Demonstrating the value of interactions between neurology and the basic sciences that underpin it, this volume considers a range of topics from the points of view of both neurobiologist and clinician and reveals how advances in our understanding have been and continue to be made. The selected topics cover a wide range of levels of organization in the nervous system. For each one, distinguished researchers from around the world have made contributions, including general reviews of normal and pathological mechanisms as well as detailed accounts of the basic processes involved. The book's coverage boasts an excellent section on the physiology and pathophysiology of central and peripheral nerve fibres. This ranges from considerations of ion channels through to mechanisms of loss of function in multiple sclerosis and strategies for restitution of function in this and other disorders of myelination. Motor control is also dealt with in depth via consideration of respiratory movements – a vital system so frequently ignored in other texts. Also of particular note are the chapters on neuronal plasticity, cell death and axonal regeneration; these are active areas in neuroscience, where new knowledge will almost certainly revolutionize neurological treatments in years to come.

In dedicating this book to Tom Sears, retiring Professor at London's Institute of Neurology (Queen Square), it is hoped that, as Professor Sears has done in the past, inspiration will be given to the next generation of neurologists to pursue research at the most fundamental level possible as well as encouraging young neuroscientists to take a deeper interest in pathological processes.

THE NEUROBIOLOGY OF DISEASE

THE NEUROBIOLOGY OF DISEASE

Contributions from Neuroscience to Clinical Neurology

Edited by
H. BOSTOCK, P.A. KIRKWOOD AND A.H. PULLEN
Institute of Neurology, Queen Square, London, UK

CAMBRIDGE UNIVERSITY PRESS
Cambridge, New York, Melbourne, Madrid, Cape Town,
Singapore, São Paulo, Delhi, Tokyo, Mexico City

Cambridge University Press
The Edinburgh Building, Cambridge CB2 8RU, UK

Published in the United States of America by Cambridge University Press, New York

www.cambridge.org
Information on this title: www.cambridge.org/9780521342384

First published 1996
First paperback edition 2011

A catalogue record for this publication is available from the British Library

Library of Congress Cataloguing in Publication data

The neurobiology of disease: contributions from neuroscience to clinical neurology/edited by
H. Bostock, P. A. Kirkwood, and A. H. Pullen.
p. cm.
Includes bibliographical references and index.
ISBN 0 521 45132 9
1. Nervous system – Diseases. 2. Neurophysiology.
I. Bostock, H. II. Kirkwood, P.A. III. Pullen, A. H.
[DNLM: 1. Nervous System – physiology. 2. Nervous System Diseases – physiopathology.
WL 102 N4945216 1996]
RC347.N473 1996
616.8–dc20 95–33564 CIP
DNLM/DLC
for Library of Congress

ISBN 978-0-521-45132-1 Hardback
ISBN 978-0-521-34238-4 Paperback

Professor T.A. Sears

Dedication

This volume is dedicated to Tom Sears, who has recently retired from the chair of the Sobell Department of Neurophysiology in the Institute of Neurology in London. Nearly all his scientific career has been made in this Institute or its sister organization, the National Hospital for Neurology and Neurosurgery (Queen Square).

After graduating in physiology from University College London he moved to the National Hospital, where he collaborated with several clinical neurologists. His independent career became firmly established after a period of study with Sir John Eccles in Canberra in the early 1960s, where he performed seminal work using intracellular recordings from respiratory motoneurones. By emphasizing the Sherringtonian, integrative role of the motoneurone, these studies transformed how we think about the way in which the nervous system deals with the command signals for respiratory movements and, indeed, for movements in general. The theme of respiration as a model motor control system has been central to his research, which has expanded to embrace respiratory mechanics, the chemical control of breathing, human stretch reflexes, cerebellar and olivary influences on motor control and a host of other related topics.

However, his interests in neuroscience have gone well beyond motor control and he developed quite separate lines of work in the physiology and pathophysiology of nerve conduction, synaptic plasticity, neural degeneration and development, with a particular interest in the motoneurone. An essential element in this catholic approach was the early addition of an anatomical section to his neurophysiological laboratory, equipped to undertake basic histological and ultrastructural studies, later incorporating *in vitro*, immunocytochemical and *in situ* hybridization technology.

His work in all fields has involved collaboration with other distinguished scientists from across the world and, most importantly, the training of young clinical neurologists in scientific methods.

His abilities were recognized by the Institute of Neurology, which created the Department of Neurophysiology for him in 1968, the department being established by London University in 1975 as the Sobell Department. He is now Emeritus Professor.

He has been an editor of many scientific journals, including chairman of the *Journal of Physiology (London)*, and his distinction was rewarded by an honorary doctorate from the University of Aix, Marseille and the Presidency of the European Neuroscience Association.

Contents

Contributors

M.J. Aminoff
Department of Neurology, School of Medicine, University of California, San Francisco, CA 94143-0114, USA

P. Andersen
Department of Neurophysiology, Institute of Basic Medical Sciences, University of Oslo, Postbox 1104 Blindern, 0317 Oslo, Norway

M.D. Baker
Sobell Department of Neurophysiology, Institute of Neurology, Queen Square, London WC1N 3BG, UK

D.A. Bayliss
Department of Pharmacology, University of Virginia, Health Science Center, Box 448, Jordan Hall, Charlottesville, VA 22908, USA

A.J. Berger
Department of Physiology & Biophysics, University of Washington School of Medicine, Seattle, WA 98195, USA

A.L. Bianchi
Département de Physiologie et Neurophysiologie, Faculté des Sciences et Techniques de Saint Jérôme, 13397 Marseille cedex 20, France

W. Burke
Department of Physiology, University of Sydney, Sydney, NSW 2006, Australia

W.E. Crill
Department of Physiology and Biophysics, SJ-40, Room G424 HSB, University of Washington School of Medicine, Seattle, WA 98195, USA

J. Diamond
Montreal Neurological Institute, McGill University, 3801 rue University, Montreal, Quebec, Canada H3A 2B4

T.E. Dick
Department of Medicine, Case Western Reserve University, Cleveland, OH 44106–5000, USA

A.F. DiMarco
Case Western Reserve University, Pulmonary Division, MetroHealth Medical Center, 3395 Scranton Road, Cleveland, OH 44109, USA

E. Di Pasquale
'Biologie des Rythmes et du Développement', Département de Physiologie et Neurophysiologie, URA1832, Faculté Saint Jérôme, 13397 Marseille cedex 20, France

R. Durbaba
Department of Physiology, Charing Cross and Westminster Medical School, Fulham Palace Road, London W6 8RF, UK

B. Duron
Laboratoire de Neurophysiologie, URA 1331 CNRS 10, rue Frédéric Petit, 80036 Amiens cedex, France

T.E. Feasby
Foothills Hospital, 1403, 29 Street NW, Calgary, Alberta, Canada

H.-J. Freund
Neurologische Klinik, Medizinische Einrichtungen der Universität Düsseldorf, Moorenstrasse 5, 40225 Düsseldorf, Germany

C. Gestreau
Département de Physiologie et Neurophysiologie, Faculté des Sciences et Techniques de Saint Jérôme, 13397 Marseille cedex 20, France

A. Gloster
Montreal Neurological Institute, McGill University, 3801 rue University, Montreal, Quebec, Canada H3A 2B4

M. Goldman
Pulmonary Function Lab, W 111 B, West Los Angeles VA Medical Center, Wilshire & Sawtelle Blvds, Los Angeles, CA 90073, USA

D.S. Goodin
Department of Neurology, School of Medicine, University of California, San Francisco, CA 94143-0114, USA

M. Gorassini
Division of Neuroscience, University of Alberta, Edmonton, Alberta, Canada 26G 2H7

P. Grafe
Department of Physiology, University of Munich, Pettenkoferstrasse 12, 80336 Munich, Germany

M. Green
Respiratory Muscle Laboratory, Royal Brompton National Heart and Lung Hospital, London SW3 6NP, UK

L. Grélot
Département de Physiologie et Neurophysiologie, Faculté des Sciences et Techniques de Saint Jérôme, 13397 Marseille cedex 20, France

G. Hilaire
'Biologie des Rythmes et du Développement', Département de Physiologie et Neurophysiologie, URA 1832, Faculté Saint Jérôme, 13397 Marseille cedex 20, France

R.S. Howard
The Harris Unit, National Hospital for Neurology and Neurosurgery, Queen Square, London WC1N 3BG, UK

A. Iggo
Department of Preclinical Veterinary Sciences, University of Edinburgh, Summerhall, Edinburgh EH9 1QH, UK

J.G.R. Jefferys
Department of Physiology, The Medical School, University of Birmingham, Birmingham B15 2TT, UK

V. Jensen
Department of Neurophysiology, Institute of Biomedical Sciences, University of Oslo, Postbox 1104 Blindern, 0317 Oslo, Norway

I.P. Johnson
Department of Anatomy and Developmental Biology, Royal Free Hospital School of Medicine, Rowland Hill Street, London NW3 2PF, UK

T. Konishi
Department of Neurology, Utano National Hospital, 8 Ondoyama-cho, Narutaki, Kyoto 616, Japan

C. Krieger
Division of Neurology, Department of Medicine, Vancouver Hospital and Health Sciences Centre (UBC), 2211 Wesbrook Mall, Vancouver BC, Canada V6T 2B5

W.I. McDonald
Department of Clinical Neurology, Institute of Neurology, Queen Square, London WC1N 3BG, UK

S. Milano
Département de Physiologie et Neurophysiologie, Faculté des Sciences et Techniques de Saint Jérôme, F-13397 Marseille cedex 20, France

G.H. Mills
Respiratory Muscle Laboratory, Royal Brompton National Heart and Lung Hospital, London SW3 6NP, UK

R. Monteau
'Biologie des Rythmes et du Développement', Département de Physiologie et Neurophysiologie, URA 1832, Faculté Saint Jérôme, 13397 Marseille cedex 20, France

D. Morin
Laboratoire des Neurosciences de la Motricité, URA 339, Université Bordeaux I, Avenue des Facultés, 33405 Talence cedex, France

E. Moser
Department of Neurophysiology, Institute of Biomedical Sciences, University of Oslo, Postbox 1104 Blindern, 0317 Oslo, Norway

J.B. Munson
Department of Neuroscience, University of Florida College of Medicine, Box J-244, JHM Health Science Center, Gainesville, FL 32610-0244, USA

J. Newsom-Davis
Neurosciences Unit, Institute of Molecular Medicine, John Radcliffe Hospital, Headington, Oxford OX3 9DU, UK

J.G. Nicholls
Biocenter, Klingelbergstrasse 70, CH-4056 Basel, Switzerland

H. Nishimura
Department of Neurosurgery, Yamaguchi Red Cross Hospital, Yohatababa 53–1, Yamaguchi City, Yamaguchi ken, 753 Japan

N. Nisimaru
Department of Physiology, Oita Medical University, Oita 879-55, Japan

J. Ochoa
Department of Neurology, Good Samaritan Hospital and Medical Center, 1040 NW 22nd Ave, Suite NSC-460, Portland, OR 97210, USA

M.P. Pender
Department of Medicine, Clinical Sciences Building, Royal Brisbane Hospital, Brisbane, Queensland 4029, Australia

A. Prochazka
Division of Neuroscience, University of Alberta, Edmonton, Alberta, Canada 26G 2H7

A.H. Pullen
Sobell Department of Neurophysiology, Institute of Neurology, Queen Square, London WC1N 3BG, UK

G. Reid
Physiologisches Institut, Universitätskrankenhaus Eppendorf, Martinistrasse 52, D-20246 Hamburg, Germany

J.M. Ritchie
Department of Pharmacology, Yale University School of Medicine, 333 Cedar Street, New Haven CT 06510, USA

J.R. Romaniuk
Case Western Reserve University, Pulmonary and Critical Care Medicine, MetroHealth Medical Center, 2500 MetroHealth Drive, Cleveland, OH 44109–1998, USA

J. Rosenbluth
Department of Physiology, New York University Medical Center, 400 E 34th Street, New York, NY 10016, USA

C.P. Seers
Sobell Department of Neurophysiology, Institute of Neurology, Queen Square, London WC1N 3BG, UK

E. Shen
Shanghai Brain Research Institute, Academia Sinica, 319 Yueyang Road, Shanghai 200031, China

K.J. Smith
Department of Neurology, UMDS (Guy's Campus), London Bridge, London SE1 9RT, UK

G.S. Supinski
Case Western Reserve University, Pulmonary and Critical Care Medicine, MetroHealth Medical Center, 2500 MetroHealth Drive, Cleveland, OH 44109–1998, USA

A. Taylor
Department of Physiology, Charing Cross and Westminster Medical School, Fulham Palace Road, London W6 8RF, UK

J. Taylor
Prince of Wales Medical Research Institute, High Street, Randwick, Sydney, NSW 2031, Australia

Z. Varga
Biocenter, Klingelbergstrasse 70, 4056 Basel, Switzerland

F. Viana
Laboratorium voor Fysiologie, Katholieke Universiteit Leuven, Campus Gasthuisberg, 3000 Leuven, Belgium

S.G. Waxman
Department of Neurology, PO Box 3333, Yale University Medical Center, 333 Cedar Street, New Haven, CT 06510, USA

R.H. Westgaard
Division of Organization and Work Science, Norwegian Institute of Technology, 7034 Trondheim, Norway

W.D. Willis, Jr
Marine Biomedical Institute, University of Texas Medical Branch, 200 University Boulevard, Suite 608, Galveston, TX 77555-0843, USA

Preface

Clinical neurology and the neurosciences can interact in various ways. Investigators often start from a given disease or clinical condition and ask 'What is the cause of this disorder?', with the hope that having identified the cause the way will be open to devise a remedy. Though apparently straightforward, this approach can be fraught with difficulty. The aetiology of neurological conditions is often so obscure that it is not evident which discipline, whether biochemistry, electrophysiology, immunology or epidemiology, is going to give the critical lead, so all must be supported. Only for inherited conditions has this approach been conspicuously successful, where recent advances in genetics and molecular biology have in several cases provided a quick path to the first stage of identifying the cause.

An alternative approach is to put science first: to ask fundamental questions about the workings of the nervous system, but to ensure that the scientific questions are always relevant to neurological ones. Not only may the science then help the neurologist, but clinical observations may help illuminate the science. This is the approach exemplified by most of the contributors to this book, and is also the approach of Tom Sears, to whom the book is dedicated. This alternative approach is well illustrated by his work on nerve conduction and demyelination, which underpins several of the chapters. In pioneering experiments with McDonald, using focal experimental demyelination with diphtheria toxin, he first showed that individual central nerve fibres could remain intact through a demyelinated region, with unimpaired conduction above and below the lesion, while transmission of impulses through the lesion was blocked or proceeded with a reduced safety factor. These experiments on the pathophysiology of demyelination were of obvious clinical significance, but also provided access to a fundamental question about axonal physiology.

Huxley & Stämpfli (1949) had shown that conduction in isolated myelinated axons was saltatory: the inward membrane currents of excitation occurred only at the nodes of Ranvier. This showed that the internodal membrane was not excited, but not whether it was actually *excitable*, a question which could only be answered by demyelination. Sears therefore (with Rasminsky and later with Bostock) adapted Huxley and Stämpfli's technique to investigate demyelinated axons, taking advantage of the anatomy of rat spinal roots to record longitudinal currents from undissected but functionally isolated single axons. For some demyelinated fibres conduction was always saltatory (Rasminsky & Sears, 1972), but in smaller fibres, or at temperatures below 37 °C, continuous conduction could sometimes be recorded over one or more demyelinated internodes, at velocities about one-twentieth of normal (Bostock & Sears, 1978). These studies helped to open investigation of a range of basic questions concerning the distribution of sodium channels within the membrane and the mechanisms of regulation of this distribution (see chapters by Ritchie, Waxman). On the clinical side, these results helped to resolve the puzzling lack of correlation between degree of demyelination and conduction failure, and to explain the remarkable delays in visual and other evoked responses in multiple sclerosis. Neither the cause of demyelinating disease was revealed, nor a cure, but by clarifying the pathophysiology of conduction failure and slowing, this work led to further studies on how to overcome the conduction failure by prolonging the action potential, by reducing pump activity, or by promoting remyelination (see chapters by McDonald, Smith, Rosenbluth). The important lesson here is that the starting point for these insights in both neurobiology and clinical neurology was a basic scientific question about saltatory conduction, which was relevant to the clinical consequences of demyelination.

Tom Sears worked on many other problems, always with a clinical aspect at least at the back of his mind. This book was inspired by a symposium held in London to mark his retirement. The symposium brought together a group of distinguished scientists and neurologists, all of whom either worked directly with him at some stage of their careers, or were strongly influenced by his published work. Many were educated in their scientific approach by him, including us, the three editors of this volume, for which we are profoundly grateful.

The symposium which acted as a spur to production of this volume was made possible by the generous support of the Guarantors of *Brain*, Pfizer Central Research, The Spinal Cord Research Trust (Paralyzed Veterans of America), The Multiple Sclerosis Society of Great Britain and Northern Ireland, The Wellcome Trust, Digitimer Ltd, Lilly Industries Ltd, The Institute of Neurology, The Sobell Foundation and Merck, Sharp & Dohme Research Laboratories.

We are indebted to Julie Savvides, Kully Sunner and Peter Humphreys (Sobell Department) for their encouragement and help in organizing the symposium. We further relied a great deal on Julie Savvides during the preparation of this book, for which we would also like to acknowledge the advice and assistance of the editorial staff of Cambridge University Press.

London
May 1995

H.B.
P.A.K.
A.H.P.

References

Bostock, H. & Sears, T.A. (1978). The internodal axonal membrane: electrical excitability and continuous conduction in segmental demyelination. *Journal of Physiology (London)*, **280**, 273–301.
Huxley, A.F. & Stämpfli, R. (1949). Evidence for saltatory conduction in peripheral myelinated nerve fibres. *Journal of Physiology (London)* **108**, 315–339.
Rasminsky, M. & Sears, T.A. (1972). Internodal conduction in undissected demyelinated nerve fibres. *Journal of Physiology (London)*, **227**, 323–350.

Part I
Physiology and pathophysiology of nerve fibres

Ion channels and their roles in nerve are becoming more widely studied as new techniques are developed, the number of identified channel types increases, and evidence for the involvement of channels in disease processes accumulates. The first seven chapters deal with ion channels and ion exchange mechanisms from widely differing viewpoints. Ritchie reviews the distribution of Na^+ and K^+ channels, both along axons and between axons and supporting cells, while Waxman focuses on the node of Ranvier, and other ion movements that are important there, in addition to the sodium influx responsible for action potentials. Baker and Konishi consider different K^+ channels in Schwann cells that may be involved in maintaining the immediate environment of the internodal axolemma. This comprises the vast majority of the axon membrane and, as the following chapter by Reid indicates, is very far from the passive cable once assumed. Species differences are important for some aspects of nerve and channel function: the lack of fast K^+ channels at mammalian, as against frog nodes (Ritchie) is the best known, but there is also a surprising difference between K^+ channels reported in mouse and rabbit Schwann cells, the latter lacking inward rectification (Baker, Konishi). Konishi's elucidation of the factors affecting expression of this conductance may be relevant to the species difference. Ultimately we may have to study channel behaviour in human axons directly to be sure that the work on animal preparations is relevant to human physiology and disease, and pioneering results with this approach are described by Reid. Pathological alterations of channels are described by Grafe in a new model of diabetic neuropathy, and by Newsom-Davis in diseases affecting the neuromuscular junction.

The last five chapters in this section deal with different aspects of demyelinating diseases, mostly related directly or indirectly to multiple sclerosis (MS). The pathophysiology of demyelination is reviewed in the human peripheral nervous system by Feasby, in central nerve fibres by Smith, and by Pender in experimental autoimmune encephalomyelitis (normally regarded as the best available animal model of MS), in which the lesions can be primarily central or peripheral, depending on the immune challenge. MS itself is discussed in the chapters by Smith and McDonald. Finally, the use of glial transplants to overcome demyelination or amyelination is reviewed by Rosenbluth.

1

Ion channels in normal and pathophysiological mammalian peripheral myelinated nerve

J.M. RITCHIE

Department of Pharmacology, Yale University School of Medicine, New Haven, Connecticut, USA

The ionic basis of the nerve impulse was well established over four decades ago by Hodgkin & Huxley (1952); and its application to myelinated nerve soon followed (Frankenhaeuser & Huxley, 1964). Huxley & Stämpfli (1949) had already provided clear evidence that conduction in peripheral myelinated nerve fibres was saltatory; and Rushton (1951) in a seminal theoretical analysis had made general predictions about the properties of myelinated fibres, particularly how these change as the fibre diameter changes. These predictions correspond extremely well with the situation that prevails in real axons.

One question remained unanswered, namely the nature of the axolemma under the myelin. Was the internodal axolemma similar to the nodal axolemma, i.e. capable of conducting but not doing so because of the insulating myelin sheath; or were the internodal electrophysiological properties quite different? The earliest study on conduction in single demyelinated fibres (Rasminsky & Sears, 1972) failed to resolve the question of whether demyelinated axons can conduct impulses in a continuous (as opposed to saltatory) manner. However, with improvements in technique, Bostock & Sears (1978) showed clearly that single undissected myelinated fibres in perfused ventral roots of normal rats treated with diptheria toxin to produce demyelination could indeed conduct impulses – but in a continuous manner, at less than one-twentieth of the velocity expected for normal stretches of the same fibre. This provided unequivocal evidence for the presence of Na^+ channels in the now demyelinated internodal axon. However, the question whether these channels were present *normally* remained unresolved. And the fact remains that extensive demyelination produced experimentally, particularly in the acute stage, is commonly associated with conduction block.

Nodal and internodal sodium channels

The essential question raised by the demyelination experiments was whether these Na^+ channels underlying the continuous conduction resulted from a remodelling of the internodal axon, or whether they were already in place.

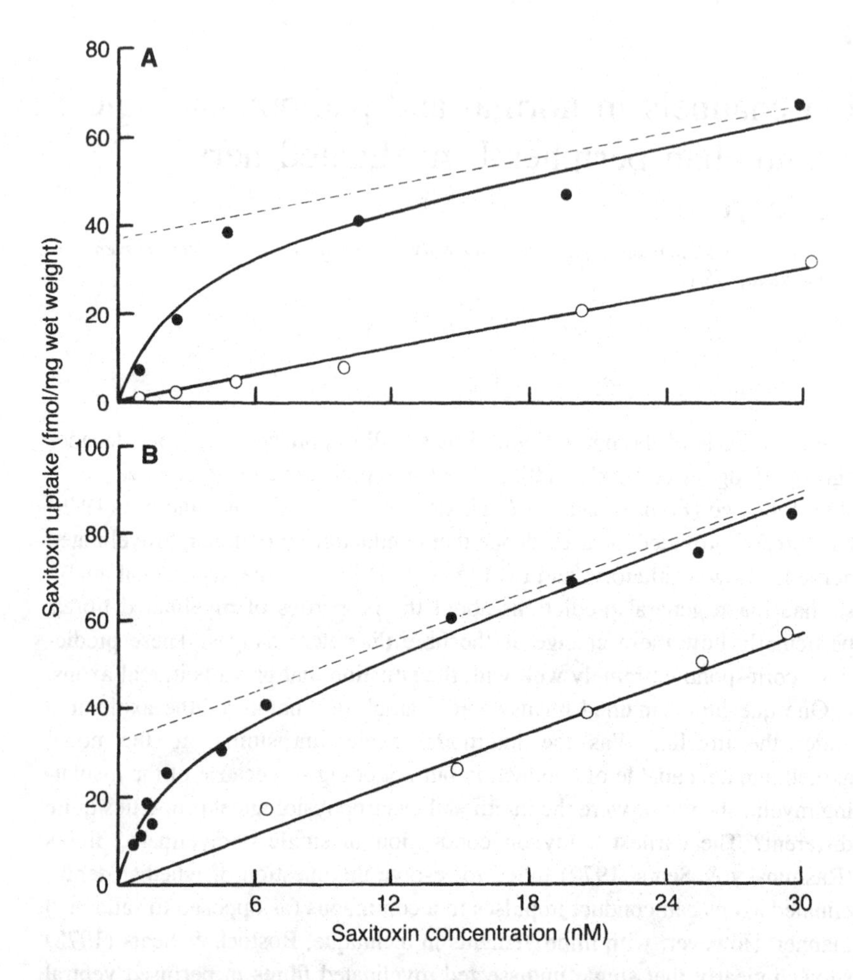

Fig. 1.1. Total (filled circles) and linear (open circles) components of the uptake of [^{3}H] saxitoxin by the myelinated fibres of desheathed rabbit sciatic nerve; either intact (A) or after homogenization (B). From Ritchie (1986), with permission.

An answer was sought experimentally by Ritchie & Rogart (1977) using radiolabelled saxitoxin as a specific marker for Na^+ channels. Arguing that extracellular saxitoxin would gain access only to the nodal axolemma and not to the internodal axolemma under the myelin, they compared the saturable uptake of saxitoxin in normal and homogenized rabbit sciatic nerve (Fig. 1.1). The answer was clear. In spite of the fact that in the homogenized preparation the axolemma exposed to the saxitoxin was now increased by 2–3 orders of magnitude, there was no statistically significant increase in the saturable binding of saxitoxin. Two conclusions were drawn from this experiment. The first was that the nodal density of Na^+ channels was extraordinarily high, being apparently about 10 000/μm^2 (but see later). The second (based on a statistical analysis of the 150 observations embodied in Fig. 1.1) was that if Na^+ channels were indeed present in the internodal axolemma, their density had to be less than 25/μm². This latter value for the upper limit has recently been confirmed electrophysiologically (Chiu & Schwarz, 1987; see also Ritchie, 1988).

Nodal and internodal potassium channels

That the non-uniform distribution of Na^+ channels described above might be accompanied by a non-uniform distribution of voltage-dependent K^+ channels was first suggested by the observation that K^+ channel blocking agents, such as tetraethylammonium (TEA) ions and 4-aminopyridine (4-AP), have little or no effect on the mammalian A-fibre action potential yet greatly prolong the action potential in demyelinated fibres (Bostock, Sherratt & Sears, 1978; Sears, Bostock & Sherratt, 1978; Sherratt, Bostock & Sears, 1980). Voltage-clamp experiments confirmed that there is an inhomogeneous distribution of the fast delayed rectifier channel that in non-myelinated nerve, and in *non-mammalian* myelinated nerve, is involved in the rapid repolarization

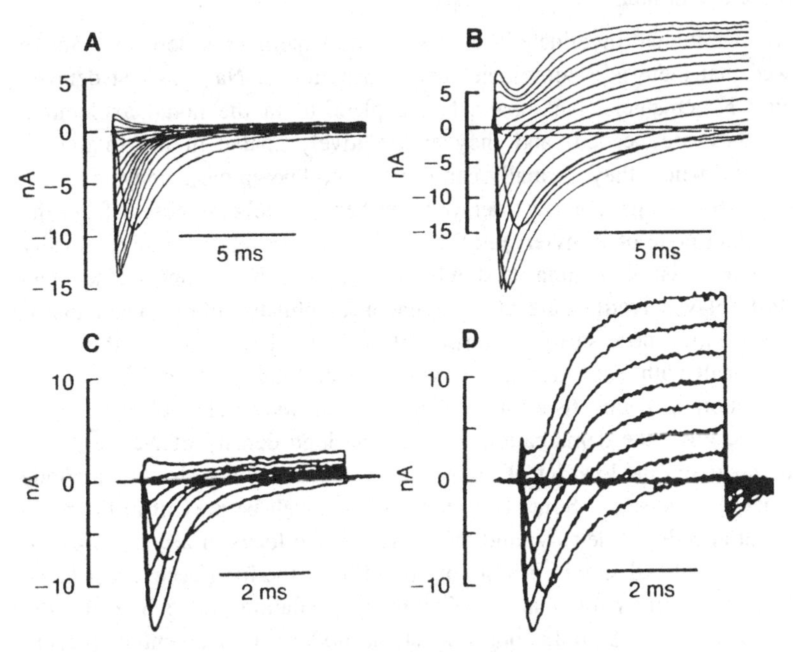

Fig. 1.2. Ionic currents in response to step depolarizations of voltage-clamped nodes of Ranvier from the sciatic nerve. Upper panel: comparison of the normal currents in rabbit (A) and frog (B) nodes. Lower panel: comparison of currents taken before (C) and 90 minutes after (D) a rabbit node had been exposed to collagenase. Data from Chiu *et al.* (1979) and Chiu & Ritchie (1980), replotted.

phase of the action potential. Thus, Chiu *et al.* (1979) showed that in rabbit myelinated nerve this phase of outward current is virtually absent (Fig. 1.2A), unlike the case of frog myelinated nerve where the inward Na^+ currents on voltage-clamp are followed by an outward delayed K^+ current (Fig. 1.2B) that can be blocked by TEA. This latter observation had originally been made by Horackova, Nonner & Stämpfli (1968) but not subsequently investigated by them in detail.

That fast delayed rectifier currents are, however, present normally in the paranodal membrane was shown in experiments where paranodal demyelination was brought about *acutely* by one of a variety of procedures (Chiu & Ritchie, 1981). When the paranodal seal between the myelin and axolemma is suddenly broken, there is an abrupt appearance of outward K^+ current (compare Fig. 1.2C and D). Fast delayed rectifiers are thus clearly present in the mammalian paranodal region. Subsequent experiments showed that these currents are also found in frog and rabbit internodal axolemma (Chiu & Ritchie, 1981, 1982; Ritchie & Chiu, 1981).

Function of the inhomogeneous distribution of axonal sodium and potassium channels

In the mammalian myelinated nerve fibre the experiments described above showed that there is a complementary distribution of Na^+ and fast delayed rectifier K^+ channels. Na^+ channels are plentiful in the nodal axolemma, where they are needed; and they are relatively sparse in the internodal axolemma, where they normally can perform no known electrophysiological function. By comparison, fast delayed rectifier channels are absent from the mammalian node of Ranvier; but they are present in relatively high density in the internodal axolemma. It should be noted that K^+ channels other than the fast delayed rectifier are also present in myelinated fibres. Their distribution does not necessarily conform with that just described; and this issue will be dealt with elsewhere in this volume (see Chapters 2 and 5).

The inhomogeneous distribution of K^+ channels may well subserve several functions. First, the demarcation between the high density of Na^+ channels in the node and the low density in the internodal membrane in the paranode is unlikely to be sharp. Some Na^+ channels may well be present in the paranode but at reduced density; and these may be the locus of a depolarization that is delayed and slowly developing because of the high access resistance in the region of the paranodal seal. Computer simulation studies indeed show that such a response could long outlast the nodal action potential, placing the nodal membrane at risk of being re-excited. The presence of K^+ channels in the paranodal axolemma, by preventing or reducing the extent of this slow depolarization, would prevent such ectopic generation of impulses and minimize repetitive firing. More importantly, these K^+ channels, together with the other K^+ conductances present in the internodal axolemma, also control the internodal resting potential. This is important because it is now clear that the maintenance of an adequate *nodal* membrane potential requires that the *internode* also has a substantial resting potential (Bohm & Straub, 1961; Barrett & Barrett, 1982; Chiu & Ritchie, 1984).

Given that there are advantages for the presence of internodal K^+ channels, what are the possible advantages of their absence in the node? The first answer is that they will permit a higher frequency of firing in the axon, which may be advantageous in the mammal. This is because the presence of a delayed potassium conductance necessarily means that for some time after the impulse, the persisting high potassium conductance makes it more difficult to set up a subsequent impulse due to the prolonged refractoriness. A second advantage is that the presence of nodal K^+ channels increases the energetic cost of an action potential (Ritchie, 1985). This is because the presence of K^+ channels leads to a faster repolarization of the membrane so that during the later stages of the action potential, larger inward Na^+ currents flow through the still incompletely inactivated Na^+ channels; more Na^+ ions have therefore to be actively extruded during recovery. For this reason, the absence of nodal K^+ channels, together with the fact that in the mammal there is a speeding up of inactivation kinetics by a factor of 2–3 (Chiu *et al.*, 1979), the metabolic cost of an action potential in a mammalian fibre might be as little as 15% of the corresponding cost in a frog myelinated fibre of the same diameter and at the same temperature.

Although it thus became established that the internodal axolemma contained both Na^+ channels (but at low density) and K^+ channels, it remains unclear to what extent continuous conduction in demyelinated fibres relies only on the channels present in the now exposed axolemma, and to what extent there is a remodelling (insertion of new Na^+ channels, spreading of erstwhile nodal channels). But whatever happens in the acute phase of demyelination, it is clear that some remodelling does occur during the recovery phase. Thus, fibres that have remyelinated following demyelination (and fibres that have regenerated following nerve section) exhibit a 2- to 3-fold increase in saturable saxitoxin binding, indicating an increase in the number of Na^+ channels per unit length of axon (Ritchie, Rang & Pellegrino, 1981; Ritchie, 1982). The increase in the number of Na^+ channels probably does not reflect an increase in nodal channel density, merely the fact that remyelinated and regenerated fibres have about 3 times as many nodes per unit length as they do normally.

In addition to this involvement of Na^+ channels, for the first several months of recovery after demyelination or crush, the action potential is considerably broadened by 4-AP, indicating a contribution of K^+ currents in the repolarization phase of the action potential. As recovery proceeds, this sensitivity to 4-AP disappears (Ritchie, 1982). The sensitivity to 4-AP need not signify the insertion of *nodal* K^+ channels, but merely that the paranodal seal, normally denying access to the internodal K^+ channels, is not perfect (as is also the case in developing myelinated axons).

Extraneuronal sodium and potassium channels

Completely unexpected was the finding that Wallerian degeneration of the sciatic nerve was not accompanied by the expected complete loss of saxitoxin binding. In the rat, about 7% of the binding remained; but in the rabbit sciatic nerve, following an initial fall after nerve section, there was a maintained 2- to 3-fold increase in the number of Na^+ channels in the degenerated stump (determined by saxitoxin binding), which was clearly axon-free (Ritchie & Rang, 1983). Patch-clamp experiments on rabbit cultured Schwann cells showed that Schwann cells, which proliferate during Wallerian degeneration, express Hodgkin–Huxley type Na^+ (Fig. 1.3) and delayed rectifier K^+ channels (Chiu, Shrager & Ritchie, 1984; Shrager, Chiu & Ritchie, 1985). Subsequent experiments (Bevan *et al.*, 1985) showed that cultured rat astrocytes similarly express both Na^+ and K^+ channels; and both Schwann cells and astrocytes express voltage-gated Cl^- channels (Bevan *et al.*, 1985; Howe & Ritchie, 1988). The species difference in the extent of Schwann cell expression of Na^+ channels (rabbit versus rat) remains unexplained.

At least three types of delayed rectifier K^+ channel are expressed by rabbit Schwann cells. The fastest activating (type 1) is blocked by 4-AP and by α-dendrotoxin. A slower channel (type 2) is blocked by 4-AP but not by α-dendrotoxin. A third channel that under normal conditions activates only at very positive potentials is blocked by neither 4-AP nor

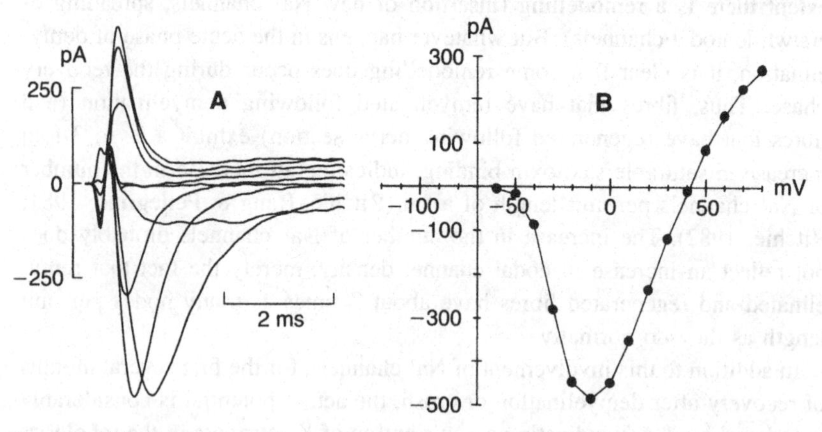

Fig. 1.3. (A) Na^+ currents in a voltage-clamped rabbit cultured Schwann cell. (B) Current–voltage relationship obtained from the records in (A). From Howe & Ritchie (1989), with permission.

α-dendrotoxin. All three are blocked by TEA (Baker, Howe & Ritchie, 1993; Howe & Ritchie, 1988). These channels are described in more detail by Baker (see Chapter 3).

Whereas K^+ channels in a variety of tissues have been implicated in a diversity of physiological functions, the presence of Na^+ channels in satellite cells is less easy to account for. It is clear, however, that their presence was the main confounding factor in the calculation of nodal Na^+ channel density mentioned earlier, which gave much too high a value (Ritchie & Rogart, 1977). Indeed, if the Na^+ channel density at the node is 1000–2000/μm^2, the saxitoxin binding experiments can now be reinterpreted to mean that about half the Na^+ channels present in a normal rabbit sciatic nerve trunk are extraneuronal, i.e. in Schwann cells. In cultured Schwann cells, the density is about 30/μm^2, but their density *in vivo*, as well as their distribution along the Schwann cell plasmalemma, are unknown. It is clear, however, that Schwann cell Na^+ channels are not an artifact of cell culture. Not only were they first described as a result of Wallerian degeneration *in vivo* (Ritchie & Rang, 1983) but immunostaining with antibodies to the Na^+ channel clearly reveals their presence *in situ* in Schwann cells of normal mammalian myelinated nerve – particularly in the region of the cell body and in the microvilli that the Schwann cell sends down to the axolemma in the nodal regions (Ritchie *et al.*, 1990).

These voltage-dependent Na^+ and K^+ channels show a considerable plasticity depending on the method of dissociation of the cells and on their developmental age. Thus, when examined under similar conditions of cell culture, the electrophysiological behaviour of Schwann cells cultured from a predominantly myelinated (sciatic) and from a predominantly non-myelinated (vagus) mammalian nerve, from either neonatal or adult tissue, is similar (Howe & Ritchie, 1990). However, Chiu (1988), looking at explants and acutely dissociated Schwann cells, has suggested that voltage-gated Na^+ currents are found only in Schwann cells associated with non-myelinated axons. Furthermore, Ritchie (1988) has shown that all Na^+ currents are abolished for several days after treatment with proteolytic enzymes. Similarly, in astrocytes Barres *et al.* (1990) have shown that replacement of their normal dissociation procedure by a 'tissue print' method radically changes the expression of different kinds of ion channel in cultured type 1 astrocytes. Finally, Sontheimer and his colleagues have shown that expression of Na^+ and K^+ channels by astrocytes is dramatically influenced by the presence or absence of neurones; and furthermore, large changes in the electrophysiological and pharmacological (tetrodotoxin sensitivity) parameters occur with developmental changes in astrocytes maintained in culture for several days or obtained at different times postnatally (Sontheimer & Ritchie, 1994).

The functional significance of these voltage-gated channels in the satellite cell of the nervous system remains unclear. Some of the kinds of K^+ channel (five have been identified so far) may be involved in mitosis: for K^+ channel blocking agents (such as TEA, 4-AP or quinidine) interfere with Schwann cell proliferation (Chiu & Wilson, 1989) as they are indeed known to do in lymphocytes (Chandy *et al.*, 1984) and fibroblasts (Gray *et al.*, 1986). They may also be involved in the maintenance of the extracellular ionic concentrations, especially of K^+ ions (Bevan *et al.*, 1985). But the role of the satellite cell Na^+ channels remains mysterious. One suggestion for the Na^+ channels (Gray & Ritchie, 1985), which could equally apply to the K^+ channels, is that they are transferred from the Schwann cell system of microvilli or at the corresponding system at the astrocytic node (perhaps both normally and during remodelling after demyelination). This suggestion is attractive, at least from the point of view of economy of supply (Ritchie, 1988); but even nearly a decade after it was first suggested it remains just a speculation.

Acknowledgements

This work was supported in part by grants NS08304 and NS12327 from the USPHS.

References

Baker, M., Howe, J.R. & Ritchie, J.M. (1993). Two types of 4-aminopyridine-sensitive potassium current in rabbit Schwann cells. *Journal of Physiology (London)*, **464**, 321–342.

Barres, B.A., Koroshetz, W.J., Chun, L.L.Y. & Corey, D.P. (1990). Ion channel expression by white matter glia: the type-1 astrocyte. *Neuron*, **5**, 527–544.

Barret, E.F. & Barret, J.N. (1982). Intracellular recording from myelinated axons: mechanism of the depolarizing afterpotential. *Journal of Physiology (London)*, **323**, 117–144.

Bevan, S., Gray, P.T.A. & Ritchie, J.M. (1984). A calcium-activated cation-selective channel in rat cultured Schwann cells. *Proceedings of the Royal Society of London, Series B*, **222**, 349–355.

Bevan, S., Chiu, S.Y., Gray, P.T.A. & Ritchie, J.M. (1985). The presence of voltage-gated sodium, potassium and chloride channels in rat cultured astrocytes. *Proceedings of the Royal Society of London, Series B*, **225**, 299–313.

Bohm, H.W. & Straub, R.W. (1961). Der Effekt von Tetraethylammonium an Kalium-depolarisierten markhaltigen Nervenfasern. *Pflügers Archiv*, 274, S28–29.

Bostock, H. & Sears, T.A. (1978). The internodal axon membrane: electrical excitability and continuous conduction in segmental demyelination. *Journal of Physiology (London)*, **280**, 273–301.

Bostock, H., Sears, T.A. & Sherratt, R.M. (1981). The effects of 4-aminopyridine and tetraethylammonium ions on normal and demyelinated mammalian nerve fibres. *Journal of Physiology (London)*, **313**, 301–315.

Bostock, H., Sherratt, R.M. & Sears, T.A. (1978). Overcoming conduction failure in demyelinated nerve fibres by prolonging action potentials. *Nature*, **274**, 385–387.

Chandy, K.G., De Coursey, T.E., Cahalan, M.D., McLaughlin, C. & Gupta, S. (1984). Voltage gated K channels are required for human T lymphocyte activation. *Journal of Experimental Medicine*, **160**, 369–385.

Chiu, S.Y. (1988). Changes in excitable membrane properties in Schwann cells of adult rabbit sciatic nerve following nerve transection. *Journal of Physiology (London)*, **396**, 173–188.

Chiu, S.Y. & Ritchie, J.M. (1980). Potassium channels in nodal and internodal axonal membrane of mammalian myelinated fibres. *Nature*, **284**, 170–171.

Chiu, S.Y. & Ritchie, J.M. (1981). Evidence for the presence of potassium channels in the paranodal region of acutely demyelinated mammalian single nerve fibres. *Journal of Physiology (London)*, **313**, 415–437.

Chiu, S.Y. & Ritchie, J.M. (1982). Evidence for the presence of potassium channels in the internode of frog myelinated nerve fibres. *Journal of Physiology (London)*, **322**, 485–501.

Chiu, S.Y. & Ritchie, J.M. (1984). On the physiological role of internodal potassium channels and the security of conduction in myelinated nerve. *Proceedings of the Royal Society of London, Series B*, **220**, 415–422.

Chiu, S.Y., Ritchie, J.M., Rogart, R.B. & Stagg, D. (1979). A quantitative description of membrane currents in rabbit myelinated nerve fibres. *Journal of Physiology (London)*, **292**, 149–166.

Chiu, S.Y. & Schwarz, W. (1987). Sodium and potassium currents in acutely demyelinated internodes of rabbit sciatic nerves. *Journal of Physiology (London)*, **391**, 631–649.

Chiu, S.Y., Shrager, P. & Ritchie, J.M. (1984). Neuronal-type Na^+ and K^+ channels in rabbit cultured Schwann cells. *Nature*, **311**, 156–157.

Chiu, S.Y. & Wilson, G.F. (1989). The role of potassium channels in Schwann cell proliferation in Wallerian degeneration of explant rabbit sciatic nerves. *Journal of Physiology (London)*, **408**, 199–122.

Frankenhaeuser, B. & Huxley, A.F. (1964). The action potential in the myelinated fibre of *Xenopus laevis* as computed on the basis of voltage clamp data. *Journal of Physiology (London)*, **171**, 302–315.

Gray, P.T.A., Chiu, S.Y., Bevan, S. & Ritchie, J.M. (1986). Ion channels in rabbit cultured fibroblasts. *Proceedings of the Royal Society of London, Series B*, **227**, 1–16.

Gray, P.T.A. & Ritchie, J.M. (1985). Ion channels in Schwann and glial cells. *Trends in Neuroscience*, **8**, 411–416.

Hodgkin, A.L. & Huxley, A.F. (1952). A quantitative description of membrane current and its application to conduction and excitation in nerve. *Journal of Physiology (London)*, **117**, 500–544.

Horackova, M., Nonner, W. & Stämpfli, R. (1968). Action potentials and voltage clamp currents of single rat Ranvier nodes. *Proceedings of the International Union of Physiological Sciences*, **7**, 198.

Howe, J.R. & Ritchie J.M. (1988). Two types of potassium current in rabbit cultured Schwann cells. *Proceedings of the Royal Society of London, Series B*, **235**, 19–27.

Howe, J.R. & Ritchie, J.M. (1989). Cation and anion channels in mammalian Schwann and glial cells. In *Peripheral Nerve Development and Regeneration: Recent Advances and Clinical Applications*, ed. E. Scarpini, M.G. Fiori, D. Pleasure & G. Scarlato, pp. 67–73. Padua: Liviana Press.

Howe, J.R. & Ritchie, J.M. (1990). Sodium currents in Schwann cells from myelinated and non-myelinated nerve of neonatal and adult rabbits. *Journal of Physiology (London)*, **425**, 169–210.

Huxley, A.F. & Stämpfli, R. (1949). Evidence for saltatory conduction in peripheral myelinated nerve fibres. *Journal of Physiology (London)*, **108**, 315–339.

Rasminsky, M. & Sears, T.A. (1972). Internodal conduction in undissected demyelinated nerve fibres. *Journal of Physiology (London)*, **227**, 323–350.
Ritchie, J.M. (1982). Sodium and potassium channels in regenerating and developing mammalian myelinated nerve. *Proceedings of the Royal Society of London, Series B*, **215**, 273–287.
Ritchie, J.M. (1985). A note on the mechanism of resistance to anoxia and ischaemia in pathophysiological mammalian myelinated nerve. *Journal of Neurology, Neurosurgery and Psychiatry*, **48**, 274–277.
Ritchie, J.M. (1986). The distribution of sodium and potassium channels in mammalian myelinated nerve. In *Ion Channels in Neural Membranes*, ed. J.M. Ritchie, D. Keynes & L. Bolis, pp. 105–122. New York: Alan Liss.
Ritchie, J.M. (1988). Sodium-channel turnover in rabbit cultured Schwann cells. *Proceedings of the Royal Society of London, Series B*, **233**, 423–430.
Ritchie, J.M., Black, J.A., Waxman, S.G. & Angelides, K.J. (1990). Sodium channels in the cytoplasm of Schwann cells. *Proceedings of the National Academy of Sciences, USA*, **87**, 9290–9294.
Ritchie, J.M. & Chiu, S.Y. (1981). Distribution of sodium and potassium channels in mammalian myelinated nerve. *Advances in Neurology*, **31**, 329–342.
Ritchie, J.M. & Rang, H.P. (1983). Extraneuronal saxitoxin binding sites in rabbit myelinated nerve. *Proceedings of the National Academy of Sciences, USA*, **80**, 2803–2807.
Ritchie, J.M., Rang, H.P. & Pellegrino, R. (1981). Sodium and potassium channels in demyelinated and remyelinated peripheral nerve. *Nature*, **294**, 257–259.
Ritchie, J.M. & Rogart, R.B. (1977). The density of sodium channels in mammalian nerve fibres and the nature of the axonal membrane under the myelin sheath. *Proceedings of the National Academy of Sciences, USA*, **74**, 211–215.
Rushton, W.A.H. (1951). A theory of the effects of fibre size in medullated nerve. *Journal of Physiology (London)*, **115**, 101–122.
Sears, T.A., Bostock, H. & Sherratt, R.M. (1978). The pathophysiology of demyelination and its implication for the systematic treatment of multiple sclerosis. *Neurology*, **28**, 21–26.
Sherratt, R.M., Bostock, H. & Sears, T.A. (1980). The effects of 4-aminopyridine on normal and demyelinated mammalian nerve fibres. *Nature*, **283**, 570–572.
Shrager, P., Chiu, S.Y. & Ritchie, J.M. (1985). Voltage-dependent sodium and potassium channels in mammalian cultured nerve cells. *Proceedings of the National Academy of Sciences, USA*, **82**, 948–952.
Sontheimer, H. & Ritchie, J.M. (1994). Voltage-gated sodium and calcium ion expression by satellite cells. In *Neuroglial Cells*, ed. H. Kettenmann & B.R. Ransom. London: Oxford University Press.

2

Molecular anatomy of the node of Ranvier: newer concepts

STEPHEN G. WAXMAN

Department of Neurology, Yale University School of Medicine, New Haven, and Center for Neuroscience Research, VA Hospital, West Haven, Connecticut, USA

An important chapter in neuroscience was opened up by Tom Sears and his colleagues (see e.g. Rasminsky & Sears, 1972; Bostock & Sears, 1976, 1978) when they carried out their beautiful analyses, using the external longitudinal current recording method, of nodal and internodal transmembrane currents in myelinated and demyelinated ventral root fibres, by implication beginning to define the molecular anatomy of the node of Ranvier. Since that time, the sequestration of voltage-sensitive Na^+ channels in the axon membrane at the node has been further examined using a variety of techniques including saxitoxin-binding (Ritchie & Rogart, 1977), cytochemical methods (Quick & Waxman, 1977a), freeze-fracture (Rosenbluth, 1976), immuno-electron microscopy (Black *et al.*, 1989), nodal voltage clamp (Chiu & Ritchie, 1981, 1982; Neumcke & Stämpfli, 1982) and single channel patch clamp (Vogel & Schwarz, 1995); and the expression of various types of K^+ channels in myelinated axons has been studied using electrophysiological and pharmacological methods (Waxman & Ritchie, 1993; Vogel & Schwarz, 1994). This chapter will discuss some of the newer aspects of the molecular anatomy of the mammalian node of Ranvier, with emphasis on Na^+ channels, the Na^+–Ca^{2+} exchanger, and the diffusion barrier that accumulates intra-axonal Na^+ in a limited space below the axon membrane.

Clustering of sodium channels in the nodal axon membrane

Fig. 2.1 is an electron micrograph which shows the node of Ranvier in the optic nerve of a normal rat following immunostaining with antibody 7493 (Black *et al.*, 1989). This antibody is directed against the alpha subunit of the rat brain Na^+ channel (Elmer *et al.*, 1990). Na^+ channel immunoreactivity is dense in the axon membrane of the node, but is absent in the internodal axon membrane under the myelin sheath. Saxitoxin-binding studies and voltage-clamp analyses also demonstrate a high density of Na^+ channels in the nodal axon membrane, approximately $1000/\mu m^2$ (Neumcke & Stämpfli, 1982; Shrager, 1989; Vogel & Schwarz, 1994; Ritchie, 1995); in contrast, these

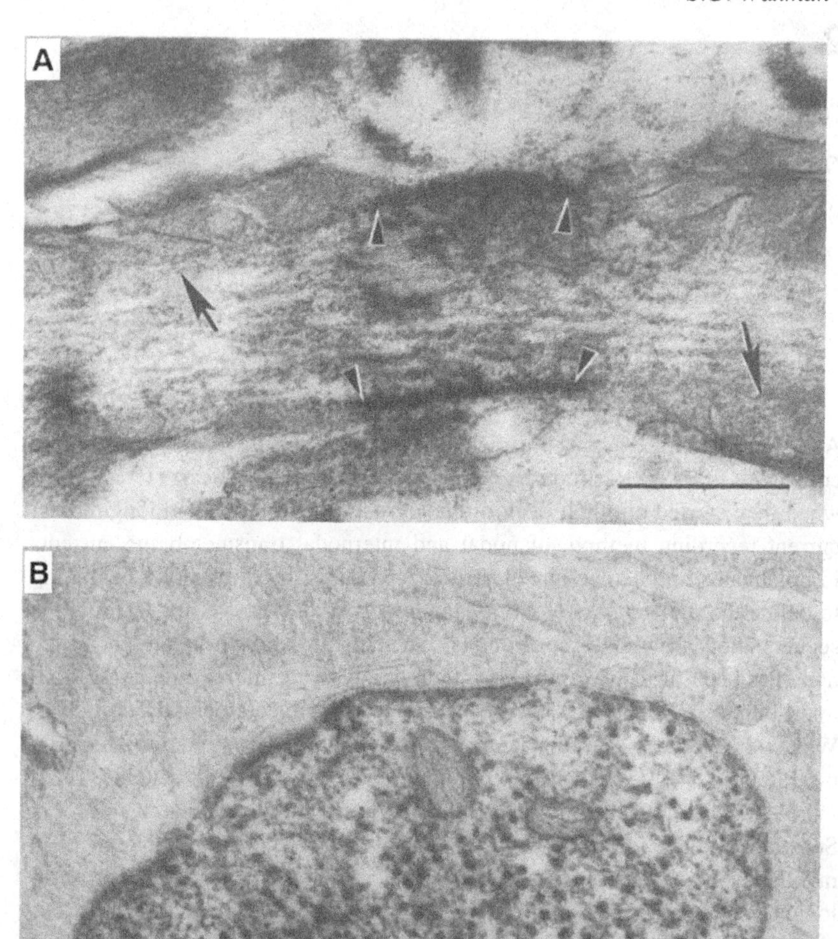

Fig. 2.1. (A) Immuno-electron micrograph showing Na^+ channel immunoreactivity of the axon membrane at the node of Ranvier (between arrowheads). The internodal axon membrane (arrows) does not stain. Adult rat optic nerve. Scale bar represents 0.5 μm. Modified from Black *et al.* (1989). (B) Na^+ channel immunoreactivity distributed along the formerly internodal axon membrane in a spinal cord axon, examined 30 days after demyelination with ethidium bromide followed by X-irradiation. Scale bar represents 1 μm. Modified from Black *et al.* (1991).

studies suggest a low Na^+ channel density ($<25/\mu m^2$) in the internodal axon membrane (Ritchie & Rogart, 1977; Shrager,1989). The clustering of Na^+ channels in the axon membrane at presumptive nodes begins early in development, prior to the formation of compact myelin or mature axo-glial paranodal junctions; at the time of development of these initial nodal precursors, however, the axon is invariably associated with abutting or loosely ensheathing glial cells (Waxman & Foster, 1980; Waxman, Black & Foster, 1982; Waxman & Ritchie, 1993).

Early in development, prior to myelination, action potential conduction is supported by a low density of Na^+ channels which are distributed apparently uniformly along the fibres. An example is provided by premyelinated axons in the neonatal rat optic nerve, where a Na^+ channel density of approximately $2/\mu m^2$ supports conduction of single action potentials; this probably reflects the high input impedance of these axons, which are very small, having a mean diameter of 0.22 μm (Waxman *et al.*, 1989).

The freeze-fracture method has been particularly useful for developmental studies on axons. This method provides *en face* views of the membrane at high magnification, and permits quantification of the density of intramembranous particles, which represent protein and glycoprotein molecules associated with the membrane; Rosenbluth (1976) presented data which suggest that large (>10 nm) particles in the external face (E-face) of the axon membrane are related to Na^+ channels. Judging from the freeze-fracture data, Na^+ channel density appears to increase substantially in premyelinated axons as they mature prior to the formation of compact myelin, approaching the values seen in normal non-myelinated axons (Black, Foster & Waxman, 1982). On the basis of a comparison with non-myelinated axons, it appears likely that Na^+ channel density in premyelinated optic nerve axons approaches a level close to $100/\mu m^2$ by the time myelination occurs. This raises the question of how Na^+ channel density is modulated so that it falls to adult levels in the internodal axon membrane in myelinated fibres.

Developmental studies using the freeze-fracture method provide some hints about this, and suggest that, following formation of compact myelin, there is a *suppression* of Na^+ channels in the underlying internodal axon membrane. These studies (Black *et al.*, 1985, 1986) utilized E-face intramembranous particle density as a measure of Na^+ channel deployment. During normal development, there is a reduction in E-face intramembranous particle density, from a value of approximately $300/\mu m^2$ in premyelinated axons just prior to myelination, to $170/\mu m^2$ in the internodal axon membrane following myelination in 19-day-old controls. This observation provides a morphological correlate for the expected drop in Na^+ channel density. Using these data from normally developing axons as a baseline for comparison, the E-face intramembranous particle density in the membrane of spinal cord axons deprived of glial cell contact (by irradiation at the time of gliogenesis) was then studied. Notably, in the glial-cell-deprived spinal cord where myelination does not occur, the E-face intramembranous particle density in glial-cell-deprived axons *increases* slightly (rather than decreasing as in normal age-matched internodal membrane), to approximately 440/μm^2, a value similar to that in normal unmyelinated axons in the control spinal cord (Fig. 2.2). This suggests that Na^+ channel density increases slightly, from normal premyelinated levels, in these glial cell-deprived axons. Consistent with this interpretation, paraplegia is not observed in developing rats after X-irradiation of the lumbar spinal cord (Black *et al.*, 1985).

Despite the paucity of glial cells in the X-irradiated spinal cord, some axons are myelinated by oligodendrocytes or Schwann cells, several weeks later than usual. Accompanying this delayed myelination, E-face intramembranous particle density decreases to 180/μm^2, a value similar to that of the normal internodal membrane (Black *et al.*, 1986). This reduction in E-face intramembranous particles is specific to myelinated axons, and does not occur in neighbouring fibres that remain unmyelinated, demonstrating that it is not a general or systemic effect. On the basis of these findings, it has been suggested that maturation involves a suppression of internodal Na^+ channel expression, which occurs consequent to formation of compact myelin (Waxman, 1987).

How myelination leads to the suppression of Na^+ channels in the underlying axon membrane is not understood. It is possible that contact-mediated cell interactions are involved, since Na^+ channels are heavily glycosylated (Trimmer & Agnew, 1989) and might interact with glial-cell-associated molecules; K^+ channels are modulated by the neural cell adhesion molecule N-CAM (Sontheimer *et al.* 1991a). Alternatively, the reduced internodal Na^+ channel density after myelination could reflect a decreased voltage gradient through the axon membrane as a result of the voltage drop in the overlying myelin. Although the half-life of Na^+ channels in the axon membrane is not known (and may be different for premyelinated, nodal and internodal membrane), it is known that the half-life of Na^+ channels in neuroblastoma cells and Schwann cells is 26 hours and 2.2 days, respectively (Waechter, Schmidt & Catterall, 1983; Ritchie, 1988). A reduction in the rate of Na^+ channel incorporation into the axon membrane after myelination could account for a decrease in the internodal Na^+ channel density over the next several days since there is presumably continuing channel turnover. Brady (1993) has proposed that there are localized changes (phosphorylation/dephosphorylation) in the axonal cytoskeleton as a result of myelination, and has suggested that these changes might affect the targeting of ion channels.

Suppression of Na^+ channel expression in the axon membrane as a result of myelination might provide an explanation for the appearance, in chronically demyelinated axons, of increased Na^+ channel densities. Fig. 2.1B is an immuno-electron micrograph showing an axon in the rat spinal cord stained

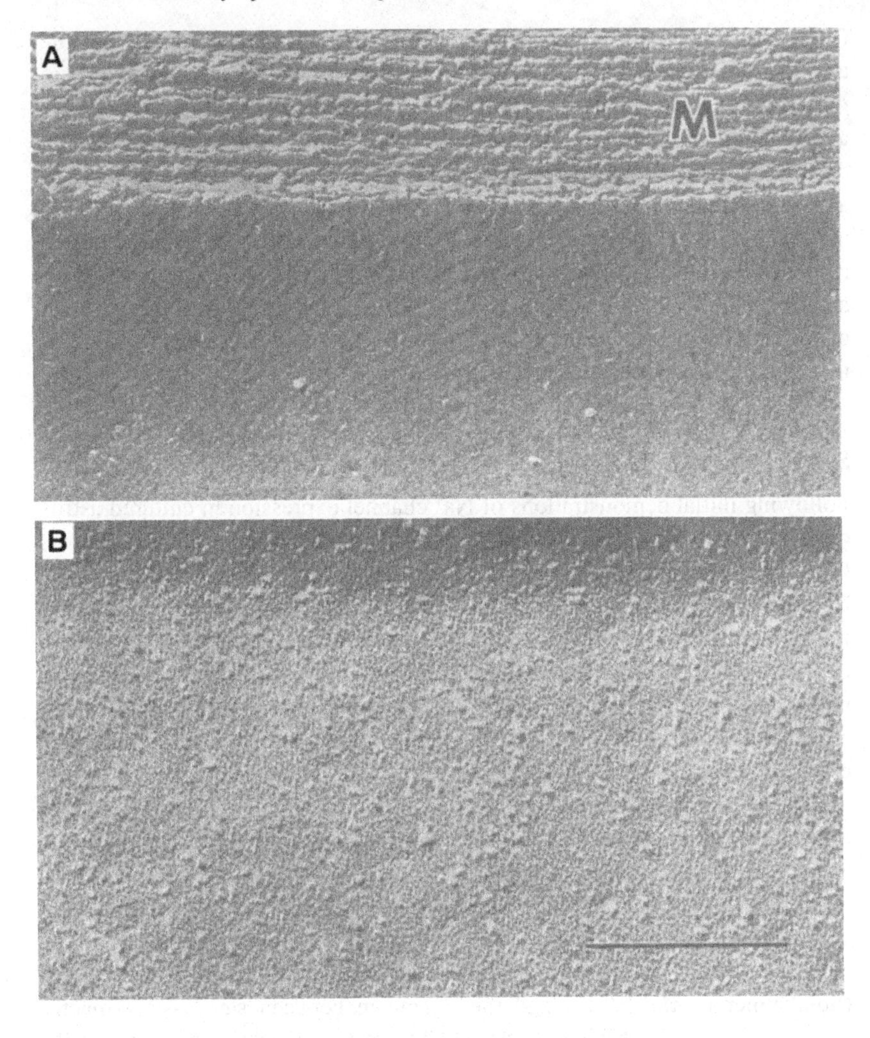

Fig. 2.2. (A) Freeze-fracture electron micrograph showing E-face of the axon membrane within a myelinated internode in 19-day-old rat optic nerve. M, myelin. The paucity of large intramembranous particles corresponds to the low Na^+ channel density. (B) In an age-matched glial-cell-deprived axon, which is not myelinated, there is a much higher density of E-face intramembranous particles. It is likely that Na^+ channel expression is suppressed in association with myelin formation. Scale bar represents 0.25 μm. Modified from Black *et al.* (1986).

for Na^+ channels, 30 days following demyelination with ethidium bromide and X-irradiation (a manoeuvre which prevents remyelination). As shown in this figure, there is distinct Na^+ channel immunoreactivity in regions of the axon membrane that were previously internodal. The increased Na^+ channel density in these regions provides a molecular correlate for conduction of action potentials in demyelinated axons which was apparent in field potential recordings through the lesion (Black *et al.*, 1991). Although the mechanisms responsible for the increase in axonal Na^+ channel density after demyelination are not understood, it is possible that it occurs as a result of a derepression of channel expression, reflecting a constitutive deployment of Na^+ channels that is unmasked when the myelin is damaged.

Sodium channels in astrocytes at the node of Ranvier

Following initial demonstrations of Na^+ channel expression in cultured astrocytes (Bevan *et al.*, 1985; Nowak, Ascher & Berwald-Netter, 1987; Barres, Chun & Corey, 1989), it has become clear that Na^+ channels are expressed by astrocytes *in situ*. Thus, immuno-ultrastructural studies demonstrate distinct Na^+ channel immunoreactivity in perinodal astrocytes fixed by perfusion of intact tissue (Black *et al.*, 1989). Electrophysiological studies using the patch-slice method (Blanton, LoTurco & Kriegstein, 1989) have demonstrated the presence of Na^+ currents, similar to those seen in cultured astrocytes, in astrocytes *in situ* within brain slices (Sontheimer & Waxman, 1993). These immunocytochemical and electrophysiological results *in situ* obviate any concern that might have been raised about astrocyte Na^+ channels being an artifact of cell culture.

Several factors, in addition to the low density of Na^+ channels in the astrocyte membrane, militate against the generation of action potentials in these cells. These include the high $K^+ : Na^+$ conductance ratio in most astrocytes (Sontheimer *et al.*, 1991b) and the mismatch between steady-state inactivation for astrocyte Na^+ channels and the resting potentials of these cells (Sontheimer & Waxman, 1992).

The function of Na^+ channels in perinodal astrocytes has not been determined. A possible clue to their role may lie in their distribution. Na^+ channel immunoreactivity in perinodal astrocytes is not confined to the plasma membrane; in contrast to the Na^+ channel immunoreactivity in the axon, which is mostly associated with the nodal axon membrane, Na^+ channel staining in perinodal astrocytes is largely cytoplasmic (Black *et al.*, 1989). Saxitoxin-binding studies provide evidence for a cytoplasmic pool of Na^+ channels, or channel precursors, in Schwann cells (the only glial cells studied by this method to date). Ritchie *et al.* (1990) found that approximately 50% of the total [^{3}H]saxitoxin binding in cultured rat Schwann cells is located in the cytoplasmic compartment; differences between the equilibrium dissociation constant (K_D) in the cytoplasmic pool and in intact cells (where saxitoxin can bind only to Na^+ channels inserted in the cell membrane) suggest a precursor–product relationship, as previously reported in growth cones (Wood, Strichartz & Pfenninger, 1989).

The presence of a cytoplasmic pool might provide a reservoir of channels or precursors that could be transferred to the adjacent nodal axon, as suggested in the hypothesis of Ritchie and co-workers (Bevan *et al.*, 1985; Gray & Ritchie, 1985) who speculated that glial cells might function as subsidiary sites for the synthesis of ion channels, subsequently transferring them (via a mechanism that is not understood) to adjacent axons. Although there is biophysical evidence indicating that Na^+ channels in some astrocytes have properties similar to those in axons (Barres *et al.*, 1989; Sontheimer *et al.*, 1991b), transfer of ion channels between cells has not been demonstrated.

Sontheimer (1993) has proposed an alternative function for astrocytic Na^+ channels, i.e. that they provide a return pathway for Na which facilitates the function of Na^+/K^+-ATPase. According to this hypothesis, even small numbers of Na^+ channels in the astrocyte membrane, at densities too low to support action potential electrogenesis, provide a 'return pathway' for Na^+ in parallel with other pathways such as ligand-gated channels and Na^+-coupled transporters, which permits maintenance of the intracellular Na^+ levels necessary for operation of the ATPase. This hypothesis suggests that Na^+ channels participate in a feedback loop modulating ionic homeostasis: depolarization of the astrocyte membrane by increased extracellular K^+ would increase the open-probability of the astrocytic Na^+ channels, providing a larger substrate for Na^+/K^+-ATPase operation, thereby facilitating re-uptake of K^+ following high-frequency action potential activity. Flux studies (Sontheimer *et al.*, 1994) provide support for this proposal, which is currently under study.

The sodium–calcium exchanger

Even though voltage-dependent Ca^{2+} channels have not been demonstrated in the axon membrane of myelinated fibres, recent studies have indicated that irreversible anoxic injury in the rat optic nerve, a CNS white matter tract, is critically dependent on the presence of extracellular Ca^{2+} (Stys *et al.*, 1990a; Ransom, Stys & Waxman, 1990a). The degree of anoxic injury of the optic nerve, measured by electrophysiological methods, is dependent on the transmembrane Ca^{2+} gradient (Stys *et al.*, 1990a). Pharmacological studies demonstrate that anoxic injury in the optic nerve is not dependent on Ca^{2+} influx via voltage-sensitive Ca^{2+} channels (Stys, Ransom & Waxman, 1990b), consistent with earlier observations which demonstrated that Ca^{2+} channels do not support action potential electrogenesis in the optic nerve (Foster, Connors & Waxman, 1982). Moreover, glutamate- or aspartate-gated Ca^{2+} channels do not mediate anoxic injury in the optic nerve (Ransom *et al.*, 1990b).

It is now clear that Ca^{2+} influx in the anoxic optic nerve is mediated by reverse operation of the Na^+–Ca^{2+} exchanger, an antiporter which mediates the coupled fluxes of Na^+ and Ca^{2+} (Stys, Waxman & Ransom, 1991, 1992). Blockade of Na^+ channels in the anoxic optic nerve with tetrodotoxin or saxitoxin provides substantial protection from anoxia, demonstrating that Na^+ channels are involved in the anoxic cascade. Perfusion of the anoxic optic nerve with zero-Na^+ (choline-substituted) solution also provides significant protection from anoxia, suggesting that coupled Na^+ and Ca^+ fluxes are involved in anoxic injury. Exposure of the anoxic optic nerve to benzamil and bepridil, which inhibit the Na^+–Ca^{2+} exchanger, results in substantially improved recovery from anoxia (Stys *et al.*, 1991, 1992).

The findings in the optic nerve indicate that a complex cascade of events is triggered by anoxia in CNS white matter tracts. As a result of depletion of ATP and failure of Na^+/K^+-ATPase, there is an increase in extracellular K^+ (Ransom *et al.*, 1992). Na^+ influx, through Na^+ channels, produces a rise in intracellular Na^+. As a result of the increased intracellular Na^+ and depolarization, the Na^+–Ca^{2+} exchanger operates in the reverse mode, bringing Ca^{2+} into the intracellular compartment (Stys *et al.*, 1991, 1992). Ultrastructural studies, which demonstrate severe damage to cytoskeletal elements (such as microtubules and neurofilaments) in the axoplasm as a result of anoxia, indicate that Ca^{2+} does, indeed, enter the axon during anoxia (Waxman *et al.*, 1992). Thus, we have hypothesized (Waxman, Ransom & Stys, 1991) that the Na^+–Ca^{2+} exchanger is present in the axon membrane, possibly co-localized with Na^+ channels at the node of Ranvier.

Activity of the Na^+–Ca^{2+} exchanger in the nodal axon membrane, in the forward mode, could provide a mechanism for extrusion of Ca^{2+} ions that leak into the axoplasm via the large number of nodal Na^+ channels, which are not perfectly selective and exhibit a finite permeability to Ca^{2+} (Meves & Vogel, 1973; Hille, 1984). Moreover, localization of the Na^+–Ca^{2+} exchanger at the node would extrude Ca^{2+} ions into the nodal gap which communicates with the extracellular space at large, rather than into the periaxonal space under the myelin or the attenuated space between the axon and myelin-forming cell at the paranode, where Ca^{2+} extruded from the axon could damage the myelin-forming cell (Schlaepfer, 1977; Smith & Hall, 1988).

Might the nodal Na^+–Ca^{2+} exchanger ever function, under normal physiological circumstances, in the reverse mode, importing Ca^{2+} in exchange for Na^+? This type of reverse exchange has, in fact, been observed in myocardial cells where it triggers further Ca^{2+} release from the sarcoplasmic reticulum (Leblanc & Hume, 1990). Persistent depolarization (Kocsis, Ruiz & Waxman, 1983) and increased intracellular Na^+ at the node (Bergmann, 1970), which are observed during high-frequency activity in some axons, both favour reverse exchange. Thus we may speculate that as a result of repetitive activity in some axons, the Na^+–Ca^{2+} exchanger might import Ca^{2+}

into the subnodal axoplasm, where it could elicit Ca^{2+}-induced Ca^{2+} release and/or activate second-messenger cascades.

Multiple sodium conductances

In at least some types of sensory neurones, several types of Na^+ channels with different kinetics and voltage dependence are expressed (Caffrey *et al.*, 1992; Rizzo, Waxman & Kocsis, 1993). Within the optic nerve, there is now evidence for the presence of a non-activating Na^+ conductance, in addition to rapidly activating and rapidly inactivating voltage-dependent Na^+ channels. Stys *et al.* (1991, 1992) speculated that a non-inactivating Na^+ conductance mediates the pathological Na^+ influx in the anoxic optic nerve. Recently, experiments using a gap recording method have demonstrated that application of tetrodotoxin to optic nerves, either resting or following depolarization in solutions containing 15 or 40 mM K^+, results in hyperpolarizing shifts in membrane potential. The conductance responsible for these shifts is present at rest where it contributes to the resting potential, and persists in nerves depolarized sufficiently to completely abolish classical transient Na^+ currents (Stys *et al.*, 1993). The tetrodotoxin sensitivity of this conductance suggests that it reflects the presence of persistent (non-inactivating) Na^+ channels in optic nerve axons (Stys *et al.*, 1993). These channels may participate in setting the threshold of the axon.

Whether this persistent Na^+ conductance represents a homogeneous population of Na^+ channels some of which do not inactivate at depolarized membrane potentials, or a distinct subpopulation of Na^+ channels, is not yet clear (see also Chapter 18). 'Window' currents can produce a persistent Na^+ conductance (Sontheimer & Waxman, 1992). Alternatively, Alzheimer, Schwindt & Crill (1993) have suggested that a uniform population of Na^+ channels might generate both transient and persistent Na^+ currents by switching between gating modes. A third possibility is the presence of a distinct subpopulation of non-inactivating Na channels in the axon membrane. Irrespective of the molecular substrate, Na^+ conductance at the node of Ranvier consists of two parallel tetrodotoxin-sensitive pathways, one of which is non-inactivating.

An axoplasmic diffusion barrier for sodium

Finally, recent work has refocused attention on the possibility of a diffusion barrier which sequesters ions in the axoplasm subjacent to the nodal part of the axon membrane. It is well established that, after ultrastructural staining with uranyl salts, an 'undercoating' can be observed extending into the axoplasm just beneath the nodal axon membrane (Peters, 1966; Berthold & Rydmark, 1983). Following aldehyde fixation as shown in Fig. 2.3, externally

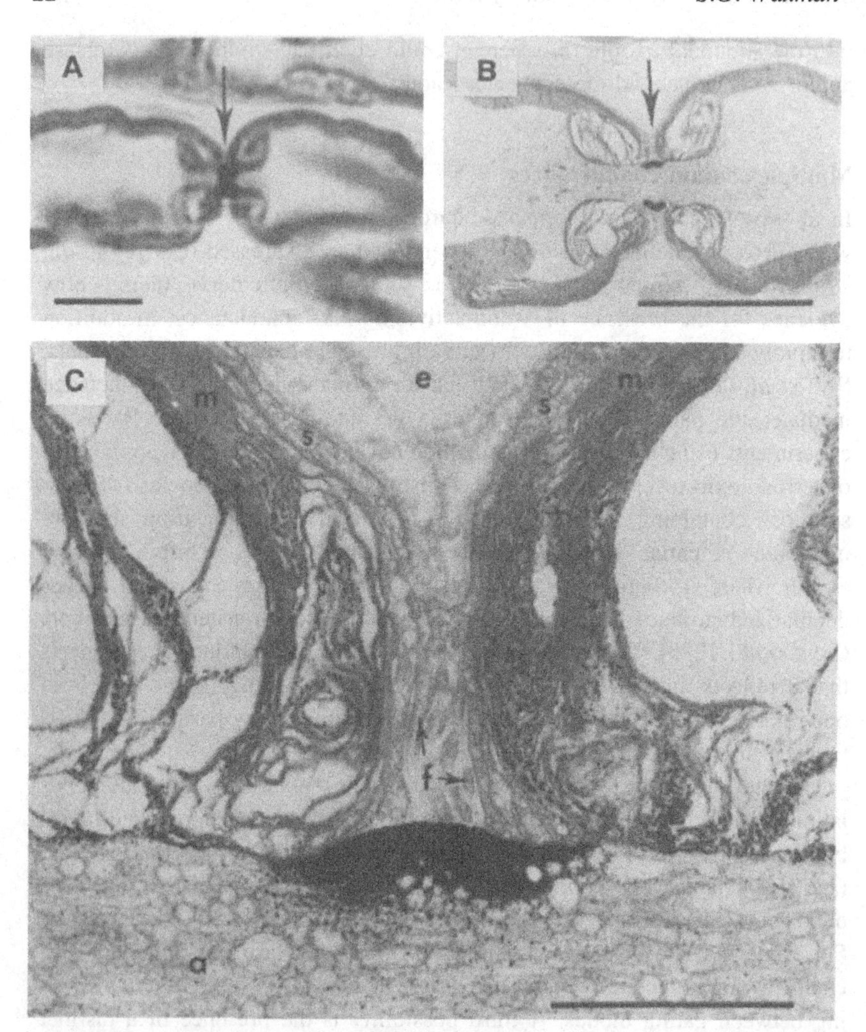

Fig. 2.3. Ferrocyanide staining in an axon from rat sciatic nerve after osmium tetroxide fixation. (A) Light microscopy of 3 μm section reveals dense nodal staining. Scale bar represents 10 μm. (B) After thin sectioning, electron microscopy demonstrates a cytoplasmic localization of the stain. Scale bar represents 10 μm. (C) Higher magnification indicates that the stain is confined to a domain less than 0.25 μm thick on the cytoplasmic side of the nodal axon membrane. a, axon; f, finger-like Schwann cell processes; s, Schwann cell; e, perinodal extracellular space; m, myelin. Scale bar represents 1 μm. Modified from Quick & Waxman (1977b).

applied ferrocyanide can cross the nodal axolemma, but does not diffuse freely through the bulk of the axoplasm, and remains confined to a limited compartment extending less than 0.25 μm into the cytoplasm beneath the axon membrane at the node (Quick & Waxman, 1977b).

Electrophysiological results provide evidence for an accumulation of Na^+

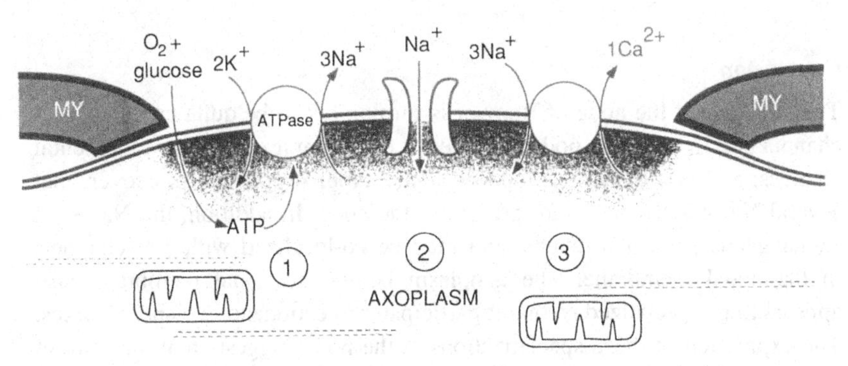

Fig. 2.4. Model proposing the co-localization of Na^+ channels (2), Na^+–Ca^{2+} exchanger (3) and Na^+/K^+-ATPase (1) in the axon membrane at the node. A diffusion barrier confines cations to a 'fuzzy space' (stippling) subjacent to the nodal axon membrane. MY, myelin. Modified from Waxman *et al.* (1991).

on the inner surface of the nodal axon membrane following electrical activity. Thus, voltage-clamp studies on nodes of Ranvier have demonstrated that the Na^+ equilibrium potential is altered as a result of repetitive activity; the predicted Na^+ concentration at the inner surface of the nodal axon membrane is much higher than the concentration that would occur with free diffusion of Na^+ within the axoplasm, suggesting that the concentration of Na^+ on the inner surface of the axon membrane increases as a result of a diffusion barrier (Bergmann, 1970). Similarly, the presence of a 'fuzzy space' subjacent to the inner surface of the cell membrane has been postulated in cardiac myocytes, where Na^+ ions are required to trigger Ca^{2+} influx via reverse operation of the Na^+–Ca^{2+} exchanger (Lederer *et al.*, 1990).

Although the presence of a diffusion barrier which collects Na^+ in the nodal axoplasm has not yet been definitively demonstrated, barriers to the diffusion of other cations have been observed within neurones. For example non-uniform cytoplasmic distribution of Ca^{2+} ions has been demonstrated during stimulated Ca^{2+} influxes in neurones (Connor, 1986; Lipscombe *et al.*, 1988). The presence of a 'fuzzy space' in the axoplasm, which retards diffusion of cations away from the node of Ranvier, would tend to accumulate Na^+ ions in the cytoplasmic domain beneath the Na^+-channel-rich axon membrane, and would thus facilitate extrusion via the Na^+/K^+-ATPase which has been demonstrated immunocytochemically (Ariyasu, Nichol & Ellisman, 1985) at the node of Ranvier. Co-localization of Na^+ channels, Na^+/K^+-ATPase, the Na^+–Ca^{2+} exchanger and a 'fuzzy space' at the node is shown diagrammatically in the hypothetical model in Fig. 2.4 (Waxman *et al.*, 1991). The availability of dyes that permit optical measurement of Na^+ and

Ca^{2+} levels within cells may provide, in the near future, a demonstration of the nodal 'fuzzy space'.

Conclusion

The structure of the node of Ranvier is turning out to be quite complex. Na^+ channel clustering at the node is the result of a complex set of developmental changes, and is a dynamic property of the axon. It appears, moreover, that several Na^+ conductances are present at the node. In addition, the Na^+–Ca^{2+} exchanger is present in axons, and may be co-localized with Na^+ channels in the nodal membrane. The axoplasm below the nodal membrane also appears to be specialized so that it participates in cation storage and/or fluxes. The expression of these specializations at the node suggests that our current understanding of nodal physiology is probably incomplete. It is likely that, over the next few years, functional and pathological correlates of the molecular organization of the node of Ranvier will become better understood.

Acknowledgements

Research in the author's laboratory has been supported in part by Grants from the NINCDS and the National Multiple Sclerosis Society, and by the Medical Research Service, Department of Veterans Affairs.

References

Alzheimer, C., Schwindt, P.C. & Crill, W.E. (1993). Modal gating of Na^+ channels as a mechanism of persistent Na^+ current in pyramidal neurones from rat and cat sensorimotor cortex. *Journal of Neuroscience*, **13**, 660–673.

Ariyasu, R.G., Nichol, J.A. & Ellisman, M.H. (1985). Localization of sodium/potassium adenosine triphosphatase in multiple cell types of the murine nervous system with antibodies raised against the enzyme from kidney. *Journal of Neuroscience*, **5**, 2581–2596.

Barres, B.A., Chun, L.L.Y. & Corey, D.P. (1989). Glial and neuronal forms of the voltage-dependent sodium channels: characteristics and cell-type distribution. *Neuron*, **2**, 1375–1388.

Bergmann, C. (1970). Increase of sodium concentration near the inner surface of the nodal membrane. *Pflügers Archiv*, **317**, 287–302.

Berthold, C. & Rydmark, M. (1983). Electron microscopic serial section analysis of nodes of Ranvier in lumbosacral spinal roots of the cat: ultrastructural organization of nodal compartments in fibres of different sizes. *Journal of Neurocytology*, **12**, 475–505.

Bevan, S., Chiu, S.Y., Gray, P.T.A. & Ritchie, J.M. (1985). The presence of voltage-gated sodium, potassium and chloride channels in rat cultured astrocytes. *Proceedings of the Royal Society of London, Series B*, **225**, 229–313.

Black, J.A., Felts, P., Smith, K.J., Kocsis, J.D. & Waxman, S.G. (1991). Distribution of sodium channels in chronically demyelinated spinal cord axons: immuno-ultrastructural localization and electrophysiological observations. *Brain Research*, **544**, 59–70.

Black, J.A., Foster, R.E. & Waxman, S.G. (1982). Rat optic nerve: freeze-fracture studies during development of myelinated axons. *Brain Research*, **250**, 1–10.

Black, J.A., Friedman, B., Waxman, S.G., Elmer, L.W. & Angelides, K.J. (1989). Immuno-ultrastructural localization of sodium channels at nodes of Ranvier and perinodal astrocytes in rat optic nerve. *Proceedings of the Royal Society of London, Series B*, **238**, 38–57.

Black, J.A., Sims, T.J., Waxman, S.G. & Gilmore, S.A. (1985). Membrane ultrastructure of developing axons in glial cell deficient rat spinal cord. *Journal of Neurocytology*, **14**, 79–104.

Black, J.A., Waxman, S.G., Sims, T.J. & Gilmore, S.A. (1986). Effects of delayed myelination by oligodendrocytes and Schwann cells on the macromolecular structure of axonal membrane in rat spinal cord. *Journal of Neurocytology*, **15**, 745–762.

Blanton, M.G., LoTurco, J.J. & Kriegstein, A.R. (1989). Whole-cell recordings from neurones in slices of reptilian and mammalian cerebral cortex. *Journal of Neuroscience Methods*, **30**, 203–210.

Bostock, H. & Sears, T.A. (1976). Continuous conduction in demyelinated mammalian nerve fibres. *Nature*, **263**, 786–787.

Bostock, H. & Sears, T.A. (1978). The internodal axon membrane: electrical excitability and continuous conduction in segmental demyelination. *Journal of Physiology (London)*, **280**, 273–301.

Brady, S.T. (1993). Axonal dynamics and regeneration. In *Neuroregeneration*, ed. A. Gorio, pp. 7–36. New York: Raven Press.

Caffrey, J.M., Eng, D.L., Black, J.A., Waxman, S.G. & Kocsis, J.D. (1992). Three types of sodium channels in adult rat dorsal root ganglion neurones. *Brain Research*, **592**, 283–297.

Chiu, S.Y. & Ritchie, J.M. (1981). Evidence for the presence of potassium channels in the paranodal region of acutely demyelinated mammalian nerve fibres. *Journal of Physiology (London)*, **313**, 415–437.

Chiu, S.Y. & Ritchie, J.M. (1982). Evidence for the presence of potassium channels in the internode of frog myelinated nerve fibres. *Journal of Physiology (London)*, **322**, 485–501.

Connor, J.A. (1986). Digital imaging of free calcium changes and of spatial gradients in growing processes in single, mammalian central nervous system cells. *Proceedings of the National Academy of Sciences, USA*, **83**, 6179–6183.

Elmer, L.W., Black, J.A., Waxman, S.G. & Angelides, K.J. (1990). The voltage-dependent sodium channel in mammalian CNS and PNS: Antibody characterization and immunocytochemical localization. *Brain Research*, **532**, 222–231.

Foster, R.E., Connors, B.W. & Waxman, S.G. (1982). Rat optic nerve: electrophysiological, pharmacological, and anatomical studies during development. *Developmental Brain Research*, **3**, 361–376.

Gray, P.T. & Ritchie, J.M. (1985). Ion channels in Schwann and glial cells. *Trends in Neurosciences*, **8**, 411–415.

Hille, B. (1984). *Ionic Channels of Excitable Membranes*. Sunderland: Sinaur Associates.

Kocsis, J.D., Ruiz, J.A. & Waxman, S.G. (1983). Maturation of mammalian myelinated fibres: changes in action-potential characteristics following 4-aminopyridine application. *Journal of Neurophysiology*, **50**, 449–463.

Leblanc, N. & Hume, J. (1990). Sodium current-induced release of calcium from cardiac sarcoplasmic reticulum. *Science*, **248**, 372–376.

Lederer, W.J., Niggli, E. & Hadley, R.W. (1990). Sodium–calcium exchange in excitable cells: fuzzy space. *Science*, **248**, 283.

Lipscombe, D., Madison, D.V., Poenie, M., Reuter, H., Tsien, R.W. & Tsien, R.Y. (1988). Imaging of cytosolic Ca^{2+} transients arising from Ca^{2+} stores and Ca^{2+} channels in sympathetic neurones. *Neuron*, **1**, 355–365.

Meves, H. & Vogel, W.J. (1973). Calcium inward currents in internally perfused giant axons. *Journal of Physiology (London)*, **235**, 225–265.

Neumcke, B. & Stämpfli, R. (1982). Sodium currents and sodium-current fluctuations in rat myelinated nerve fibres. *Journal of Physiology (London)*, **329**, 163–184.

Nowak, L., Ascher, P. & Berwald-Netter, Y. (1987). Ionic channels in mouse astrocytes in culture. *Journal of Neuroscience*, **7**, 101–109.

Peters, A. (1966). The node of Ranvier in the central nervous system. *Quarterly Journal of Experimental Physiology*, **51**, 229–236.

Quick, D.C. & Waxman, S.G. (1977a). Specific staining of the axon membrane at nodes of Ranvier with ferric ion and ferrocyanide. *Journal of the Neurological Sciences*, **31**, 1–11.

Quick, D.C. & Waxman, S.G. (1977b). Ferric ion, ferrocyanide, and inorganic phosphate as cytochemical reactants at peripheral nodes of Ranvier. *Journal of Neurocytology*, **6**, 555–570.

Ransom, B.R., Stys, P.K. & Waxman, S.G. (1990a). The pathophysiology of anoxic injury in central nervous system white matter. *Stroke*, **21** (Suppl. III), III-52–57.

Ransom, B.R., Walz, W., Davis, P.K. & Carlini, W.G. (1992). Anoxia-induced changes in extracellular K^+ and pH in mammalian central white matter. *Journal of Cerebral Blood Flow and Metabolism*, **12**, 593–602.

Ransom, B.R., Waxman, S.G. & Davis, P.K. (1990b). Anoxic injury of CNS white matter: protective effect of ketamine. *Neurology*, **40**, 1399–1404.

Rasminsky, M. & Sears, T.A. (1972). Internodal conduction in undissected demyelinated nerve fibres. *Journal of Physiology (London)*, **227**, 323–350.

Ritchie, J.M. (1988). Sodium-channel turnover in rabbit cultured Schwann cells. *Proceedings of the Royal Society of London, Series B*, **233**, 423–430.

Ritchie, J.M. (1995). Physiology of axons. In *The Axon*, ed. S.G. Waxman, J.D. Kocsis & P. Stys, pp. 68–96. Oxford University Press.

Ritchie, J.M., Black, J.A., Waxman, S.G. & Angelides, K.J. (1990). Sodium channels in the cytoplasm of Schwann cells. *Proceedings of the National Academy of Sciences, USA*, **87**, 9290–9294.

Ritchie, J.M. & Rogart, R.B. (1977). The density of sodium channels in mammalian myelinated nerve fibres and the nature of the axonal membrane under the myelin sheath. *Proceedings of the National Academy of Science, USA*, **74**, 211–215.

Rizzo, M.A., Waxman, S.G. & Kocsis, J.D. (1993). Heterogeneity of voltage-dependent Na^+ conductance in dorsal root ganglion neurons of adult rat. *Society for Neuroscience Abstracts*, **19**, 1528.

Rosenbluth, J. (1976). Intramembranous particle distribution at the node of Ranvier and adjacent axolemma in myelinated axons of the frog brain. *Journal of Neurocytology*, **5**, 731–745.

Schlaepfer, W.W. (1977). Structural alterations of peripheral nerve induced by the calcium ionophore A23187. *Brain Research*, **136**, 1–9.

Shrager, P. (1989). Sodium channels in single demyelinated mammalian axons. *Brain Research*, **483**, 149–154.

Smith, K.J. & Hall, S.M. (1988). Peripheral demyelination and remyelination initiated by the calcium-selective ionophore ionomycin: *in vivo* observations. *Journal of Neurological Science*, **83**, 37–53.

Sontheimer, H. (1993). Glia, as well as neurons, express a diversity of ion channels. *Canadian Journal of Physiology and Pharmacology*, **70**, S223-S238.

Sontheimer, H., Fernandez-Marques, E., Ullrich, N., Pappas, C.A. & Waxman, S.G. (1994). Astrocyte Na^+ channels are required for maintenance of Na^+/K^+-ATPase activity. *Journal of Neuroscience*, **14**, 2464–2476.

Sontheimer, H., Kettermann, H., Schachner, M. & Trotter, J. (1991a). The neural cell adhesion molecule (N-CAM) modulates K^+ channels in cultured glial precursor cells. *European Journal of Neuroscience*, **3**, 230–236.

Sontheimer, H., Ransom, B.R., Cornell-Bell, A.H., Black, J.A. & Waxman, S.G. (1991b). Na^+-current expression in rat hippocampal astrocytes *in vitro*: alterations during development. *Journal of Neurophysiology*, **65**, 3–19.

Sontheimer, H. & Waxman, S.G. (1992). Ion channels in spinal cord astrocytes *in vitro*. II. Biophysical and pharmacological analysis of Na^+ currents. *Journal of Neurophysiology*, **68**, 1000–1011.

Sontheimer, H. & Waxman, S.G. (1993). Expression of voltage-activated ion channels by astrocytes and oligodendrocytes in the hippocampal slice. *Journal of Neurophysiology*, **70**, 1863–1873.

Stys, P.K., Ransom, B.R. & Waxman, S.G. (1990b). Effects of polyvalent cations and dihydropyridine calcium channel blockers on recovery of CNS white matter from anoxia. *Neuroscience Letters*, **115**, 293–299.

Stys, P.K., Ransom, B.R., Waxman, S.G. & Davis, P.K. (1990a). Role of extracellular calcium in anoxic injury of mammalian central white matter. *Proceedings of the National Academy of Sciences, USA*, **87**, 4212–4216.

Stys, P.K., Sontheimer, H., Ransom, B.R. & Waxman, S.G. (1993). Non-inactivating, TTX-sensitive Na^+ conductance in rat optic nerve axons. *Proceedings of the National Academy of Sciences, USA*, **90**, 6976–6981.

Stys, P.K., Waxman, S.G. & Ransom, B.R. (1991). Na^+–Ca^{2+} exchanger mediates Ca^{2+} influx during anoxia in mammalian central nervous system white matter. *Annals of Neurology*, **30**, 375–380.

Stys, P.K., Waxman, S.G. & Ransom, B.R. (1992). Ionic mechanisms of anoxic injury in mammalian CNS white matter: role of Na^+ channels and Na^+–Ca^{2+} exchanger. *Journal of Neuroscience*, **12**, 430–439.

Trimmer, J.S. & Agnew, W.S. (1989). Molecular diversity of voltage-sensitive sodium channels. *Annual Review of Physiology*, **51**, 401–418.

Vogel, W. & Schwarz, J.R. (1995). Voltage-clamp studies in frog, rat, and human axons: macroscopic and single currents. In *The Axon*, ed. S.G. Waxman, J.D. Kocsis & P.K. Stys, pp. 257–280. Oxford: Oxford University Press.

Waechter, C.J., Schmidt, J.W. & Catterall, W.A. (1983). Glycosylation is required for maintenance of functional sodium channels in neuroblastoma cells. *Journal of Biological Chemistry*, **258**, 5117–5123.

Waxman, S.G. (1987). Molecular neurobiology of the myelinated nerve fibre: ion-channel distributions and their implications for demyelinating diseases. In *Molecular Neurobiology in Neurology and Psychiatry*, ed. E.R. Kandel, pp. 7–37. New York: Raven Press.

Waxman, S.G., Black, J.A. & Foster, R.E. (1982). Freeze-fracture heterogeneity of the axolemma of premyelinated fibres in the CNS. *Neurology*, **32**, 418–421.

Waxman, S.G., Black, J.A., Kocsis, J.D. & Ritchie, J.M. (1989). Low density of sodium channels supports action potential conduction in axons of neonatal rat optic nerve. *Proceedings of the National Academy of Sciences, USA*, **86**, 1406–1410.

Waxman, S.G., Black, J.A., Stys, P.K. & Ransom, B.R. (1992). Ultrastructural concomitants of anoxic injury and early post-anoxic recovery in rat optic nerve. *Brain Research*, **574**, 105–119.

Waxman, S.G. & Foster, R.E. (1980). Development of the axon membrane during differentiation of myelinated fibres in spinal nerve roots. *Proceedings of the Royal Society of London, Series B*, **209**, 441–446.

Waxman, S.G., Ransom, B.R. & Stys, P.K. (1991). Non-synaptic mechanisms of calcium-mediated injury in CNS white matter. *Trends in Neurosciences*, **14**, 461–468.

Waxman, S.G. & Ritchie, J.M. (1993). Molecular dissection of the myelinated axon. *Annals of Neurology*, **33**, 121–136.

Wood, M.R., Strichartz, G. & Pfenninger, K.H. (1989). Compartmentalization of Na$^+$ channels in nerve growth cones: conversion during plasmalemmal insertion. *Journal of Cell Biology*, **109**, 335.

3

Delayed rectifier type potassium currents in rabbit and rat axons and rabbit Schwann cells

MARK D. BAKER

Department of Pharmacology, Yale University School of Medicine, New Haven, Connecticut, USA

Introduction

Impulse transmission along a myelinated axon is saltatory, where only the axonal membrane at the nodes of Ranvier is subject to the full action potential. As a consequence of the cytoarchitecture of myelinated nerve, a node of Ranvier can be charged and discharged with a time constant of $\ll$1 ms as its capacity is small (1–2 pF), and its apparent leakage conductance is large (about 20 nS). Delayed rectification is thus not necessary for rapid nodal repolarization. Na^+ channel inactivation and the cessation of Na^+ ion influx allows rapid repolarization with outward current flow in the leakage pathway. Pharmacological blockade of *Xenopus* or *Rana* nodal delayed rectifier with, for example, tetraethylammonium (TEA) ions (Schmidt & Stämpfli, 1966) or extracellular gallamine (Smith & Schauf, 1981; modelled by Frankenhaeuser & Huxley, 1964) causes only a slight prolongation of the action potential (~0.3 ms at 20 °C). The few rapidly activating K^+ channels found at mammalian nodes do not contribute significantly to repolarization (rabbit: Chiu *et al.*, 1979; rat: Brismar, 1980).

The large leakage conductance of the node of Ranvier, as measured in nodal voltage-clamp experiments, is now believed to be part of a current pathway under or through the myelin sheath that allows the internodal axolemma to repolarize the node (Barrett & Barrett, 1982; Blight, 1985; Baker *et al.*, 1987). The movement of charge in this circuit from the node to the internodal capacity during an action potential gives rise to the depolarizing (or negative) after-potential, and an associated phase of increased excitability to applied currents.

Delayed rectifier potassium channels are under the myelin

The presence of delayed rectifier K^+ channels under the myelin in rabbit nerve was first demonstrated in voltage clamp by Chiu & Ritchie (1981). Active K^+ current was revealed by paranodal demyelination while performing a nodal voltage clamp. As the series resistance provided by the myelin was reduced by superfused lysolethicin, unmistakable rapidly activating, voltage-dependent K^+ currents were revealed. After the removal of the myelin, the paranodal axolemma (like the node) fully experienced the imposed voltage-clamp steps, and the paranodal K^+ channels were thus activated. Similarly, Ritchie & Chiu (1981) also found that delayed rectifier K^+ channels were present under compact myelin, where the myelin was again removed by lysolethicin treatment. Voltage-clamp evidence for paranodal and internodal K^+ channels in rat nerve was subsequently obtained by Röper & Schwarz (1989). The apparent threshold for activation and the kinetics of internodal K^+ currents suggest that a conductance similar to G_{Kf1}, described by Dubois (1981) at the frog node, is involved in generating the currents.

Pharmacological characteristics of the delayed rectifier potassium channels

Chiu & Ritchie (1981) reported the effects of tetraethylammonium (TEA) ions on the K^+ currents revealed by paranodal demyelination in rabbit nerve. The currents could be blocked by internal TEA, but they were hardly affected by external TEA (20–60 mM). Bostock, Sears & Sherratt (1981) examined the effects of 4-aminopyridine (4-AP) and TEA on the action currents generated by segmentally demyelinated rat spinal root axons and on C-fibre compound action potentials. Five millimolar 4-AP abolished the late, outward, membrane currents produced during continuous conduction through a demyelinated internode, indicating the involvement of 4-AP-sensitive delayed rectifier K^+ channels in the repolarization phase of the pathological action potential. Low millimolar concentrations of 4-AP also substantially widened the C-fibre compound action potential whereas external TEA ions had little effect on action potential duration, even at 50 mM.

The channels generating the internodal delayed rectifier current in rabbit and rat axons may therefore lack the external TEA binding site that confers sensitivity to the blocker within the micromolar range. If so, they share this characteristic with several other types of delayed rectification, including that of the squid giant axon (Tasaki & Hagiwara, 1957). It can be argued that external TEA ions have some difficulty diffusing to the internodal axolemma in intact myelinated nerve, and that this might explain, at least in part, the differential effects of TEA on fast and slow electrotonus recorded from rat nerve (attributable to actions at the node and internode, respectively; Baker *et al.*, 1987). In comparison, 4-AP probably diffuses through membranes in its non-protonated form, and blocks K^+ channels under the myelin when applied extra-axonally. External TEA ions presumably diffuse more easily to the axolemma after myelin disruption or removal, which may occur as a result of the superfusion of a detergent, penetration of the myelin by a microelectrode, or death of the ensheathing Schwann cells after exposure to diphtheria toxin (Rasminsky & Sears, 1972). It is well known that an invertebrate Schwann cell does not provide a significant diffusion barrier to molecules that are much larger than TEA ions. For example, tetrodotoxin selectively blocks Na^+ currents in lobster giant axons at nanomolar concentrations (Narahashi, Moore & Scott, 1964). Taken together with the reported pharmacological characteristics of the K^+ currents generated by cultured rat dorsal root ganglion cells (Kostyuk *et al.*, 1981) these findings suggest that, like the mammalian central A current (Numann, Wadman & Wong, 1987), the mammalian peripheral nerve delayed rectifier may be very insensitive to external TEA ions.

Differences in sensitivity to TEA and other pharmacological agents have already been used to discriminate between K^+ channel types that appear similar in other respects. Recent evidence obtained by cloning and expressing mammalian brain K^+ channels in *Xenopus* oocytes has revealed that some mammalian brain K^+ channels, e.g. homomultimeric RCK1 and RCK5, which appear very similar on the basis of their electrophysiological characteristics are, in fact, different channels, coded for by different mRNAs (Stühmer *et al.*, 1989). The sensitivity of these two channel types to external TEA is very different (IC_{50} on RCK1 = 600 μM; IC_{50} on RCK5 = 129 mM). Furthermore, mutations of only a single residue in an expressed channel can alter the IC_{50} for external TEA from 600 mM to 400 μM (e.g. mutation R476Y in RMK2; Liman, Tytgat & Hess, 1992). Similarly, Hurst *et al.* (1991) have demonstrated that three amino acid residues positioned in the loop between transmembrane domains S5 and S6 are crucial for the binding of a toxin in green mamba venom, α-dendrotoxin (α-DTX), to a rat brain K^+ channel, RBK1 (sequentially homologous with RCK1). In particular the channel must apparently have a negatively charged amino acid residue at the position corresponding to Glu^{353} in order for α-DTX to have an IC_{50} for block of less than 20 nM. Thus marked differences in the sensitivity of K^+ channels to blockers and toxins have been demonstrated to be related to differences in primary amino acid structure.

Potassium channels in rabbit Schwann cells

Rabbit Schwann cells in culture express voltage-activated Na^+ and K^+ channels (Chiu, Shrager & Ritchie,1984; Shrager, Chiu & Ritchie, 1985), which Chiu *et al.* (1984) provocatively described as of 'neuronal type'. The puzzle as to why Schwann cells should express Na^+ channels, normally associated with electrical excitability and the generation and transmission of action potentials, remains unresolved. The characteristic is also, interestingly, shared by astrocytes (e.g. Bevan *et al.*, 1985). One suggestion is that satellite cells make channels for *transfer* to the neighbouring axon (Gray & Ritchie, 1985), and in this way, the burden of manufacturing channels for insertion into a

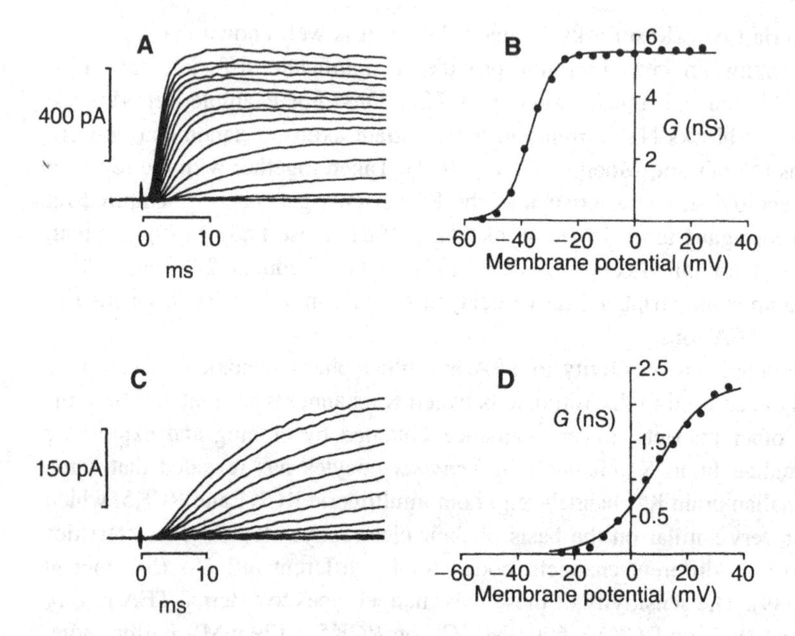

Fig. 3.1. (A) Family of outward K^+ currents recorded from a Schwann cell that appeared to generate only type I current over the range of membrane potentials presented. The currents were evoked by stepping the membrane potential, at time zero, from a pre-pulse potential of −120 mV. (B) Conductance versus membrane potential (*G–E*) plot derived from recordings in (A). (C) Family of type II currents, recorded from another Schwann cell that did not generate type I current. (D) *G–E* plot derived from recordings presented in (C). For plots in (B) and (D) values of mean current were obtained between 35 and 40 ms and converted to a conductance assuming E_K = −86 mV. Membrane potential values have been corrected for residual series-resistance errors. From Baker, Howe & Ritchie (1993).

long axon could, at least, be shared between the neurone and its satellite cells. However, an intercellular channel transfer mechanism remains in the realm of speculation, and the transfer hypothesis stands alone only because no other plausible scheme has been put forward for the role of Na^+ channels (in particular) in satellite cells.

In order to determine whether rabbit Schwann cell K^+ channels conform to the 'neuronal type', we have recently studied the characteristics of whole cell K^+ currents and single K^+ channels in rabbit Schwann cells (Howe, Baker & Ritchie, 1992; Baker, Howe & Ritchie, 1993). The Schwann cell K^+ currents appear to be classifiable into categories similar to those for K^+ currents described at amphibian nodes of Ranvier (Dubois, 1981; Bräu *et al.*, 1990). The currents include two types that are sensitive to 4-AP (type I and type II; Fig. 3.1) both of which deactivate rapidly (with normal extracellular $[K^+]$) at potentials more negative than their thresholds for activation. Type I current is selectively blocked by nanomolar concentrations of α-DTX, with an apparent K_D of 1.3 nM, whereas type II current is insensitive to concentrations up to 500 nM (Baker *et al.*, 1993).

One of the fast deactivating K^+ currents at *Xenopus* nodes, K_{f1}, has been shown by Bräu *et al.* (1990) to be blocked selectively by nanomolar concentrations of α-DTX. This same toxin also blocks rapidly activating, voltage-activated K^+ channels in mammalian dorsal root ganglion cells where the IC_{50} for block is reported to be close to 2 nM (Stansfeld & Feltz, 1988). Patch-clamp experiments performed at the paranodal loops of an intact Schwann cell associated with an axon (Wilson & Chiu, 1990) have provided evidence for channels with the same unitary conductance and similar activation kinetics as the putative type I channels in cultured Schwann cells (Howe *et al.*, 1992), suggesting that Schwann cells *in situ* express type I channels. It might therefore be concluded that the Schwann cell K^+ channels do indeed appear to be of 'neuronal type', particularly as it has been suggested that dendrotoxin sensitivity is probably an exclusive property of K^+ channels in neuronal membranes (Dreyer, 1990). The possibility is thus raised that not only Na^+ channels but also K^+ channels might be manufactured for transfer from Schwann cells to axons.

However, this interpretation may be incorrect. While our present knowledge of the characteristics of the macroscopic rabbit type II Schwann cell current suggests that it is similar in its voltage dependence, deactivation kinetics, inactivation kinetics and pharmacology to amphibian axonal K_{f2} currents (Baker & Ritchie, 1993), two clear differences have so far emerged between the rabbit axonal fast delayed rectifier and Schwann cell type I currents. Firstly, their apparent activation thresholds differ by 15–20 mV (external pH 7.2), and secondly, the Schwann cell current is blocked by micromolar concentrations of external TEA (K_D ~200 μM; Howe & Ritchie, 1988), while the axonal delayed rectifier is apparently much less sensitive to this ion applied externally (discussed above). The first difference may be ephemeral, in as much as the apparent voltage dependence of channel gating might be expected to depend on the constitution and surface charge of the particular cell membrane into which it is inserted. However, the pharmacological difference suggests a difference in amino acid sequence between the axonal K_{f1} and Schwann cell type I channels, and also that type I K^+ currents are generated by channels that are not found in axons in significant numbers. If substantiated by further experiments, this suggests that if channel transfer occurred from Schwann cell to axon, type I channels would have to be excluded from the process.

The distribution of K^+ channels in mammalian axons has been investigated quite thoroughly (for review see e.g. Black, Kocsis & Waxman, 1990), and their different physiological functions have been substantially clarified over the last decade (e.g. Baker *et al.*, 1987; Kocsis *et al.*, 1987; Baker & Bostock, 1989). However, one challenge which so far has not received sustained attention is to provide evidence as to the physiological and pathophysiological functioning of the myelinated axon associated Schwann cell, in relation to its possible involvement in controlling the periaxonal environment. For example, do the classes of K^+ channel expressed by Schwann cells help maintain the ionic environment around an axon? Knowledge gained of the biophysical and pharmacological attributes of Schwann cell ion channels *in vitro* may help us to elucidate their roles *in vivo*.

References

Baker, M. & Bostock, H. (1989). Depolarization changes the mechanism of accommodation in rat and human motor axons. *Journal of Physiology (London)*, **411**, 546–561.

Baker, M., Bostock, H., Grafe, P. & Martius, P. (1987). Function and distribution of three types of rectifying channel in rat spinal root myelinated axons. *Journal of Physiology (London)*, **383**, 45–67.

Baker, M., Howe, J.R. & Ritchie, J.M. (1993). Two types of 4-aminopyridine-sensitive potassium current in rabbit Schwann cells. *Journal of Physiology (London)*, **464**, 321–342.

Baker, M.D. & Ritchie, J.M. (1993). Characteristics of type II K^+ current in rabbit Schwann cells. *Abstracts of the XXXII Congress of the International Union of Physiological Sciences*, 204.2/P.

Barrett, E.F. & Barrett, J.N. (1982). Intracellular recording from vertebrate myelinated axons: mechanism of the depolarizing afterpotential. *Journal of Physiology (London)*, **323**, 117–144.

Bevan, S., Chiu, S.Y., Gray, P.T.A. & Ritchie, J.M. (1985). The presence of voltage-gated sodium, potassium and chloride channels in rat cultured astrocytes. *Proceedings of the Royal Society of London, Series B*, **225**, 299–313.

Black, J.A., Kocsis, J.D. & Waxman, S.G. (1990). Ion channel organization of the myelinated fibre. *Trends in Neurosciences*, **13**, 48–54.

Blight, A.R. (1985). Computer simulation of action potentials and afterpotentials in mammalian myelinated axons: the case for a lower resistance myelin sheath. *Neuroscience*, **15**, 13–31.

Bostock, H., Sears, T.A. & Sherratt, R.M. (1981). The effects of 4-aminopyridine and tetraethylammonium ions on normal and demyelinated mammalian nerve fibres. *Journal of Physiology (London)*, **313**, 301–315.

Bräu, M.E., Dreyer, F., Jonas, P., Repp, H. & Vogel, W. (1990). A K^+ channel in *Xenopus* nerve fibres selectively blocked by bee and snake toxins: binding and voltage-clamp experiments. *Journal of Physiology (London)*, **420**, 365–385.

Brismar, T. (1980). Potential clamp analysis of membrane current in rat myelinated nerve fibres. *Journal of Physiology (London)*, **298**, 171–184.

Chiu, S.Y. & Ritchie, J.M. (1981). Evidence for the presence of potassium channels in the paranodal region of acutely demyelinated mammalian single nerve fibres. *Journal of Physiology (London)*, **313**, 415–437.

Chiu, S.Y., Ritchie, J.M., Rogart, R.B. & Stagg, D. (1979). A quantitative description of membrane currents in rabbit myelinated nerve. *Journal of Physiology (London)*, **292**, 149–166.

Chiu, S.Y., Shrager, P. & Ritchie, J.M. (1984). Neuronal-type Na^+ and K^+ channels in rabbit cultured Schwann cells. *Nature*, **311**, 156–157.

Dubois, J.-M. (1981). Evidence for the existence of three types of potassium channels in the frog Ranvier node membrane. *Journal of Physiology (London)*, **318**, 297–316.

Dreyer, F. (1990). Peptide toxins and potassium channels. *Reviews of Physiology, Biochemistry and Pharmacology*, **115**, 93–137.

Frankenhaeuser, B. & Huxley, A.F. (1964). The action potential in the myelinated nerve fibre of *Xenopus laevis* as computed on the basis of voltage clamp data. *Journal of Physiology (London)*, **171**, 302–315.

Gray, P.T.A. & Ritchie, J.M. (1985). Ion channels in Schwann and glial cells. *Trends in Neurosciences*, **8**, 411–415.

Howe, J.R., Baker, M. & Ritchie, J.M. (1992). On the block of outward potassium current in rabbit Schwann cells by internal sodium ions. *Proceedings of the Royal Society of London, Series B*, **249**, 309–316.

Howe, J.R. & Ritchie, J.M. (1988). Two types of potassium current in rabbit cultured Schwann cells. *Proceedings of the Royal Society of London, Series B*, **235**, 19–27.

Hurst, R.S., Busch, A.E., Kavanaugh, M.P., Osborne, P.B., North, R.A. & Adelman, J.P. (1991). Identification of amino-acid residues involved in dendrotoxin block of rat voltage-dependent potassium channels. *Molecular Pharmacology*, **40**, 572–576.

Kocsis, J.D., Eng, D.L., Gordon, T.R. & Waxman, S.G. (1987). Functional differences between 4-aminopyridine and tetraethylammonium-sensitive potassium channels in myelinated axons. *Neuroscience Letters*, **75**, 193–198.

Kostyuk, P.G., Veselovsky, N.S., Fedulova, S.A. & Tsyndrenko, A.Y. (1981). Ionic currents in the somatic membrane of rat dorsal root ganglion neurones. III. Potassium currents. *Neuroscience*, **6**, 2439–2444.

Liman, E.R., Tytgat, J. & Hess, P. (1992). Subunit stoichiometry of a mammalian K^+ channel determined by construction of multimeric cDNAs. *Neuron*, **9**, 861–871.

Narahashi, T., Moore J.W. & Scott, W.R. (1964). Tetrodotoxin blockage of sodium conductance increase in lobster giant axons. *Journal of General Physiology*, **47**, 965–974.

Numann, R.E., Wadman, W.J. & Wong, R.K.S. (1987). Outward currents of single hippocampal cells obtained from the adult guinea-pig. *Journal of Physiology (London)*, **393**, 331–353.

Rasminsky, M. & Sears, T.A. (1972). Internodal conduction in undissected demyelinated nerve fibres. *Journal of Physiology (London)*, **227**, 323–350.

Ritchie, J.M. & Chiu, S.Y. (1981). Distribution of sodium and potassium channels in mammalian myelinated nerve. In *Demyelinating Disease: Basic and Clinical Electrophysiology*, ed. S.G. Waxman & J.M. Ritchie, pp. 329–342. New York: Raven Press.

Röper, J. & Schwarz, J.R. (1989). Heterogeneous distribution of fast and slow potassium channels in myelinated rat nerve fibres. *Journal of Physiology (London)*, **416**, 93–110.

Schmidt, H. & Stämpfli, R. (1966). Die Wirkung von Tetraäthylammoniumchlorid auf den einzelnen Ranvierschen Schnürring. *Pflügers Archiv*, **287**, 311–325.

Shrager, P., Chiu S.Y. & Ritchie, J.M. (1985). Voltage-dependent sodium and potassium channels in mammalian cultured Schwann cells. *Proceedings of the National Academy of Sciences, USA*, **82**, 948–952.

Smith, K.J. & Schauf, C.L. (1981). Effects of gallamine triethiodide on membrane currents in amphibian and mammalian peripheral nerve. *Journal of Pharmacology and Experimental Therapeutics*, **217**, 719–726.

Stansfeld, C. & Feltz, A. (1988). Dendrotoxin-sensitive K^+ channels in dorsal root ganglion cells. *Neuroscience Letters*, **93**, 49–55.

Stühmer, W., Ruppersberg, J.P., Schröter, K.H., Sakmann, B., Stocker, M., Giese,

K.P., Perschke, A., Baumann, A. & Pongs, O. (1989). Molecular basis of functional diversity of voltage-gated potassium channels in mammalian brain. *EMBO Journal*, **8**, 3235–3244.
Tasaki, I. & Hagiwara, S. (1957). Demonstration of two stable potential states in the squid giant axon under tetraethylammonium chloride. *Journal of General Physiology*, **40**, 859–885.
Wilson, G.F. & Chiu, S.Y. (1990). Ion channels in axon and Schwann cell membranes at paranodes of mammalian myelinated fibres studied with patch-clamp. *Journal of Neuroscience*, **10**, 3263–3274.

4

Axonal signals for potassium channel expression in Schwann cells

TETSURO KONISHI

Department of Neurology, Utano National Hospital, Kyoto, Japan

Introduction

Following a report of voltage-gated Na^+ and K^+ currents in cultured Schwann cells (Chiu, Shrager & Ritchie, 1984), various kinds of voltage-gated ionic channels have been found in glial cells in the peripheral and central nervous systems (Barres, Chun & Corey, 1990). Among these ionic channels in glial cells, inwardly rectifying potassium (K_{ir}) channels are important for the regulation of the potassium microenvironment in the nervous system by potassium siphoning (Newman, Frambach & Odette, 1984; Konishi, 1990) or spatial potassium buffering (Orkand, Nicholls & Kuffler, 1966). This chapter focuses on the mechanism of the expression of functional K_{ir} channels in mouse Schwann cells in relation to axonal contact, intracellular cAMP and neuronal activity, and discusses the physiological significance of Schwann cells in potassium regulation.

Axonal contact

In cultured Schwann cells obtained from dissociated sciatic nerves of neonatal mice, only voltage-gated outward currents were recorded with the whole-cell patch-clamp technique during depolarizing voltage steps (Konishi, 1989). The equilibrium potentials of the tail currents indicated that these outward currents were carried by potassium ions (Konishi, 1989). They were eliminated by bath application of quinine, but were not affected by external barium (Konishi, 1990). In freshly dissociated Schwann cells from neonatal sciatic nerves, both myelinating and non-myelinating cells showed barium-sensitive inward currents during hyperpolarizing voltage steps, which were disclosed by subtracting a record in a solution containing barium from a record in standard solution (Fig. 4.1) (Konishi, 1992). The incidence of cells showing such K_{ir} currents decreased within a few days after transection of sciatic nerves of neonatal mice (Fig. 4.1). The elimination of K_{ir} currents from Schwann cells in nerve segments distal to the transected site or in

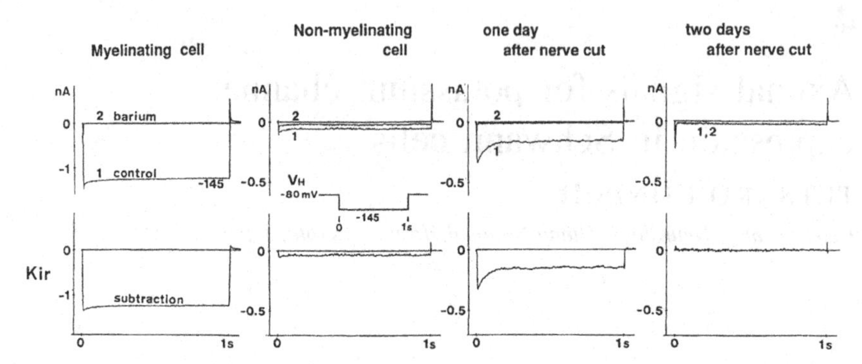

Fig. 4.1. Left half: Current records elicited in myelinating and non-myelinating Schwann cells from 2-day-old mouse by −145 mV, 1 s hyperpolarizing voltage steps (holding potential (HP) −80 mV) before (1) and after (2) external application of 2 mM $BaCl_2$. Lower tracings: Inward K^+ currents (K_{ir}) were obtained by subtracting tracing 2 from tracing 1. Right half: Currents in Schwann cells obtained 1 and 2 days after nerve transection in neonatal mouse. Inward K^+ currents were absent in a cell 2 days after nerve transection. From Konishi (1992) with permission of *Brain Research*.

cultured conditions (Konishi, 1992) suggested that axonal contact plays an important role in the expression of these currents in Schwann cells.

Intracellular cAMP

The expression in Schwann cells of galactocerebroside (GC), which is a major component of the myelin sheath, disappears from cultured rat Schwann cells within a few days after the start of culture and is restored by an increase in Schwann cell cAMP by adding membrane-permeable cAMP analogue to the culture medium (Sobue & Pleasure, 1984). Schwann cells lose GC expression in transected nerves and restore it after nerve regeneration (Jessen, Mirsky & Morgan, 1987), suggesting that axon–Schwann cell interactions lead to an elevation of intracellular cAMP levels and thus to Schwann cell differentiation. To see whether the expression of functional K_{ir} channels might also be induced by an increase in Schwann cell cAMP, cAMP analogues or forskolin were added to the culture medium.

cAMP or chlorophenylthio(CPT)-cAMP, which is a membrane-permeable cAMP analogue, was added to the culture medium 4 days after the start of culture, when K_{ir} currents had virtually disappeared from the Schwann cells. Whole-cell recordings were made in the standard solution, 5 days after incubation with the drugs. Barium-sensitive K_{ir} currents during hyperpolarizing voltage steps were observed in cells incubated with CPT-cAMP but not in cells incubated with the same concentration of cAMP (Fig. 4.2).

The inward rectification in cells co-cultured with CPT-cAMP was obvious

cAMP

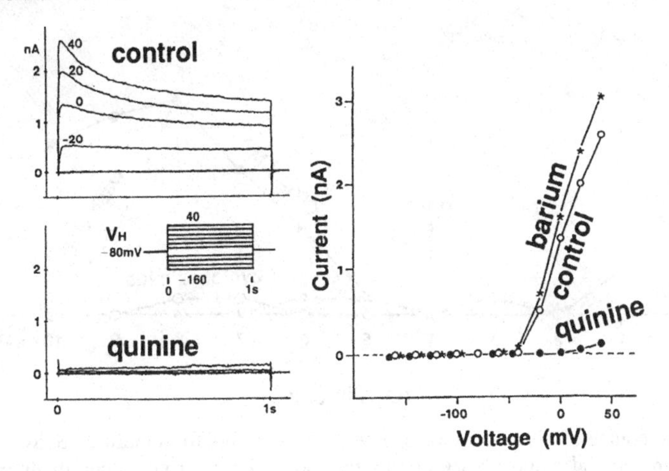

CPT-cAMP

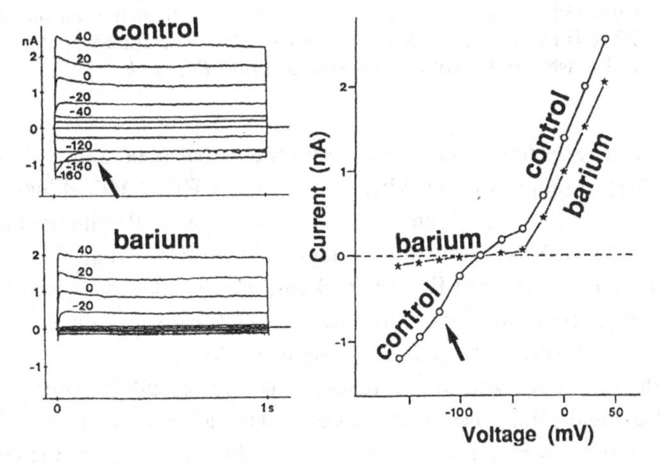

Fig. 4.2. Recordings of currents, and current–voltage plots of peak currents, obtained from a cultured Schwann cell after 5 days of incubation with 1 mM cAMP (upper panel) or 1 mM chlorophenylthio (CPT)-cAMP (lower panel) before (control) and after bath application of 2 mM $BaCl_2$ (barium). In a cell with cAMP, 2 mM quinine hydrochloride was added to a solution containing 2 mM barium (quinine). Currents were elicited by 20 mV voltage steps (HP −80 mV); some voltage steps are indicated in the figure. Hyperpolarization-activated inward currents (arrow), which are barium sensitive, were elicited in a cell co-cultured with CPT-cAMP but were not observed in a cell with cAMP. From Konishi (1992) with permission of *Brain Research*.

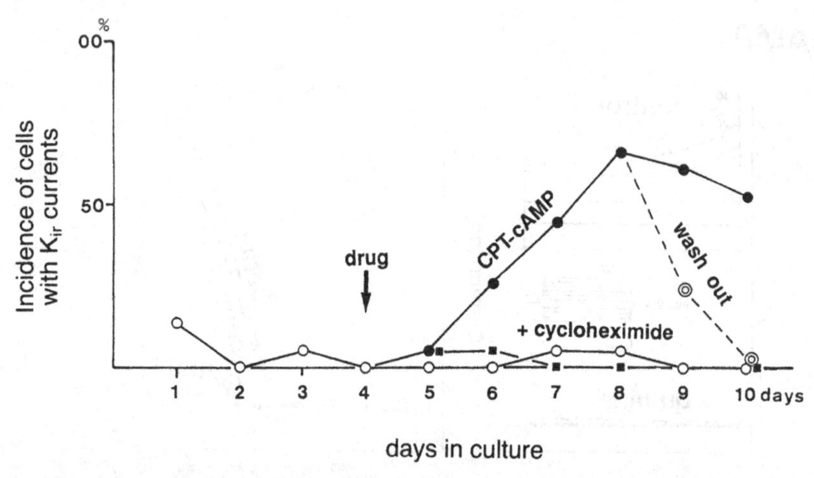

Fig. 4.3. Time course of disappearance of inward K^+ currents from cultured Schwann cells (open circles) and reappearance or suppression of these currents upon treatment of the cultures with 1 mM CPT-cAMP (filled circles) or both 1 mM CPT-cAMP and cycloheximide (1 μg/ml) (filled squares) at 96 hours of culture (arrow). The removal of CPT-cAMP from the medium after a 96 hour application caused rapid elimination of inward K^+ currents from the cultured cells within 2 days (double circles). The percentage of cells with inward K^+ currents (ordinate) was obtained from the total number of 16–20 cells (30 cells at 24 hours). Abscissa: days of culture from the start of plating. From Konishi (1992) with permission of *Brain Research*.

when whole-cell recordings were made in high external potassium solution (Konishi, 1992). Among various kinds of agents, CPT-cAMP, dibutyryl-(DB)-cAMP and forskolin, which are known to increase the intracellular cAMP level, induced K_{ir} currents after 4 days of incubation (Konishi, 1992). The concentrations of DB-cAMP and forskolin which induced K_{ir} currents were similar to the concentrations known to induce GC expression in cultured rat Schwann cells (Sobue, Shuman & Pleasure, 1986).

The incidence of cells with K_{ir} currents elicited by co-culture with 1 mM CPT-cAMP gradually increased and peaked 4 days after addition of CPT-cAMP (filled circles in Fig. 4.3). The K_{ir} currents disappeared from the cells within 2 days of removal of cAMP from the medium (double circles in Fig. 4.3), suggesting that continuously elevated intracellular cAMP is necessary for K_{ir} expression, as it is for the cAMP-dependent induction of GC (Sobue *et al.*, 1986). The simultaneous application of cycloheximide (1 μg/ml), an inhibitor of protein synthesis, with 1 mM CPT-cAMP suppresssed K_{ir} currents for up to 6 days of incubation (filled squares in Fig. 4.3).

These results show that cAMP is an intracellular second messenger for the expression of the functional K_{ir} channels, as well as for the expression of myelin-specific proteins in Schwann cells.

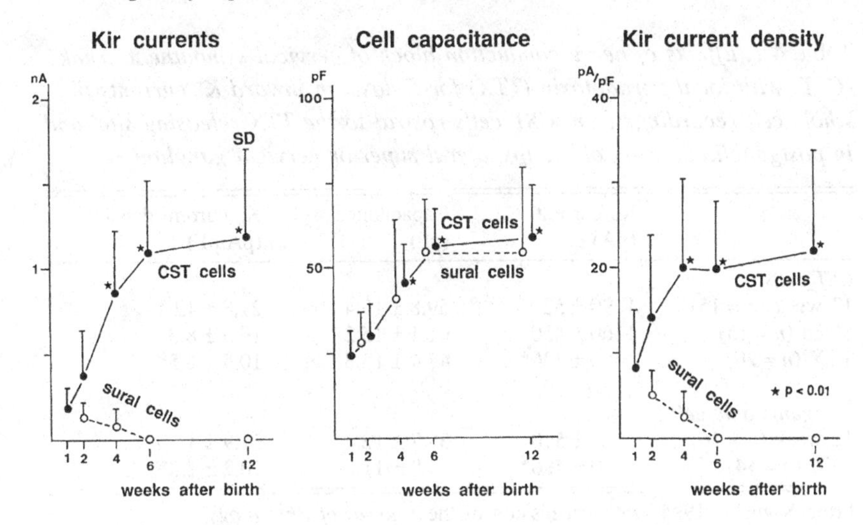

Fig. 4.4. Changes with development of the magnitude of inward K^+ currents, cell capacitance and current density of inward K^+ currents in whole-cell recordings in sural cells 2–12 weeks after birth (open circles) and in cervical sympathetic trunk (CST) cells 1–12 weeks after birth (filled circles). Vertical bars indicate standard deviations of the mean; stars for CST cells show significant differences of mean values in comparison with CST cells in 1-week-old mice (P<0.01 by Student's t-test). From Konishi (1994) with permission of the *Journal of Physiology*.

Neuronal activity

In adult mouse, K_{ir} currents were not recorded in freshly dissociated non-myelinating Schwann cells from the sural nerve (sural cells) but were observed in non-myelinating cells from the cervical sympathetic trunk (CST cells) (Konishi, 1994). Developmental changes in the magnitude of the K_{ir} currents (measured by the subtraction method described in Fig. 4.1), cell capacitance and K_{ir} current density were investigated in freshly dissociated sural cells and CST cells from 1- to 12-week-old mice (Fig. 4.4). In sural cells, K_{ir} currents decreased in magnitude and were virtually undetectable in cells more than 6 weeks after birth. In contrast, the magnitude of K_{ir} currents increased in CST cells in parallel with an increase in cell capacitance 1–6 weeks after birth, while the K_{ir} current density increased for 1–4 weeks after birth and then stayed constant for up to 12 weeks.

To determine the factors responsible for the increase in K_{ir} current density with development in CST cells, single K_{ir} channel currents were recorded in CST cells in the cell-attached configuration with 154 mM K^+ in the pipette. The single channel conductance (measured between 0 and −100 mV relative to the resting potential) was around 30 pS and did not change with development (Konishi, 1994). The steady-state open channel probability of single K_{ir}

Table 4.1. *Effects of nerve conduction block of cervical sympathetic trunk (CST) with local tetrodotoxin (TTX) for 5 days on inward K^+ currents in whole-cell recordings from CST cells rostral to the TTX releasing site, and in postganglionic cells of the ipsilateral superior cervical ganglion*

	K_{ir} current (pA)	Capacitance (pF)	K_{ir} current density (pA/pF)
CST cells			
12 week ($n = 15$)	1190 ± 520	59.8 ± 15.4	21.8 ± 12.1
Sham ($n = 15$)	1160 ± 450	61.1 ± 15.2	19.9 ± 8.3
TTX ($n = 20$)	490 ± 330*	47.4 ± 15.8	10.5 ± 5.5*
Postganglionic cells			
12 week ($n = 16$)	1720 ± 510	55.7 ± 12.4	31.9 ± 11.0
TTX ($n = 14$)	650 ± 480*	63.9 ± 11.6	10.2 ± 7.2*

From Konishi (1994) with permission of the *Journal of Physiology*.
Asterisks indicate a significant difference of the mean current density in TTX-treated mice from that of 12-week-old mice or sham-operated mice ($P < 0.01$). Values are the mean ± SD.

channels in CST cells decreased with membrane hyperpolarization, as has been described for K_{ir} channels in rat Schwann cells (Wilson & Chiu, 1990). The constancy of unitary conductance and open channel probability of the single K_{ir} channels suggests that the increase in K_{ir} current density observed in whole-cell recordings with development is due to increased numbers of functional K_{ir} channels.

The absence of K_{ir} currents in adult sural cells suggests that the expression of functional K_{ir} channels in adult CST cells might be controlled by undetermined signals from neurones other than axonal contact. Since only 16–20% of the total population of unmyelinated fibres in sural nerves are known to be sympathetic efferents (Chad *et al.*, 1983; Baron, Jänig & Kollmann, 1988), the expression of functional K_{ir} channels in adult CST cells might be related to the level of sympathetic activity. To test this hypothesis, sympathetic neuronal activity of CST in adult mice was modified chronically by local tetrodotoxin (TTX) application. A glass tube containing TTX was implanted with its open end close to one side of the CST at the mid-cervical portion. Successful TTX block of the CST was shown by a narrowing of the ipsilateral palpebral fissure and retraction of the eyeball. After 5 days of successful TTX block, a segment of CST rostral to the site of the open end of the capillary tube and ipsilateral inferior postganglionic branches of the superior cervical ganglion were excised for whole-cell recordings.

After 5 days of chronic conduction block, the magnitude and density of K_{ir} currents in whole-cell recordings in CST cells (TTX) were significantly lower than in cells with control tubes (sham) or in cells from 12-week-old mice (control) (Table 4.1). A similar significant decrease in the magnitude and density of K_{ir} currents was observed in non-myelinating cells in the inferior postganglionic branches of the superior cervical ganglion (postganglionic cell). These cells are located well away from the TTX releasing site and were thus unaffected by any direct TTX effects (Table 4.1).

These results show that neuronal activity regulates the expression of functional K_{ir} channels in Schwann cells. Although this study does not identify the signalling mechanisms responsible for the activity-related regulation of K_{ir} channel expression, candidates for the intercellular signals include potassium ions released by active axons and taken up by Schwann cells, or depolarization by potassium accumulation, or trophic factors released from active axons. A process mediated by intracellular second messengers activated by neuronal activities might be responsible for the expression of functional K_{ir} channels. A study of gene expression of messenger RNA for K_{ir} channel protein would show whether K_{ir} channel protein synthesis is involved in the activity-dependent regulation of the expression of functional K_{ir} channels in Schwann cells. It should also be determined whether synthesis of regulatory proteins, post-translational modulation of channel proteins by phosphorylation or other modulatory factors are involved.

Regulation of the potassium microenvironment by Schwann cells

In CST, asynchronous sympathetic neuronal activity would cause a potassium gradient in a single non-myelinating Schwann cell as follows (Fig. 4.5). When impulses are conducting in the shaded axon, potassium would accumulate in the periaxonal space of this axon during neuronal activity (Frankenhaeuser & Hodgkin, 1956). If the potassium concentration in the periaxonal space increases to 40 mM, as shown in Fig. 4.5, the current–voltage relation of K^+ currents in the cell membrane facing the periaxonal space of the active axon would shift to the right, towards depolarization. Although K_{ir} channels determine the membrane potential near the potassium equilibrium potential (Konishi, 1990), depolarization of the membrane potential in this region would not reach the degree predicted by the new potassium equilibrium potential, due to the clamping of the membrane potential near the resting level by K_{ir} channels in the rest of the cell membrane, where external potassium concentration is low (cross in Fig. 4.5). The potassium gradient in a single cell causes potassium entry into the Schwann cell cytoplasm through K_{ir} channels facing the periaxonal space with a high potassium concentration, as has been demonstrated in neonatal Schwann cells (Konishi, 1991). In fact, an increase in glial intracellular K^+ during neuronal activity has been demonstrated in a variety of preparations (Coles & Tsacopoulos, 1979; Ballanyi, Grafe & Ten Bruggencate, 1987). The presence of outward currents elicited during depolarizing voltage steps in whole-cell recordings

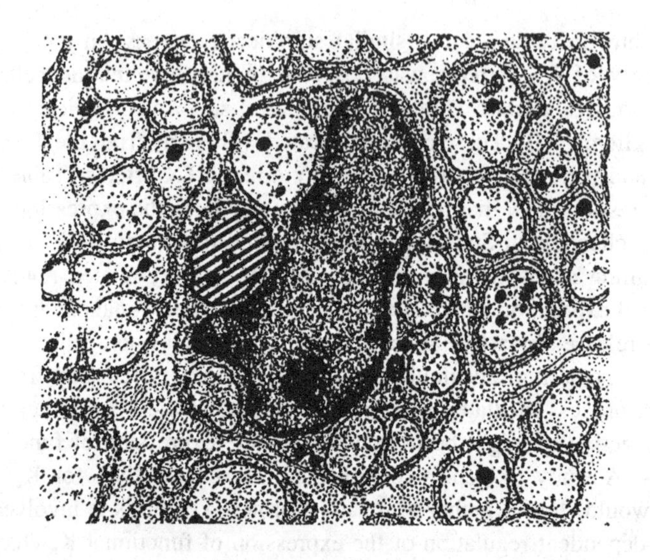

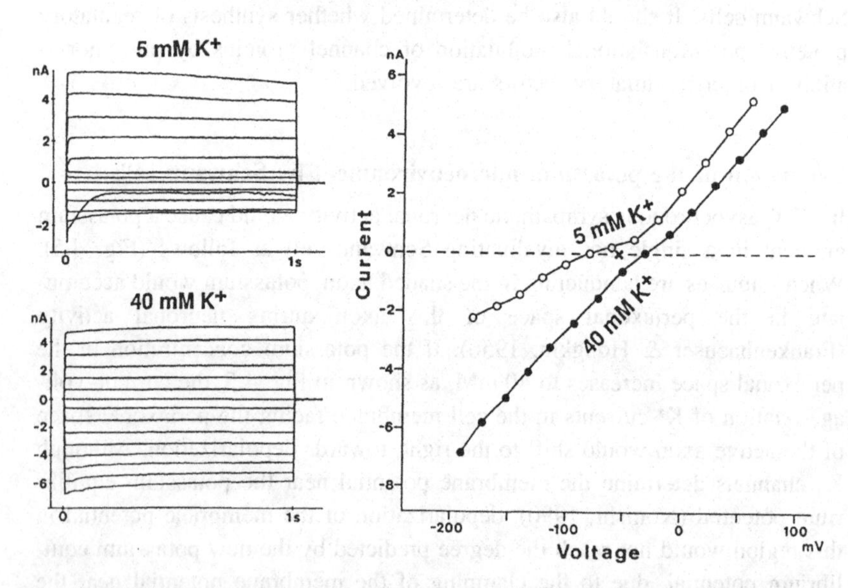

Fig. 4.5. Digitised electron micrograph of non-myelinating Schwann cells in cervical sympathetic trunk (CST) and recordings of currents and their current–voltage plots of peak currents in whole-cell recordings obtained from a CST cell of 12-week-old mouse under 5 mM or 40 mM K^+ solution. Currents were elicited by 20 mV voltage steps from a holding potential of zero current potential (79 mV in 5 mM K^+ and –32 mV in 40 mM K^+). Electron micrograph reproduced from Aguayo *et al.* (1976) with permission of the *Journal of Neuropathology and Experimental Neurology.*

in CST cells (Fig. 4.5) suggests that excess potassium in these cells would be released into the general extracellular space and/or into the periaxonal space with a low potassium concentration by a mechanism called potassium siphoning (Newman *et al.*, 1984; Konishi, 1990). The potassium released from Schwann cells would be taken into axons by the neuronal Na^+/K^+ pump. Thus non-myelinating Schwann cells in sympathetic nerves play important roles, not only in the regulation of the potassium concentration around active axons, but also in returning excess potassium to neurones.

Conclusion

In summary, axonal contact and neuronal activity play an important role in the expression of functional K_{ir} channels in Schwann cells in which cAMP may be involved as an intracellular second messenger for axon–Schwann cell interactions.

References

Aguayo, A.J., Bray, G.M., Terry, L.C. & Sweezey, E. (1976). Three dimensional analysis of unmyelinated fibres in normal and pathologic autonomic nerves. *Journal of Neuropathology and Experimental Neurology*, **35**, 136–151.

Ballanyi, K., Grafe, P. & Ten Bruggencate, G. (1987). Ion activities and potassium uptake mechanisms of glial cells in guinea-pig olfactory cortex slices. *Journal of Physiology (London)*, **382**, 159–174.

Baron, R., Jänig, W. & Kollmann, W. (1988). Sympathetic and afferent somata projecting in hindlimb nerve and the anatomical organization of the lumbar sympathetic nervous system of the rat. *Journal of Comparative Neurology*, **275**, 460–468.

Barres, B.A., Chun, L.L.Y. & Corey, D.P. (1990). Ion channels in vertebrate glia. *Annual Review of Neuroscience*, **275**, 460–468.

Chad, D., Bradley, W.G., Rasool, C., Good, P., Reichlin, S. & Zivin, J. (1983). Sympathetic postganglionic unmyelinated axons in the rat peripheral nervous system. *Neurology*, **33**, 841–847.

Chiu, S.Y., Shrager, P. & Ritchie, J.M. (1984). Neuronal-type Na^+ and K^+ channels in rabbit cultured Schwann cells. *Nature*, **311**, 156–157.

Coles, J.A. & Tsacopoulos, M. (1979). Potassium activity in photoreceptors, glial cells and extracellular space in the drone retina: changes during photostimulation. *Journal of Physiology (London)*, **290**, 525–549.

Frankenhaeuser, B. & Hodgkin, A.L. (1956). The after-effects of impulses in the giant fibres of *Loligo*. *Journal of Physiology (London)*, **131**, 341–376.

Jessen, K.R., Mirsky, R. & Morgan, L. (1987). Axonal signals regulate the differentiation of non-myelin-forming Schwann cells: an immunohistochemical study of galactocerebroside in transected and regenerating nerves. *Journal of Neuroscience*, **7**, 3362–3369.

Konishi, T. (1989). Voltage-dependent potassium channels in mouse Schwann cells. *Journal of Physiology (London)*, **411**, 115–130.

Konishi, T. (1990). Voltage-gated potassium currents in myelinating Schwann cells in the mouse. *Journal of Physiology (London)*, **431**, 123–139.

Konishi, T. (1991). Potassium channel-dependent changes in the volume of developing mouse Schwann cells. *Brain Research*, **565**, 57–66.
Konishi, T. (1992). cAMP-mediated expression of inwardly rectifying potassium channels in cultured mouse Schwann cells. *Brain Research*, **594**, 197–204.
Konishi, T. (1994). Activity-dependent regulation of inwardly rectifying potassium currents in non-myelinating Schwann cells in mice. *Journal of Physiology (London)*, **474**, 193–202.
Newman, E.A., Frambach, D.A. & Odette, L.L. (1984). Control of extracellular potassium levels by retinal glial cell K^+ siphoning. *Science*, **225**, 1174–1175.
Orkand, R.K., Nicholls, J.G. & Kuffler, S.W. (1966). The effect of nerve impulses on the membrane potential of glial cells in the central nervous system of amphibia. *Journal of Neurophysiology*, **29**, 788–806.
Sobue, G. & Pleasure, D. (1984). Schwann cell galactocerebroside induced by derivatives of adenosine 3′,5′-monophosphate. *Science*, **224**, 72–74.
Sobue, G., Shuman, S. & Pleasure, D. (1986). Schwann cell responses to cyclic AMP: proliferation, changes in shape, and appearance of surface galactocerebroside. *Brain Research*, **362**, 23–32.
Wilson, G.F. & Chiu, S.Y. (1990). Potassium channel regulation in Schwann cells during early developmental myelinogenesis. *Journal of Neuroscience*, **10**, 1615–1625.

5

Ion channels in human axons

GORDON REID

Sobell Department of Neurophysiology, Institute of Neurology, London, UK.

Some puzzling differences between rat and human myelinated nerve fibres emerged from work in this department some years ago. Rat nerve was being used to investigate the origin of the spontaneous activity which seems to underlie paraesthesiae, a common and disturbing symptom in neuropathies. It proved to be surprisingly difficult to induce rat nerve to become spontaneously active, although this is easily done in human nerve by manoeuvres such as ischaemia. A related finding is that rat nerve fibres accommodate more than those of humans to long stimuli. The electrotonic behaviour of nerve fibres differs between the two species in ways which suggest different populations of K^+ channels (Bostock & Baker, 1988).

To try to understand these differences, single-channel and multi-channel patch clamping were applied to human axons, followed by voltage clamping of the node of Ranvier. Much of this work was done in collaboration with groups in Giessen, Hamburg and Munich. The results show that the ion channels in human axons are very similar to those in rat axons, as are the action potential and membrane currents in the node. The species differences are still not understood, but may result from differences in ion channel distribution or density, probably in the paranode or internode rather than the node. This chapter will summarize the information about ion channels in human peripheral myelinated axons which arose from this work, and relate this to descriptions of channels in other species.

The background

Firstly, to provide a context, it is necessary to review axonal excitability briefly, since developments during the past 15 years have shown that it is more complex and interesting than previously thought. More detailed surveys can be found in several reviews (see, for example, Vogel & Schwarz, 1995; Black, Kocsis & Waxman, 1990).

A simple model of myelinated nerve

A simple and useful model of a myelinated nerve fibre consists of short segments of active axon membrane at the nodes of Ranvier, connected by passive internodes insulated by myelin. This is sufficient to explain saltatory conduction (Fitzhugh, 1962). Action potentials in frog nodes of Ranvier can be modelled using very few currents, with one type of Na^+ current generating the upstroke of the action potential, one type of (fast) K^+ current involved in repolarization, and a passive 'leak' conductance setting the resting potential (Frankenhaeuser & Huxley, 1964). This simple representation spawned a fertile research programme on the frog node of Ranvier during the 1960s and 1970s (reviewed in Stämpfli & Hille, 1976). It has since been revised in several ways.

Some problems with the early model

More ion channels exist than was at first thought

In frog and rat axons the fast K^+ current consists of at least two components with different voltage dependence (I_{Kf1} and I_{Kf2}), and a slow K^+ current (I_{Ks}) is also present (Dubois, 1981; Röper & Schwarz, 1989; Corrette *et al.*, 1991). Single-channel recordings in both species reveal three types of voltage-dependent K^+ channel, each with similarities to one of the macroscopic currents. F (fast) channels, which activate (open) and deactivate (close) with fast kinetics, behave in a similar way to I_{Kf2}; I (intermediate) channels, which activate with fast kinetics, but deactivate more slowly than F channels, behave like I_{Kf1}; and S (slow) channels, which activate and deactivate more slowly than either of the other types, may produce I_{Ks} (Jonas *et al.*, 1989; Safronov, Kampe & Vogel, 1993). Experiments in intact axons (see below) have clarified the roles of these K^+ currents.

Single-channel recording has identified several other channel types in frog axons: a large conductance Ca^{2+}-activated K^+ channel; a K^+ channel closed by ATP; a 'flickery' K^+ channel, active around the resting potential; and a Na^+-activated K^+ channel (Vogel & Schwarz, 1995). The first of these has also been described in rat axons (Kampe, Safronov & Vogel, 1992), as has a Cl^- channel (Strupp & Grafe, 1991). In addition, *in vivo* recordings of electrotonus in rat and human axons reveal a hyperpolarization-activated conductance (Baker *et al.* 1987; Bostock & Baker, 1988).

The internode is more than a passive cable

The fact that ion channels are not confined to the node of Ranvier but are present throughout the paranode and internode (see Chiu & Ritchie, 1984), seems at first sight paradoxical since these channels are normally covered by myelin. However, the myelin sheath does not exclude communication between the outside surface of the internodal membrane and the extracellular space. They are connected by a high-resistance pathway, which, along with the large capacitance of the internodal axon, produces a long time constant (Barrett & Barrett, 1982). This resistive pathway probably takes several routes, partly through the paranodal area, which has a remarkably complex structure (Berthold & Rydmark, 1983).

The density of each ion channel type differs between nodal, paranodal and internodal membrane, and between species. In all species, Na^+ channels are concentrated at the node, and their density is low in the paranode and internode. In frog nodes, fast K^+ currents (I_{Kf1} and I_{Kf2}) are also large, almost as large as the nodal Na^+ currents, but in rat and rabbit nerve fibres fast K^+ currents in the node are very small (Brismar, 1980; Chiu *et al.*, 1979). In rat axons, the highest apparent density of fast K^+ currents occurs in the paranode, immediately adjacent to the node but under the myelin, and their density is low in the node and internode. Most of the small K^+ current in rat nodes is slow (I_{Ks}), and the slow K^+ current appears to be concentrated in the node, with a lower density in the paranode and internode (Röper & Schwarz, 1989).

A less simple model of a myelinated fibre

Most of the experiments which have helped to clarify the roles of these ion channels have been done in intact axons rather than in isolated patches or nodes, since the complex structure of myelinated fibres makes it difficult to predict what the channels do from their characteristics alone. Such experiments can be difficult to interpret because of the uncertain access of test solutions or channel-blocking drugs to internodal channels.

The simple model describes adequately the role of the nodal Na^+ channels in generating the upstroke of the action potential. Repolarization after an action potential and the generation of the resting potential have turned out to be more complex, as the role of the internode in these processes has been recognized. As a rough approximation, because of its long time constant, the internodal axonal membrane has little involvement in electrical events on a short time scale, such as the action potential. On a longer time scale, the connection between the internodal membrane and the node is very good and the internode exerts a considerable influence on nodal excitability. Under normal conditions, the resting potential is probably generated mainly by channels in the internode (Chiu & Ritchie, 1984; Baker *et al.*, 1987).

Because of their different distribution in the axonal membrane, the role of the fast K^+ channels is different in frog and rat axons, although the channels themselves are almost identical. In frog axons, the large fast K^+ currents in the node are important in repolarizing the membrane after the action potential (Weller *et al.*, 1985). In mammalian axons, however, the small nodal K^+

currents are not involved in repolarization; blocking them with tetraethylammonium (TEA) produces almost no change in the action potential (Schwarz & Eikhof, 1987).

Repolarization in mammalian axons depends mostly on the leak conductance (Chiu *et al.*, 1979). The leak recorded in voltage-clamped nodes was at first assumed to be through nodal channels, but recordings from intact axons do not support the idea of a substantial nodal leak conductance. Instead, they indicate that most of the leak occurs in an external pathway, probably the access resistance to the internode mentioned above (Baker *et al.*, 1987). During the action potential, the axonal membrane in the internode stays close to the resting potential, and most of the change in potential is across the myelin (Barrett & Barrett, 1982). After the action potential, as the nodal conductance falls due to inactivation of the nodal Na^+ channels, it is probably mostly current flow from internode to node outside the axon that brings the node back towards the internodal resting potential.

The major role of the paranodal fast K^+ channels in rat axons seems to be to limit the depolarizing after-potential (DAP) which results from the small depolarization of the internodal axon membrane during the action potential (Barrett & Barrett, 1982). Blocking these channels with 4-aminopyridine (4-AP) increases the DAP, which then re-excites the node, producing bursts of action potentials after a single short stimulus (Baker *et al.*, 1987). The slow K^+ current appears to be important in accommodation; blocking it with TEA induces repetitive activity in response to a long stimulus (Baker *et al.*, 1987).

The hyperpolarization-activated conductance may limit the hyperpolarization resulting from increased activity of the Na^+/K^+-ATPase, which might otherwise block conduction (Baker *et al.*, 1987). The channels which have so far been found only in single-channel recordings have no definite function, but two (the ATP-sensitive and Ca^{2+}-activated K^+ channels) would probably open only at times of metabolic stress and may therefore also help to maintain normal excitability under abnormal conditions (Jonas *et al.*, 1991)

Patch-clamped human axons

To investigate the ion channels of human axons directly, the first approach we used was patch clamping, based on the method introduced in *Xenopus* by Jonas *et al.* (1989). Enzyme treatment removes the connective tissue and loosens the axo-glial junctions, allowing single axons with retracted myelin to be patch clamped. The preparation of human axons is much more difficult than in frog because of the density of the connective tissue around the axons, but a successful procedure was eventually developed which allowed recordings to be made (Reid *et al.*, 1991; Scholz *et al.*, 1993). Nerves have been obtained from graft operations, from post-mortem examinations and from sural nerve biopsies.

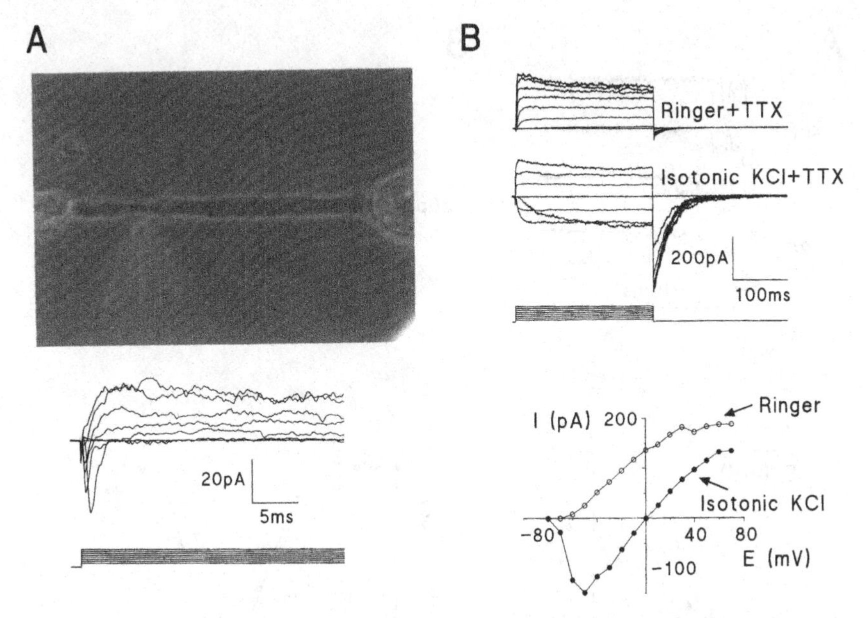

Fig. 5.1. (A) Above: A typical human axon prepared for patch clamping. A patch pipette points towards the patch site at the node. The diameter of this axon in the internode is about 10 μm, and the length of the demyelinated section is about 130 μm. Below: Currents in Ringer solution, from an outside-out patch made near the node. Pulses to between –90 mV and +50 mV were applied after 50 ms prepulses to –130 mV to remove Na^+ channel inactivation; the holding potential (E_h) was –90 mV. All currents in Figs. 5.1–5.4 are shown after capacitance and leakage correction. (B) Above: Currents in a different outside-out patch, in Ringer and isotonic KCl; Na^+ currents were blocked with 300 nM TTX. E_h was –100 mV; pulses to between –80 mV and +60 mV. Below: The shift in reversal potential of the late current in isotonic KCl demonstrates its K^+ selectivity.

Fig 5.1A (upper part) shows a typical human axon prepared for patch clamping. In 20 experiments in human axons, the inclusion of Lucifer Yellow in the patch pipette (Wilson & Chiu, 1990) indicated an axonal origin in every case, and it is therefore likely that other patches not so verified were also from axonal membrane. After enzyme treatment, a resting potential of about –40 mV to –60 mV can still be recorded from some axons; whether any still conduct has not been tested, but enzymatically dissociated rat axons do (Levy *et al.*, 1985).

Currents in a typical multi-channel patch from a site at or very close to the node are shown in Fig. 5.1A (lower part). Virtually all the current in such patches is contributed by a tetrodotoxin (TTX)-sensitive Na^+ current, and a voltage-dependent K^+ current (Fig. 5.2B). Even patches from sites which appear to be exactly at the node contain many fast-activating K^+ channels, which are scarce in intact human nodes (see below); since the width of the nodal membrane is probably similar to that of the patch pipette, it may be

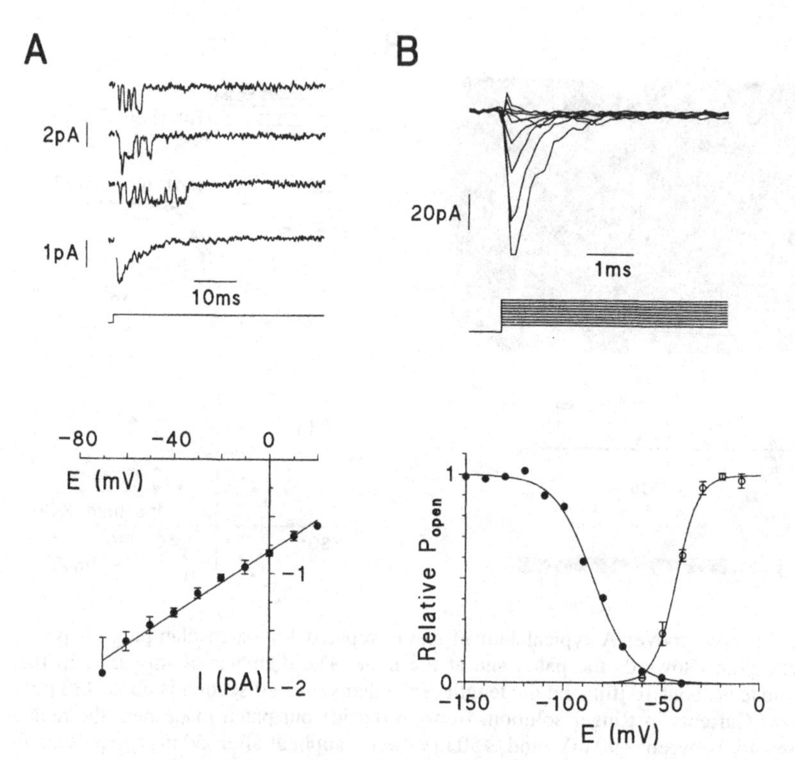

Fig. 5.2. (A) Above: Single Na^+ channel currents in an outside-out patch with three channels; pulse protocol as Fig. 5.1A, pulses to –60 mV are shown. Three individual records and the average of 24 such records are shown. Below: Current–voltage relation of single Na^+ channels from five outside-out patches (the error bars in all figures show the mean ± SEM). The single-channel conductance is about 14 pS, and the extrapolated reversal potential is close to the Na^+ equilibrium potential. (B) Above: Na^+ currents in a large outside-out patch from near the node. Pulse protocol as in Fig. 5.1A; pulses to between –80 mV and +100 mV shown. Below: Voltage dependence of Na^+ current inactivation (filled circles) and activation (open circles). Activation was estimated by extrapolation to the beginning of the pulse of a double exponential fit to the falling phase of the current (see text). Continuous lines show fits of the Boltzmann function $f(E) = 1/\{1 + [(E_{50} - E)/k]\}$ with the following values: activation: $E_{50} - 43$ mV, $k = 5.5$ mV; inactivation: $E_{50} = -85$ mV, $k = -9.3$ mV.

impossible to make patches from exclusively nodal membrane, and it is likely that most apparently nodal patches also contain some paranodal membrane. The term 'near-nodal' is therefore used. To make the recording of individual current components easier, solutions are chosen to block other currents (Scholz *et al.*, 1993).

Sodium channels (Scholz *et al.*, 1993)

Na^+ currents are found at high density (peak currents 40–60 pA) at near-nodal sites. At sites more distant from the node, smaller numbers of Na^+ channels are seen, and sometimes none. Single Na^+ channels are shown in Fig. 5.2A (upper part); the averaged current shows kinetics typical of voltage-dependent Na^+ currents, and the single-channel conductance (Fig. 5.2A, lower part) is about 14 pS.

A recording from a near-nodal patch with a large number of Na^+ channels is shown in Fig. 5.2B (upper part). The current has similar kinetics to the averaged current shown in Fig. 5.2A, and the time course of inactivation in single- and multi-channel recordings is often better fitted by two exponential components than by one. The maximum inward current occurs at potentials around −20 mV to −30 mV. The voltage dependence of activation and inactivation is shown in Fig. 5.2B (lower part). Human axonal sodium channels appear similar to those in *Xenopus* (Jonas *et al*., 1989) and rat (Hermsteiner, Kampe & Vogel, 1991).

Voltage-dependent 'delayed rectifier' potassium channels (Scholz *et al.* 1993)

Voltage-dependent K^+ currents, activated by depolarization, are found at moderately high density at all patch sites, whether paranodal, internodal or near-nodal. As in frog and rat, several different voltage-dependent K^+ channel types are present, and are normally found together in the same patch (Fig. 5.3A).

A few patches contain only one or very few channels of a single type, allowing single-channel properties to be studied. The commonest K^+ channel type, found in nearly every patch (Fig. 5.3B–D; also seen in Fig. 5.3A) normally has a conductance of 30–35 pS (inward current in isotonic KCl), although the range is from about 20 to 40 pS. Its conductance, its voltage dependence of activation, its deactivation kinetics and its pharmacology are virtually identical to those of the rat and *Xenopus* I (intermediate) channel, and it has been given the same name.

Like those in *Xenopus* and rat, the human I channel can be blocked by the green mamba snake toxin α-dendrotoxin (DTX), allowing the remaining channels in a patch to be studied. In one example (Fig. 5.4A), blocking I channels with DTX reveals a 10 pS channel with slow kinetics (also seen in Fig. 5.3A) which is similar to the rat and frog S (slow) channel. A few patches have also shown a 50 pS channel similar to the rat and frog F (fast) channel, and some human patches have shown a slow channel type, not so far reported in other species, with a larger conductance than the S channel (Scholz *et al*., 1993).

In patches with a mixed population of channels, components of K^+ current can be separated (if the currents are large enough) using their different kinetics of deactivation. 'Tail currents', the currents recorded as ion channels close on returning to the holding potential, can often be described as the sum of several exponential components (Dubois, 1981). Three components with

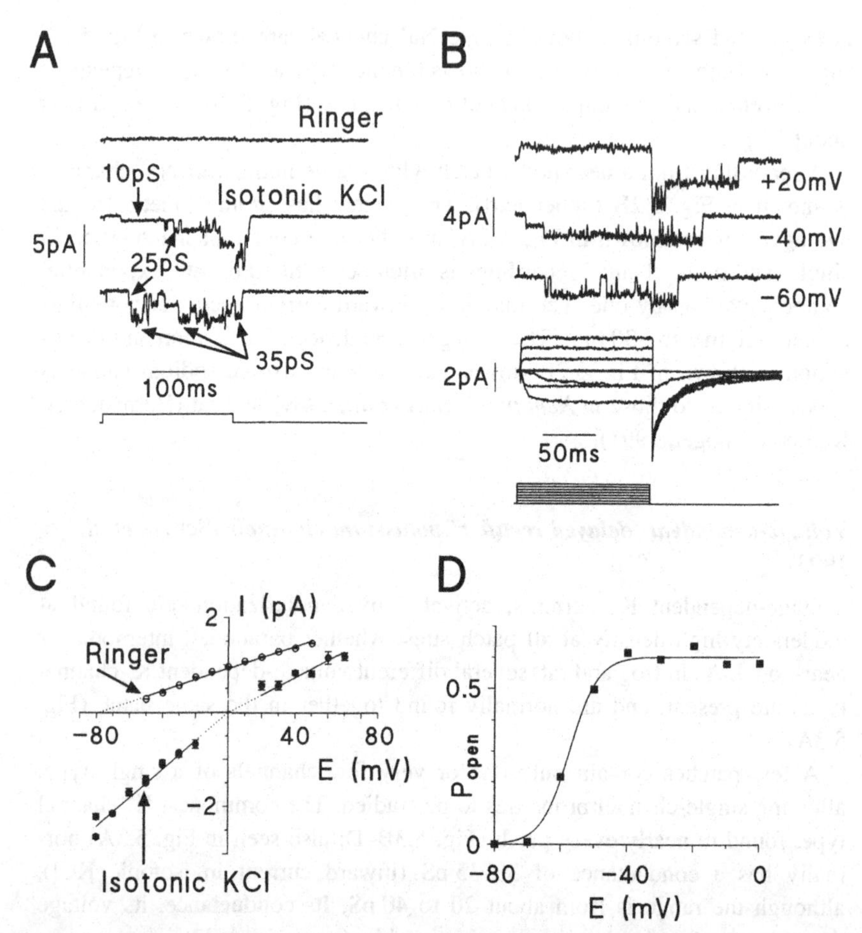

Fig. 5.3. (A) K^+ channels of several types in an outside-out patch in Ringer solution and isotonic KCl (with 300 nM TTX), in response to pulses to −60 mV; E_h = −100 mV. (B) Currents in an inside-out patch which contained three I channels, with isotonic KCl + TTX in the pipette; pulses to between −60 mV and +60 mV, E_h = −80 mV. Three individual records at different potentials are shown, and the average (n = 90). (C) Current–voltage relation of single I channels in two outside-out patches, measured in Ringer solution and isotonic KCl. The single-channel conductance in Ringer solution is 12 pS, and in isotonic KCl the conductance for inward current is 30 pS and for outward current 18 pS. (D) Voltage dependence of activation of single I channels, measured from the averaged currents shown in (A). The fitted line is a Boltzmann function (see Fig. 5.2B): E_{50} = −57 mV, k = 4.2 mV.

different deactivation kinetics – fast, intermediate and slow – can be identified (Fig. 5.4B, upper part). The intermediate and fast components activate at different potentials (Fig. 5.4B, lower part). The voltage dependence of activation of the intermediate component is similar to that of the human I channel, and to the rat I channel (Safronov *et al.*, 1993) and I_{Kf1} current

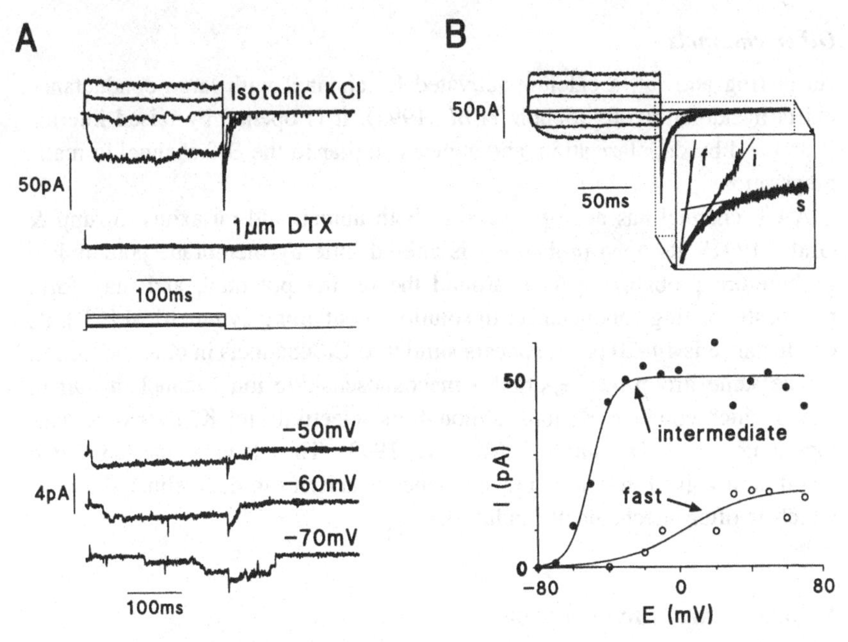

Fig. 5.4. (A) Above: Currents in a large outside-out patch in isotonic KCl; pulses to –70 mV, –50 mV, +20 mV and +50 mV, $E_h = -100$ mV. Currents are shown before and after addition of 1 μM α-dendrotoxin. Below: Currents in the same patch in isotonic KCl + 1 μM α-dendrotoxin, at higher gain. Pulses to potentials shown; $E_h = -100$ mV. Slow 10 pS channels are visible. (B) Above: Currents in a large outside-out patch in isotonic KCl, in response to pulses to –60 mV to +60 mV; $E_h = -100$ mV. Inset: Semilogarithmic plot of the averaged tail currents in this patch, fitted with three exponential components (time constants: fast (f) 2.35 ms, intermediate (i) 11.7 ms and slow (s) 101 ms). Below: Voltage dependence of activation of the fast and intermediate components of the tail currents above. The fitted lines are Boltzmann functions (see Fig. 5.2B), with the following values: intermediate: $E_{50} = -50$ mV, $k = 6.0$ mV, amplitude 50 pA; fast: $E_{50} = +3.4$ mV, $k = 18.3$ mV, amplitude 20 pA.

(Corrette *et al.*, 1991), while the voltage dependence of the fast component is like that of the F channel and the I_{Kf2} current in rat. The fast component in most human patches is smaller than has been reported in rat (up to 30% of total current). Longer pulses, with longer tails, allow the slow component to be measured accurately. It has a similar voltage dependence to the S channel in rat (Safronov *et al.*, 1993) though not to the macroscopic slow current (Röper & Schwarz, 1989; Schwarz, Reid & Bostock, 1995), and may correspond to more than one channel type, since it can be recorded in patches which have no S channels.

Direct comparisons of F and I channel currents in multi- and single-channel patches from rat and human axons show close similarities between the two species in voltage dependence, kinetics and pH sensitivity (Quasthoff, Mitrović & Grafe, 1993).

Other channels

As in frog and rat, a calcium-activated K^+ channel with large conductance exists in human axons (Scholz *et al.*, 1993). It is opened by raised internal $[Ca^{2+}]$ and by depolarization, and appears similar to the BK channel in many preparations.

A Cl^- channel has been recorded in both human and rat axons (Strupp & Grafe, 1991). Its open probability is altered little by membrane potential; it is therefore probably open at around the resting potential, and may form part of the resting conductance. In solutions containing symmetrical $[Cl^-]$, its conductance is 14–20 pS. It appears similar to Cl^- channels in other neurones.

The same group has reported a mechanosensitive ion channel in human axons which conducts K^+ ions, although its selectivity for K^+ over other cations is not clear (Quasthoff & Mitrović, 1993). They have suggested that it could be involved in generating spontaneous activity in demyelinated axons, which is often mechanically induced.

Voltage-clamped human axons

Recently, the macroscopic nodal voltage-clamp technique has been applied to human axons (Schwarz *et al.*, 1993, 1995). An action potential from an ulnar nerve axon is shown in Fig. 5.5A (upper part). It has a similar amplitude to that in rat axons (Schwarz & Eikhof, 1987). A long stimulus pulse elicits a train of three or four action potentials (Fig. 5.5B, upper part). Voltage-clamp currents from another axon are shown in Fig. 5.5C (upper part). A TTX-sensitive inward Na^+ current is followed by a small, fast-activating, outward K^+ current which is reduced by 4-AP. These voltage-dependent conductances are superimposed on a time- and voltage-independent 'leak' conductance. In three axons which were partially demyelinated by stretch during dissection, fast K^+ currents were large, presumably due to paranodal channels. When 'tail currents' in these axons were recorded in isotonic KCl, the fast K^+ current could be separated into two components, apparently similar to I_{Kf1} and I_{Kf2} in rat (Corrette *et al.*, 1991), with most of the current being I_{Kf1}-like. A slow K^+ current, similar to I_{Ks} in rat (Röper & Schwarz, 1989), was found in all human axons in isotonic KCl.

These current components have been analysed quantitatively (Reid, Bostock & Schwarz, 1993; Schwarz *et al.*, 1995) and used to generate a mathematical model based on that of Frankenhaeuser & Huxley (1964). There are four components: a Na^+ current; a small fast K^+ current of the I_{Kf1} type (the very small I_{Kf2} is ignored); a slow K^+ current; and a voltage- and time-independent leak. The model generates realistic single action potentials and repetitive activity (lower parts of Fig. 5.5A and B), and calculated membrane currents (lower part of Fig. 5.5C) are similar to those recorded. As in rat

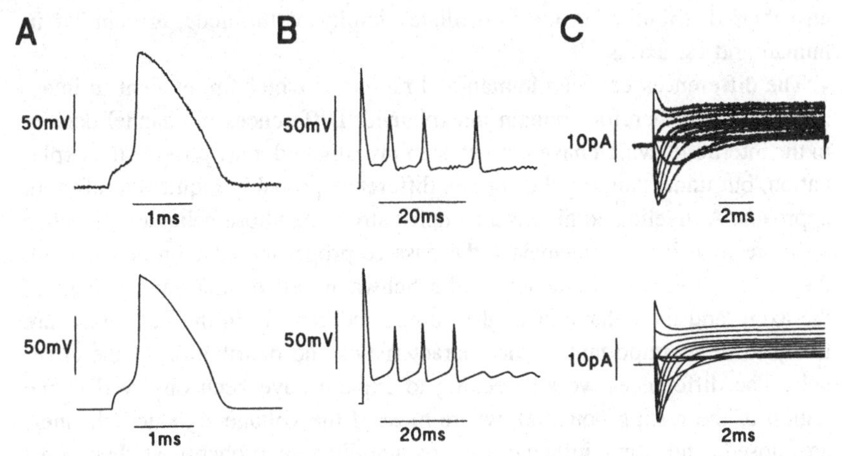

Fig. 5.5. Upper half: (A) Action potential recorded in a human node of Ranvier, in response to a brief current stimulus. (B) A train of action potentials elicited by a long current pulse in the same fibre. (C) Voltage-clamp currents in a different human node. Pulses to between −78 mV and +62 mV were applied after 50 ms prepulses to −118 mV to remove Na^+ channel inactivation; the holding potential (E_h) was −78 mV. The Na^+ currents have been reduced to about half their original amplitude by applying 3 nM TTX, in order to reduce the clamp error caused by series resistance. Lower half: Action potential, repetitive activity and voltage-clamp currents produced by the human node model under conditions corresponding to those above.

(Schwarz & Eikhof, 1987), the fast K^+ current has very little effect on the repolarization phase of the modelled single action potential. Removal of the slow K^+ current from the model produces prolonged repetitive activity similar to that induced in rat axons by TEA, which blocks the slow K^+ current (Baker *et al.*, 1987).

Conclusions

The characteristics of the ion channels in human axons appear to be essentially the same as those in the other species which have been studied. Although channels exist in other species which have not so far been found in human axons, this probably reflects only the relative difficulty of obtaining and recording from human nerve. Similarly, it would be unsafe to conclude that findings at the single-channel level in human axons which have not been reported in other species represent a genuine species difference.

The macroscopic voltage-clamp recordings support this conclusion. Currents in intact nodes of Ranvier are similar to those in rat nodes, while recordings from partially demyelinated axons, like the patch-clamp recordings, show that the paranodes of human axons contain a high density of fast K^+ channels. This suggests that not only the channels themselves, but also their distribution in the immediate vicinity of the node, are similar in human and rat axons.

The differences between human and rat axons which are evident in intact nerves *in vivo* therefore remain unexplained. Differences in channel density in the internode, which have not yet been investigated, may provide the explanation, but understanding the species difference probably requires a different approach. A myelinated fibre is a complex structure whose behaviour depends on more than its ion channels – the passive properties of axon and myelin, the largely unknown influence of the Schwann cell on the environment of the axon, and the behaviour of the pumps and carriers in the membrane are likely to be as important as the characteristics and distribution of the channels. The differences we are seeking to explain have been observed in the region of the resting potential, where most of the voltage-activated channels are closed, and other influences on excitability are probably at their most significant.

Mathematical modelling allows some possibilities to be explored (Bostock, Baker & Reid, 1991; Schwarz *et al.*, 1995), but lack of information imposes limitations on this approach. Interpreting the meaning of the single-channel and voltage-clamp recordings reviewed here, and determining where the species differences might lie, requires experiments on intact human axons. Much can be done *in vivo* (Bostock *et al.*, 1991), but channel blockers cannot be used, and only potentials in the region of the resting potential can be investigated, which limits the information that can be obtained. *In vitro* recording from fresh whole human nerve fascicles, under the same conditions as have been used for rat nerve, offers perhaps the best hope of allowing channels to be studied in their natural environment.

Acknowledgements

I am grateful to Hugh Bostock, Allison Hunter and Jürgen Schwarz for comments on the manuscript, and to them and other collaborators for many interesting and provocative discussions. None of the work in this laboratory would have been possible without the enthusiastic collaboration of Mr Rolfe Birch, Royal National Orthopaedic Hospital, London, who has given us tissue from some 65 patients over the past 3 years. Work in this laboratory was supported by the Medical Research Council, and the collaborations by the British Council and the Deutsche Akademische Austauschdienst.

References

Baker, M., Bostock, H., Grafe, P. & Martius, P. (1987). Function and distribution of three types of rectifying channel in rat spinal root myelinated axons. *Journal of Physiology (London)*, **383**, 45–67.

Barrett, E.F. & Barrett, J.N. (1982.) Intracellular recording from vertebrate myelinated axons: mechanism of the depolarising afterpotential. *Journal of Physiology (London)*, **323**, 117–144.

Berthold, C.-H. & Rydmark, M. (1983). Anatomy of the paranode–node–paranode region in the cat. *Experientia*, **39**, 964–976

Black, J.A., Kocsis, J.D. & Waxman, S.G. (1990). Ion channel organization of the myelinated fiber. *Trends in Neurosciences*, **13**, 48–54.

Bostock, H. & Baker, M. (1988). Evidence for two types of potassium channel in human motor axons *in vivo*. *Brain Research*, **462**, 354–358.

Bostock, H., Baker, M. & Reid, G. (1991). Changes in excitability of human motor axons underlying post-ischaemic fasciculations: evidence for two stable states. *Journal of Physiology (London)*, **441**, 537–557.

Brismar, T. (1980). Potential clamp analysis of membrane currents in rat myelinated nerve fibres. *Journal of Physiology (London)*, **298**, 171–184.

Chiu, S.Y. & Ritchie, J.M. (1984). On the physiological role of internodal potassium channels and the security of conduction in myelinated nerve fibres. *Proceedings of the Royal Society of London, Series B*,, **220**, 415–422.

Chiu, S.Y., Ritchie, J.M., Rogart, R.B. & Stagg, D. (1979). A quantitative description of membrane currents in rabbit myelinated nerve. *Journal of Physiology (London)*, **292**, 149–166.

Corrette, B.J., Repp, H., Dreyer, F. & Schwarz, J.R. (1991). Two types of fast K^+ channels in rat myelinated nerve fibres and their sensitivity to dendrotoxin. *Pflügers Archiv*, **418**, 408–416.

Dubois, J.M. (1981). Evidence for the existence of three types of potassium channels in the frog Ranvier node membrane. *Journal of Physiology (London)*, **318**, 297–316.

Fitzhugh, R. (1962). Computation of impulse initiation and saltatory conduction in a myelinated nerve fibre. *Biophysical Journal*, **2**, 11–21.

Frankenhaeuser, B. & Huxley, A.F. (1964). The action potential in the myelinated nerve fibre of *Xenopus laevis* as computed on the basis of voltage clamp data. *Journal of Physiology (London)*, **171**, 302–315.

Hermsteiner, M., Kampe, K. & Vogel, W. (1991). Mammalian and amphibian single sodium channels recorded from the membrane of peripheral nerve fibres. *Pflügers Archiv*, **418**, R29.

Jonas, P., Brau, M.E., Hermsteiner, M. & Vogel, W. (1989). Single-channel recording in myelinated nerve fibres reveals one type of Na channel but different K channels. *Proceedings of the National Academy of Sciences, USA*, **86**, 7238–7242.

Jonas, P., Koh, D.-S., Kampe, K., Hermsteiner, M. & Vogel, W. (1991). ATP-sensitive and Ca-activated K channels in vertebrate axons: novel links between metabolism and excitability. *Pflügers Archiv*, **418**, 68–73.

Kampe, K., Safronov, B. & Vogel, W. (1992). A Ca-activated and three voltage-dependent K channels identified in mammalian peripheral nerve. *Pflügers Archiv* , **420**, R28.

Levy, W.J., Spagnolia, T., Rumpf, R. & York, D.H. (1985). A method for *in vitro* enzymatic dissociation of nerve roots and peripheral nerves from adult mammals. *Journal of Neuroscience Methods*, **14**, 281–291.

Quasthoff, S. & Mitrović, N. (1993). A mechanosensitive K^+ channel with fast gating kinetics on human axons blocked by gadolinium ions. *Pflügers Archiv*, **422**, R20.

Quasthoff, S., Mitrović, N. & Grafe, P. (1993). Different sensitivity of F and I channel currents in rat and human axons to changes in cytoplasmic pH. *Pflügers Archiv*, **422**, R20.

Reid, G., Bostock, H. & Schwarz, J.R. (1993). Quantitative description of action

potentials and membrane currents in human node of Ranvier. *Journal of Physiology (London)*, **467**, 247P.
Reid, G., Kampe, K., Scholz, A., Birch, R., Bostock, H. & Vogel, W. (1991). Single channel currents in internodes of human axons. *European Journal of Neuroscience* , Supplement **4**, 254.
Röper, J. & Schwarz, J.R. (1989). Heterogeneous distribution of fast and slow potassium channels in myelinated rat nerve fibres. *Journal of Physiology (London)*, **416**, 93–110.
Safronov, B.V., Kampe, K. & Vogel, W. (1993). Single voltage-dependent potassium channels in rat peripheral nerve membrane. *Journal of Physiology (London)*, **460**, 675–691.
Scholz, A., Reid, G., Vogel, W. & Bostock, H. (1993). Ion channels in human axons. *Journal of Neurophysiology*, **70**, 1274–1279.
Schwarz, J.R. & Eikhof, G. (1987). Na currents and action potentials in rat myelinated fibres at 20 and 37 °C. *Pflügers Archiv*, **409**, 569–577.
Schwarz, J.R., Reid, G., Birch, R. & Bostock, H. (1993). Action potentials and membrane currents in human node of Ranvier. *Pflügers Archiv*, **422**, R18.
Schwarz, J.R., Reid, G. & Bostock, H. (1995). Action potentials and membrane currents in the human node of Ranvier. *Pflügers Archiv*, **430**, 283–292.
Stämpfli, R. & Hille, B. (1976). Electrophysiology of the peripheral myelinated nerve. In *Frog Neurobiology*, ed. R. Llinas & W. Precht. Berlin: Springer.
Strupp, M. & Grafe, P. (1991). A chloride channel in rat and human axons. *Neuroscience Letters*, **133**, 237–240.
Vogel, W. & Schwarz, J.R. (1995). Voltage-clamp studies in frog, rat and human axons: macroscopic and single channel currents. In: *The Axon*, ed. S.G. Waxman, J.D. Kocsis & P.K. Stys, pp. 257–280. Oxford: Oxford University Press.
Weller, U., Bernhardt, U., Siemen, D., Dreyer, D., Vogel, W. & Habermann, E. (1985). Electrophysiological and neurobiochemical evidence for the blockade of a potassium channel by dendrotoxin. *Naunyn-Schmiedebergs Archives of Pharmacology*, **330**, 77–83.
Wilson, G.F. & Chiu, S.Y. (1990). Ion channels in axon and Schwann cell membranes at paranodes of mammalian myelinated fibers studied with patch clamp. *Journal of Neuroscience*, **10**, 3263–3274.

6

An *in vitro* model of diabetic neuropathy: electrophysiological studies

P. GRAFE

Department of Physiology, University of Munich, Munich, Germany

Introduction

Diabetic neuropathy is characterized by alterations in axonal excitability which can lead to either 'positive' or 'negative' symptoms (paraesthesiae and dysaesthesiae as compared to hypaesthesia and anaesthesia). At present, little is known about the mechanisms underlying such pathological changes in nerve function. Modification of axonal membrane conductances might be one factor involved. Hyperglycaemia and hypoxia are regarded as most important for the development of diabetic neuropathy (Low, 1987). This chapter reviews a series of studies in which the combined effects of hyperglycaemia and hypoxia have been tested on isolated peripheral myelinated axons. As compared with *in vivo* experiments using diabetic animals, such an *in vitro* model of diabetic neuropathy has enabled us to explore directly the effects of factors possibly involved in the pathogenesis of this disease on electrophysiological axonal parameters.

The *in vitro* nerve preparation

The effects of hyperglycaemic hypoxia were investigated using electrophysiological recordings from isolated rat dorsal and ventral roots. After isolation (Schneider *et al*., 1992), the spinal roots were incubated at room temperature for 30 minutes to about 8 hours in solutions with different concentrations of D-glucose or other hexoses. Afterwards these nerves were transferred to the experimental organ bath used to record compound action potentials, extracellular direct current (DC) potentials, and electrotonus (Marsh *et al*., 1987; Schneider *et al*., 1992, 1993b). It consisted of a three-chambered Plexiglas bath, into which the spinal root was sealed with silicone grease. The central compartment (volume 1.5 ml) was continuously perfused at 14 ml/min with normoxic or hypoxic buffer. The nerve end in one of the lateral compartments was drawn into a suction electrode which was used for the application of current pulses. The solution in this lateral compartment was identical to the one in the central compartment, but it was not made hypoxic. The K^+ concentration in the second lateral compartment was elevated by 30 mM, and the potential difference was recorded across the resistance between this compartment and the central compartment of the organ bath.

A model of diabetic neuropathy

Peripheral nerves in patients with diabetes mellitus differ in their sensitivity to ischaemia as compared with those in control persons with normal glucose metabolism. These differences are: (*a*) resistance to ischaemia and (*b*) liability to ischaemic lesions with a prevalence of functional deficits in sensory nerve fibres (Thomas & Tomlinson, 1993). All these characteristic abnormalities were also found in our experiments on isolated rat spinal roots exposed to normo- or hyperglycaemic hypoxia (Fig. 6.1; Strupp *et al.*, 1991; Schneider *et al.*, 1992). Therefore, we consider the effects of hyperglycaemic hypoxia on isolated peripheral nerves as an *in vitro* model of diabetic neuropathy.

Fig. 6.1 summarizes observations made on 75 ventral or dorsal rat spinal roots, which had been incubated in either 2.5, 12.5, or 25 mM D-glucose before, during and after hypoxia. All roots were exposed to hypoxia for 30 minutes at 36 °C. Normo- as well as hyperglycaemic hypoxia depolarized the axons and decreased the peak height of compound action potentials. However, in particular in ventral spinal roots, both membrane depolarization and the decline in the amplitude of the compound action potential were strongly reduced during exposure to hypoxia in 12.5 or 25 mM D-glucose. This is a clear indication for resistance to hypoxia. Electrophysiological damage due to hyperglycaemic hypoxia, on the other hand, was revealed in the experiments on dorsal spinal roots. Neither membrane potential nor action potential amplitude recovered completely after hypoxia in solutions containing high concentrations of glucose.

In another series of experiments, spinal roots incubated in 25 mM D-glucose were simultaneously exposed to both hypoxia and inhibition of glycolysis (Fig. 6.1; Schneider, Niedermeier & Grafe, 1993a). Glycolysis was inhibited by 10 mM iodoacetate (IAA), an inhibitor of the enzyme phosphoglyceraldehyde dehydrogenase. IAA strongly enhanced the effects of hypoxia; action potential amplitude and extracellular DC potential of both ventral and dorsal roots responded very sensitively to hypoxia. Furthermore, IAA improved post-hypoxic functional recovery in spite of the augmented response during hypoxia. On average, peak height of dorsal spinal roots incubated in 25 mM D-glucose recovered to only 38.2 ± 3.3 % (mean ± SEM; n = 13) of the pre-hypoxic level, however, the recovery was nearly complete (88.6 ± 3.4%; n = 5) when hypoxia was induced with simultaneous inhibition of glycolysis.

The observations made using IAA indicate that glycolysis is responsible

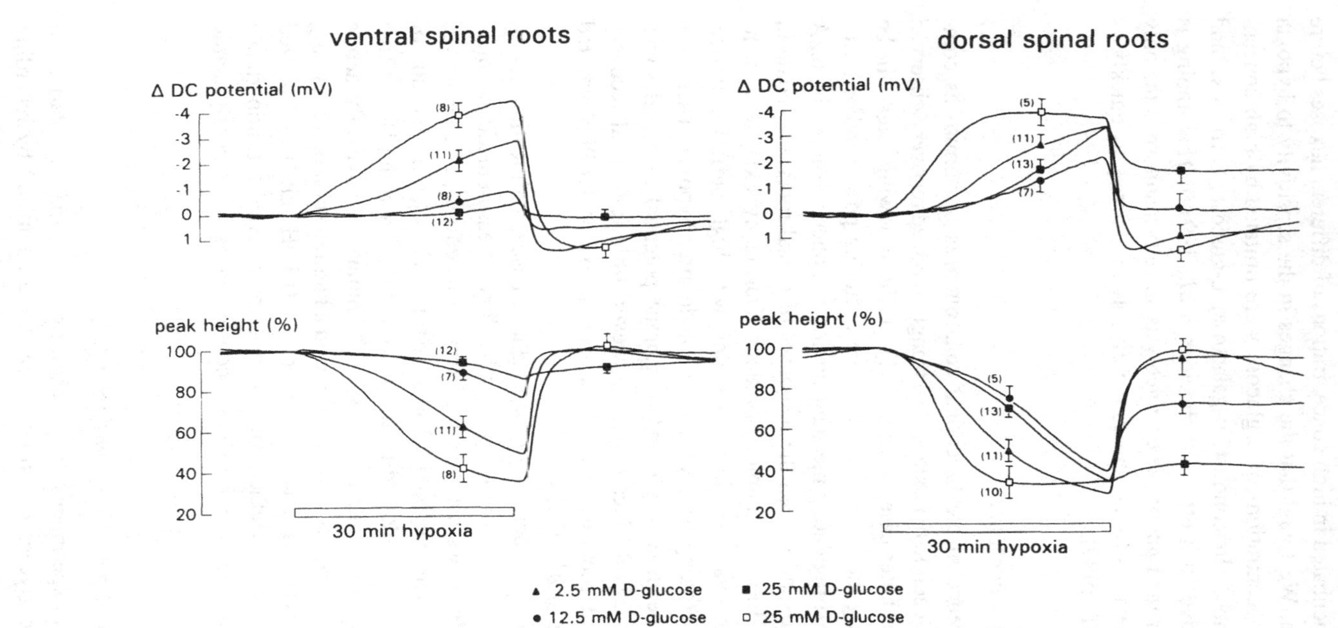

Fig. 6.1. Changes in the peak height of compound nerve action potentials and of the extracellular direct current (DC) potential before, during and after hypoxia. Dorsal and ventral spinal roots had been incubated for several hours in 2.5, 12.5 or 25 mM D-glucose before the exposure to hypoxia. In some of the spinal roots, the response to both hyperglycaemic hypoxia plus inhibition of glycolysis by means of iodoacetate (IAA, 10 mM) was tested. Data were digitized and averaged (mean ± SEM; number of roots given in brackets). Only one exposure to hypoxia was performed on each spinal root. Adapted from Schneider *et al.* (1993a).

for (*a*) the development of resistance to hypoxia and (*b*) post-hypoxic functional deficits in sensory fibres of dorsal spinal roots. This view is supported by the effects of D-mannose, tested in another series of experiments in which spinal roots were incubated in high concentrations of different hexoses before and during hypoxia. We found that the changes in the sensitivity to hypoxia induced by high concentrations of D-glucose were imitated by high concentrations of D-mannose. In contrast, D-galactose, L-glucose, D-fructose and L-fucose did not have such effects (Schneider *et al.*, 1993a). This finding is further evidence for the importance of glycolysis since D-mannose is the only hexose which, in nervous tissue, enters the glycolytic pathway as quickly as D-glucose (Sokoloff, 1989).

The importance of bicarbonate

One striking difference in the functional consequences of normo- as compared with hyperglycaemic hypoxia is the strongly reduced recovery observed after hypoxia in the latter case (see Fig. 6.1). In the following, data will be summarized showing the importance of tissue buffering power and/or bicarbonate-dependent pH regulating mechanisms for this phenomenon. Fig. 6.2A illustrates a comparison between the responses of dorsal roots to hypoxia in bicarbonate- and HEPES-containing solutions. During hypoxia, less membrane depolarization was seen in the solution with high buffering power; however, the most striking observation was made after hypoxia. There was good recovery of the peak height and membrane potential in dorsal spinal roots buffered with 25 mM bicarbonate; however, little functional recovery was found in dorsal roots exposed to hypoxia in 6 mM HEPES-buffered medium (Schneider *et al.*, 1992).

The importance of bicarbonate for the extent of tissue acidification during hyperglycaemic hypoxia was also revealed by direct measurements of interstitial pH. Fig. 6.2B shows data obtained from isolated rat peroneal nerves exposed to hyperglycaemic hypoxia. The interstitial pH (pH_e) in the centre of such isolated nerves was followed by means of recordings with pH-sensitive microelectrodes (Strupp *et al.*, 1991). After 30 minutes of hyperglycaemic hypoxia, pH_e decreased by about 0.1 pH units in a bathing solution containing 25 mM bicarbonate whereas an acid-going shift of 0.3 pH units was observed in 6 mM HEPES. Taken together, these data clearly reveal the importance of tissue acidification for the functional consequences of hyperglycaemic hypoxia.

pH-dependent axonal potassium conductance

In a recent series of experiments, the mechanisms of hypoxic depolarization in normal and hyperglycaemic sensory axons were compared by recording

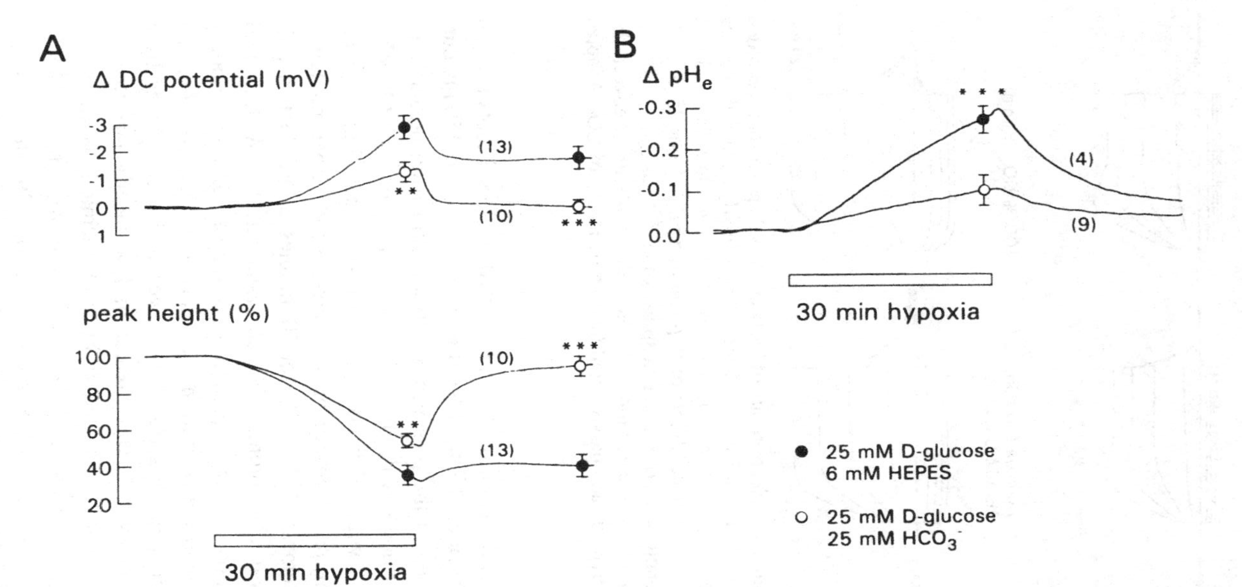

Fig. 6.2. (A) Changes in the peak height of compound nerve action potentials and of the extracellular direct current (DC) potential before, during and after hypoxia. Dorsal spinal roots incubated in a solution with 25 mM D-glucose were exposed to hypoxia in saline buffered with either bicarbonate (25 mM HCO_3^-/ 5% CO_2) or HEPES (6 mM). Note the much better post-hypoxic functional recovery of the dorsal spinal roots in high bicarbonate solution. $^{**}P < 0.01$; $^{***}P < 0.001$. (B) Changes in interstitial pH (pH_e) recorded by pH-sensitive microelectrodes in the centre of isolated rat peroneal nerves. These nerves were exposed to hypoxia in saline buffered with either bicarbonate (25 mM HCO_3^-/5% CO_2) or HEPES (6 mM). Note the much smaller hypoxia-induced extracellular acidification in high bicarbonate solution. $^{***}P < 0.001$. Adapted from Strupp *et al.* (1991) and Schneider *et al.* (1992).

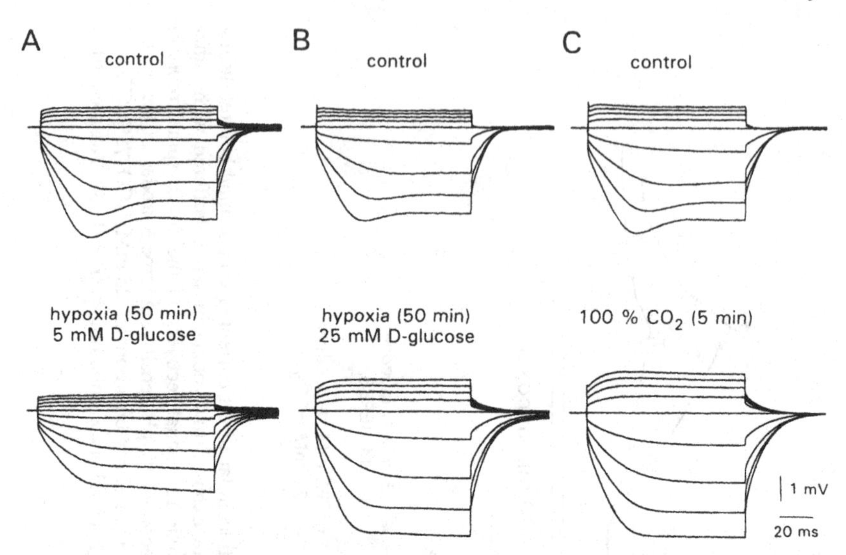

Fig. 6.3. Examples of changes in electrotonus of isolated dorsal spinal roots recorded in tetrodotoxin-containing solutions. Illustrated are extracellularly recorded changes in voltage (*E*) induced by current pulses of 150 ms duration (current amplitude ±10 μA). Note the opposite effects of hypoxia in 5 mM (A) or 25 mM (B) D-glucose on the input conductance of the dorsal spinal roots. The spinal root of (B) was also exposed to a high CO_2 concentration (C). Note the similarities between the effect of hyperglycaemic hypoxia and passive acidification by means of CO_2. Adapted from Grafe *et al.* (1994).

demarcation potentials and electrotonus from isolated rat dorsal roots *in vitro* (Grafe, Bostock & Schneider, 1994). Electrotonic responses to 150 ms current pulses were recorded in the presence of tetrodotoxin from dorsal roots incubated for at least 3 hours with either normal (5 mM) or high (25 mM) D-glucose solutions, and with either normal (25 mM) or low (5 mM) bicarbonate to alter buffering power and/or bicarbonate-dependent pH regulation mechanisms. On replacement of O_2 by N_2 for 50 minutes, all the roots depolarized, but the changes in electrotonus differed systematically. With normal glucose, the depolarization was accompanied by an increase in input conductance, whether bicarbonate concentration was normal or reduced. For the hyperglycaemic roots the depolarization was slower, and accompanied by a *fall* in input conductance, which was greater in low bicarbonate than in normal bicarbonate. The changes induced by hyperglycaemic hypoxia in low bicarbonate could be mimicked by exposure of the root for 5 min to 100% CO_2. Examples of these observations are given in Fig. 6.3. Similar, but more profound changes in electrotonus and input conductance could be produced by a combination of 3 mM tetraethylammonium chloride and 3 mM 4-aminopyridine, to block both fast and slow K^+ channels (not illustrated; see also Baker *et al.*, 1987; Waxman & Ritchie, 1993). The similarity in the effects of hyperglycaemic hypoxia, passive acidification by high CO_2 and K^+ channel blockers on the electrotonus indicate that hyperglycaemic hypoxia reduces the input conductance of spinal root axons by inhibition of pH-sensitive K^+ channels. In order to test this idea further, experiments were performed using the patch-clamp technique on axonal membrane patches.

In these studies, multi-channel current recordings from excised, inside-out axonal membrane patches (Jonas *et al.*, 1989; Safronov, Kampe & Vogel, 1993) were used to study the effects of cytoplasmic acidification on voltage-dependent K^+ conductances with fast (F channels) and intermediate (I channels) kinetics of deactivation. It was found that F channels were blocked by small changes in cytoplasmic pH (50% inhibition at pH 6.9). I channels, on the other hand, were less sensitive to intra-axonal acidification (Schneider *et al.*, 1993b). Other types of axonal K^+ channel have been described in recent publications (Jonas *et al.*, 1991; Koh *et al.*, 1992). Their possible modulation by cytoplasmic acidification has not yet been studied. However, inhibition of these conductances may also contribute to the changes seen in electrotonus during hyperglycaemic hypoxia.

Conclusions

Effects of hyperglycaemic hypoxia on isolated peripheral nerves resemble observations made in peripheral nerves of human diabetics. Using such an *in vitro* model of diabetic neuropathy, we found that hyperglycaemic hypoxia inhibits the potassium conductance of axons by cytoplasmic acidification. This alteration in membrane conductance may contribute to the abnormal excitability of peripheral nerves seen in diabetic neuropathy. 'Positive' symptoms may be related to enhanced repetitive firing after inhibition of K^+ channels; 'negative' symptoms may be due to membrane depolarization after block of K^+ channels involved in the generation of the resting potential. A method to quantify the axonal potassium conductance in human axons *in vivo* has been developed by Bostock & Baker (1988). Our data indicate that this technique of 'threshold tracking' should be used to explore axonal potassium conductance in peripheral nerves of human diabetics.

Acknowledgements

I should like to thank my colleagues U. Schneider, M. Strupp, S. Quasthoff, R. Jund, W. Niedermeier, N. Mitrović (all from the Department of Physiology, University of Munich) and H. Bostock (Institute of Neurology, University of London) for their valuable collaboration. Ms C. Müller provided excellent laboratory and secretarial assistance. This project was supported by grants from the Deutsche Forschungsgemeinschaft (SFB 220, B1).

References

Baker, M., Bostock, H., Grafe, P. & Martius, P. (1987). Function and distribution of three types of rectifying channel in rat spinal root myelinated axons. *Journal of Physiology (London)*, **383**, 45–67.

Bostock, H. & Baker, M. (1988). Evidence for two types of potassium channel in human motor axon *in vivo*. *Brain Research*, **462**, 354–358.

Grafe, P., Bostock, H. & Schneider, U. (1994). The effects of hyperglycaemic hypoxia on rectification in rat dorsal root axons. *Journal of Physiology (London)*, **480**, 297–307.

Jonas, P., Brau, M.E., Hermsteiner, M. & Vogel, W. (1989). Single-channel recording in myelinated nerve fibres reveals one type of Na channel but different K channels. *Proceedings of the National Academy of Sciences, USA*, **86**, 7238–7242.

Jonas, P., Koh, D.S., Kampe, K., Hermsteiner, M. & Vogel, W. (1991). ATP-sensitive and Ca-activated K-channels in vertebrate axons: novel links between metabolism and excitability. *Pflügers Archiv*, **418**, 68–73.

Koh, D.S., Jonas, P., Brau, M.E. & Vogel, W. (1992). A TEA-insensitive flickering potassium channel active around the resting potential in myelinated nerve. *Journal of Membrane Biology*, **130**, 149–162.

Low, P.A. (1987). Recent advances in the pathogenesis of diabetic neuropathy. *Muscle and Nerve*, **10**, 121–128.

Marsh, S.J., Stansfeld, C.E., Brown, D.A., Davey, R. & McCarthy, D. (1987). The mechanism of action of capsaicin on sensory C-type neurones and their axons *in vitro*. *Neuroscience*, **23**, 275–289.

Safronov, B.V., Kampe, K. & Vogel, W. (1993). Single voltage-dependent potassium channels in rat peripheral nerve membrane. *Journal of Physiology (London)*, **460**, 675–691.

Schneider, U., Jund, R., Nees, S. & Grafe, P. (1992). Differences in sensitivity to hyperglycemic hypoxia of isolated rat sensory and motor nerve fibers. *Annals of Neurology*, **31**, 605–610.

Schneider, U., Niedermeier, W. & Grafe, P. (1993a). The paradox between resistance to hypoxia and liability to hypoxic damage in hyperglycemic peripheral nerves: evidence for glycolysis involvement. *Diabetes*, **42**, 981–987.

Schneider, U., Quasthoff, S., Mitrović, N. & Grafe, P. (1993b). Hyperglycaemic hypoxia alters after-potential and fast K^+ conductance of rat axons by cytoplasmic acidification. *Journal of Physiology (London)*, **465**, 679–697.

Sokoloff, L. (1989). Circulation and energy metabolism of the brain. In *Basic Neurochemistry*, ed. G. Siegel, B. Agranoff, R.W. Albers & P. Molinoff, pp. 565–590. New York: Raven Press.

Strupp, M., Jund, R., Schneider, U. & Grafe, P. (1991). Glucose availability and sensitivity to anoxia of isolated rat peroneal nerve. *American Journal of Physiology*, **261**, E389-E394.

Thomas, P.K. & Tomlinson, D.R. (1993). Diabetic and hypoglycemic neuropathy. In *Peripheral neuropathy*, ed. P.J. Dyck, P.K. Thomas, J.W. Griffin, P.A. Low & J.F. Podulso, pp. 1219–1250. Philadelphia: W.B. Saunders.

Waxman, S.G. & Ritchie, J.M. (1993). Molecular dissection of the myelinated axon. *Annals of Neurology*, **33**, 121–136.

7

Autoimmunity at the neuromuscular junction

JOHN NEWSOM-DAVIS

Neurosciences Group, Institute of Molecular Medicine, John Radcliffe Hospital, Oxford, UK

The neuromuscular junction is arguably the best understood synapse in the human nervous system. Building on the pioneering studies in the frog (see Katz, 1969), Elmqvist *et al* (1964) and Elmqvist & Lambert (1968) defined the principal properties of nerve-to-muscle transmission in man, based on observations made on biopsied human intercostal muscle. At that time, the neuromuscular junction was believed to be the site of the disorder in myasthenia gravis (MG), but its physiological mechanism was not known until the studies of Elmqvist *et al.* (1964) revealed a reduction in the amplitudes of the miniature endplate potentials and of the endplate potentials. The later discovery of MG's antibody-mediated nature was made possible by the earlier physiological studies. Subsequently, it has become clear that MG is not the only autoimmune disorder affecting the neuromuscular junction. Its vulnerability may be because the ion channels and cell surface molecules concerned in synaptic transmission, unlike those at synapses elsewhere in the nervous system, lie outside the blood–brain barrier and thus lack the protection from systemic immune attack that the latter affords.

This chapter will first describe the background for the autoimmune process in MG, a postsynaptic disorder, before describing two other disorders distinct from MG in which the targets for immune attack are presynaptic.

Myasthenia gravis

That MG might be an autoimmune disease was first suggested by Simpson (1960) who deduced this from clinical clues: its association with other autoimmune disease (notably of the thyroid) and the transfer of the disorder transiently to babies born to myasthenic mothers. But the validation of this hypothesis depended on a crucial, and apparently unrelated, discovery: the isolation of a snake toxin (α-bungarotoxin) that specifically bound muscle acetylcholine receptors (Chang & Lee, 1962). Using this toxin to affinity purify acetylcholine receptor (AChR) from the electric organ of electric eel, Patrick & Lindstrom (1973) immunized rabbits with the primary aim of characterizing AChRs by raising experimental antibodies. The experiment, however, also induced muscle paralysis in the rabbits, that was subsequently shown to be due to loss of muscle AChRs caused by the induced AChR antibodies. The disorder was termed experimental autoimmune MG.

The question of whether human MG was presynaptic or postsynaptic in origin (a considerable controversy at that time) was resolved by the finding that AChRs were strikingly reduced in number at the postsynaptic membrane in MG (Fambrough, Drachman & Satyamurti, 1973; Ito *et al.*, 1978), and that this reduction could fully account for the observed physiological changes. Evidence that a circulating factor caused the AChR loss was provided by the detection of anti-AChR antibodies in 85% of MG patients (Lindstrom *et al.*, 1976), by the clinical response to plasma exchange that removes circulating antibodies (Pinching, Peters & Newsom-Davis, 1976), and the finding that MG immunoglobulins injected into mice could reproduce the pathophysiological changes of the human disease (Toyka *et al.*, 1977).

These experiments established unequivocally the cause for the muscle weakness in MG. They also met the criteria for regarding a disorder as antibody-mediated: detection of antibodies to a defined antigen, and induction of the disease in animals by immunization with the purified antigen or by injection of patients' plasma or immunoglobulins ('passive transfer'). Importantly, these discoveries made available a sensitive and highly specific diagnostic test for MG in the anti-AChR antibody assay, and a means of monitoring disease progress. Furthermore, the immune-mediated nature of MG led to the increasing and successful use of immunosuppressive therapies in the management of the condition.

Subsequent cloning and sequencing of the separate genes coding for the five subunits of the human AChR (see Beeson *et al.*, 1993) has allowed the investigation of the binding sites for the antibodies and also for the T cells on which antibody production by B cells depends. Such knowledge may in due course make possible selective control of the abnormal immune response, for example by inactivating the relevant helper T cells as we have shown in one experimental paradigm (Nicolle *et al.*, 1994).

Lambert–Eaton myasthenic syndrome

Lambert–Eaton myasthenic syndrome (LEMS), a disorder of neuromuscular transmission often associated with small cell lung cancer (SCLC), was first shown to be physiologically distinct from MG by Lambert and his colleagues (Eaton & Lambert, 1957; Lambert *et al.*, 1961). Later studies with Elmqvist established that the disorder was due to a reduced nerve-evoked quantal release of acetylcholine (*m*), values for *m* typically being <10 compared with 50 or more in controls (Elmqvist & Lambert, 1968; Lambert & Elmqvist, 1971). Clinical features pointing to an autoimmune origin for the condition were the occurrence of other immunological diseases in its non-SCLC-linked form and the improvement in muscle strength following plasma exchange treatment (Lang *et al.*, 1981). Its antibody-mediated nature was first shown experimentally by passive transfer experiments in which mice were injected daily with LEMS IgG. The principal physiological abnormalities of LEMS were reproduced in the mice, namely reduced quantal content of the endplate potential (Lang *et al.*, 1981) and morphological changes characterized by paucity and disorganization of active zone particles (Fukunaga *et al.*, 1983). These, and further experiments, indicated that voltage-gated Ca^{2+} channels (VGCCs) or structures closely associated with them were targets for the IgG autoantibodies in LEMS (Vincent, Lang & Newsom-Davis, 1989). Cross-linking of the VGCCs by divalent antibody leads to their downregulation (Nagel *et al.*, 1988). Anti-VGCC antibodies can be detected in the serum of many patients by means of a radio-immunoassay based on the use of the marine snail toxin ω-conotoxin to label VGCCs (Sher *et al.*, 1989), and titres appear to correlate with disease severity in longitudinal studies on individual patients (Leys *et al.*, 1991).

There is strong evidence that, in the 60% of LEMS patients with SCLC, the tumour itself triggers the autoantibody response. First, SCLC is believed to be of neuroectodermal origin and might therefore share antigenic determinants with the peripheral nervous system. Second, the tumour cells show voltage-gated Ca^{2+} flux when briefly depolarized by high external K^{+} concentrations (Roberts *et al.*, 1985). Third, SCLC cell lines exposed to LEMS IgG, whether from SCLC-LEMS or non-SCLC-LEMS patients, show a highly significant reduction in K^{+}-stimulated Ca^{2+} influx compared with cells grown in control IgG (Roberts *et al.*, 1985). Fourth, SCLCs from patients with LEMS show a significant increase in macrophage infiltration of the tumour compared with non-LEMS tumours (Morris *et al.*, 1992). Finally, treatment of the tumour by resection or local radiotherapy can lead to a complete recovery of the neurological disorder (Chalk *et al.*, 1990).

Thus LEMS in its SCLC-associated form appears to provide a model for other paraneoplastic neurological disorders, for example subacute sensory neuropathy or subacute cerebellar degeneration. In each of these, the immune system may provide the link between the cancer and the neurological disorder. The stimulus for the autoantibody response in non-SCLC cases is unknown.

Acquired neuromyotonia

Acquired neuromyotonia (NMT) is characterized by widespread muscle twitching (myokymia), severe muscle cramps often induced by muscle contraction, pseudomyotonia and sometimes by muscle weakness. Denny-Brown & Foley (1948) were the first to demonstrate the characteristic electromyographic findings of doublet, triplet or multiplet motor unit discharges occurring at instantaneous frequencies much higher than can be produced by voluntary contraction. Isaacs (1961) established the peripheral nerve origin of the discharges by showing that they persisted in the presence of proximal peripheral nerve block, and suggested that the abnormal discharges were arising in the terminal arborization of motor nerves (Isaacs, 1967), a condition that he referred to as 'quantal squander'.

Clues to a possible autoimmune basis for NMT include its association with other autoimmune disorders or with thymoma, the presence of oligoclonal bands in the cerebrospinal fluid, and the transient improvement that can follow plasma exchange (Newsom-Davis & Mills, 1993). This suggested that a circulating antibody was increasing peripheral nerve excitability, one possible action being a reduction in K^+ conductance. Such an effect would increase the quantal content of the endplate potential. In preliminary experiments to test this, mice were injected with plasma or IgG from NMT patients or from controls, and resistance to *d*-tubocurarine was assessed *in vitro* in the mouse phrenic nerve–diaphragm preparation (Sinha *et al.*, 1991). The results showed a significant increase in tubocurarine resistance in the NMT-injected mice, consistent with an increase in quantal content. In recent similar passive transfer experiments, microelectrodes were used to measure quantal content by the direct method (in the presence of μ-conotoxin to block the muscle action potential). The results confirmed the increase in quantal content.

The results of the physiological experiments are consistent with an antibody-mediated process that leads either to a reduction in the number of voltage-gated K^+ channels (VGKCs) at motor nerve terminals or, less likely, to a prolongation of Na^+ channel open time. To try to distinguish between these, anti-VGKC antibodies are currently being sought in the peripheral blood of NMT patients using α-dendrotoxin as a specific ligand. Antibodies at raised titre have clearly been detected in two of our cases (Vincent & Newsom-Davis, unpublished data). These findings appear to establish an antibody-mediated autoimmune aetiology for acquired neuromyotonia in some patients.

Conclusions

The neuromuscular synapse, rich in ion channels concerned with neuromuscular transmission, is a favoured target for autoimmune attack. It is vulnerable, too, to neurotoxins, whose availability has greatly aided the identification of the target antigens. Recognition of the distinct autoantibody-mediated mechanisms underlying myasthenia gravis, the Lambert–Eaton syndrome and acquired neuromyotonia has offered new therapeutic possibilities. Finally, the neuromuscular junction may be an informative site at which to study other peripheral nerve disorders in which immune processes appear to be implicated (Roberts *et al.*, 1994).

References

Beeson, D., Brydson, M., Betty, M., Jeremiah, S., Povey, S., Vincent A. & Newsom-Davis, J. (1993). Primary structure of the human muscle acetylcholine receptor: cDNA cloning of the gamma and epsilon subunits. *European Journal of Biochemistry*, **215**, 229–238.

Chalk, C.H., Murray, N.M., Newsom-Davis, J., O'Neill, J.H. & Spiro, S.G. (1990). Response of the Lambert–Eaton myasthenic syndrome to treatment of associated small-cell lung carcinoma. *Neurology*, **40**, 1552–1556.

Chang, C.E. & Lee, C.Y. (1962). Isolation of neurotoxins from the venom of *Bungarus multicinctus* and their modes of neuromuscular blocking action. *Archives of Pharmacodynamic Therapy*, **144**, 241–257.

Denny-Brown, D. & Foley, J.M. (1948). Myokymia and the benign fasciculation of muscular cramps. *Transactions of the Association of American Physicians*, **61**, 88–96.

Eaton, L.M. & Lambert, E.H. (1957). Electromyography and electric stimulation of nerves in diseases of motor unit: observations on myasthenic syndrome associated with malignant tumors. *Journal of the American Medical Association*, **163**, 1117–1124.

Elmqvist, D., Hofmann, W.W., Kugelberg, J. & Quastel, D.M.J. (1964). An electrophysiological investigation of neuromuscular transmission in myasthenia gravis. *Journal of Physiology (London)*, **174**, 417–434.

Elmqvist, D. & Lambert, E.H. (1968). Detailed analysis of neuromuscular transmission in a patient with the myasthenic syndrome sometimes associated with bronchogenic carcinoma. *Mayo Clinic Proceedings*, **43**, 689–713.

Fambrough, D.M., Drachman, D.B. & Satyamurti, S. (1973). Neuromuscular junction in myasthenia gravis: decreased acetylcholine receptors. *Science*, **182**, 293–295.

Fukunaga, H., Engel, A.G., Lang, B., Newsom-Davis, J. & Vincent, A. (1983). Passive transfer of Lambert–Eaton myasthenic syndrome with IgG from man to mouse depletes the presynaptic membrane active zones. *Proceedings of the National Academy of Sciences USA*, **80**, 7636–7640.

Isaacs, H. (1961). A syndrome of continuous muscle-fibre activity. *Journal of Neurology, Neurosurgery and Psychiatry*, **24**, 319–325.

Isaacs, H. (1967). Continuous muscle fibre activity in an Indian male with additional evidence of terminal motor fibre abnormality. *Journal of Neurology, Neurosurgery and Psychiatry*, **30**, 126–133.

Ito, Y., Miledi, R., Vincent, A. & Newsom-Davis, J. (1978). Acetylcholine receptors and end-plate electrophysiology in myasthenia gravis. *Brain*, **101**, 345–368.

Katz, B. (1969). *The Release of Neural Transmitter Substances*. Liverpool: Liverpool University Press.

Lambert, E.H. & Elmqvist, D. (1971). Quantal components of end-plate potentials in the myasthenic syndrome. *Annals of the New York Academy of Sciences*, **183**, 183–199.

Lambert, E.H., Rooke, E.D., Eaton, L.M. & Hodgson, C.H. (1961). Myasthenic syndrome occasionally associated with bronchial neoplasm: neurophysiologic studies. In *Myasthenia Gravis*, ed. H.R. Viets, pp. 362–410. Springfield, IL: Charles C. Thomas.

Lang, B., Newsom-Davis, J., Wray, D., Vincent, A. & Murray, N.M.F. (1981). Autoimmune aetiology for myasthenic (Eaton–Lambert) syndrome. *Lancet*, **ii**, 224–226.

Leys, K., Lang, B., Johnston, I. & Newsom-Davis, J. (1991). Calcium channel autoantibodies in the Lambert–Eaton myasthenic syndrome. *Annals of Neurology*, **29**, 307–314.

Lindstrom, J.M., Seybold, M.E., Lennon, V.A., Whittingham, S. & Duane, D.D. (1976). Antibody to acetylcholine receptor in myasthenia gravis: prevalence, clinical correlates and diagnostic value. *Neurology*, **26**, 1054–1059.

Morris, C.S., Esiri, M.M., Marx, A. & Newsom-Davis, J. (1992). Immunocytochemical characteristics of small cell lung carcinoma associated with the Lambert–Eaton myasthenic syndrome. *American Journal of Pathology*, **140**, 839–845.

Nagel, A., Engel, A.G., Lang, B., Newsom-Davis, J. & Fukuoka, T. (1988). Lambert–Eaton myasthenic syndrome IgG depletes presynaptic membrane active zone particles by antigenic modulation. *Annals of Neurology*, **24**, 552–558.

Newsom-Davis, J. & Mills, K.R. (1993). Immunological associations of acquired neuromyotonia (Isaacs' syndrome): report of 5 cases and literature review. *Brain*, **116**, 453–469.

Nicolle, M.W., Nag, B., Sharma, S.D., Willcox, N., Vincent, A., Ferguson, D.J.P & Newsom-Davis, J. (1994). Specific tolerance to an acetylcholine receptor epitope induced *in vitro* in myasthenia gravis $CD4^+$ lymphocytes by soluble major histocompatibility complex class II-peptide complexes. *Journal of Clinical Investigation*, **93**, 1361–1369.

Patrick, J. & Lindstrom, J. (1973). Autoimmune response to acetylcholine receptor. *Science*, **180**, 871–872.

Pinching, A.J., Peters, D.K. & Newsom-Davis, J. (1976). Remission of myasthenia gravis following plasma exchange. *Lancet*, **ii**, 1373–1376.

Roberts, A., Perera, S., Lang, B., Vincent, A. & Newsom-Davis, J. (1985). Paraneoplastic myasthenic syndrome IgG inhibits $^{45}Ca^{2+}$ flux in a human small cell carcinoma line. *Nature*, **317**, 737–739.

Roberts, M., Willison, H., Vincent, A. & Newsom-Davis, J. (1994). Serum factor in Miller–Fisher variant of Guillain–Barré syndrome and neurotransmitter release. *Lancet*, **343**, 454–455.

Sher, E., Canal, N., Piccolo, G., Gotti, C., Scoppetta, C., Evoli, A. & Clementi, F. (1989). Specificity of calcium channel autoantibodies in Lambert–Eaton myasthenic syndrome. *Lancet*, **ii**, 640–643.

Simpson, J.A. (1960). Myasthenia gravis: a new hypothesis. *Scottish Medical Journal*, **5**, 419–439.

Sinha, S., Newsom-Davis, J., Mills, K., Byrne, N., Lang, B. & Vincent, A. (1991). Autoimmune aetiology for acquired neuromyotonia (Isaacs' syndrome). *Lancet*, **338**, 75–77.

Toyka, K.V., Drachman, D.B., Griffin, D.E., Pestronk, A., Winkelstein, J.A., Fischbeck, K.H. & Kao, I. (1977). Myasthenia gravis: study of humoral immune mechanisms by passive transfer to mice. *New England Journal of Medicine*, **296**, 125–131.

Vincent, A., Lang, B. & Newsom-Davis, J. (1989). Autoimmunity to the voltage-gated calcium channel underlies the Lambert–Eaton myasthenic syndrome, a paraneoplastic disorder. *Trends in Neurosciences*, **12**, 496–502.

8

Immunopathology and pathophysiology of experimental autoimmune encephalomyelitis

M.P. PENDER

Department of Medicine, The University of Queensland, Brisbane, Queensland, Australia

Introduction

Experimental autoimmune encephalomyelitis (EAE) is an inflammatory demyelinating disease of the central nervous system (CNS), and is widely studied as an animal model of the human CNS demyelinating diseases, including multiple sclerosis (Raine, 1984). EAE can be induced by inoculation with whole CNS tissue, purified myelin basic protein (MBP) or myelin proteolipid protein (PLP), together with adjuvants. It may also be induced by the passive transfer of T cells specifically reactive to these myelin antigens. EAE may have either an acute or a chronic relapsing course. Acute EAE closely resembles the human disease acute disseminated encephalomyelitis, while chronic relapsing EAE resembles multiple sclerosis. EAE is also the prototype for T-cell-mediated autoimmune disease in general. This chapter will focus on the immunopathology and pathophysiology of EAE, which are the subjects of investigation in my laboratory.

Immunopathology

In EAE, activated $CD4^+$ T cells specific for myelin antigens cross the blood–brain barrier and enter the CNS parenchyma. Macrophages, $CD4^+$ T cells of other specificities, and a limited number of B cells and $CD8^+$ T cells also then enter the CNS (Traugott *et al.*, 1981; McCombe *et al.*, 1992). It is generally held that myelin, rather than the oligodendrocyte, is the primary target of the autoimmune attack in EAE (Itoyama & Webster, 1982; Moore, Traugott & Raine, 1984), and that the myelin damage is initiated by macrophages that have been activated by cytokines secreted by the invading T cells. However, it remains possible that the oligodendrocyte may be the primary target of an autoimmune attack by cytotoxic T cells in EAE. We have recently discovered that apoptosis (programmed cell death) occurs in the CNS in EAE (Pender *et al.*, 1991). On the basis of morphological findings obtained by conventional histological techniques, we suggested that some of the apoptotic cells may be oligodendrocytes, but this has not been proven by immunocytochemistry. As apoptosis is the mechanism of target cell death induced by cytotoxic T cells, apoptosis of oligodendrocytes in EAE may occur as a result of specific T cell cytotoxicity and may contribute to the demyelination. It would be predicted that such oligodendrocyte apoptosis would be followed by phagocytosis of the apoptotic oligodendrocyte and the myelin it supported.

In our initial report describing the occurrence of apoptosis in the CNS in EAE, we also suggested that some of the apoptotic cells were haematogenous cells (lymphocytes or possibly macrophages) (Pender *et al.*, 1991). Using a pre-embedding immunocytochemical technique we subsequently demonstrated that T cell apoptosis occurs in the spinal cord of rats with acute EAE induced by inoculation with MBP (Pender *et al.*, 1992). About half the apoptotic cells in the spinal cord were found to be labelled by monoclonal antibodies specific for T cells. We proposed that the apoptotic elimination of encephalitogenic T cells in the CNS may contribute to the subsidence of inflammation and the clinical recovery from attacks of EAE. Apoptosis of immature T cells in the thymus is the mechanism of clonal deletion of autoreactive cells during the normal neonatal development of immunological self-tolerance. Apoptosis of mature T cells may play a role in peripheral (extrathymic) tolerance. Using standard electron microscopy we have also demonstrated that macrophage apoptosis occurs in the spinal cord in EAE, and suggest that this contributes to the downregulation of this autoimmune disease (K.B. Nguyen, P.A. McCombe & M.P. Pender, unpublished data).

Pathophysiology

To determine how EAE interferes with the function of the nervous system and produces the neurological signs, we have used combined electrophysiological and light and electron microscopic histological techniques. The aims have been to determine: (1) the sites of nerve conduction abnormalities, their relationship to neurological signs, and the cause of these conduction abnormalities in acute EAE and chronic relapsing EAE; (2) the mechanism of clinical recovery; and (3) the mechanism for persistent neurological deficits.

Sites of nerve conduction abnormalities in acute EAE and relationship to neurological signs

The site of nerve conduction abnormalities in EAE varies with the species of animal and with the sensitizing inoculum. It is important to emphasize that abnormalities can occur in the peripheral nervous system (PNS), especially the spinal roots, as well as in the CNS, and that the PNS involvement may predominate.

EAE induced by inoculation with whole CNS tissue

Rabbits with acute EAE induced by inoculation with whole CNS tissue develop neurological signs about 20 days after inoculation. The signs consist of lateral splaying and ataxia of the hindlimbs, followed by similar involvement of the forelimbs. Hindlimb and forelimb weakness later develop in some animals. In rabbits with this form of acute EAE, localized conduction block occurs in a high proportion of the large-diameter afferent fibres in the dorsal root ganglia (Pender & Sears, 1982, 1984). Such conduction block is present in the lumbosacral and thoracic dorsal root ganglia and correlates well with the predilection of these ganglia to demyelination. The functional deafferentation resulting from the conduction block in the dorsal root ganglia readily accounts for the postural abnormalities, ataxia and areflexia that occur in rabbits with this form of EAE.

Lewis rats with acute EAE induced by inoculation with whole CNS tissue develop neurological signs about 11 days after inoculation. The signs consist of tail, hindlimb and occasionally forelimb weakness and paralysis, without ataxia. In rats with this form of EAE, the involvement of the dorsal root ganglia is much less extensive than in rabbits and is largely restricted to the sacrococcygeal dorsal root ganglia (Pender & Sears, 1986). Conduction block in Aδ fibres in the caudal dorsal root ganglia is likely to be partly responsible for the impairment of tail nociception that also occurs in these rats (Pender, 1986a). In rats with hindlimb weakness, there is conduction block in the large- and small-diameter myelinated fibres at the ventral root exit zones of the lumbar spinal cord (Pender, 1986b, 1988a). The CNS part of the spinal cord ventral root exit zone is a site of predilection for demyelination in these animals. Conduction block in this region is an important cause of the hindlimb weakness. As prominent demyelination of the ventral root exit zone also occurs in rabbits with acute EAE induced by inoculation with whole CNS tissue, it is likely that the limb weakness that also develops in these animals is due to conduction block in this region, but this has not been studied electrophysiologically (Pender & Sears, 1984). In both the rabbit and the rat with this form of EAE, there is some demyelination in other regions of the spinal cord, and in the dorsal and ventral roots, which may also lead to conduction abnormalities and contribute to the neurological signs; however, the PNS distal to the dorsal root ganglion is spared (Pender & Sears, 1984, 1986; Pender, 1988a).

EAE induced by inoculation with MBP

Rats with acute EAE induced by inoculation with MBP (MBP-EAE) develop neurological signs similar to those of rats with acute EAE induced by inoculation with whole CNS tissue; however, rats with acute MBP-EAE have less extensive conduction block at the ventral root exit zone (Pender, 1988b). The main sites of demyelination in rats with acute MBP-EAE are the ventral and dorsal spinal roots (Pender, 1988c). Conduction block occurs in the ventral and dorsal roots in these animals, especially in the more caudal roots (Pender, 1986a, 1988b; J.B. Chalk, P.A. McCombe & M.P. Pender, unpublished data). The conduction block in the ventral roots and in the ventral root exit zone accounts for the tail and hindlimb weakness (Pender, 1988b) while conduction block in Aδ fibres in the sacrococcygeal dorsal roots is an important cause of the impairment of tail nociception in these rats (Pender, 1986a).

A similar predilection of the spinal roots for demyelination also occurs in rats with acute EAE induced by the passive transfer of MBP-specific T cells, although we have not performed electrophysiological studies in these animals (Pender, Nguyen & Willenborg, 1989). Heininger *et al.* (1989) have demonstrated conduction abnormalities in the spinal cord dorsal columns and in the spinal roots of rats with EAE induced by the passive transfer of MBP-specific T line cells; they attributed these abnormalities to demyelination, especially paranodal demyelination. Rats with MBP-EAE have a characteristic ascending progression of the motor and sensory deficits (Simmons *et al.*, 1982; Pender, 1986a). This can be at least partly accounted for by the caudally increasing length of the spinal roots. In a study on male albino rats, Waibl (1973) showed that the lengths of the spinal roots increase progressively from the first thoracic root (2 mm) to the third coccygeal root (59 mm). Thus the probability of many lesions and of a high proportion of fibres undergoing conduction block in an entire root increases progressively in a caudal direction (Pender, 1986a). A tendency for the more caudal roots to be more severely affected at any one level may also contribute to the ascending progression of neurological signs (Pender *et al.*, 1989). In rats with MBP-EAE, there is mild involvement of the dorsal root ganglia and sparing of the PNS distal to the dorsal root ganglia (Pender, 1988b; Pender *et al.*, 1989). These rats have limited demyelination in regions of the CNS other than the ventral root exit and dorsal root entry zones; however, significant conduction block occurs in the lumbosacral dorsal columns (J.B. Chalk, P.A. McCombe & M.P. Pender, unpublished data).

EAE induced by inoculation with PLP

Rats with acute PLP-EAE have neurological signs similar to those of rats with acute MBP-EAE; however, demyelination and nerve conduction abnormalities are restricted to the CNS in rats with acute PLP-EAE (J.B. Chalk, P.A. McCombe, R. Smith & M.P. Pender, unpublished results). Conduction in the dorsal and ventral spinal roots is normal although conduction abnormalities occur at the dorsal root entry zone of the spinal cord; these abnormalities can be explained by demyelination in the CNS part of this transitional zone. As there is also prominent demyelination in the CNS part of the ventral root exit zone, conduction block in the ventral root exit zone is likely to be an important cause of tail and hindlimb weakness in rats with PLP-EAE, but this region has not been assessed electrophysiologically. Demyelination is also present in other regions of the spinal cord, and significant conduction block occurs in the dorsal columns (J.B. Chalk, P.A. McCombe, R. Smith & M.P. Pender, unpublished data).

Explanation for the different distributions of conduction block and demyelination in different forms of EAE

The involvement of the PNS when rats are inoculated with purified CNS MBP is explained by the fact that the P_1 protein from the PNS is identical to CNS MBP (Brostoff & Eylar, 1972; Greenfield *et al.*, 1973). The preferential involvement of the spinal roots with sparing of the peripheral nerves may be due to the reduced blood–nerve barrier in the roots (Olsson, 1968) and the almost 3-fold higher concentration of P_1 in the spinal root than in the peripheral nerve in the rat (Greenfield *et al.*, 1973).

Sparing of the PNS in PLP-EAE is expected because of the apparent absence of PLP in the PNS (Finean, Hawthorne & Patterson, 1957; Folch, Lees & Carr, 1958). The distribution of conduction block and demyelination in EAE induced by inoculation with whole CNS tissue can be partly explained as the combination of the lesions of MBP-EAE and of PLP-EAE. However, whole CNS tissue contains not only MBP and PLP but also other myelin antigens such as galactocerebroside, myelin-associated glycoprotein and myelin/oligodendrocyte glycoprotein. None of the latter antigens has been shown to be capable of inducing EAE when inoculated with adjuvants. However, antibodies to myelin/oligodendrocyte glycoprotein, which is restricted to the CNS, augment CNS demyelination when administered to rats with MBP-EAE (Linington *et al.*, 1988) and may contribute to the CNS demyelination in EAE induced by inoculation with whole CNS tissue. Antibodies to galactocerebroside can induce CNS demyelination *in vivo* after intraneural injection (Sergott *et al.*, 1984) and may also contribute to CNS demyelination in this form of EAE. The greater involvement of the dorsal root ganglia in rats with EAE induced by inoculation with whole CNS tissue than in rats with MBP-EAE indicates that immune responses to other antigens, for example galactocerebroside, are involved. The greater involvement of the dorsal root ganglia in the rabbit than in the rat with EAE induced by inoculation with whole CNS tissue probably reflects an interspecies difference in the immune response to specific myelin antigens rather than a difference in the dorsal root ganglia, as extensive demyelination and conduction block occur in the dorsal root ganglia of rats with experimental autoimmune neuritis induced by inoculation with PNS myelin (Stanley, McCombe &

Pender, 1992). The selective involvement of the dorsal root ganglion can be explained by the deficient blood–nerve barrier in this region (Waksman, 1961; Jacobs, MacFarlane & Cavanagh, 1976).

Cause of conduction block in EAE

Because of reports of absent or minimal demyelination in some animals with neurological signs of acute EAE or the first attack of chronic relapsing EAE, it has been suggested that the signs of EAE are not due to demyelination (Hoffman, Gaston & Spitler, 1973; Lassmann & Wisniewski, 1979; Panitch & Ciccone, 1981; Raine *et al.*, 1981; Simmons *et al.*, 1981, 1982; Kerlero de Rosbo *et al.*, 1985). Furthermore, it has been asserted that the clinical recovery in rats with acute EAE is too rapid to be accounted for by remyelination, and that demyelination is therefore not responsible for the neurological signs (Panitch & Ciccone, 1981; Simmons *et al.*, 1981). Thus the neurological signs of EAE have been attributed to other factors such as oedema (Paterson, 1976; Simmons *et al.*, 1982; Kerlero de Rosbo *et al.*, 1985) or an impairment of monoaminergic neurotransmission (White, 1984).

The reports of absent or minimal demyelination in some animals with EAE can be explained by a failure to examine the PNS and the lumbar, sacral and coccygeal segments of the spinal cord and/or the use of insensitive histological techniques (Pender, 1987). It is well established that primary demyelination in the absence of inflammation can cause nerve conduction block (McDonald, 1963; McDonald & Sears, 1970). Hence, when conduction block is localized to sites of inflammatory demyelination in EAE, as described above, it is reasonable to conclude that the demyelination is the cause of the conduction block. Other electrophysiological features characteristic of primary demyelination are also found in these lesions, namely: slowing of conduction in large-diameter fibres; normal conduction in unmyelinated fibres; the aggravation of conduction block by heating and its amelioration by cooling; rate-dependent conduction block; the restoration of conduction by the K^+ channel blocking agent 4-aminopyridine, and by ouabain, which inhibits the electrogenic Na^+ pump (Pender & Sears, 1982, 1984; Stanley & Pender, 1991).

The preservation of conduction in unmyelinated fibres in inflamed and demyelinated dorsal root ganglia in which conduction is blocked in the majority of large-diameter myelinated fibres indicates that the conduction block is not due to an action of cytokines or other factors on the axonal membrane. The temporal association of restoration of conduction with remyelination (see below) also indicates that demyelination is responsible for the conduction block. Oedema is unlikely to contribute to the neurological signs of EAE except when it occurs in a confined space and leads to vascular compression and secondary ischaemia, for example in the optic canal.

Mechanism of clinical recovery

Lewis rats with acute EAE rapidly recover from the neurological signs. For example, rats with MBP-EAE develop distal tail weakness 10 days after inoculation. Over the next 2–3 days the tail weakness ascends to involve the whole tail, and hindlimb weakness also develops. Clinical recovery usually commences 3–5 days after the onset of neurological signs and is complete by 20 days after inoculation (Pender, 1988b). Clinical recovery is associated with a restoration of conduction in the CNS and in the PNS (Pender, 1989). The onset of recovery correlates well with the ensheathment of CNS and PNS demyelinated fibres by oligodendrocytes and Schwann cells respectively (Pender, 1989; Pender *et al.*, 1989). Such ensheathment commences in the PNS 3 days after the onset of neurological signs and in the CNS 5 days after the onset of signs. Compact myelin formation commences in the PNS 5 days after neurological onset and in the CNS 6 days after neurological onset, and rapidly progresses thereafter. There is evidence from studies on non-inflammatory models of PNS demyelination that nerve conduction may be restored in demyelinated fibres in the early stages of repair before the formation of compact myelin lamellae (Bostock & Sears, 1978; Smith & Hall, 1980; Smith, Bostock & Hall, 1982). In the lysophosphatidyl choline model, restoration of conduction occurred when demyelinated fibres became closely associated with debris-free Schwann cells (Smith & Hall, 1980; Smith *et al.*, 1982). Hence, it is likely that the initial restoration of conduction in the PNS and CNS in EAE is a consequence of ensheathment of the axons by Schwann cells and oligodendrocytes respectively. Whether the restoration of conduction results from a reorganization of axonal ion channels (Bostock & Sears, 1978) or from the insulating effect of the investing Schwann and oligodendrocyte processes (Shrager & Rubinstein, 1990) is unclear. The subsequent rapid formation of compact myelin would make the restoration of conduction more secure (Smith, Blakemore & McDonald, 1981).

Chronic relapsing EAE

In Lewis rats with chronic relapsing EAE induced by inoculation with whole CNS tissue and treatment with low-dose cyclosporin A, there is demyelination and conduction block in the CNS and the PNS during clinical attacks (Pender *et al.*, 1990; Stanley & Pender, 1991). Conduction block in the PNS afferent pathway can be explained by demyelination in the dorsal root ganglia and dorsal roots. In the CNS, conduction block and prominent demyelination were demonstrated in the lumbosacral dorsal columns. In most rats during clinical episodes the cerebral somatosensory evoked potential was reduced in amplitude and prolonged in latency, which could be accounted for by conduction block in the CNS and PNS components of the afferent pathway (Stanley & Pender, 1991). In occasional rats with episodes of EAE, however, this potential was markedly increased in amplitude, which might have been due to conduction block of descending pathways that normally inhibit synaptic transmission in the afferent pathway. Demyelination was also present at the dorsal root entry and ventral root exit zones of the spinal cord, as well as in other regions of the CNS, and was also present in the ventral roots in the PNS. There was minimal involvement of the PNS distal to the dorsal root ganglia. In the later stages of clinically active disease (>28 days after inoculation) there was extensive spinal cord demyelination but minimal PNS demyelination (Pender *et al.*, 1990). During clinical remission there was restoration of conduction in the CNS and the PNS, and remyelination by oligodendrocytes and Schwann cells respectively (Pender *et al.*, 1990; Stanley & Pender, 1991).

Axonal damage in EAE

In well-established clinical remission from chronic relapsing EAE, there is residual conduction failure in the CNS and the PNS but minimal persistent or continuing demyelination (Pender *et al.*, 1990; Stanley & Pender, 1991). This residual conduction failure can be accounted for by CNS and PNS axonal degeneration and axonal loss which were demonstrated morphologically. Axonal damage and degeneration are well-recognized features of hyperacute EAE (Lampert, 1967; Hansen & Pender, 1989) and may occur to a limited extent in acute EAE (Lampert & Kies, 1967; Pender, 1989); thus they may also contribute to the neurological deficits in these forms of EAE.

Conclusions

The neurological signs of acute EAE and chronic relapsing EAE can generally be explained by demyelination-induced conduction block in the CNS and/or the PNS. The sites of demyelination and conduction abnormalities vary with the species inoculated and with the sensitizing antigens. In the rat, inoculation with MBP results in spinal cord and prominent spinal root involvement whereas immunization with PLP results in disease restricted to the CNS. Clinical recovery from attacks of EAE is readily explained by ensheathment and remyelination of CNS and PNS demyelinated axons by oligodendrocytes and Schwann cells respectively. Axonal damage and loss are likely to have an important role in producing persistent neurological deficits.

Acknowledgements

This work was supported by the National Health and Medical Research Council of Australia and the National Multiple Sclerosis Society of Australia.

References

Bostock, H. & Sears, T.A. (1978). The internodal axon membrane: electrical excitability and continuous conduction in segmental demyelination. *Journal of Physiology (London)*, **280**, 273–301.

Brostoff, S.W. & Eylar, E.H. (1972). The proposed amino acid sequence of the P1 protein of rabbit sciatic nerve myelin. *Archives of Biochemistry and Biophysics*, **153**, 590–598.

Finean, J.B., Hawthorne, J.N. & Patterson, J.D.E. (1957). Structural and chemical differences between optic and sciatic nerve myelins. *Journal of Neurochemistry*, **1**, 256–259.

Folch, J., Lees, M. & Carr, S. (1958). Studies of the chemical composition of the nervous system. *Experimental Cell Research (Supplement)*, **5**, 58–71.

Greenfield, S., Brostoff, S., Eylar, E.H. & Morell, P. (1973). Protein composition of myelin of the peripheral nervous system. *Journal of Neurochemistry*, **20**, 1207–1216.

Hansen, L.A. & Pender, M.P. (1989). Hypothermia due to an ascending impairment of shivering in hyperacute experimental allergic encephalomyelitis in the Lewis rat. *Journal of the Neurological Sciences*, **94**, 231–240.

Heininger, K., Fierz, W., Schafer, B., Hartung, H.P., Wehling, P. & Toyka, K.V. (1989). Electrophysiological investigations in adoptively transferred experimental autoimmune encephalomyelitis in the Lewis rat. *Brain*, **112**, 537–552.

Hoffman, P.M., Gaston, D.D. & Spitler, L.E. (1973). Comparison of experimental allergic encephalomyelitis induced with spinal cord, basic protein, and synthetic encephalitogenic peptide. *Clinical Immunology and Immunopathology*, **1**, 364–371.

Itoyama, Y. & Webster, H.D. (1982). Immunocytochemical study of myelin-associated glycoprotein (MAG) and basic protein (BP) in acute experimental allergic encephalomyelitis (EAE). *Journal of Neuroimmunology*, **3**, 351–364.

Jacobs, J.M., MacFarlane, R.M. & Cavanagh, J.B. (1976). Vascular leakage in the dorsal root ganglia of the rat, studied with horseradish peroxidase. *Journal of the Neurological Sciences*, **29**, 95–107.

Kerlero de Rosbo, N., Bernard, C.C.A., Simmons, R.D. & Carnegie, P.R. (1985). Concomitant detection of changes in myelin basic protein and permeability of blood–spinal cord barrier in acute experimental autoimmune encephalomyelitis by electroimmunoblotting. *Journal of Neuroimmunology*, **9**, 349–361.

Lampert, P. (1967). Electron microscopic studies on ordinary and hyperacute experimental allergic encephalomyelitis. *Acta Neuropathologica*, **9**, 99–126.

Lampert, P.W. & Kies, M.W. (1967). Mechanism of demyelination in allergic encephalomyelitis of guinea pigs: an electron microscopic study. *Experimental Neurology*, **18**, 210–223.

Lassmann, H. & Wisniewski, H.M. (1979). Chronic relapsing experimental allergic encephalomyelitis: effect of age at the time of sensitization on clinical course and pathology. *Acta Neuropathologica*, **47**, 111–116.

Linington, C., Bradl, M., Lassmann, H., Brunner, C. & Vass, K. (1988). Augmentation of demyelination in rat acute allergic encephalomyelitis by circulating mouse monoclonal antibodies directed against a myelin/oligodendrocyte glycoprotein. *American Journal of Pathology*, **130**, 443–454.

McCombe, P.A., Fordyce, B.W., de Jersey, J., Yoong, G. & Pender, M.P. (1992). Expression of CD45RC and Ia antigen in the spinal cord in acute

experimental allergic encephalomyelitis: an immunocytochemical and flow cytometric study. *Journal of the Neurological Sciences*, **113**, 177–186.
McDonald, W.I. (1963). The effects of experimental demyelination on conduction in peripheral nerve: a histological and electrophysiological study. II. Electrophysiological observations. *Brain*, **86**, 501–524.
McDonald, W.I. & Sears, T.A. (1970). The effects of experimental demyelination on conduction in the central nervous system. *Brain*, **93**, 583–598.
Moore, G.R., Traugott, U. & Raine, C.S. (1984). Survival of oligodendrocytes in chronic relapsing experimental autoimmune encephalomyelitis. *Journal of the Neurological Sciences*, **65**, 137–145.
Olsson, Y. (1968). Topographical differences in the vascular permeability of the peripheral nervous system. *Acta Neuropathologica*, **10**, 26–33.
Panitch, H. & Ciccone, C. (1981). Induction of recurrent experimental allergic encephalomyelitis with myelin basic protein. *Annals of Neurology*, **9**, 433–438.
Paterson, P.Y. (1976). Experimental allergic encephalomyelitis: role of fibrin deposition in immunopathogenesis of inflammation in rats. *Federation Proceedings*, **35**, 2428–2434.
Pender, M.P. (1986a). Ascending impairment of nociception in rats with experimental allergic encephalomyelitis. *Journal of the Neurological Sciences*, **75**, 317–328.
Pender, M.P. (1986b). Conduction block due to demyelination at the ventral root exit zone in experimental allergic encephalomyelitis. *Brain Research*, **367**, 398–401.
Pender, M.P. (1987). Demyelination and neurological signs in experimental allergic encephalomyelitis. *Journal of Neuroimmunology*, **15**, 11–24.
Pender, M.P. (1988a). The pathophysiology of acute experimental allergic encephalomyelitis induced by whole spinal cord in the Lewis rat. *Journal of the Neurological Sciences*, **84**, 209–222.
Pender, M.P. (1988b). The pathophysiology of myelin basic protein-induced acute experimental allergic encephalomyelitis in the Lewis rat. *Journal of the Neurological Sciences*, **86**, 277–289.
Pender, M.P. (1988c). Demyelination of the peripheral nervous system causes neurologic signs in myelin basic protein-induced experimental allergic encephalomyelitis: implications for the etiology of multiple sclerosis. *Annals of the New York Academy of Sciences*, **540**, 732–734.
Pender, M.P. (1989). Recovery from acute experimental allergic encephalomyelitis in the Lewis rat: early restoration of nerve conduction and repair by Schwann cells and oligodendrocytes. *Brain*, **112**, 393–416.
Pender, M.P., McCombe, P.A., Yoong, G. & Nguyen, K.B. (1992). Apoptosis of αβ T lymphocytes in the nervous system in experimental autoimmune encephalomyelitis: its possible implications for recovery and acquired tolerance. *Journal of Autoimmunity*, **5**, 401–410.
Pender, M.P., Nguyen, K.B., McCombe, P.A. & Kerr, J.F.R. (1991). Apoptosis in the nervous system in experimental allergic encephalomyelitis. *Journal of the Neurological Sciences*, **104**, 81–87.
Pender, M.P., Nguyen, K.B. & Willenborg, D.O. (1989). Demyelination and early remyelination in experimental allergic encephalomyelitis passively transferred with myelin basic protein-sensitized lymphocytes in the Lewis rat. *Journal of Neuroimmunology*, **25**, 125–142.
Pender, M.P. & Sears, T.A. (1982). Conduction block in the peripheral nervous system in experimental allergic encephalomyelitis. *Nature*, **296**, 860–862.
Pender, M.P. & Sears, T.A. (1984). The pathophysiology of acute experimental allergic encephalomyelitis in the rabbit. *Brain*, **107**, 699–726.

Pender, M.P. & Sears, T.A. (1986). Involvement of the dorsal root ganglion in acute experimental allergic encephalomyelitis in the Lewis rat: a histological and electrophysiological study. *Journal of the Neurological Sciences*, **72**, 231–242.

Pender, M.P., Stanley, G.P., Yoong, G. & Nguyen, K.B. (1990). The neuropathology of chronic relapsing experimental allergic encephalomyelitis induced in the Lewis rat by inoculation with whole spinal cord and treatment with cyclosporin A. *Acta Neuropathologica*, **80**, 172–183.

Raine, C.S. (1984). Biology of disease. Analysis of autoimmune demyelination: its impact upon multiple sclerosis. *Laboratory Investigation*, **50**, 608–635.

Raine, C.S., Traugott, U., Farooq, M., Bornstein, M.B. & Norton, W.T. (1981). Augmentation of immune-mediated demyelination by lipid haptens. *Laboratory Investigation*, **45**, 174–182.

Sergott, R.C., Brown, M.J., Silberberg, D.H. & Lisak, R.P. (1984). Antigalactocerebroside serum demyelinates optic nerve *in vivo*. *Journal of the Neurological Sciences*, **64**, 297–303.

Shrager, P. & Rubinstein, C.T. (1990). Optical measurement of conduction in single demyelinated axons. *Journal of General Physiology*, **95**, 867–890.

Simmons, R.D., Bernard, C.C., Ng, K.T. & Carnegie, P.R. (1981). Hind-limb motor ability in Lewis rats during the onset and recovery phases of experimental autoimmune encephalomyelitis. *Brain Research*, **215**, 103–114.

Simmons, R.D., Bernard, C.C., Singer, G. & Carnegie, P.R. (1982). Experimental autoimmune encephalomyelitis: an anatomically-based explanation of clinical progression in rodents. *Journal of Neuroimmunology*, **3**, 307–318.

Smith, K.J., Blakemore, W.F. & McDonald, W.I. (1981). The restoration of conduction by central remyelination. *Brain*, **104**, 383–404.

Smith, K.J., Bostock, H. & Hall, S.M. (1982). Saltatory conduction precedes remyelination in axons demyelinated with lysophosphatidyl choline. *Journal of the Neurological Sciences*, **54**, 13–31.

Smith, K.J. & Hall, S.M. (1980). Nerve conduction during peripheral demyelination and remyelination. *Journal of the Neurological Sciences*, **48**, 201–219.

Stanley, G.P., McCombe, P.A. & Pender, M.P. (1992). Focal conduction block in the dorsal root ganglion in experimental allergic neuritis. *Annals of Neurology*, **31**, 27–33.

Stanley, G.P. & Pender, M.P. (1991). The pathophysiology of chronic relapsing experimental allergic encephalomyelitis in the Lewis rat. *Brain*, **114**, 1827–1853.

Traugott, U., Shevach, E., Chiba, J., Stone, H.J. & Raine, C.S. (1981). Autoimmune encephalomyelitis: simultaneous identification of T and B cells in the target organ. *Science*, **214**, 1251–1253.

Waibl, H. (1973). Zur Topographie der Medulla spinalis der Albinoratte (*Rattus norvegicus*). *Advances in Anatomy, Embryology and Cell Biology*, **47**, 5–42.

Waksman, B.H. (1961). Experimental study of diphtheritic polyneuritis in the rabbit and guinea pig. III. The blood–nerve barrier in the rabbit. *Journal of Neuropathology and Experimental Neurology*, **20**, 35–77.

White, S.R. (1984). Experimental allergic encephalomyelitis: effects on monoaminergic neurotransmission. In *Brainstem Control of Spinal Cord Function*, ed. C.D. Barnes, pp. 257–281. Orlando, FL: Academic Press.

9

Pathophysiology of human demyelinating neuropathies

T.E. FEASBY

Department of Clinical Neurosciences, University of Calgary, Calgary, Alberta, Canada

Introduction

This chapter will review our understanding of the pathophysiology of the human demyelinating neuropathies with particular emphasis on acute Guillain–Barré syndrome (GBS). Abnormalities of conduction in peripheral nerves are secondary to changes in the axon or its myelin sheath. Those affecting the axon include neuronal or axonal degeneration, axonal shrinkage and, theoretically, alterations of ion channels. The last are usually pharmacological or toxicological. Abnormalities of myelin include demyelination, both segmental and paranodal, and reduced thickness of the sheath. The two major pathological lesions, axonal degeneration and demyelination, are sufficient to explain most of the clinical and physiological deficits in GBS.

Demyelination and conduction block

The first inference of conduction block was by Erb (1876) who studied focal traumatic nerve lesions. He observed that, in some patients, faradic stimulation could activate the nerve distal to a lesion. He postulated that the nerve fibres had not degenerated. Just a few years later, Gombault (1881) demonstrated myelin sheath abnormalities at the site of focal nerve lesions but these observations were not linked to those of Erb.

Seddon (1943), in his studies of war nerve injuries, classified them into three categories: neurotmesis, axonotmesis and neurapraxia. The first two were characterized by transection of the whole nerve and the axons respectively, with degeneration of the distal portions of the axons. In neurapraxia, however, the nerve trunk remained intact, the distal axons did not degenerate and recovery occurred after a short interval. The pathological explanation for neurapraxia was provided a year later by Denny-Brown & Brenner (1944a, b) in their studies of nerve compression in cats. Temporary compression of nerves by a tourniquet or a metal clip produced transient block of conduction through the compressed region. Conduction was restored over the next few weeks. Microscopy at the time of conduction block revealed demyelination with preserved axons.

W.I. McDonald (1963) studied diphtheria-toxin-induced demyelination in cat dorsal roots. He demonstrated conduction block directly in the region of the dorsal root ganglia during antidromic conduction by stimulating the dorsal roots proximally. Cragg & Thomas (1964) studied experimental allergic neuritis (EAN), an autoimmune demyelinating peripheral nerve disease, in guinea pigs. They recorded conduction over successively longer distances and showed progressive diminution in the size of the compound muscle action potential, the hallmark of conduction block.

Single-fibre studies of demyelination

The saltatory nature of conduction in myelinated fibres was proved by Tasaki & Takeuchi (1941) and Huxley & Stämpfli (1949), who showed that excitation occurred only at nodes of Ranvier. The internodal axon with its myelin sheath acts as a passive cable. The insulating properties of the myelin sheath prevent significant transmembrane current flow in the internode, confining it to the nodes. Tasaki (1953), in an isolated frog myelinated fibre, dissolved the internodal myelin with saponin, a detergent. He noted a large increase in the internodal transmembrane current and, eventually, conduction block.

Rasminsky & Sears (1972) devised an ingenious method of studying conduction in single mammalian myelinated nerve fibres *in vivo*, based on the method of Huxley & Stämpfli (1949). The preparation involved peripheral stimulation of single nerve fibres and measurement of the external longitudinal currents from single axons from intact dorsal or ventral roots. Demyelination was induced by the topical application of diphtheria toxin. They found that slowed conduction, which is characteristic of demyelination, was due to slowing of internodal conduction caused by internodal transmembrane current leak, delaying the activation of the next node. Internodal conduction times could increase up to 30-fold, but in cases of excessive leak, presumably due to more complete demyelination, conduction block occurred. Subsequent studies by Bostock and colleagues (reviewed in Bostock, 1993) showed that conduction block could be produced by paranodal as well as segmental demyelination. Bostock & Sears (1978), using an improved technique, made the interesting observation that the demyelinated internodal axonal membrane can support continuous conduction. This appears to be of more importance in the human central nervous system than the peripheral nervous system.

Studies of human conduction block

Austin (1958) described a case of chronic inflammatory demyelinating polyneuropathy with many relapses. He noted that recovery from the relapses was often very rapid and could not be explained on the basis of axonal regeneration. He observed that the axons were largely preserved in the nerve biopsy, that there was little atrophy or fibrillation of muscle and that stimulation of motor nerves distally evoked greater contractions than did voluntary efforts. He explained these features on the basis of 'proximal conduction block' analogous to that described by Denny-Brown & Brenner (1944a, b).

Bannister & Sears (1962) described a case of acute GBS with extensive physiological studies. They concluded that 'failure to conduct impulses over some 50 cm of nerve could not have been due to Wallerian degeneration' and 'the nerve fibres which showed early recovery and on which clinical recovery was clearly dependent retained a structural integrity such that Wallerian degeneration did not take place'. They speculated that 'the myelin sheath loses its insulating properties or that there is actual loss of myelin. The effect of this would be to increase the ''leakage'' current through the internode (Huxley & Stämpfli, 1949) thus preventing adjacent nodes from being excited. The sort of process envisaged here may be likened to the effects of experimental ''demyelination'' caused by saponin applied to the myelin sheath of single isolated nerve fibres by Tasaki (1953)'.

The experimental work of Denny-Brown & Brenner (1944a, b) had shown that tourniquet compression of nerve produced demyelination and conduction block. Gilliatt (1980a, b) and colleagues extended these findings and further delineated the nature of the myelin lesion. Clinical studies later showed that conduction block in human tourniquet paralysis signified a good prognosis (Rudge, 1974; Bolton & McFarlane, 1978). Conduction block was also observed to be important in both pressure (Trojaborg, 1970) and entrapment neuropathies (Brown *et al.*, 1976).

To explore the pathological basis of conduction block in human neuropathies, we compared the nerve conduction findings in a series of 15 polyneuropathy patients with the results of quantitative teased fibres from their nerve biopsies (Feasby *et al.*, 1985). Ten patients had either definite or probable conduction block and five did not. Those with conduction block had an average of 50% of the fibres with demyelination compared with 5% in those without block.

Conduction in human polyneuropathies

The most important human 'demyelinating' polyneuropathy is acute Guillain–Barré syndrome (GBS). Despite comments by Bannister & Sears (1962) and Lambert & Mulder (1964) about conduction block in GBS, little interest was shown in this phenomenon for many years and investigators concentrated on other features of demyelination such as slowed conduction velocities, prolonged terminal latencies, temporal dispersion of the compound action potential and delayed 'F' responses (McLeod *et al.*, 1976). However, none of these easily measurable abnormalities, despite their diagnostic value, explained the often profound clinical deficits of GBS.

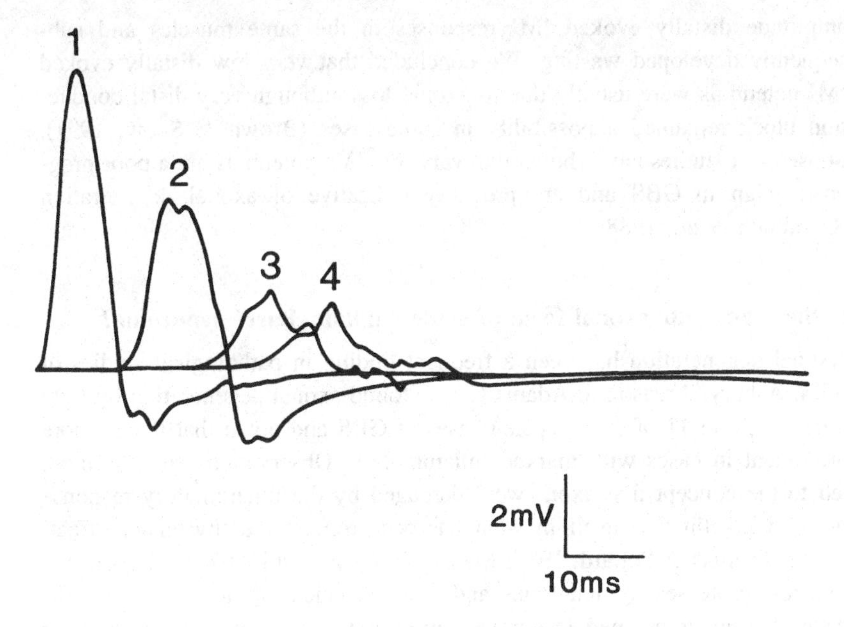

Fig. 9.1. Median nerve conduction study showing conduction block and temporal dispersion in a patient with acute Guillain–Barré syndrome. The recording was from the abductor pollicis brevis muscle, with stimulation at the motor point (1), wrist (2), elbow (3) and axilla (4). Recordings by Dr W.F. Brown.

In 1984 W.F. Brown and I published a systematic neurophysiological study of 25 patients with acute GBS (Brown & Feasby, 1984). We sought evidence of conduction block by comparing the size of the compound muscle action potential with stimulation of motor nerves at both distal and proximal sites. Our criteria were: (1) a decline in the peak–peak amplitude of the maximum 'M' potential of >20% between the proximal and distal stimulation sites, and (2) an increase in the duration of the negative peak of the 'M' potential of no more than 15%. We also sought evidence of denervation using electromyography.

In 13 of 19 GBS patients studied in the first 2 weeks of illness, evidence of conduction block was found in at least one motor nerve. In 15 of 20 motor nerves with demonstrable conduction block, the drop in amplitude between proximal and distal stimulation points was >50%. Conduction block was sometimes distributed uniformly over the length of the nerve but often seemed to be more pronounced distally (Fig. 9.1) (see also Brown & Snow, 1991; van der Meche *et al.*, 1988) or at points of common nerve entrapment. Serial studies often revealed progression and then resolution of conduction block, paralleling the clinical course.

Electromyography showed that denervation of muscle was present in 64% of patients. Those who had significant denervation often had very low amplitude distally evoked 'M' responses in the same muscles and subsequently developed wasting. We concluded that very low distally evoked 'M' potentials were usually due to axonal loss although very distal conduction block remained a possibility in some cases (Brown & Snow, 1991). Subsequent studies have shown that very low 'M' potentials are a poor prognostic sign in GBS and are probably indicative of axonal degeneration (Cornblath *et al.*, 1988).

Is there an acute axonal form of acute Guillain–Barré syndrome?

Axonal degeneration has been a frequent finding in pathological studies of GBS. Asbury, Arnason & Adams (1969) found axonal degeneration by light microscopy in 11 of 19 autopsied cases of GBS and noted that it was more prominent in cases with marked inflammation. Observations such as these led to the concept that axons were damaged by the inflammatory response or by demyelination itself in what has been termed the 'bystander effect' (King, Thomas & Pollard, 1977; Madrid & Wisniewski, 1977). Oxygen free radicals, proteases, lymphokines and other products of inflammatory cells have all been implicated in demyelination and could cause axonal damage (Hartung *et al.*, 1993).

We recently had the opportunity to study serial nerve biopsies in a patient with severe acute GBS (Feasby *et al.*, 1993). The first biopsy, from the deep and superficial nerves at day 15, showed profuse inflammation and demyelination. The second biopsy, from the opposite side at day 75, showed complete loss of large axons. The patient's recovery was very slow and incomplete. This example showed that severe inflammatory demyelination can lead to severe axonal degeneration.

We have also examined the relationship between inflammation, demyelination and axonal degeneration in experimental allergic neuritis, the experimental model of GBS (Hahn *et al.*, 1988, 1991). We found that the severity of the disease correlated with the dose of the antigen. More importantly, we found that the severity of axonal degeneration correlated with the amount of inflammation and demyelination.

In contrast with the above, we have studied eight patients with GBS who had very severe axonal degeneration with no physiological or pathological evidence of demyelination or inflammation (Feasby *et al.*, 1986, 1993). These patients all had very acute severe clinical disease. Six required artificial ventilation, two died and five of six had very delayed and incomplete recovery. Motor nerves were inexcitable as early as 3 days after the onset of the disease.

Histological sampling included an autopsy at day 28 on one patient, a nerve biopsy at day 17 followed by an autopsy at 19 months on one patient, and nerve biopsies on days 18 and 29 on two other patients. The biopsies and the early autopsy all showed the same findings: severe axonal loss, continuing axonal degeneration and a striking lack of inflammation or demyelination. The autopsy at 19 months showed severe loss of axons in distal nerves and significant regeneration in the ventral roots, indicative of previous axonal injury (Feasby *et al.*, 1993).

These latter cases raise the question of whether there is a separate form of GBS characterized by 'primary' axonal degeneration (Thomas, 1992; Dyck, 1993). It has been argued that we could have missed the presence of inflammation and demyelination in our cases because the biopsies and autopsies were too late or because of insufficient sampling, particularly distally (Honavar *et al.*, 1991; Berciano *et al.*, 1993). These arguments might have been persuasive if the clinical and pathological findings in our cases had been minor or focal. In fact, they were severe and extensive. Terminal nerve inexcitability revealed the distal extent of the lesion. The autopsy finding of marked ventral root lesions showed that even the most proximal part of the peripheral nervous system was affected. We contend that secondary axonal degeneration should be proportional to the amount of inflammation and demyelination, as it is in EAN (Hahn *et al.*, 1988, 1991). Thus, in cases of such severe axonal degeneration, it would have been difficult to miss evidence of inflammation and demyelination.

We concluded that there may be two forms of acute GBS. The typical form is characterized by inflammatory demyelination and frequently shows secondary axonal degeneration, sometimes severe. A second form of GBS may be a 'primary' axonal variety, characterized by axonal degeneration without inflammation or demyelination. Recent support for this concept has come from studies of an acute paralytic disease of the peripheral nervous system of young Chinese which also seems to be an acute primary axonal form of GBS (McKhann *et al.*, 1993).

Conclusion

Clinical neurophysiological findings in human neuropathies reflect the underlying pathological changes, demyelination and axonal degeneration. Conduction block, the most significant effect of demyelination, explains much of the acute and chronic paralysis of many human neuropathies. The combination of physiological, pathological and clinical studies has allowed the subdivision of GBS into two distinct pathophysiological categories.

Acknowledgements

I owe a special debt of gratitude to my teacher Professor T.A. Sears. He triggered my interest in demyelination and taught me many important lessons. I am also grateful to many colleagues for collaboration and support, particularly Dr A.F. Hahn and Dr W.F. Brown. This work has been supported by the Medical Research Council of Canada and the Muscular Dystrophy Association of Canada.

References

Asbury, A.K., Arnason, B.G. & Adams, R.D. (1969). The inflammatory lesion in idiopathic polyneuritis: its role in pathogenesis. *Medicine*, **48**, 173–215.

Austin, J.H. (1958). Recurrent polyneuropathies and their corticosteroid treatment, with five-year observations of a placebo-controlled case treated with corticotrophin, cortisone and prednisone. *Brain*, **81**, 157–191.

Bannister, R.G. & Sears, T.A. (1962). The changes in nerve conduction in acute idiopathic polyneuritis. *Journal of Neurology, Neurosurgery and Psychiatry*, **25**, 321–328.

Berciano, J., Coria, F., Monton, F., Calleja, J., Figols, J. & LaFarga, M. (1993). Axonal form of Guillain-Barré syndrome: evidence for macrophage-associated demyelination. *Muscle and Nerve*, **16**, 744–751.

Bolton, C.F. & McFarlane, R.M. (1978). Human pneumatic tourniquet paralysis. *Neurology*, **28**, 787–793.

Bostock, H. (1993). Impulse propagation in experimental neuropathy. In *Peripheral Neuropathy*, 3rd edn, ed. P.J. Dyck & P.K. Thomas, pp. 109–120. Philadelphia: W.B. Saunders.

Bostock, H. & Sears, T.A. (1978). The internodal axon membrane: electrical excitability and continuous conduction in segmental demyelination. *Journal of Physiology (London)*, **280**, 273–301.

Brown, W.F. & Feasby, T.E. (1984). Conduction block and denervation in Guillain–Barré polyneuropathy. *Brain*, **107**, 219–239.

Brown, W.F., Ferguson, G.G., Jones, M.W. & Yates, S.K. (1976). The location of conduction abnormalities in human entrapment neuropathies. *Canadian Journal of Neurological Sciences*, **3**, 111–122.

Brown, W.F. & Snow, R. (1991). Patterns and severity of conduction abnormalities in Guillain–Barré syndrome. *Journal of Neurology, Neurosurgery and Psychiatry*, **54**, 768–774.

Cornblath, D.R., Mellits, E.D., Griffin, J.W., McKhann, G.M., Albers, J.W., Miller, R.G., Feasby, T.E., Quaskey, S.A. and the Guillain–Barré Syndrome Study Group. (1988). Motor conduction studies in Guillain–Barré syndrome: description and prognostic value. *Annals of Neurology*, **23**, 354–359.

Cragg, B.G. & Thomas, P.K. (1964). Changes in nerve conduction in experimental allergic neuritis. *Journal of Neurology, Neurosurgery and Psychiatry*, **27**, 106–115.

Denny-Brown, D. & Brenner, C. (1944a). Paralysis of nerve induced by direct pressure and by tourniquet. *Archives of Neurology and Psychiatry*, **51**, 1–26.

Denny-Brown, D. & Brenner, C. (1944b). Lesion in peripheral nerve resulting from compression by spring clip. *Archives of Neurology and Psychiatry*, **52**, 1–19.

Dyck, P.J. (1993). Is there an axonal variety of GBS? *Neurology*, **43**, 1277–1280.

Erb, W.H. (1876). Diseases of the peripheral cerebro-spinal nerves. In *Cyclopaedia of the Practice of Medicine*, ed. H. von Ziemssen. New York: Wood.

Feasby, T.E., Brown, W.F., Gilbert, J.J. & Hahn, A.F. (1985). The pathological basis of conduction block in human neuropathies. *Journal of Neurology, Neurosurgery and Psychiatry*, **48**, 239–244.

Feasby, T.E., Gilbert, J.J., Brown, W.F., Bolton, C.F., Hahn, A.F., Koopman, W.J. & Zochodne, D.W. (1986). An acute axonal form of Guillain–Barré polyneuropathy. *Brain*, **109**, 1115–1126.

Feasby, T.E., Hahn, A.F., Brown, W.F., Bolton, C.F., Gilbert, J.J. & Koopman, W.J. (1993). Severe axonal degeneration in acute Guillain–Barré syndrome: evidence of two different mechanisms? *Journal of the Neurological Sciences*, **116**, 185–192.

Gilliatt, R.W. (1980a). Acute compression block. In *The Physiology of Peripheral Nerve Disease*, ed. A.J. Sumner, pp. 287–315. Philadelphia: W.B. Saunders.

Gilliatt, R.W. (1980b). Chronic nerve compression and entrapment. In *The Physiology of Peripheral Nerve Disease*, ed. A.J. Sumner, pp. 316–339. Philadelphia: W.B. Saunders.

Gombault, M. (1881). Contribution à l'étude anatomique de la nevrite parenchymateuse subaique et chronique: nevrite segmentaire peri-axile. *Archives Neurologique (Paris)*, **1**, 11.

Hahn, A.F., Feasby, T.E., Steele, A., Lovgren, D.S. & Berry, J. (1988). Demyelination and axonal degeneration in Lewis rat experimental allergic neuritis depend on the myelin dosage. *Laboratory Investigation*, **59**, 115–125.

Hahn, A.F., Feasby, T.E., Wilkie, L. & Lovgren, D. (1991). P_2-peptide induced experimental allergic neuritis: a model to study axonal degeneration. *Acta Neuropathologica*, **82**, 60–65.

Hartung, H.-P. (1993). Immune-mediated demyelination. *Annals of Neurology*, **33**, 563–567.

Honavar, M., Tharakan, J.K.J., Hughes, R.A.C., Leibowitz, S. & Winer, J.B. (1991). A clinico-pathological study of the Guillain–Barré syndrome: nine cases and a literature review. *Brain*, **114**, 1245–1270.

Huxley, A.F. & Stämpfli, R. (1949). Evidence for saltatory conduction in peripheral myelinated nerve fibres. *Journal of Physiology (London)*, **108**, 315–339.

King, R.H.M., Thomas, P.K. & Pollard, J.D. (1977). Axonal and dorsal root ganglion cell changes in experimental allergic neuritis. *Neuropathology and Applied Neurobiology*, **3**, 471–486.

Lambert, E.H. & Mulder, D.W. (1964). Nerve conduction in the Guillain–Barré syndrome. *Electroencephalography and Clinical Neurophysiology*, **17**, 86–93.

Madrid, R.E. & Wisniewski, H.M. (1977). Axonal degeneration in demyelinating disorders. *Journal of Neurocytology*, **6**, 103–117.

McDonald, W.I. (1963). The effects of experimental demyelination on conduction in peripheral nerve: a histological and electrophysiological study. II. Electrophysiological observations. *Brain*, **86**, 501–523.

McKhann, G.M., Cornblath, D.R., Griffin, J.W., Ho, T.W., Li, C.Y., Jiang, Z., Wu, H.S., Zhaori, G., Liu, Y., Jou, L.P., Liu, T.C., Gao, C.Y., Mao, J.Y., Blaser, M.J., Mishu, B. & Asbury, A.K. (1993). Acute motor axonal neuropathy: a frequent cause of acute flaccid paralysis in China. *Annals of Neurology*, **33**, 333–342.

McLeod, J.G., Walsh, J.C., Prineas, J.W. & Pollard, J.D. (1976). Acute idiopathic polyneuritis: a clinical and electrophysiological follow-up study. *Journal of the Neurological Sciences*, **27**, 145–162.

Rasminsky, M. & Sears, T.A. (1972). Internodal conduction in undissected demyelinated nerve fibres. *Journal of Physiology (London)*, **227**, 323–350.

Rudge, P. (1974). Tourniquet paralysis with prolonged conduction block: an electro-physiological study. *Journal of Bone and Joint Surgery*, **56B**, 716–720.

Seddon, H.J. (1943). Three types of nerve injury. *Brain*, **66**, 236–288.

Tasaki, I. (1953). *Nervous Transmission*. Springfield, IL: Charles C. Thomas.

Tasaki, I. & Takeuchi, T. (1941). Der am Ranvierschen Knoten entschende Aktiansstran und seine Bedeutung für die Erregnungsleitung. *Pflügers Archiv*, **244**, 686–711.

Thomas, P.K. (1992). The Guillain–Barré syndrome: no longer a simple concept. *Journal of Neurology*, **239**, 361–362.
Trojaborg, W. (1970). Rate of recovery in motor and sensory fibres of the radial nerve: clinical and electrophysiological aspects. *Journal of Neurology, Neurosurgery and Psychiatry*, **33**, 625–638.
van der Meché, F.G.A., Meulstee, J., Vermeulen, M. & Kievet, A. (1988). Patterns of conduction failure in the Guillain–Barré syndrome. *Brain*, **111**, 405–416.

10

Conduction properties of central demyelinated axons: the generation of symptoms in demyelinating disease

KENNETH J. SMITH

Department of Neurology and Division of Anatomy and Cell Biology, United Medical and Dental Schools of Guy's and St Thomas' Hospitals, London, UK

The loss of myelin from central axons is a prominent feature of the lesion of multiple sclerosis (MS), and a direct cause of the several conduction deficits which lead, in turn, to the symptoms associated with central demyelinating disease. This review of the conduction deficits will focus on experimentally demyelinated central axons, since in experimental lesions it is usually possible to determine the morphology of the axons with some certainty, whereas this is rarely possible in human demyelinating disease.

The conduction properties of central axons passing through a region of demyelination were reliably described for the first time by McDonald and Sears in their landmark studies of 1969 and 1970 (McDonald & Sears, 1969b, 1970a, b). These authors found that conduction along axons passing through long experimental demyelinating lesions (>5 mm in length) was often blocked at the site of the lesion, but that conduction could sometimes continue if the lesions were small. Where conduction occurred it was usually abnormal, and proceeded with a locally reduced velocity and a prolonged refractory period of transmission (RPT); the axons were also incapable of conducting trains of impulses at high frequency. These conduction abnormalities remain the hallmark of conduction in axons passing through demyelinating lesions in either the central or peripheral nervous systems.

Conduction block

Conduction block is believed to be the dominant cause of the most distressing symptoms of MS, such as paralysis, blindness and numbness (McDonald, 1975, 1986; Halliday & McDonald, 1977; Ulrich & Groebke-Lorenz, 1983; Waxman, 1988), and it was also the first conduction abnormality to be described in experimentally demyelinated central axons (McDonald & Sears, 1969a) (Fig. 10.1).

The conduction block was found to be restricted to the site of the lesion, with conduction continuing in an apparently normal manner along the morphologically unaffected portions of the axon on either side (McDonald &

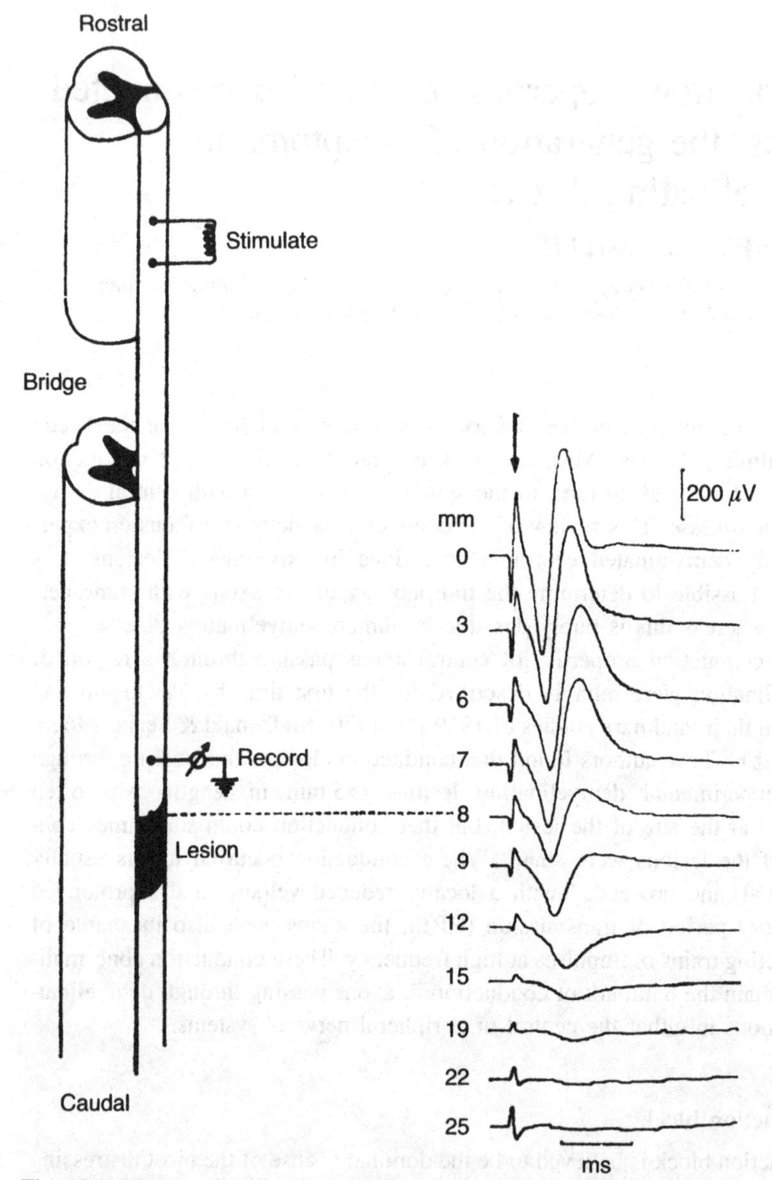

Fig. 10.1. Diagram showing the recording arrangements by which the series of averaged compound action potentials on the right were obtained. The potentials were recorded from the cat dorsal columns at progressive distances from the rostral site of stimulation. At the demyelinating lesion situated in the dorsal columns the potential becomes dominantly positive (downward deflection), indicating conduction block. The positivity declines in an approximately exponential manner with distance, and virtually no axons conduct through the lesion. Reprinted from McDonald & Sears (1970a) by permission of Oxford University Press.

Sears, 1969a, b, 1970a). McDonald and Sears induced demyelination by the injection of diphtheria toxin into the cat dorsal columns. This lesion contained axons with a variety of different types and degrees of demyelination (Harrison *et al.*, 1972a–c) and so it was not possible in the first experiments to determine the type of demyelination which characterized the blocked axons. Several more recent studies, however, have found that the complete block of conduction is a common feature of lesions in which the demyelination is largely segmental, i.e. where myelin is lost in units of complete internodes. Such lesions include those induced by lysophosphatidyl choline (LPC) (Smith, Blakemore & McDonald, 1979, 1981) and ethidium bromide (EBr) (Blakemore, 1982; Felts & Smith, 1992). In lesions where myelin damage is less extensive it is difficult to predict whether conduction will continue or not.

There are several factors associated with segmental demyelination which contribute to conduction block, of which the most important appears to be the time since demyelination was established: freshly segmentally demyelinated axons routinely exhibit conduction block (e.g. Felts & Smith, 1992), probably because the density of Na^+ channels in newly exposed internodal axolemma is insufficient to support conduction (Ritchie & Rogart, 1977; Shrager, 1989). Secondly, the loss of the myelin sheath alters the cable properties of the axon, an effect which results in what has been described as an 'impedance mismatch' at the site of demyelination (Waxman, 1977, 1978; Waxman & Foster, 1980): loss of the myelin sheath greatly increases the capacitance of the demyelinated internode, so that much more local current is required to depolarize the axolemma to its firing threshold. These factors serve to reduce the safety factor for conduction several-fold, so that in demyelinated axons it is typically near unity. Where the safety factor is even fractionally above 1, conduction will continue (albeit with a reduced velocity, etc.) and there will probably be few symptoms, but if the safety factor is fractionally below 1 then conduction will be blocked and symptoms may be expected. Many MS patients appear to have large numbers of demyelinated axons with a safety factor near unity, and so even very small changes in the safety factor can have profound consequences in terms of the severity of the symptoms presented (see, for example, 'Effects of temperature' below).

Another consideration which can determine the presence or absence of conduction block is the geometry of the internode preceding the demyelinated region, since this can dictate that the safety factor will be below unity and thus that conduction will be blocked even if the demyelinated axolemma is excitable. For example, if the length of this internode is normal and relatively long, and yet has undergone partial thickness demyelination (i.e. thinning of the myelin sheath), then the increased capacitance and conductance along the internode can so reduce the local current available to depolarize the demyelinated axolemma that permanent conduction block ensues. Thus in diseases such as MS, partial thickness demyelination may result in permanent conduction block, whereas complete demyelination can at least offer the opportunity for a later restoration of conduction. The prevalence in MS of stable internodes which have undergone partial thickness demyelination is not known.

Whereas partial thickness demyelination can be disadvantageous, short internodes (resulting, for example, from partial length or paranodal demyelination or from remyelination) can promote the excitation of the demyelinated axolemma (Waxman & Brill, 1978): a short internode preceding the demyelinated region will reduce the opportunity for local currents to be diminished, thereby increasing the safety factor for conduction. In some circumstances, short internodes may therefore be expected to compensate for other factors which may not be optimal for conduction in a particular axon.

The biophysical properties of the demyelinated axolemma are also critical elements determining whether conduction will be possible, but relatively little is known about these properties. The importance of a sufficient population of Na^+ channels has been mentioned, and although central myelinated axons certainly contain internodal K^+ channels (Kapoor *et al.*, 1993b), the types, distribution and density of these channels are uncertain and it is therefore impossible to predict the effect they may have on conduction. Similarly, it is difficult to predict the effects of changes in the distribution and density of the Na^+/K^+-ATPase and other pumps and ion exchangers, although there is evidence that the Na^+/K^+-ATPase can induce conduction block during periods of intense activity (see 'Transmission of impulse trains'). The biophysical properties should be illuminated by current patch-clamp investigations of demyelinated axolemma. The various channels, etc., found along normal and pathological axons are the subject of an excellent recent review (Waxman & Ritchie, 1993) and are discussed in other chapters in this volume.

The potential role that the presence of inflammation and inflammatory mediators may play in conduction block is uncertain. Notably, a profound re-expression of previous symptoms is acutely associated with antibody-mediated killing of T lymphocytes (and perhaps other cells) in MS patients, and this effect has tentatively been linked with the release of inflammatory mediators (D.A.S. Compston, personal communication). Indeed, several other observations suggest that inflammation may impair conduction (Youl *et al.*, 1991; McDonald *et al.*, 1992), and various cytokines have now been reported to affect ionic currents in neural tissues (e.g. Sawada, Hara & Maeno, 1991; Brinkmeier *et al.*, 1992; Szucs *et al.*, 1992; ffrench-Mullen & Plata-Salamán, 1992; Plata-Salamán & ffrench-Mullen, 1992). This is an interesting area for future research. The role that putative circulating 'neuroelectric blocking factors' may have in modulating conduction in central demyelinated axons is reviewed elsewhere (Smith, 1994).

Conduction in demyelinated axons

McDonald and Sears found that some of the axons which were able to conduct impulses through smaller diphtheria toxin lesions did so with certain conduction deficits (McDonald & Sears, 1969b, 1970a), suggesting that despite the presence of conduction the axons had indeed been affected by the lesion. However, the lesion induced by diphtheria toxin produces many types of myelin abnormality, including segmental demyelination, paranodal demyelination (Harrison *et al.*, 1972c), partial thickness demyelination (i.e. myelin thinning) (Harrison *et al.*, 1972b, c), and, in longer-term lesions (after 19 days), repair of some axons by remyelination (Harrison *et al.*, 1972b). Thus, while it was reasonable to suppose that the conducting, pathological axons were among those affected by the lesion, McDonald and Sears were careful to point out that the type and degree of the insult remained uncertain: it was also possible that the axons may have been affected directly by the diphtheria toxin (McDonald & Sears, 1970a); certainly a number of axons underwent degeneration.

Several more homogeneous experimental demyelinating lesions have since been developed for electrophysiological study (Smith *et al.* 1979, 1981; Black *et al.*, 1991; Felts & Smith, 1992), but the inability to correlate particular conduction deficits with specific morphological abnormalities has been a persistent problem. However, this problem has recently been overcome through studies using the EBr lesion where the conduction properties of single axons have been determined in detail, and then the particular axons have been labelled by the iontophoresis of horseradish peroxidase so that they can be identified subsequently at light and electron microscope levels (P. A. Felts and K. J. Smith, unpublished observations). These studies have established that segmentally demyelinated central axons can conduct, and that they do so with the familiar properties of a prolonged latency and RPT (Fig. 10.2), and a reduced ability to conduct impulse trains.

Most of the conducting, labelled axons have been found to have a patchy or continuous ensheathment by the cellular processes of glial or Schwann cells, raising the possibility that ensheathment of demyelinated axons, even without remyelination, can promote the restoration of conduction. However, although 18 of 20 axons examined to date have been at least partially ensheathed, conduction has also been observed in two axons which exhibited lengths of demyelination over which the axolemma was entirely free of glial contacts for at least 100 μm. These axons demonstrate that patchy or continuous glial ensheathment is not an absolute requirement for conduction, at least over distances of 100 μm.

This was an unexpected finding since several previous morphological studies had indicated that such axolemmae may lack the requisite density of Na^+ channels. For example, an immunocytochemical study of the EBr lesion revealed the presence of Na^+ channel immunoreactivity only in demyelinated

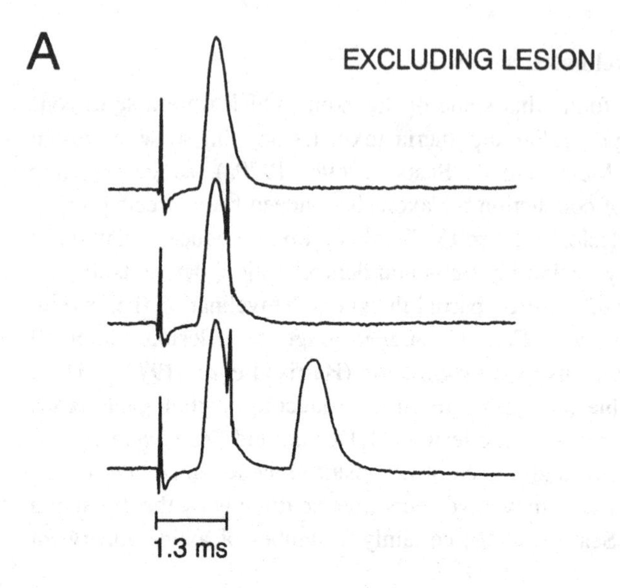

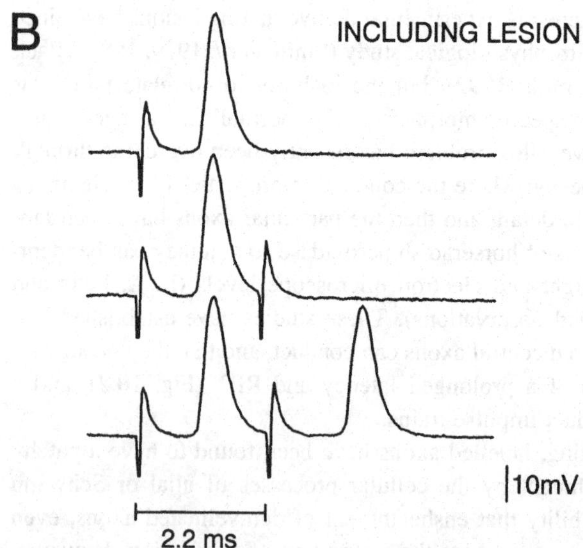

Fig. 10.2. Series of action potentials recorded intra-axonally from the same site along an axon which has been segmentally demyelinated by the injection of ethidium bromide into the rat dorsal column. The conduction distance is similar for all records, but the lesion is included in the conduction pathway in (B) alone. In each series of three records, the first record shows the response to a single supramaximal stimulus, and the second and third records show responses to paired stimuli whose interval straddles the refractory period of transmission (RPT) of the axon. Note the prolongation of the latency and RPT when the lesion is included in the conduction pathway.

axons which were ensheathed by glial or Schwann cell processes (Black *et al.*, 1991), and other ultrastructural studies showed that node-like axonal specializations formed along demyelinated central axons only at sites where the axons were contacted by glial processes (Blakemore & Smith, 1983; Rosenbluth & Blakemore, 1984; Rosenbluth, Tao-Cheng & Blakemore, 1985). This paradox may be partially explained by the fact that the antibody used in the immunocytochemical study was effective at detecting only relatively high concentrations of Na^+ channels (e.g. >100 channels/μm^2), whereas it is known that much lower densities are still compatible with conduction (see above). Alternatively, it may be that conduction can proceed despite the presence of short lengths of inexcitable membrane, as appears to happen in some peripheral demyelinated axons (Smith, Bostock & Hall, 1982). Whether conduction can continue in the presence of long lengths (e.g. >500 µm) of naked demyelinated axon awaits further study.

The fact that conduction can occur in segmentally demyelinated central axons can help to explain the common occurrence of remissions in MS, and also the frequent finding that not all lesions in MS are associated with the production of symptoms, even when the lesions are located in pathways such as the optic nerve where prominent symptoms may be expected (Ghatak *et al.*, 1974; Wisniewski *et al.*, 1976). The observation that conduction can occur in both continuously and patchily ensheathed demyelinated axons is also of interest with respect to the appearance of MS lesions upon MRI examination. Most MS lesions can be categorized as either 'closed' or 'open' on MRI criteria, and detailed ultrastructural study has revealed the presence in these lesions of either continuously or patchily ensheathed demyelinated axons respectively (Barnes *et al.*, 1991). MRI examination would be all the more valuable if it were able to discriminate symptomatic from asymptomatic lesions, and this may be possible if, for example, the open lesions tended to be associated with conduction block and were thereby typically symptomatic. However, the fact that conduction can occur in both patchily and continuously ensheathed axons indicates that both open and closed lesions may be clinically silent under some circumstances. Whether one type of lesion is more likely to produce symptoms remains to be determined.

Since a major cause of the symptoms in diseases such as MS and optic neuritis is conduction block in central demyelinated axons (McDonald, 1975; Halliday & McDonald, 1977; Waxman, 1988), it follows that the restoration of conduction to the demyelinated axons may be expected to contribute to the remissions commonly seen in these diseases. Thus a symptomatic therapy for the disease might be based on the use of pharmacological agents which would promote the restoration of conduction in blocked axons, and Sears' group, in particular, has established the feasibility of this approach (Schauf & Davis, 1974; Sears, Bostock & Sherratt, 1978; Bostock, Sherratt & Sears, 1978; Sears & Bostock, 1981; Davis & Schauf, 1981; Kocsis, Black &

Waxman, 1993; Waxman, 1993), which has led directly to successful clinical trials (Stefoski *et al.*, 1991; van Diemen *et al.*,1992a). Other pharmacological strategies have also been tested clinically (Davis *et al.*, 1970; Gilmore, Kasarskis & McAllister, 1985; Kaji, Happel & Sumner, 1990). Some of these approaches are discussed briefly under 'Effects of temperature' and 'Transmission of impulse trains'.

Conduction slowing

Although normal myelinated fibres often have conduction velocities exceeding 50 m/s, it is clear that the conduction velocity is reduced to only a few metres per second along the demyelinated region (McDonald & Sears, 1970a; P.A. Felts & K.J. Smith, unpublished observations). Thus conduction along even a few demyelinated internodes can impose a prominent increase in latency (McDonald, 1974). These latency increases, possibly coupled with a reduction in the total number of axons conducting, lead to significant delays and other alterations in the visual, somatosensory and brainstem auditory evoked potentials (Halliday, McDonald & Mushin, 1973; Halliday & McDonald, 1977; McDonald & Halliday, 1977; Small, Matthews & Small, 1978; Matthews & Small, 1979; Chiappa, 1988; Hume & Waxman, 1988; Swanson, 1989; Onofrj *et al.*, 1990; Hallpike, 1992), and these changes have been found to be of significant diagnostic value (Halliday, McDonald & Mushin, 1972; Halliday & McDonald, 1977; Regan, Murray & Silver, 1977).

Although conduction slowing leads to prominent delays in evoked potentials, homogeneous slowing appears to have relatively little effect on the signs and symptoms seen on clinical presentation: for example, MS patients often fail to perceive any deficit in vision even when there are gross delays in the visual evoked potential (Halliday *et al.*, 1973). However, when the slowing is asymmetric, as in patients with unilateral optic neuritis, the conduction slowing can result in subtle signs and symptoms, such as the Pulfrich phenomenon (Rushton, 1975) (e.g. the fact that a pendulum swung from side to side may be described by the patient as traversing a curved trajectory). Where conduction is delayed to differing degrees in different somatosensory axons it may also interfere with vibration sensitivity (Gilliatt & Willison, 1962; McDonald, 1975).

Refractory period of transmission

McDonald and Sears found that some axons passing through the central diphtheria toxin lesion were unable to transmit the second impulse of closely spaced pairs of impulses, although they could do so where the lesion was excluded from the pathway (McDonald & Sears, 1970a). They introduced the term 'refractory period of transmission' (RPT) to describe this deficit and they distinguished this term from the 'refractory period' which strictly describes properties of the axon at the stimulating cathode (Paintal, 1966). The RPT is defined as the maximum interval between two supramaximal stimuli at which the action potential arising from the second stimulus just fails to be propagated through the lesion. McDonald and Sears found that whereas the range of RPTs was 0.4–1.1 ms (mean 0.86 ms) in normal cat dorsal column fibres, this was prolonged to 1.2–4.2 ms in most axons passing through a demyelinating lesion (McDonald & Sears, 1970a). More recent studies utilizing labelled axons have confirmed the prolongation of RPT (from 0.55–1.4 ms (mean 0.83 ms, $n = 21$) in the unaffected portions of rat dorsal column axons, to 1.0–27.0 ms (mean 3.76 ms) in the same axons where a demyelinating lesion was included in the conduction pathway), and have also established that the prolongation is a property of axons which are segmentally demyelinated (P.A. Felts & K.J. Smith, unpublished observations). The factors underlying the increase in RPT have recently been examined in computer models of partially demyelinated axons (Quandt & Davis, 1992).

Transmission of impulse trains

Demyelinated central axons exhibit several deficits with regard to the conduction of impulse trains. McDonald & Sears (1970a) found that the maximum frequency for the faithful transmission of impulses through the lesion was decreased, and this is, in part, a natural consequence of the prolongation of the RPT. Demyelinating central lesions can therefore act as filters, transforming trains of impulses at high frequency into trains at low frequency: in motor axons this effect can be expected to contribute to weakness, and in sensory axons it may be expected subtly to impair sensation. McDonald and Sears also found that axons affected by the lesion tended to accumulate refractoriness with repeated activation such that an axon might be able faithfully to conduct the first three or four impulses in response to a stimulus train at 400–500 Hz, but then be able only to conduct impulses in response to alternate stimuli, and eventually to every third or fourth stimulus (Fig. 10.3) (McDonald & Sears, 1970a; see also Kaji, Suzumura & Sumner, 1988). The accumulation of refractoriness may well contribute to the increasing weakness observed upon sustained muscular contraction in MS patients (McDonald, 1975), and possibly to the 'fading out' of vision reported by some MS patients if the gaze is fixated on a point for several seconds (Waxman, 1981).

Another cause of increasing weakness may be the fact that demyelinated central axons can also acquire intermittent periods of complete conduction block during the passage of impulse trains. Thus an axon may faithfully follow a stimulus at 200 Hz for several seconds but then enter a short period of complete conduction block before returning to the faithful transmission

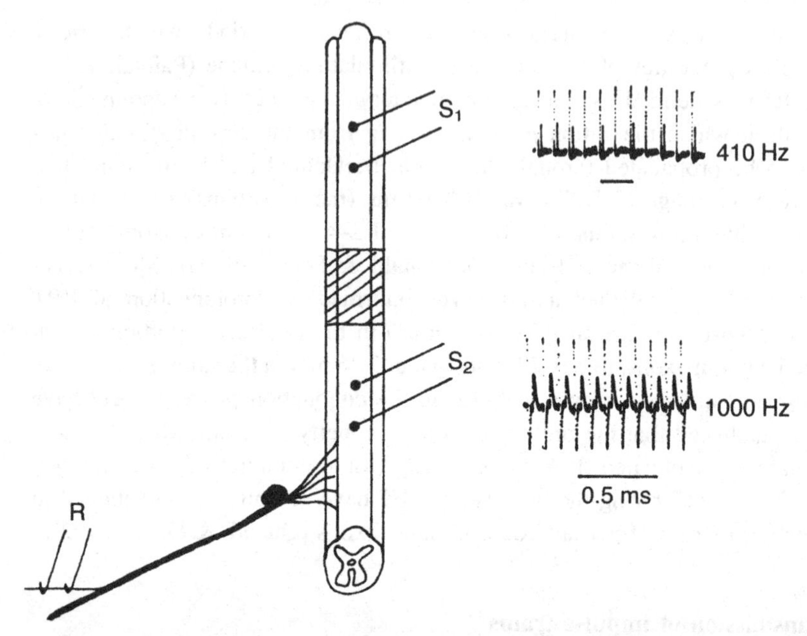

Fig. 10.3. Records of activity in a single unit isolated from an intercostal nerve caudal to an experimental demyelinating lesion (hatched area) in the cat dorsal column. The lower record shows that when the conduction pathway excluded the lesion the unit was able faithfully to follow supramaximal stimuli (S_2) applied at 1000 Hz, but when the lesion was included in the conduction pathway (upper record) the unit was only able to follow stimuli at 410 Hz (S_1): indeed, after the first few responses the unit was only able to conduct the impulses arising from alternate stimuli. In this figure the action potentials are the smaller, thicker lines immediately following each stimulus artifact, which appears as a dotted line). Reprinted from McDonald & Sears (1970a) by permission of Oxford University Press.

of impulses at 200 Hz (R. Kapoor, P.A. Felts & K.J. Smith, unpublished observations). In recordings made at the site of demyelination (but not at other sites along the axon) we have found that the periods of conduction block are coincident with periods of membrane hyperpolarization (Fig. 10.4; R. Kapoor, P.A. Felts & K.J. Smith, unpublished observations), suggesting that the block may be due to increased activity of the axolemmal Na^+/K^+-ATPase (or Na^+ pump) as has been observed in peripheral demyelinated axons (Bostock & Grafe, 1985). Pump activity would be stimulated by the gradual accumulation of $[K^+]_o$ and, especially, $[Na^+]_i$ during the period of high-frequency transmission along the demyelinated portion, and since the pump is electrogenic its activity would result in a membrane hyperpolarization which would, in turn, raise the threshold for excitation of the demyelinated axolemma. Since demyelinated axons already have a reduced safety factor, such a raised threshold might easily result in the conduction block observed. The block would be relieved as the normal ion balance was

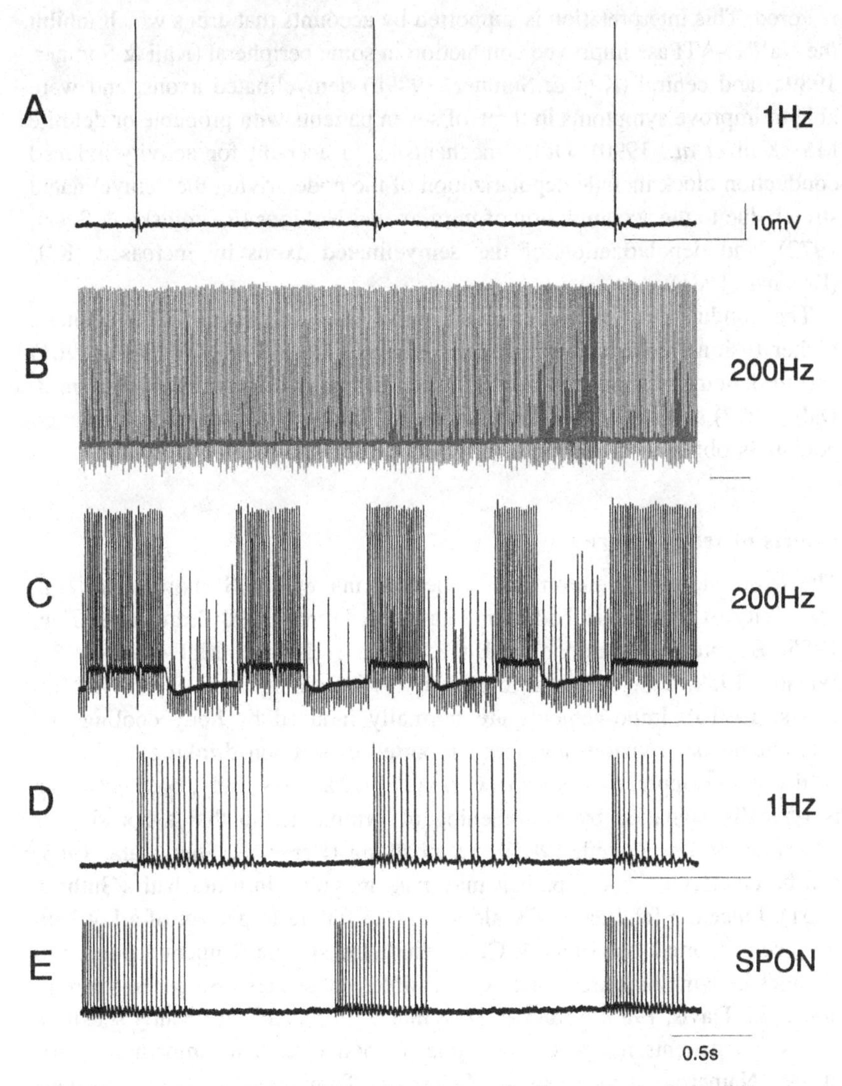

Fig. 10.4. A series of records obtained intra-axonally from one demyelinated sensory axon at the margin of a lesion in the rat dorsal column, including the lesion in the conduction pathway. (A) Single electric stimuli at 1 Hz initially result in single impulses which traverse the lesion to the recording electrode. (B) The axon can initially faithfully follow stimulation at 200 Hz, but after several seconds of such high-frequency stimulation (C) the train of impulses is periodically interrupted by short periods of complete conduction block, after which faithful conduction of the 200 Hz train resumes: the short vertical lines seen during the periods of conduction block are stimulus artifacts. The periods of conduction block are associated with membrane hyperpolarization. Following the period of high-frequency stimulation (D) the passage of a single action potential now evokes short bursts of additional, ectopic impulses, believed to arise at the site of the lesion. Furthermore (E), in the absence of any deliberate stimulation, the axon developed continuing activity, generating bursts of impulses believed to arise at the site of demyelination. (R. Kapoor, P.A. Felts and K.J. Smith, unpublished observations.)

restored. This interpretation is supported by accounts that drugs which inhibit the Na^+/K^+-ATPase improved conduction in some peripheral (Kaji & Sumner, 1989a) and central (Kaji & Sumner, 1989b) demyelinated axons, and were able to improve symptoms in three of seven patients with probable or definite MS (Kaji *et al.*, 1990). Other mechanisms to account for activity-induced conduction block include depolarization of the node driving the demyelinated stretch due to the accumulation of intra-axonal Na^+ ions (Rasminsky & Sears, 1972), and depolarization of the demyelinated axons by increased $[K^+]_o$ (Brismar, 1981).

The conduction deficits described above may contribute to the reduced flicker fusion frequency observed in patients (Titcombe & Willison, 1961) and to deficits apparent in visual (Milner, Regan & Heron, 1974; Celesia & Daly, 1977) and somatosensory (Sclabassi, Namerow & Enns, 1974) evoked potentials obtained in response to repetitive stimuli.

Effects of temperature

The expression of symptoms in some patients with MS shows a striking sensitivity to changes in body temperature (Uhthoff, 1890; Edmund & Fog, 1955; Boynton, Garramone & Buca, 1959; Nelson & McDowell, 1959; Watson, 1959; Galvin, Regan & Heron, 1976; Symington *et al.*, 1977; Matthews, 1991b). Improvements are normally induced by body cooling and improvements in vision have been reported even upon drinking a glass of cold water (Hopper, Matthews & Cleland, 1972). Conversely, body warming is typically signalled by a worsening of symptoms, so that a hot shower (Waxman & Geschwind, 1983) or sunbathing (Berger & Sheremata, 1985) can be deleterious, and a patient may risk drowning in a hot bath (Guthrie, 1951). Indeed, a fatal case of scalding due to failure to get out of a hot bath has been reported (Waxman & Geschwind, 1983). The temperature-induced changes in symptoms are typically reversible (but see also Berger & Sheremata, 1983; Davis, 1985; Berger & Sheremata, 1985), and sufficiently sensitive that some patients notice changes due to their circadian temperature variations (Namerow, 1968; Davis, Michael & Tomaszewski, 1973). Furthermore, since the effects of warming are relatively specific for central demyelinating diseases, it has been possible to base a diagnostic test for MS upon the phenomenon: the 'hot bath test' (Davis, 1970; Matthews, 1991a).

The sensitivity of the symptoms to temperature changes arises from a combination of the reduced safety factor for conduction in demyelinated axons, and the effects of temperature on action potential duration (Schoepfle & Erlanger, 1941; Paintal, 1966; Koles & Radminsky, 1972; Schauf & Davis, 1974). Temperature increases shorten the duration of action potentials, curtailing the flow of depolarizing current generated by the node driving the demyelinated axolemma (Schauf & Davis, 1974; Davis & Schauf, 1981).

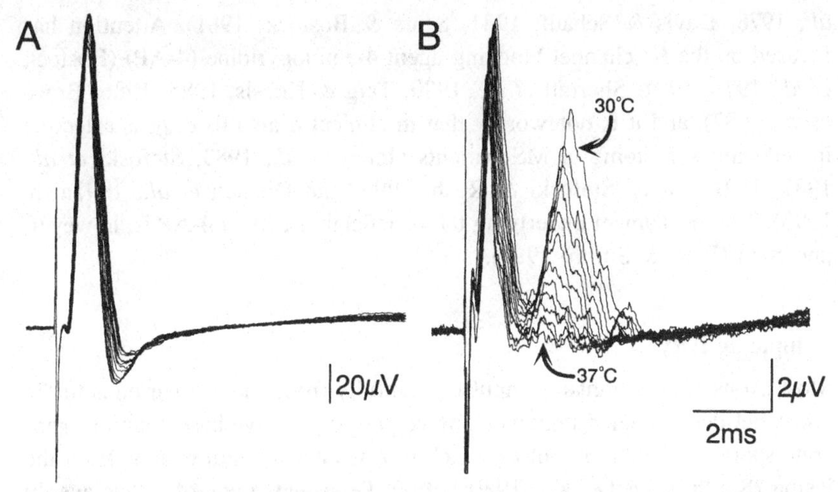

Fig. 10.5. Two series of superimposed, monophasic compound action potentials recorded *in vitro* from a normal rat dorsal column (A) and one containing an experimental demyelinating lesion (B). The delayed peak in (B) is due to conduction in segmentally demyelinated axons. The stimulating and recording conditions remained constant throughout the experiment, but the different records were obtained as the central length of the column (which contained the demyelinated lesion in B) was progressively warmed from 30 °C to normal body temperature (37 °C). Note that the temperature changes had little effect on conduction in the normal axons in (A) and (B), but that many more demyelinated axons were able to conduct at the cooler temperatures: virtually no demyelinated axons were able to conduct through the lesion at body temperature (P.A. Felts and K.J. Smith, unpublished observations). Reprinted from Smith (1994) with permission.

Where the safety factor is only just above unity, this reduction is likely to result in conduction block, accounting for the deleterious effects of increases in body temperature. Conversely, cooling lengthens the action potential and is likely to restore conduction to axons which have a safety factor just below unity at normal body temperature.

These effects of temperature have been examined in detail in peripheral demyelinated axons (Rasminsky, 1973; Davis *et al.*, 1975; Sears *et al.*, 1978; Pencek *et al.*, 1980; Sears & Bostock, 1981; Pender & Sears, 1984) and these studies have now been extended to the central demyelinating lesion induced by the intraspinal injection of EBr (Fig. 10.5). Notably, in this model virtually no demyelinated axons are able to conduct at normal body temperature, despite the presence in the preparation of many axons capable of conducting when cooled.

The prominent beneficial effects of cooling patients with MS have prompted several investigators, including Sears' group, to search for pharmacological agents which may similarly prolong nerve impulses in the hope that they may form an effective symptomatic therapy (reviewed in: Sears *et al.*, 1978; Davis & Schauf, 1981; Sears & Bostock, 1981). Attention has focused on the K^+ channel blocking agent 4-aminopyridine (4-AP) (Bostock *et al.*, 1978, 1981; Sherratt *et al.*, 1980; Targ & Kocsis, 1985, 1986; Bowe *et al.*, 1987), and it is noteworthy that in clinical trials this drug is effective in reducing symptoms in MS patients (Jones *et al.*, 1983; Stefoski *et al.*, 1987, 1991; Davis, Stefoski & Rush, 1990; van Diemen *et al.*, 1992a, b, 1993). The mechanism underlying the beneficial effects of 4-AP is, however, uncertain (Felts & Smith, 1994).

Ectopic activity

Most axons simply transmit impulses faithfully from one of their ends to the other, but demyelinated regions along central axons have been found to generate spurious, ectopic impulses which propagate in both directions from the lesion (Smith & McDonald, 1980, 1982). Persistent trains of either evenly spaced impulses or rhythmic bursts of impulses have been observed, and since they occurred in many sensory axons simultaneously, these authors suggested that they formed a sufficient explanation for the continuous tingling paraesthesiae experienced by many patients with MS at some time during the course of their disease. Experimentally demyelinated central axons were also found to be exquisitely mechanosensitive, such that even minute deformations of the dorsal columns using an eyelash were sufficient to induce prompt increases in firing rate, or to induce discharges in axons which had previously been electrically silent. This mechanosensitivity has been advanced as an explanation of certain movement-induced symptoms in MS patients, such as Lhermitte's sign (i.e. a 'pins and needles' sensation radiating down the body upon neck flexion) in patients with demyelinating lesions in the cervical spinal cord (Lhermitte, Bollack & Nicholas, 1924; Kanchandani & Howe, 1982), and the perception of flashes of light upon eye movement in patients with demyelinating lesions in the optic nerves (Davis *et al.*, 1976).

The mechanism by which the ectopic discharges are generated is not understood. Although ephaptic interactions between adjacent axons may account for some of the activity, such interactions have so far only been demonstrated in the peripheral, amyelinated axons of the dystrophic mouse (Rasminksy 1978, 1980). However, intra-axonal recordings made in normal myelinated axons (Kapoor *et al.*, 1993b), and at a site of central demyelination (Fig. 10.6) (Kapoor, Felts & Smith, 1993a; see also Young *et al.*, 1989), have recently indicated that continuing activity can arise from inward, and thereby excitatory, K^+ currents. K^+ currents would be expected to become inward when $[K^+]_o$ is slightly raised, for example by disturbed glial function or by high-frequency electrical stimulation. Certainly, in demyelinated axons *in vitro* such stimulation can induce continuing bursts of ectopic discharges,

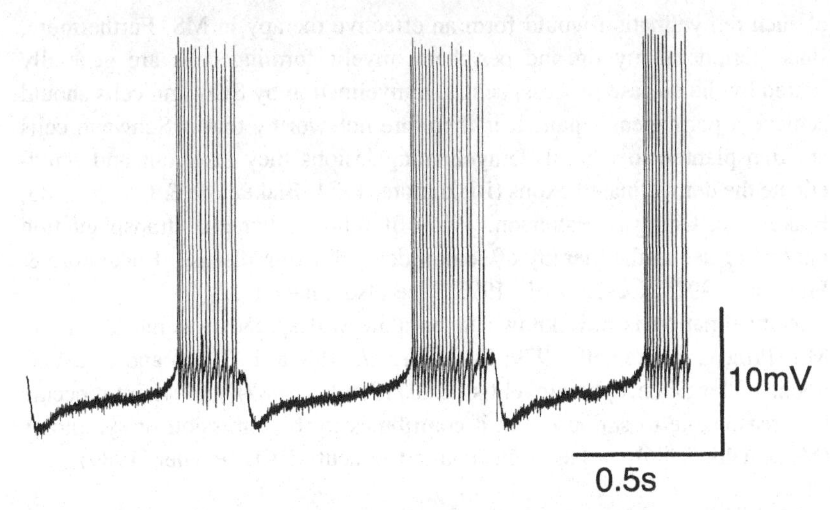

Fig. 10.6. Bursts of continuing ectopic impulses recorded intra-axonally from a sensory axon close to the site of an experimental demyelinating lesion in the rat dorsal column: the changes in membrane potential associated with the bursts are clearly seen. The bursts result from a plateau of depolarization which is terminated by a period of hyperpolarization believed to arise from activity of the Na^+/K^+-ATPase, or Na^+ pump (R. Kapoor, P.A. Felts and K.J. Smith, unpublished observations). Reprinted from Smith (1994) with permission.

and bursts of impulses initiated by the passage of a single action potential (fig. 10.4D, E).

Conduction in remyelinated central axons

Although the central diphtheria toxin lesion revealed many of the fundamental conduction properties of central demyelinated axons, extensive remyelination does not occur (Harrison, McDonald & Ochoa, 1972a) and so the study of central remyelinated axons awaited the development of new lesions. In the central demyelinating lesion produced using lysophosphatidyl choline (LPC, lysolecithin) (Hall, 1972; Blakemore *et al.*, 1977) many of the affected axons are repaired by oligodendrocyte remyelination, and these axons have been found to conduct with normal or near normal velocity and RPT: their ability faithfully to transmit high-frequency trains is also restored (Smith *et al.*, 1979, 1981). A later study established that remyelination of central demyelinated axons by Schwann cells was similarly effective in restoring secure conduction (Felts & Smith, 1992), even though Schwann cells are the myelinating cell of the peripheral nervous system.

The ability of Schwann cells to effect both morphological and electrophysiological repair of central demyelinated axons suggests that the induction of such remyelination would form an effective therapy in MS. Furthermore, since peripheral myelin and peripheral myelin-forming cells are generally spared by the disease process, central remyelination by Schwann cells should achieve a permanent repair. It is therefore noteworthy that if Schwann cells are transplanted to central demyelinating lesions they seek out and remyelinate the demyelinated axons (Blakemore, 1984; Blakemore & Crang, 1985; Blakemore, Crang & Patterson, 1987). It follows that glial transplantation may be of use in the therapy of central demyelinating disease (Blakemore & Franklin, 1991; Kocsis *et al.*, 1993) (see also Chapter 12).

Remyelination is now known to be quite widespread in some lesions in MS (Prineas & Connell, 1979; Prineas *et al.*, 1987a, b, 1993) and in experimental allergic encephalomyelitis (EAE) (Pender, 1987), and where it occurs it is reasonable to suppose that it contributes to the remission of symptoms (McDonald, 1986), as has been reported in acute EAE (Pender, 1989).

It is clear from the preceding account that the major conduction properties of central demyelinated axons were identified by Tom Sears in his initial studies with Ian McDonald 25 years ago. Both authors have continued to make major contributions to the study of demyelination, with Sears contributing particularly to our understanding of the electrophysiological properties of peripheral demyelinated axons. With so firm a foundation established in this field, it is hoped that future research will soon be rewarded with an effective symptomatic therapy for demyelinating diseases.

Acknowledgements

I wish to thank Dr P.A. Felts and Dr Susan Hall for their helpful comments on the text. The support of the Wellcome Trust, the Multiple Sclerosis Society, the British Council and the Special Trustees of Guy's Hospital is gratefully acknowledged.

References

Barnes, D., Munro, P.M., Youl, B.D., Prineas, J.W. & McDonald, W.I. (1991). The longstanding MS lesion: a quantitative MRI and electron microscopic study. *Brain*, **114**, 1271–1280.

Berger, J.R. & Sheremata, W.A. (1983). Persistent neurological deficit precipitated by hot bath test in multiple sclerosis. *Journal of the American Medical Association*, **249**, 1751–1753.

Berger, J.R. & Sheremata, W.A. (1985). Reply to letter by F.A. Davis. *Journal of the American Medical Association*, **253**, 203.

Black, J.A., Felts, P.A., Smith, K.J., Kocsis, J.D. & Waxman, S.G. (1991). Distribution of sodium channels in chronically demyelinated spinal cord axons: immuno-ultrastructural localization and electrophysiological observations. *Brain Research*, **544**, 59–70.

Blakemore, W.F. (1982). Ethidium bromide induced demyelination in the spinal cord of the cat. *Neuropathology and Applied Neurobiology*, **8**, 365–375.
Blakemore, W.F. (1984). Limited remyelination of CNS axons by Schwann cells transplanted into the sub-arachnoid space. *Journal of the Neurological Sciences*, **64**, 265–276.
Blakemore, W.F. & Crang, A.J. (1985). The use of cultured autologous Schwann cells to remyelinate areas of persistent demyelination in the central nervous system. *Journal of the Neurological Sciences*, **70**, 207–223.
Blakemore, W.F., Crang, A.J. & Patterson, R.C. (1987). Schwann cell remyelination of CNS axons following injection of cultures of CNS cells into areas of persistent demyelination. *Neuroscience Letters*, **77**, 20–24.
Blakemore, W.F., Eames, R.A., Smith, K.J. & McDonald, W.I. (1977). Remyelination in the spinal cord of the cat following intraspinal injections of lysolecithin. *Journal of the Neurological Sciences*, **33**, 31–43.
Blakemore, W.F. & Franklin, R.J.M. (1991). Transplantation of glial cells into the CNS. *Trends in Neurosciences*, **14**, 323–327.
Blakemore, W.F. & Smith, K.J. (1983). Node-like axonal specialisations along demyelinated central nerve fibres: ultrastructural observations. *Acta Neuropathologica*, **60**, 291–296.
Bostock, H. (1984). Internodal conduction along undissected nerve fibres in experimental neuropathy. In *Peripheral Neuropathy*, ed. P.J. Dyck, P.K. Thomas, E.H. Lambert & R. Bunge, pp. 900–910. Philadelphia: W.B. Saunders.
Bostock, H. & Grafe, P. (1985). Activity-dependent excitability changes in normal and demyelinated rat spinal root axons. *Journal of Physiology (London)*, **365**, 239–257.
Bostock, H. & Sears, T.A. (1978). The internodal axon membrane: electrical excitability and continuous conduction in segmental demyelination. *Journal of Physiology (London)* , **280**, 273–301.
Bostock, H., Sears, T.A. & Sherratt, R.M. (1981). The effects of 4-aminopyridine and tetraethylammonium ions on normal and demyelinated mammalian nerve fibres. *Journal of Physiology (London)*, **313**, 301–315.
Bostock, H., Sherratt, R.M. & Sears, T.A. (1978). Overcoming conduction failure in demyelinated nerve fibres by prolonging action potentials. *Nature*, **274**, 385–387.
Bowe, C.M., Kocsis, J.D., Targ, E.F. & Waxman, S.G. (1987). Physiological effects of 4-aminopyridine on demyelinated mammalian motor and sensory fibres. *Annals of Neurology*, **22**, 264–268.
Boynton, B.L., Garramone, P.M. & Buca, J.T. (1959). Observations on the effects of cool baths for patients with multiple sclerosis. *Physical Therapy Review*, **39**, 297–299.
Brinkmeier, H., Kaspar, A., Wietholter, H. & Rudel, R. (1992). Interleukin-2 inhibits sodium currents in human muscle cells. *Pflügers Archiv.*, **420**, 621–623.
Brismar, T. (1981). Specific permeability properties of demyelinated rat nerve fibres. *Acta Physiologica Scandinavica*, **113**, 167–176.
Celesia, G.G. & Daly, R.F. (1977). Visual electroencephalographic computer analysis (VECA): a new electrophysiologic test for the diagnosis of optic nerve lesions. *Neurology*, **27**, 637–641.
Chiappa, K.H. (1988). Use of evoked potentials for diagnosis of multiple sclerosis. *Neurologic Clinics*, **6**, 861–880.
Davis, F.A. (1970). Pathophysiology of multiple sclerosis and related clinical implications. *Modern Treatment*, **7**, 890–902.

Davis, F.A. (1985). Neurological deficits following the hot bath test in multiple sclerosis [letter]. *Journal of the American Medical Association*, **253**, 203.

Davis, F.A., Becker, F.O., Michael, J.A. & Sorensen, E. (1970). Effect of intravenous sodium bicarbonate, disodium edetate (Na_2EDTA), and hyperventilation on visual and oculomotor signs in multiple sclerosis. *Journal of Neurology, Neurosurgery and Psychiatry*, **33**, 723–732.

Davis, F.A., Bergen, D., Schauf, C., McDonald, I. & Deutsch, W. (1976). Movement phosphenes in optic neuritis: a new clinical sign. *Neurology*, **26**, 1100–1104.

Davis, F.A., Michael, J.A. & Tomaszewski, J.S. (1973). Fluctuation of motor function in multiple sclerosis related to circadian temperature variations. *Diseases of the Nervous System*, **34**, 33–36.

Davis, F.A. & Schauf, C.L. (1981). Approaches to the development of pharmacological interventions in multiple sclerosis. *Advances in Neurology*, **31**, 505–510.

Davis, F.A., Schauf, C.L., Reed, B.J. & Kesler, R.L. (1975). Experimental studies of the effects of extrinsic factors on conduction in normal and demyelinated nerve. *Journal of Neurology, Neurosurgery and Psychiatry*, **39**, 442–448.

Davis, F.A., Stefoski, D. & Rush, J. (1990). Orally administered 4-aminopyridine improves clinical signs in multiple sclerosis. *Annals of Neurology*, **27**, 186–192.

Edmund, J. & Fog, T. (1955). Visual and motor instability in multiple sclerosis. *AMA Archives of Neurology and Psychiatry*, **73**, 316–323.

Felts, P.A. & Smith, K.J. (1992). Conduction properties of central nerve fibres remyelinated by Schwann cells. *Brain Research*, **574**, 178–192.

Felts, P.A. & Smith, K.J. (1994). The use of potassium channel blocking agents in the therapy of demyelinating disease. *Annals of Neurology*, **36**, 454.

ffrench-Mullen, J.M.H. & Plata-Salamán, C.R. (1992). Interleukin-1β inhibition of calcium channel currents in isolated hippocampal CA1 neurones: pharmacology and mode of action. *Society for Neuroscience Abstracts*, **18**, 433.

Galvin, R.J., Regan, D. & Heron, J.R. (1976). A possible means of monitoring the progress of demyelination in multiple sclerosis: effect of body temperature on visual perception of double light flashes. *Journal of Neurology, Neurosurgery and Psychiatry*, **39**, 861–865.

Ghatak, N.R., Hirano, A., Lijtmaer, H. & Zimmerman, H.M. (1974). Asymptomatic demyelinated plaque in the spinal cord. *Archives of Neurology*, **30**, 484–486.

Gilliatt, R.W. & Willison, R.G. (1962). Peripheral nerve conduction in diabetic neuropathy. *Journal of Neurology, Neurosurgery and Psychiatry*, **25**, 11–18.

Gilmore, R.L., Kasarskis, E.J. & McAllister, R.G. (1985). Verapamil-induced changes in central conduction in patients with multiple sclerosis. *Journal of Neurology, Neurosurgery and Psychiatry*, **48**, 1140–1146.

Guthrie, T.C. (1951). Visual and motor changes in patients with multiple sclerosis. *AMA Archives of Neurology and Psychiatry*, **65**, 437–451.

Hall, S.M. (1972). The effect of injections of lysophosphatidyl choline into white matter of the adult mouse spinal cord. *Journal of Cell Science*, **10**, 535–546.

Halliday, A.M. & McDonald, W.I. (1977). Pathophysiology of demyelinating disease. *British Medical Bulletin*, **33**, 21–27.

Halliday, A.M., McDonald, W.I. & Mushin, J. (1972). Delayed visual evoked response in optic neuritis. *Lancet*, **I**, 982–985.

Halliday, A.M., McDonald, W.I. & Mushin, J. (1973). Visual evoked response in diagnosis of multiple sclerosis. *British Medical Journal*, **4**, 661–664.

Hallpike, J. (1992). Multiple sclerosis: making the diagnosis. *Australian Family Physician*, **21**, 1407–1410.

Harrison, B.M., McDonald, W.I. & Ochoa, J. (1972a). Central demyelination produced by diphtheria toxin: an electron microscopic study. *Journal of the Neurological Sciences*, **17**, 281–291.
Harrison, B.M., McDonald, W.I. & Ochoa, J. (1972b). Remyelination in the central diphtheria toxin lesion. *Journal of the Neurological Sciences*, **17**, 293–302.
Harrison, B.M., McDonald, W.I., Ochoa, J. & Ohlrich, G.D. (1972c). Paranodal demyelination in the central nervous system. *Journal of the Neurological Sciences*, **16**, 489–494.
Hines, M. & Shrager, P. (1991). A computational test of the requirements for conduction in demyelinated axons. *Restorative Neurology and Neuroscience*, **3**, 81–93.
Hopper, C.L., Matthews, C.G. & Cleeland, C.S. (1972). Symptom instability and thermoregulation in multiple sclerosis. *Neurology*, **22**, 142–148.
Hume, A.L. & Waxman, S.G. (1988). Evoked potentials in suspected multiple sclerosis: diagnostic value and prediction of clinical course. *Journal of the Neurological Sciences*, **83**, 191–210.
Huxley, A.F. & Stämpfli, R. (1949). Evidence for saltatory conduction in peripheral myelinated nerve fibres. *Journal of Physiology (London)*, **108**, 315–339.
Jones, R.E., Heron, J.R., Foster, D.H., Snelgar, R.S. & Mason, R.J. (1983). Effects of 4-aminopyridine in patients with multiple sclerosis. *Journal of the Neurological Sciences*, **60**, 353–362.
Kaji, R., Happel, L. & Sumner, A.J. (1990). Effect of digitalis on clinical symptoms and conduction variables in patients with multiple sclerosis. *Annals of Neurology*, **28**, 582–584.
Kaji, R. & Sumner, A.J. (1989a). Ouabain reverses conduction disturbances in single demyelinated nerve fibres. *Neurology*, **39**, 1364–1368.
Kaji, R. & Sumner, A.J. (1989b). Effect of digitalis on central demyelinative conduction block *in vivo*. *Annals of Neurology*, **25**, 159–165.
Kaji, R., Suzumura, A. & Sumner, A.J. (1988). Physiological consequences of antiserum-mediated experimental demyelination in CNS. *Brain*, **111**, 675–694.
Kanchandani, R. & Howe, J.G. (1982). Lhermitte's sign in multiple sclerosis: a clinical survey and review of the literature. *Journal of Neurology, Neurosurgery and Psychiatry*, **45**, 308–312.
Kapoor, R., Felts, P.A. & Smith, K.J. (1993a). A mechanism for ectopic bursts of impulses in central demyelinated axons. *Annals of Neurology*, **34**, 271.
Kapoor, R., Smith, K.J., Felts, P.A. & Davies, M. (1993b). Internodal potassium currents can generate ectopic impulses in mammalian myelinated axons. *Brain Research*, **611**, 165–169.
Kocsis, J.D., Black, J.A. & Waxman, S.G. (1993). Pharmacological modification of axon membrane molecules and cell transplantation as approaches to the restoration of conduction in demyelinated axons. *Research Publications – Association for Research in Nervous and Mental Disease*, **71**, 265–292.
Koles, A.J. & Rasminsky, M. (1972). A computer simulation of conduction in demyelinated nerve fibres. *Journal of Physiology (London)*, **227**, 351–364.
Lhermitte, J., Bollack, J. & Nicholas, M. (1924). Les douleurs à type de décharge éléctrique consécutives à la flexion céphalique dans le sclérose en plaque. *Revue Neurologique*, **2**, 56–62.
Matthews, W.B. (1991a). Laboratory Diagnosis. In *McAlpine's Multiple Sclerosis*, ed. W.B. Matthews, pp. 189–229. Edinburgh: Churchill Livingstone.
Matthews, W.B. (1991b). Pathophysiology. In *McAlpine's Multiple Sclerosis*, ed. W.B. Matthews, pp. 231–250. Edinburgh: Churchill Livingstone.
Matthews, W.B. & Small, D.G. (1979). Serial recording of visual and

somatosensory evoked potentials in multiple sclerosis. *Journal of the Neurological Sciences*, **40**, 11–21.
McDonald, W.I. (1974). Pathophysiology of conduction in central nerve fibres. In *New Developments in Visual Evoked Potentials in the Human Brain*, ed. J.E. Desmedt, pp. 1–19. Oxford: Oxford University Press.
McDonald, W.I. (1975). Mechanisms of functional loss and recovery in spinal cord damage. In *Outcome of Severe Damage to the Central Nervous System*, ed. R. Porter & D.W. Fitzsimons, pp. 23–33. Ciba Foundation Symposium 34. Amsterdam: Elsevier.
McDonald, W.I. (1986). The pathophysiology of multiple sclerosis. In *The Diagnosis of Multiple Sclerosis*, ed. W.I. McDonald & D.H. Silberberg, pp. 112–133. London: Butterworth.
McDonald, W.I. & Halliday, A.M. (1977). Diagnosis and classification of multiple sclerosis. *British Medical Bulletin*, **33**, 4–9.
McDonald, W.I., Miller, D.H. & Barnes, D. (1992). The pathological evolution of multiple sclerosis. *Neuropathology and Applied Neurobiology*, **18**, 319–334.
McDonald, W.I. & Sears, T.A. (1969a). Effect of demyelination on conduction in the central nervous system. *Nature*, **221**, 182–183.
McDonald, W.I. & Sears, T.A. (1969b). The effects of demyelination on conduction in the central nervous system. *Transactions of the American Neurological Association*, **94**, 168–173.
McDonald, W.I. & Sears, T.A. (1970a). The effects of experimental demyelination on conduction in the central nervous system. *Brain*, **93**, 583–598.
McDonald, W.I. & Sears, T.A. (1970b). Focal experimental demyelination in the central nervous system. *Brain*, **93**, 575–582.
Milner, B.A., Regan, D. & Heron, J.R. (1974). Differential diagnosis of multiple sclerosis by visual evoked potential recording. *Brain*, **97**, 755–772.
Namerow, N.S. (1968). Circadian temperature rhythm and vision in multiple sclerosis. *Neurology*, **18**, 417–422.
Nelson, D.A. & McDowell, F. (1959). The effects of induced hyperthermia on patients with multiple sclerosis. *Journal of Neurology, Neurosurgery and Psychiatry*, **22**, 113–116.
Onofrj, M., Bazzano, S., Malatesta, G. & Gambi, D. (1990). Pathophysiology of delayed evoked potentials in multiple sclerosis. *Functional Neurology*, **5**, 301–319.
Paintal, A.S. (1966). The influence of diameter of medullated nerve fibres of cats on the rising and falling phases of the spike and its recovery. *Journal of Physiology (London)*, **184**, 791–811.
Pencek, T.L., Schauf, C.L., Low, P.A., Eisenberg, B.R. & Davis, F.A. (1980). Disruption of the perineurium in amphibian peripheral nerve: morphology and physiology. *Neurology*, **30**, 593–599.
Pender, M.P. (1987). Demyelination and neurological signs in experimental allergic encephalomyelitis. *Journal of Neuroimmunology*, **15**, 11–24.
Pender, M.P. (1988). The pathophysiology of acute experimental allergic encephalomyelitis induced by whole spinal cord in the Lewis rat. *Journal of the Neurological Sciences*, **84**, 209–222.
Pender, M.P. (1989). Recovery from acute experimental allergic encephalomyelitis in the Lewis rat: early restoration of nerve conduction and repair by Schwann cells and oligodendrocytes. *Brain*, **112**, 393–416.
Pender, M.P. & Sears, T.A. (1984). The pathophysiology of acute experimental allergic encephalomyelitis in the rabbit. *Brain*, **107**, 699–726.
Plata-Salamán, C.R. & ffrench-Mullen, J.M.H. (1992). Interleukin-1β inhibits calcium channel currents in isolated hippocampal CA1 neurones. *Society for Neuroscience Abstracts*, **18**, 433.

Prineas, J.W., Barnard, R.O., Kwon, E.E., Sharer, L.R. & Cho, E.-S. (1993). Multiple sclerosis: remyelination of nascent lesions. *Annals of Neurology*, **33**, 137–151.

Prineas, J.W. & Connell, F. (1979). Remyelination in multiple sclerosis. *Annals of Neurology*, **5**, 22–31.

Prineas, J.W., Kwon, E.E., Goldenberg, P.Z., Cho, E.-S. & Sharer, L.R. (1987a). Multiple sclerosis: destruction of proliferating oligodendrocytes and new myelin in developing lesions. *Journal of Neuropathology and Experimental Neurology*, **43**, 366.

Prineas, J.W., Kwon, E.E., Sharer, L.R. & Cho, E.-S. (1987b). Massive early remyelination in acute multiple sclerosis. *Neurology*, **37** (Supplement 1), 109.

Quandt, F.N. & Davis, F.A. (1992). Action potential refractory period in axonal demyelination: a computer simulation. *Biological Cybernetics*, **67**, 545–552.

Rasminsky, M. (1973). The effects of temperature on conduction in demyelinated single nerve fibres. *Archives of Neurology*, **28**, 287–292.

Rasminsky, M. (1978). Ectopic generation of impulses and cross-talk in spinal nerve roots of 'dystrophic' mice. *Annals of Neurology*, **3**, 351–357.

Rasminsky, M. (1980). Ephaptic transmission between single nerve fibres in the spinal nerve roots of dystrophic mice. *Journal of Physiology (London)*, **305**, 151–169.

Rasminsky, M., Kearney, R.E., Aguayo, A.J. & Bray, G.M. (1978). Conduction of nervous impulses in spinal roots and peripheral nerves of dystrophic mice. *Brain Research*, **143**, 71–85.

Rasminsky, M. & Sears, T.A. (1972). Internodal conduction in undissected demyelinated nerve fibres. *Journal of Physiology (London)*, **227**, 323–350.

Regan, D., Murray, T.J. & Silver, R. (1977). Effect of body temperature on visual evoked potential delay and visual perception in multiple sclerosis. *Journal of Neurology, Neurosurgery and Psychiatry*, **40**, 1083–1091.

Ritchie, J.M. & Rogart, R.B. (1977). Density of sodium channels in mammalian myelinated nerve fibres and nature of the axonal membrane under the myelin sheath. *Proceedings of the National Academy of Sciences, USA*, **74**, 211–215.

Rosenbluth, J. & Blakemore, W.F. (1984). Structural specializations in cat of chronically demyelinated spinal cord axons as seen in freeze-fracture replicas. *Neuroscience Letters*, **48**, 171–177.

Rosenbluth, J., Tao-Cheng, J.-H. & Blakemore, W.F. (1985). Dependence of axolemmal differentiation on contact with glial cells in chronically demyelinated lesions of cat spinal cord. *Brain Research*, **358**, 287–302.

Rushton, D. (1975). Use of the Pulfrich pendulum for detecting abnormal delay in the visual pathway in multiple sclerosis. *Brain*, **98**, 283–296.

Sawada, M., Hara, N. & Maeno, T. (1991). Analysis of a decreased Na^+ conductance by tumor necrosis factor in identified neurones of *Aplysia kurodai*. *Journal of Neuroscience Research*, **28**, 466–473.

Schauf, C.L. & Davis, F.A. (1974). Impulse conduction in multiple sclerosis: a theoretical basis for modification by temperature and pharmacological agents. *Journal of Neurology, Neurosurgery and Psychiatry*, **37**, 152–161.

Schoepfle, G.M. & Erlanger, J. (1941). The action of temperature on the excitability, spike height and configuration and the absolute refractory period observed in the responses of single medullated nerve fibres. *American Journal of Physiology*, **134**, 694–704.

Sclabassi, R.J., Namerow, N.S. & Enns, N.F. (1974). Somatosensory response to stimulus trains in patients with multiple sclerosis. *Electroencephalography and Clinical Neurophysiology*, **37**, 23–33.

Sears, T.A. & Bostock, H. (1981). Conduction failure in demyelination: is it inevitable? *Advances in Neurology*, **31**, 357–375.

Sears, T.A., Bostock, H. & Sherratt, M. (1978). The pathophysiology of demyelination and its implications for the symptomatic treatment of multiple sclerosis. *Neurology*, **28**, 21–26.

Sherratt, R.M., Bostock, H. & Sears, T.A. (1980). Effects of 4-aminopyridine on normal and demyelinated mammalian nerve fibres. *Nature*, **283**, 570–572.

Shrager, P. (1989). Sodium channels in single demyelinated mammalian axons. *Brain Research*, **483**, 149–154.

Small, D.G., Matthews, W.B. & Small, M. (1978). The cervical somatosensory evoked potential (SEP) in the diagnosis of multiple sclerosis. *Journal of the Neurological Sciences*, **35**, 211–224.

Smith, K.J., Blakemore, W.F. & McDonald, W.I. (1979). Central remyelination restores secure conduction. *Nature*, **280**, 395–396.

Smith, K.J., Blakemore, W.F. & McDonald, W.I. (1981). The restoration of conduction by central remyelination. *Brain*, **104**, 383–404.

Smith, K.J., Bostock, H. & Hall, S.M. (1982). Saltatory conduction precedes remyelination in axons demyelinated with lysophosphatidyl choline. *Journal of the Neurological Sciences*, **54**, 13–31.

Smith, K.J. (1994). Conduction properties of central demyelinated and remyelinated axons, and their relation to symptom production in demyelinating disorders. *Eye* **8**, 224–237.

Smith, K.J. & McDonald, W.I. (1980). Spontaneous and mechanically evoked activity due to central demyelinating lesion. *Nature*, **286**, 154–155.

Smith, K.J. & McDonald, W.I. (1982). Spontaneous and evoked electrical discharges from a central demyelinating lesion. *Journal of the Neurological Sciences*, **55**, 39–47.

Stefoski, D., Davis, F.A., Faut, M. & Schauf, C.L. (1987). 4-Aminopyridine improves clinical signs in multiple sclerosis. *Annals of Neurology*, **21**, 71–77.

Stefoski, D., Davis, F.A., Fitzsimmons, W.E., Luskin, S.S., Rush, J. & Parkhurst, G.W. (1991). 4-Aminopyridine in multiple sclerosis: prolonged administration. *Neurology*, **41**, 1344–1348.

Swanson, J.W. (1989). Multiple sclerosis: update in diagnosis and review of prognostic factors. *Mayo Clinic Proceedings*, **64**, 577–586.

Symington, G.R., Mackay, I.R. & Currie, T.T. (1977). Improvement in multiple sclerosis during prolonged induced hypothermia. *Neurology*, **27**, 302–303.

Szucs, A., Stefano, G.B., Hughes, T.K. & Rozsa, K.S. (1992). Modulation of voltage-activated ion currents on identified neurones of *Helix pomatia* L. by interleukin-1. *Cellular and Molecular Neurobiology*, **12**, 429–438.

Targ, E.F. & Kocsis, J.D. (1985). 4-Aminopyridine leads to restoration of conduction in demyelinated rat sciatic nerve. *Brain Research*, **328**, 358–361.

Targ, E.F. & Kocsis, J.D. (1986). Action potential characteristics of demyelinated rat sciatic nerve following application of 4-aminopyridine. *Brain Research*, **363**, 1–9.

Tasaki, I. (1953). *Nervous Transmission.* Springfield, IL: Charles C. Thomas.

Titcombe, A.F. & Willison, R.G. (1961). Flicker fusion in multiple sclerosis. *Journal of Neurology, Neurosurgery and Psychiatry*, **24**, 260–265.

Uhthoff, W. (1890). Untersuchungen über die bei der multiplen Herdsklerose vorkommenden Augenstörungen. *Archiv für Psychiatrie und Nervenkrankheiten*, **21**, 55–116.

Ulrich, J. & Groebke-Lorenz, W. (1983). The optic nerve in multiple sclerosis: a morphological study with retrospective clinico-pathological correlations. *Neuro-ophthalmology*, **3**, 149–159.

van Diemen, H.A., Polman, C.H., van Dongen, T.M., van Loenen, A.C., Nauta, J.J., van Walbeek, H.K. & Koetsier, J.C. (1992a). The effect of 4-aminopyridine on clinical signs in multiple sclerosis: a randomized, placebo-controlled, double-blind, cross-over study. *Annals of Neurology*, **32**, 123–130.

van Diemen, H.A., van Dongen, M.M., Dammers, J.W. & Polman, C.H. (1992b). Increased visual impairment after exercise (Uhthoff's phenomenon) in multiple sclerosis: therapeutic possibilities. *European Neurology*, **32**, 231–234.

van Diemen, H.A., Polman, C.H., Koetsier, J.C., van Loenen, A.C. & Nauta, J.J. (1993). 4-Aminopyridine in patients with multiple sclerosis: dosage and serum level related to efficacy and safety. *Clinical Neuropharmacology*, **16**, 195–204.

Watson, C.W. (1959). Effect of lowering of body temperature on the symptoms and signs of multiple sclerosis. *New England Journal of Medicine*, **261**, 1253–1259.

Waxman, S.G. (1977). Conduction in myelinated, unmyelinated, and demyelinated fibres. *Archives of Neurology*, **34**, 585–589.

Waxman, S.G. (1978). Prerequisites for conduction in demyelinated fibres. *Neurology*, **28**, 27–33.

Waxman, S.G. (1981). Clinicopathological correlations in multiple sclerosis and related diseases. *Advances in Neurology*, **31**, 169–182.

Waxman, S.G. (1988). Clinical course and electrophysiology of multiple sclerosis. *Advances in Neurology* **47**, 157–184.

Waxman, S.G. (1993). Molecular and cellular organization of the central nervous system: implications for new therapeutics. *Research Publications – Association for Research in Nervous and Mental Disease*, **71**, 1–21.

Waxman, S.G., Black, J.A., Kocsis, J.D. & Ritchie, J.M. (1989). Low density of sodium channels supports action potential conduction in axons of neonatal rat optic nerve. *Proceedings of the National Academy of Sciences, USA*, **86**, 1406–1410.

Waxman, S.G. & Brill, M.H. (1978). Conduction through demyelinated plaques in multiple sclerosis: computer simulations of facilitation by short internodes. *Journal of Neurology, Neurosurgery and Psychiatry*, **41**, 408–416.

Waxman, S.G. & Foster, R.E. (1980). Ionic channel distribution and heterogeneity of the axon membrane in myelinated fibres. *Brain Research*, **203**, 205–234.

Waxman, S.G. & Geschwind, N. (1983). Major morbidity related to hyperthermia in multiple sclerosis. *Annals of Neurology*, **13**, 348.

Waxman, S.G. & Ritchie, J.M. (1993). Molecular dissection of the myelinated axon. *Annals of Neurology*, **33**, 121–136.

Wisniewski, H.M., Oppenheimer, D. & McDonald, W.I. (1976). Relation between myelination and function in MS and EAE. *Journal of Neuropathology and Experimental Neurology*, **35**, 327.

Youl, B.D., Turano, G., Miller, D.H., Towell, A.D., Macmanus, D.G., Moore, S.G., Barrett, G., Kendall, B.E., Moseley, I.F., *et al.* (1991). The pathophysiology of acute optic neuritis: an association of gadolinium leakage with clinical and electrophysiological deficits. *Brain*, **114**, 2437–2450.

Young, W., Rosenbluth, J., Wojak, J.C., Sakatani, K. & Kim, H. (1989). Extracellular potassium activity and axonal conduction in spinal cord of the myelin-deficient mutant rat. *Experimental Neurology*, **106**, 41–51.

11

Mechanisms of relapse and remission in multiple sclerosis

W.I. McDONALD

Department of Clinical Neurology, Institute of Neurology, London, UK

Introduction

In the late 1960s Tom Sears and I showed that either complete or partial conduction block resulted from demyelination of central nerve fibres induced by diphtheria toxin. When conduction survived it was slow, and insecure; the refractory period of transmission (a term coined by Tom in the course of these experiments) was prolonged and the damaged fibres were unable to conduct long trains of impulses at high frequencies (McDonald & Sears, 1970).

The introduction of evoked potential methods for assessing transmission in afferent pathways in man in the 1970s then made it possible to interpret some of the clinical phenomena of demyelinating disease (and in particular of multiple sclerosis) on the basis of the earlier experimental work. Multiple sclerosis (MS) is characterized by four main pathological changes: demyelination with preservation of axons, Wallerian degeneration (scanty in the early stages, more marked later), astrocytic proliferation and varying amounts of inflammation. A variable amount of remyelination also occurs. In considering the mechanism of the conduction changes, it seemed likely that demyelination *per se* made an important contribution as it does in experimental demyelination in the peripheral nervous system (McDonald, 1963; Rasminsky & Sears, 1972). Whether this was the whole explanation remained an unexplored issue until the technical advances of the 1980s led to the application of high-resolution magnetic resonance imaging (MRI) to the study of MS. These advances allowed us to tackle the questions 'What factors contribute to relapse?' and 'What factors contribute to remission?'

MRI in multiple sclerosis

MRI shows the lesions of MS with high sensitivity. The overwhelming majority of acute lesions in the relapsing/remitting and secondary progressive forms of the disease exhibit enhancement after the injection of gadolinium-DTPA (Gd-DTPA), a substance which is normally excluded from the parenchyma of the central nervous system by the blood–brain barrier. Clinicopathological studies have shown that in MS the presence of enhancement indicates the presence of inflammation (Katz *et al.*, 1993).

The evolution of the lesion in multiple sclerosis

By combining standard serial MRI, Gd-DTPA enhancement and certain quantitative MR techniques it is possible to determine the sequence of pathological changes in the development of the new lesion (McDonald, Miller & Barnes, 1992). The earliest detectable event is an increase in permeability of the blood–brain barrier in association with inflammation (Kermode *et al.*, 1990). Oedema develops and at this stage the lesion first becomes visible on unenhanced MRI. The oedema increases to a maximum at about 4–6 weeks (Isaac *et al.*, 1988; Willoughby *et al.*, 1989). At this stage, enhancement – indicating inflammation – ceases and the oedema is absorbed to leave a smaller residual scar (Miller *et al.*, 1988; Thompson *et al.*, 1991). The study of late lesions shows that while many are characterized by the presence of gliosis and persistently demyelinated axons, many others show evidence of extensive axonal loss (Barnes *et al.*, 1991).

Demyelination

Conventional MRI does not reveal normal or abnormal myelin. However, proton magnetic resonance spectroscopy (MRS) can reveal myelin breakdown products in volumes as small as 1.3 cm^3, the size of many MS lesions. By combining MRS and Gd-DTPA enhancement and performing frequent serial examinations, Davie *et al.* (1994) have shown that myelin breakdown occurs during the inflammatory phase of the lesion: electrophysiological evidence (see below) suggests that it occurs very early in it. The evidence is compatible with the interpretation that demyelination is secondary to the inflammation, but the point has not been unequivocally established.

Relationship between MRI and functional changes

In order to explore the relationship between inflammation, conduction changes and symptomatology, we undertook a serial study of patients with acute optic neuritis – a common manifestation of MS (Youl *et al.*, 1991). Patients were recruited within 2 weeks of onset of symptoms. None had shown evidence of improvement at the time of first investigation. In addition to recording the clinical features in each patient, the optic nerves were examined by MRI with Gd-DTPA enhancement and pattern-reversal visual evoked potentials (VEP) were recorded. The patients were studied again 1 month later.

All 11 affected optic nerves had MRI-visible lesions, and all enhanced at presentation; 9 of 11 had ceased to enhance 1 month later, in keeping with the observations on cerebral lesions in MS (Miller *et al.*, 1988). There was a strong correlation between a reduction in the amplitude of the VEP (principally due to conduction block), visual loss and enhancement. After enhancement ceased there was recovery of visual accuity, and there was recovery of amplitude in the VEP towards normal, indicating reversal of conduction block. In those patients in whom an evoked potential was still recordable at presentation, it was delayed, indicating that demyelination was already present. In several patients delay was observed within 48 hours of the onset of symptoms, at a time when enhancement was present, indicating (as already mentioned) that demyelination occurs very early in the evolution of the lesion. The delay in the VEP persisted, as it is well known to do, despite the recovery of vision. The crucial difference between the two stages was the presence of inflammation in the earlier, but not in the later one.

From these observations we concluded that whatever contribution demyelination made to conduction block at the earlier time, there was an additional contribution from the inflammatory process itself. The mechanism of this effect is the subject of current investigations. It seems likely that the inflammatory cytokines contribute, since there is evidence that both interleukin 2 and tumour necrosis factor alpha affect channels in excitable membranes, though direct evidence for an effect on central nerve fibres is lacking (Brinkmeier *et al.*, 1992; Kagan *et al.*, 1992).

It is thus clear that the characteristic relapse of MS is due to conduction block, partly as a direct result of demyelination, and partly from a blocking effect of the inflammatory process itself. It is equally clear that remission is associated with the resolution of inflammation. But other factors too must be involved, since the immediate effect of demyelination is conduction block, and demyelination persists after recovery, as is demonstrated in the persistence of a delayed VEP after optic neuritis in 90% of cases. Much is now known about the compensatory mechanisms which develop in the demyelinated peripheral axon. Bostock & Sears (1978) showed that conduction can be restored, at least in smaller fibres, a few days after acute demyelination induced by diphtheria toxin. A similar result was obtained following demyelination with lysolecithin (Smith, Bostock & Hall, 1982), though whereas conduction was continuous in the former lesion, it was microsaltatory in the latter.

The membrane mechanisms underlying restoration of conduction probably involve the extension of Na^+ channels from the nodal into the demyelinated internodal axon (Waxman & Ritchie, 1985; England *et al.*, 1990; Black *et al.*, 1991). Very recently it has been convincingly demonstrated that conduction can be restored in persistently demyelinated central nerve fibres (W. Levick & W. Carroll, personal communication). The mechanism is probably similar, since Moll *et al.* (1991) have demonstrated a marked increase in tritiated saxitoxin binding (indicating an increase in Na^+ channels) at post mortem in MS plaques which are rich in axons, but not in those with severe axonal loss.

What of remyelination? In experimental animals, remyelination leads to the restoration of secure conduction (Smith, Blakemore & McDonald, 1981) and it is therefore likely that such remyelination as occurs in MS contributes to recovery. There is some evidence to suggest that this is especially likely to be the case in young patients (Kriss *et al.*, 1988) and early in the course of the disease (Prineas *et al.*, 1993). However, given the limited amount of remyelination found at post-mortem in advanced cases, it probably makes relatively little contribution in the later stages of the disease.

Conclusion

To sum up, the seminal work of Sears, beginning 25 years ago, has led to a broad understanding of how relapse and remission occur in MS. Both demyelination and inflammation contribute to the functional impairment occurring in relapse. During the inflammatory phase, compensatory changes occur of which the extension of Na^+ channels into the denuded stretches of the internodal axon are likely to be important. When inflammation ceases, these compensatory changes are expressed, conduction (albeit delayed) is restored, and the symptoms resolve. The next step in completing our understanding of the pathogenesis and pathophsyiology of the MS lesion will be to elucidate the factors which lead to the failure of the compensatory processes and the development of the irrecoverable deficit so characteristic of the advanced stages of the disease.

References

Barnes, D., Munro, P.M.G., Youl, B.D., Prineas, J.W. & McDonald, W.I. (1991). The longstanding MS lesion: a quantitative MRI and electron microscopic study. *Brain*, **114**, 1271–1280.

Black, J.A., Felts, P., Smith, K.J., Kocsis, J.D. & Waxman, S.G. (1991). Distribution of sodium channels in chronically demyelinated spinal cord axons: immuno-ultrastructural localization and electrophysiological observations. *Brain Research*, **544**, 59–60.

Bostock, H. & Sears, T.A. (1978). The internodal axon membrane: electrical excitability and continuous conduction in segmental demyelination. *Journal of Physiology (London)*, **280**, 273–301.

Brinkmeier, H., Kaspar, A., Wiethölter, H. & Rüdel, R. (1992). Interleukin-2 inhibits sodium currents in human muscle cells. *Pflügers Archiv*, **420**, 621–623.

Davie, C.A., Hawkins, C.P., Barker, G.J., Brennan, A., Tofts, P.S., Miller, D.H. & McDonald, W.I. (1994). Serial proton magnetic resonance spectroscopy in acute multiple sclerosis lesions. *Brain*, **117**, 49–58.

England, J.D., Gamboni, F., Levinson, S.R. & Finger, T.E. (1990). Changed distribution of sodium channels along demyelinated axons. *Proceedings of the National Academy of Sciences, USA*, **87**, 6777–6780.

Isaac, C., Li, D.K., Genton, M., Jardine, C., Grochowski, E., Palmer, M., Katrukoff, L.F., Oger, J. & Paty, D.W. (1988). Multiple sclerosis: a serial study using MRI in relapsing patients. *Neurology*, **38**, 1511–1515.

Kagan, B.L., Baldwin, R.L., Munoz, D. & Wisniewski, B.J. (1992). Formation of ion-permeable channels by tumor necrosis factor-α. *Science*, **255**, 1427–1430.

Katz, D., Taubenberger, J.K., Cannella, B., McFarlin, D.E., Raine, C.S. & McFarland, H.F. (1993). Correlation between MRI findings and lesion development in chronic active multiple sclerosis. *Annals of Neurology*, **34**, 661–669.

Kermode, A.G., Thompson, A.J., Tofts, P., MacManus, D.G., Kendall, B.E., Kingsley, D.P.E., Moseley, I.F., Rudge, P. & McDonald, W.I. (1990). Breakdown of the blood–brain barrier precedes symptoms and other MRI signs of new lesions in multiple sclerosis: pathogenetic and clinical implications. *Brain*, **113**, 1477–1489.

Kriss, A., Francis, D.A., Cuemdet, F., Halliday, A.M., Taylor, D.S.I., Wilson, J., Keast-Butler, J., Batchelor, J.R. & McDonald, W.I. (1988). Recovery after optic neuritis in childhood. *Journal of Neurology, Neurosurgery and Psychiatry*, **51**, 1253–1258.

McDonald, W.I. (1963). The effects of experimental demyelination on conduction in peripheral nerve: a histological and electrophysiological study. II. Electrophysiological observations. *Brain*, **86**, 501–524.

McDonald, W.I., Miller, D.H. & Barnes, D. (1992). The pathological evolution of multiple sclerosis. *Neuropathology and Applied Neurobiology*, **18**, 319–334.

McDonald, W.I. & Sears, T.A. (1970). The effects of experimental demyelination on nerve conduction in the central nervous system. *Brain*, **93**, 583–598.

Miller, D.H., Rudge, P., Johnson, G., Kendall, B.E., MacManus, D.G., Moseley, I.F., Barnes, D. & McDonald, W.I. (1988). Serial gadolinium enhanced magnetic resonance imaging in multiple sclerosis. *Brain*, **111**, 927–939.

Moll, C., Mourre, C., Lazdunsky, M. & Ulrich, J. (1991). Increase of sodium channels in demyelinated lesions of multiple sclerosis. *Brain Research*, **556**, 311–316.

Prineas, J.W., Barnard, R.O., Kwon, E.E., Sharer, L.R. & Cho, E.-S. (1993). Multiple sclerosis: remyelination of nascent lesions. *Annals of Neurology*, **33**, 137–151.

Rasminsky, M. & Sears, T.A. (1972). Internodal conduction in undissected demyelinated nerve fibres. *Journal of Physiology (London)*, **227**, 323–350.

Smith, K.J., Blakemore, W.F. & McDonald, W.I. (1981). The restoration of conduction by central remyelination. *Brain*, **104**, 383–404.

Smith, K.J., Bostock, H. & Hall, S.M. (1982). Saltatory conduction precedes remyelination in axons demyelinated with lysophosphatidyl choline. *Journal of the Neurological Sciences*, **54**, 13–31.

Thompson, A.J., Kermode, A.G., Wicks, D., MacManus, D.G., Kendall, B.E., Kingsley, D.P.E. & McDonald, W.I. (1991). Major differences in the dynamics of primary and secondary progressive multiple sclerosis. *Annals of Neurology*, **29**, 53–62.

Waxman, S.G. & Ritchie, J.M. (1985). Organization of ion channels in the myelinated nerve fibre. *Science*, **228**, 1502–1507.

Willoughby, E.W., Grochowski, E., Li, D.K.B., Oger, J., Kastrukoff, L.F. & Paty, D.W. (1989). Serial magnetic resonance scanning in multiple sclerosis: a second prospective study in relapsing patients. *Annals of Neurology*, **25**, 43–49.

Youl, B.D., Turano, G., Miller, D.H., Towell, A.D., MacManus, D.G., Moore, S.G., Jones, S.J., Barrett, G., Kendall, B.E., Moseley, I.F., Tofts, P.S., Halliday, A.M. & McDonald, W.I. (1991). The pathology of acute optic neuritis: an association of gadolinium leakage with clinical and electrophysiological deficits. *Brain*, **114**, 2437–2450.

12

Glial transplantation in the treatment of myelin loss or deficiency

J. ROSENBLUTH

Department of Physiology and Neuroscience and Rusk Institute of Rehabilitation Medicine, New York University School of Medicine, New York, USA

The purpose of this chapter is to consider what, realistically, might be achieved by glial transplantation, to review what has been done thus far in the field, and to present several current unresolved issues and controversies that are the subjects of continuing investigation.

What are the aims of glial transplantation?

When myelin in the central nervous system (CNS) has been lost as a result of disease or trauma, endogenous glial cells may not remyelinate. In those cases in which demyelination persists, glial transplantation offers the potential of repairing lesions that would otherwise remain and, perhaps, of alleviating functional deficits associated with the demyelination. These deficits fall into two categories: conduction block and spontaneous activity.

Conduction block

Although it is often assumed that demyelination inevitably results in conduction block, it has been known for more than 15 years that some fibres exhibiting experimental segmental demyelination are, in fact, able to conduct (Bostock & Sears, 1978). Signals are carried by saltatory conduction along the myelinated portion of the axon, then by continuous conduction along the demyelinated segment, then by saltatory conduction again beyond the site of demyelination. All that appears to be lost is the additional time required for continuous conduction (Fig. 12.1).

Even though frank conduction block does not occur in such fibres there are, nevertheless, other potential functional deficits. The demyelinated segment may not be able to conduct trains of impulses at high frequency, and it may fatigue quickly, since continuous conduction is metabolically more expensive than saltatory conduction (see below). Still, it is clear that the absence of myelin does not preclude conduction. Indeed, some animal mutants, e.g. the myelin-deficient (md) rat, have virtually no CNS myelin

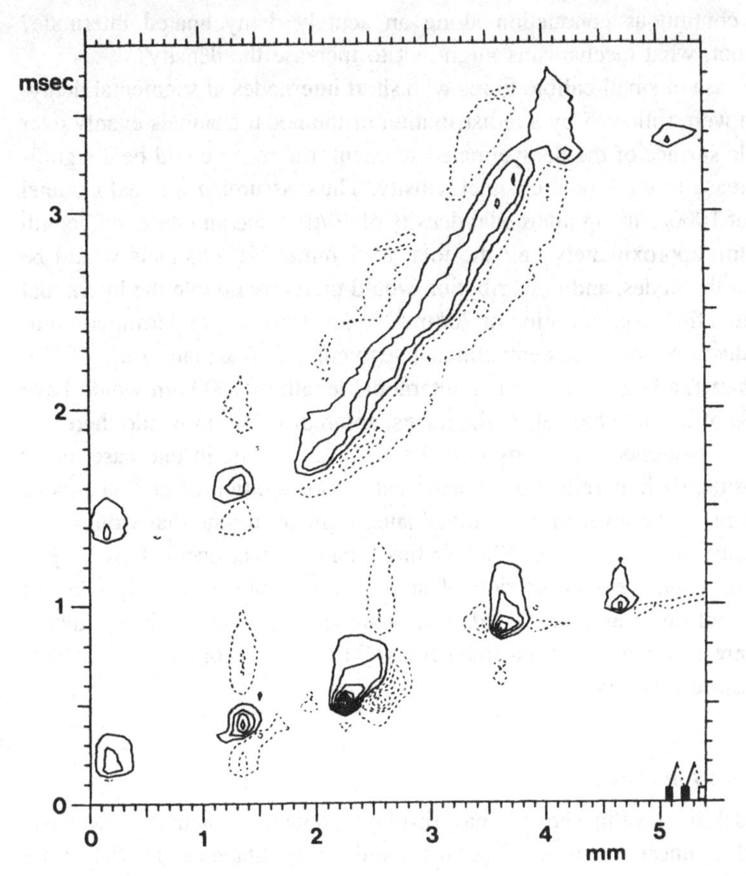

Fig. 12.1. Diagram showing inward (continuous lines) and outward (dashed lines) currents recorded along a nerve fibre treated 14 days previously with diphtheria toxin (upper record). Focal inward currents at the 0.2, 1.3 and 5 mm positions identify nodes of Ranvier. Between them, a 2 mm demyelinated segment exhibits continuous conduction. The lower record shows traces from a relatively normal fibre for comparison. From Bostock & Sears (1978).

yet are still able to carry out all basic functions without evidence of conduction block. If fibres can, therefore, conduct without myelin, what is to be gained by remyelinating them?

In fact, it is not clear what proportion of demyelinated fibres do conduct. Continuous conduction after demyelination may require a higher concentration of axolemmal Na^+ channels than is normally present underneath the myelin sheath. Although it was thought at one time that the internodal axolemma is virtually devoid of Na^+ channels, recent data show them to be present, albeit at much lower concentration than at the nodes of Ranvier (Grissmer, 1986; Chiu & Schwarz, 1987). But is the density sufficient to support continuous conduction along an acutely demyelinated internode? And, if not, what mechanisms might act to increase the density?

In the case of small-calibre fibres with short internodes, if segmental demyelination were followed by a redistribution of the nodal channels evenly over the whole surface of the demyelinated segment, the result could be a significant increase in internodal channel density. Thus, assuming a nodal channel density of $1000/\mu m^2$, an internodal density of $10/\mu m^2$ and an internodal length of 100 μm, approximately half the total axolemmal Na^+ channels would be located at the nodes, and redistribution would therefore double the internodal density to a final concentration of $20/\mu m^2$ – a level that might permit continuous conduction along the demyelinated segment (cf. Waxman *et al.*, 1989).

In contrast, a large fibre with an internodal length of 1000 μm would have only 10% of its Na^+ channels at the nodes, and redistribution would therefore increase the internodal level by only $1/\mu m^2$. Presumably, in that case, other mechanisms, such as reduction in axon calibre or addition of new channels, could increase the level in the demyelinated segment to one that would support continuous conduction. Whether this would in fact occur, how long it would take, and what proportion of such fibres would eventually conduct again are all uncertain. If, indeed, a substantial proportion of demyelinated fibres were to remain blocked, then remyelination by exogenous glial cells might restore function.

Spontaneous activity

Axons lacking myelin sheaths may display spontaneous activity, the basis for which is uncertain. In the case of the md rat, spontaneous activity is the primary neurological problem and ultimately the cause of death. Its manifestations are dramatic, consisting of generalized tonic seizures that appear when the animals are about 3 weeks old. At first, the seizures can be elicited only by vigorous stimulation – by rotating the animals, for example. Over the course of about a week, the intensity of the stimulus required to induce a seizure diminishes progressively until, ultimately, it can be triggered by the merest touch.

CNS Na^+ channel density is known to rise markedly during the first 3 postnatal weeks in the rat, the level increasing nearly 10-fold in both normal and md rat brain between days 6 and 21 (Oaklander, Pellegrino & Ritchie 1984). In normal white matter, most of the axolemmal Na^+ channels presumably become covered by myelin sheaths; only those remaining exposed at the nodes of Ranvier are active. In individual myelinated fibres, as discussed above, the proportion of total axolemmal Na^+ channels covered by myelin can range from half to more than 90%, depending upon internodal length. In the amyelinated md rat fibres, in contrast, *all* voltage-gated axolemmal

Table 12.1. *Effect of axonal activity on extracellular potassium ion concentration*

K^+ efflux per impulse over 1 cm^2 of axolemma	1×10^{-9} mmol[a]
Width of surrounding extracellular space	10–20 nm
Volume of extracellular space over 1 cm^2 of axolemma	$1–2 \times 10^{-9}$ l
Local increase in $[K^+]$[b]	0.5–1 mM per impulse

Modified from Rosenbluth (1989).
[a]From Keynes & Ritchie (1965) (rabbit vagus).
[b]Assuming no K^+ clearance.

channels are exposed, and presumably active, during conduction. The level of ion exchange associated with conduction is therefore much higher in the amyelinated axons.

The seizures that appear in the md rats at 3 weeks could thus reflect a rise in the magnitude of the currents generated by axonal activity and a decline in the threshold for excitation, associated with increasing axolemmal channel density during development, together with increasing 'compaction' of the maturing CNS, resulting in reduction of the extracellular space. These factors, plus the absence of myelin between adjacent axons, could predispose to direct ephaptic activation of quiescent amyelinated axons by neighbouring active fibres. Such indiscriminate activation of contiguous fibres could spread widely.

Alternatively, or perhaps in addition, the seizures could reflect increased K^+ efflux, corresponding to the increased Na^+ influx associated with continuous conduction, resulting in transient elevation of the extracellular K^+ concentration within the narrow extracellular space around active fibres to levels sufficient to depolarize adjacent inactive fibres to threshold. Those, in turn, would release still more K^+ as they become active.

As shown previously (Rosenbluth, 1989), a single action potential could raise extracellar $[K^+]$ by 1 mM (Table 12.1), and a train of action potentials by considerably more. Of course, mechanisms for maintaining the stability of the extracellular milieu would counteract such an increase, but these mechanisms, which are designed to deal with the relatively low levels of ionic exchange in normal white matter, may be overwhelmed by the much higher levels in myelin-deficient white matter. $[K^+]$ transients evoked in md rat dorsal columns by peripheral nerve stimulation are in fact significantly greater than those in normal controls (Young *et al.*, 1989).

In either case, axonal activity originating at a focus could spark a 'chain reaction' that spreads rapidly throughout the CNS. Thus, in md rats, modest CNS activity associated with either sensory stimulation or motor activity could trigger the generalized tonic seizures seen.

Equivalent mechanisms are thought to underlie the 'paroxysmal' phenomena seen in multiple sclerosis (MS) (Ostermann & Westerberg, 1975), though in that case spread of activity is much more limited, since the demyelination involves only discontinuous, focal regions rather than the entire white matter of the CNS.

In addition to the increased level of ionic exchange consequent upon continuous, as opposed to saltatory, conduction, an additional burden arises from evidence that some fibres lacking myelin sheaths may display a further increase in axolemmal Na^+ channel concentration beyond what is needed to support continuous conduction. Examination of freeze-fracture replicas of md rat CNS shows that the concentration of node-like membrane particles, thought to correspond to Na^+ channels, is relatively low in immature animals (~2 weeks), corresponding to the internodal particle density in normal rats, but at ~4 weeks, when seizures are prominent, particle density is much higher in some amyelinated fibres than is ever seen in normal internodal axons (Rosenbluth, 1990). As discussed above, saxitoxin-binding studies show that Na^+ channel density rises in both normal and md rat CNS during the first few postnatal weeks (Oaklander *et al.*, 1984). The freeze-fracture data suggest that the level reaches a plateau in the normals, after myelin formation is complete, but continues to rise in some of the large amyelinated fibres of the md rats.

Recent saxitoxin-binding studies of another myelin-deficient mutant, the *shiverer* mouse, have also been interpreted as showing abnormally high Na^+ channel levels in dysmyelinated fibres (Noebels, Marcon & Jalilian-Tehrani, 1991). These authors attribute the abnormality specifically to the deficiency of myelin basic protein in *shiverer*. However, since spontaneous activity, as manifested by seizures, occurs in all of the CNS myelin-deficient mutants, regardless of the biochemical defect, it seems more likely that the significant underlying factor is the myelin deficiency itself, rather than the lack of any one myelin protein.

Previous work has shown that myelin exerts a clear effect on the *distribution* of axolemmal Na^+ channels and on the counterpart particles in freeze-fracture replicas (Rosenbluth, 1981). These more recent studies suggest that myelin also exerts a regulatory effect on the overall *number* and *concentration* of axolemmal Na^+ channels, maintaining them at low density under the myelin sheath. Whether this is a direct effect of glial contact or an indirect effect, related perhaps to the electrical fields generated during activity, is not known. In either case, assuming that an increased level of axolemmal channel activity in demyelinated lesions can contribute to spontaneous or inappropriate activity, another potential goal of bringing about remyelination by glial transplantation would be to downregulate axolemmal Na^+ channels and to restrict the number active during conduction. Thus reducing the ion fluxes associated with conduction might alleviate the 'paroxysmal' phenomena and seizures associated with demyelination.

Previous studies of glial transplantation

Public attention to transplantation in the CNS has been focused on repair of neuronal deficits (Bjorklund, 1991), primarily as a potential treatment for Alzheimer's disease and Parkinsonism. Interest in the use of exogenous glial cells to repair myelin deficits, a much less visible effort, has recently begun to receive more attention (Blakemore & Franklin, 1991).

Early relevant studies were carried out *in vitro* by Wolf *et al.* (1981) who co-cultured normal optic nerve with myelin-deficient *jimpy* mouse cerebellum and reported migration of oligodendrocytes from the optic nerve into the mutant cerebellum with formation of apparently normal myelin there. Shortly thereafter, Gumpel *et al.* (1983) reported comparable success *in vivo*, transplanting fragments of normal brain into the forebrain of the dysmyelinating *shiverer* mouse mutant. There followed an extensive series of studies from Gumpel's group, as well as from several other laboratories, confirming the original observation that exogenous glia are able to form normal myelin in the CNS of congenitally myelin-deficient host animals (e.g. Lachapelle, Lapie & Gumpel, 1984; Gansmuller *et al.*, 1986; Friedman *et al.*, 1986; Kohsaka *et al.*, 1986; Duncan *et al.*, 1988; Lubetzki *et al.*, 1988; Rosenbluth, Hasegawa & Schiff, 1989; Warrington, Barbarese & Pfeiffer, 1993). These studies have been carried out using CNS fragments, partially purified oligodendrocytes and cultured glia. Transplantation has been performed on both brain and spinal cord and the host animals now include the md rat (Duncan *et al.*, 1988; Rosenbluth *et al.*, 1989), *jimpy* mouse (Lachapelle *et al.*, 1992) and *shaking* pup (Archer, Leven & Duncan 1994), in addition to the *shiverer* mouse used originally. Transplants have also been carried out in normal animals at sites of demyelinating lesions (Blakemore & Crang, 1988; Gout *et al.*, 1988). In some cases the myelin formed has been found to persist for prolonged periods.

The issue of host rejection of grafted tissue, which was a source of great concern, has not become problematic in the case of allografts and, in some cases, even when xenografts were used (Gumpel *et al.*, 1987; Archer *et al.*, 1994). It must be added, however, that all studies on the experimental animals used thus far have been short term by human standards, because of the lifespan of the animals, and it is therefore not clear whether rejection of allo- or xenografted glia would or would not be a problem for humans.

Thus, there is general agreement that glial transplantation is feasible, at least under laboratory conditions, and therefore potentially useful in the treatment of human diseases such as MS. Yet problems with this approach persist and several aspects of the subject remain controversial.

Unresolved issues in glial transplantation

How far do transplanted cells migrate?

The original studies by Gumpel's group reported widespread migration of the donor cells and foci of myelin formation at sites remote from the original site of injection into the forebrain, including cerebellum, opposite half of the forebrain and even brainstem (Lachapelle *et al.*, 1984; Gansmuller *et al.*, 1986; Baulac *et al.*, 1987). However, later studies, in which transplants were placed into the spinal cord, showed surprisingly limited spread. The difference can be accounted for partially by the stage of development of the host CNS at the injection sites. The spinal cord is further developed than the brain in neonatal rats, as used by Duncan *et al.* (1988), and in juvenile rats (Rosenbluth *et al.*, 1989) development has progressed further still.

In the spinal cord studies, cells transplanted into the dorsal columns tended

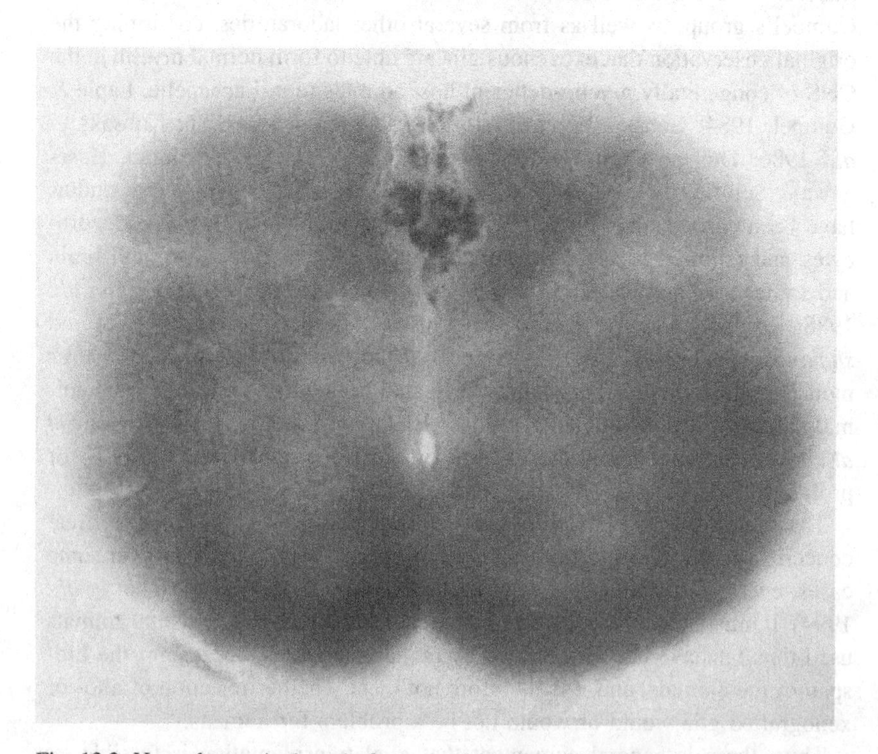

Fig. 12.2. Normal rat spinal cord after transplantation of *LacZ*+ mouse oligodendrocytes, showing reaction product confined to dorsal columns. J. Rosenbluth, R. Schiff & W.-L. Liang, unpublished data.

to remain within the dorsal columns. Although examples of myelin in lateral and ventral fibre tracts were found, these were the exception, not the rule, suggesting either that transplanted myelin-forming cells do not readily cross fibre tract boundaries or that, if they do, they are less likely to form myelin there for other reasons, perhaps related to the 'maturity' of the axons.

More recent approaches, involving transplantation of glial cells carrying the *LacZ* gene, confirm these results. Large numbers of tagged, myelin-forming cells can be demonstrated at the transplantation site in md rats, but only very few are seen outside the tract into which the cells were injected. Use of the *LacZ* marker also permits identification of transplanted cells in normal spinal cords, and here as well, the injected cells tend to form myelin primarily at the injection site (Fig. 12.2) with only occasional exceptions in other tracts.

Wider spread has been found along the *length* of fibre tracts, extending up to 8 mm in adult *shiverer* mice (Gout *et al.*, 1988) and approximately five spinal segments rostrally and caudally from the site of injection in juvenile md rats (Rosenbluth *et al.*, 1993). Use of transgenic oligodendrocytes carrying *LacZ* or *CAT* reporter genes (Friedrich & Lazzarini, 1993; Lachapelle *et al.*, 1994; Tontsch *et al.*, 1994; Rosenbluth *et al.*, 1995) provides a more sensitive indicator of their spread and shows migration of small numbers of labelled cells along even greater distances within the same tracts in the md rat spinal cord (Fig. 12.3).

Despite these successful results, it is nevertheless clear that the entire nervous system is not colonized by the transplanted cells, in the way that bone marrow might be following a graft. Indeed, the new myelin formed by the grafted glial cells was insufficient to prolong the lives of the host animals or to alleviate their tremors or seizures to any measurable degree. Regrettably, the only clear evidence, so far, of functional improvement following glial transplantation remains the demonstration of increased dorsal column conduction velocity in md rats receiving glial grafts to that tract (Utzschneider *et al.*, 1994).

Thus, focal transplantation of glia into an adult host may result in the formation of substantial amounts of myelin close to the site of transplantation but probably not elsewhere. In the case of MS, making use of modern stereotactic methods to target the transplant precisely to a lesion identified by magnetic resonance imaging (MRI), focal myelin formation could still be of value. The extent to which the transplanted glia would migrate beyond the injection site, especially in the adult brain, is uncertain, however.

Are glial grafts in the CNS immunologically 'privileged'?

As described above, glial allografts in the CNS have met with surprisingly little host reaction. These observations may not represent a general phenomenon, however. The original studies on *shiverer* mice were carried out on

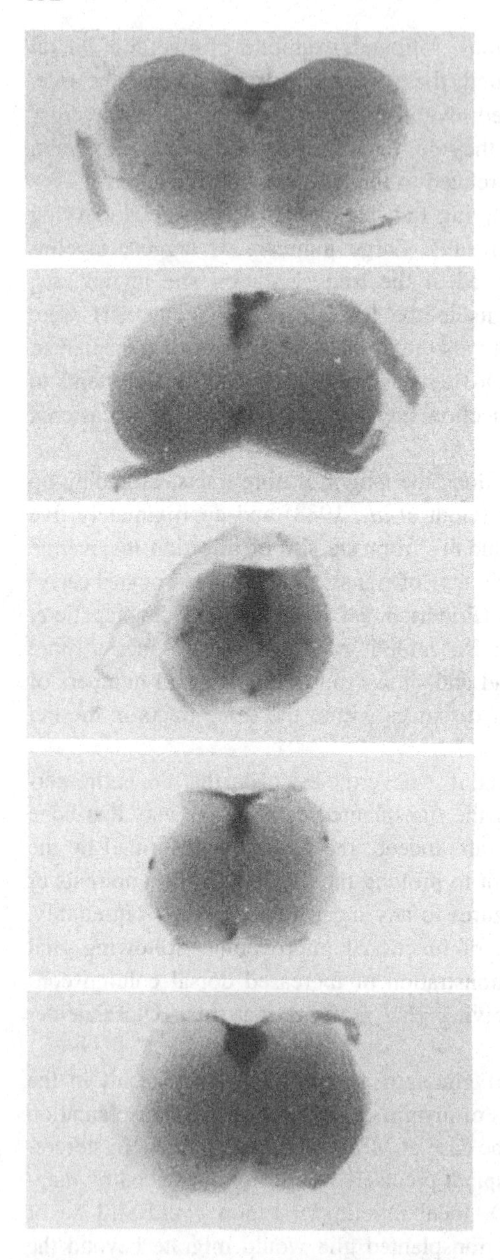

Fig. 12.3. An md rat spinal cord following transplantation of *LacZ+* mouse oligodendrocytes into thoracic dorsal columns. Reaction product is visible up into cervical levels, confined primarily to dorsal columns, but with occasional labelled cells in ventral or lateral fibre tracts. J. Rosenbluth, R. Schiff & W.-L. Liang, unpublished data.

neonatal hosts (Gumpel *et al*., 1983; Lachapelle *et al*., 1984; Gansmuller *et al*., 1986; Duncan *et al*., 1988), and the possibility exists that transplantation at that stage could have created immunological tolerance, accounting for the ability of these grafts to survive without apparent host response. However, other studies have shown prolonged survival even in older hosts (e.g. Kohsaka *et al*., 1986; Gout *et al*., 1988), suggesting either that intracerebral grafts could tolerize older hosts as well, or that the grafts are indeed in a 'privileged' location.

Recent xenograft studies raise questions about both of these interpretations. When mouse glia were transplanted into either md rat spinal cord (Rosenbluth *et al*., 1993) or normal rat spinal cord following formation of a demyelinated lesion (Crang & Blakemore, 1991), the grafts failed consistently. This could reflect a mismatch in axo-glial interaction. However, in both cases, when the host animals were immunosuppressed with cyclosporine, the grafts succeeded in forming an abundance of myelin in a high percentage of cases (Fig. 12.4).

The simplest conclusion to be drawn is that xenografts do incite a host reaction and that the CNS is not invisible to the host immune system. Whether the host responds or not may depend not only on the extent of the difference between the host and donor immunologically, but also on the technique used for transplantation and location of the transplant. The reported success of some xenografts may reflect transplantation of the donor material in such a way as to minimize its exposure to the host circulation and lymphatic system.

Thus, the available data show that the CNS is not entirely 'privileged' and that xenografts can elicit rejection. Whether long-term allografts would ultimately elicit a host immune response as well is uncertain. Even if they did, however, the xenograft studies suggest that immunosuppression is effective in retarding rejection in the CNS, as it is with transplanted peripheral organs.

In the case of MS, if a partial CNS 'privilege' permitted long-term survival of glial xenografts, this would open up the possibility of using donor sources that might not be susceptible to the same autoimmune response that caused the original lesions. Allografts, in contrast, lacking this protection, might ultimately fail. The same rationale has been used in transplantation of baboon liver, which is not susceptible to the hepatitis B virus, to human victims of that disease (Starzl *et al*., 1993). Rejection, which could be a problem in the case of peripheral organ xenografts, may be more readily surmountable in the case of CNS xenografts, however, because of the partial 'privilege' in that location.

What stage cell in the oligodendrocyte lineage is most suitable for transplantation?

The original studies showing myelin formation by transplanted glia made use of neonatal donor tissue, and it was presumed that transplanted oligodendro-

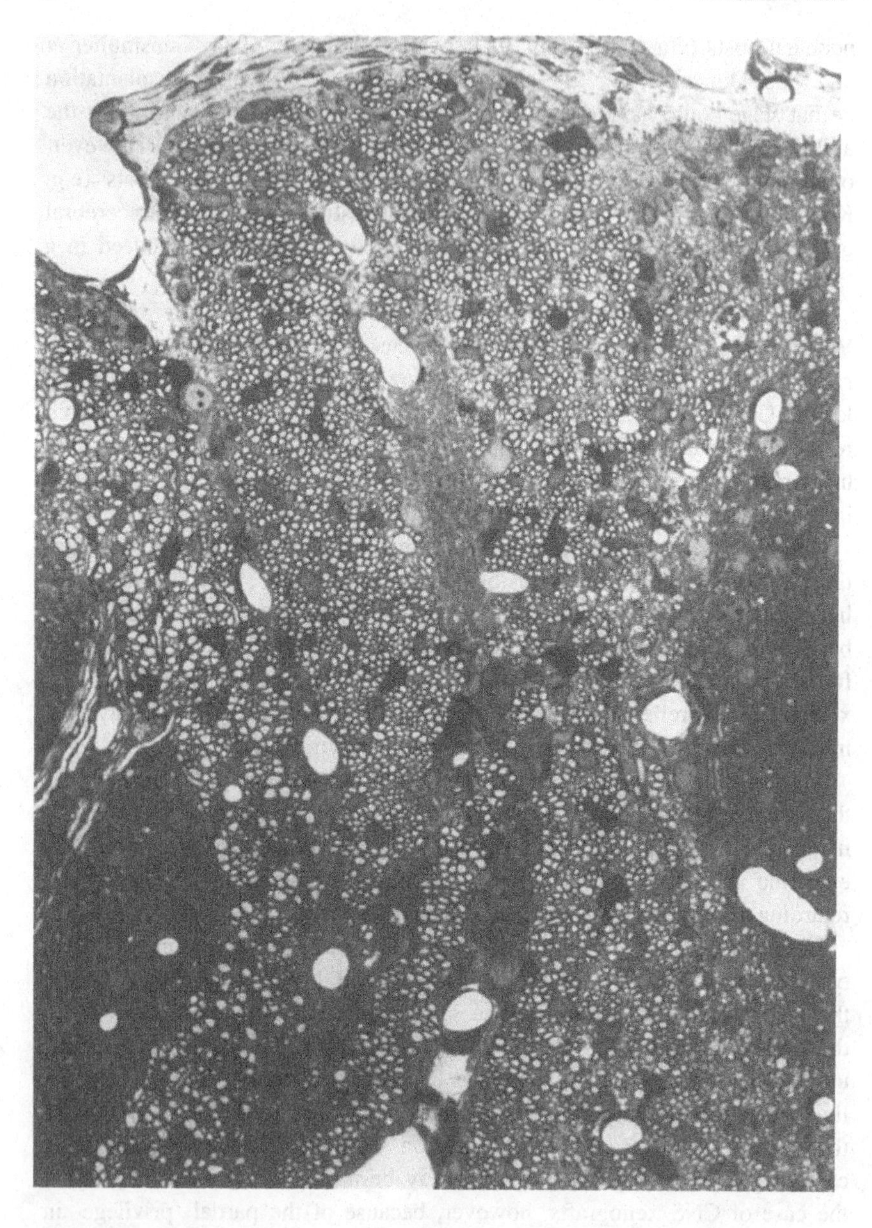

Fig. 12.4. Extensive myelination of dorsal columns in immunosuppressed md rat after transplantation of fetal mouse glia. From Rosenbluth *et al.* (1993).

cytes were the cells responsible for the myelin formed (Gumpel *et al.*, 1983; Duncan *et al.*, 1988). However, studies using fetal donors (Friedman *et al.*, 1986; Rosenbluth *et al.*, 1989) showed that myelin formation was successful in that case as well. It was shown subsequently that the rat fetal donor cells contained an abundance of oligodendrocyte precursors, identified by the A2B5 antibody, while mature oligodendrocytes, identified by the anti-GalC antibody, were entirely absent (Rosenbluth *et al.*, 1990)! Since there were no mature oligodendrocytes in the donor tissue at the time of transplantation, it must be concluded that transplanted *precursor* cells from the fetal CNS were capable of surviving and maturing in the environment of a much older host animal and of forming myelin there by ~11 days.

In addition, when a direct comparison was made between glia derived from fetal donors and postnatal donors, it was apparent that success in myelin formation declined progressively with donor age, concurrent with the decline in the proportion of oligodendrocyte precursor cells present. Glia derived from adult donors contain large numbers of mature oligodendrocytes, but few precursors. It was proposed, accordingly, that precursors, rather than fully differentiated oligodendrocytes, could be the cells primarily responsible for myelin formation after transplantation (Rosenbluth *et al.*, 1990).

These studies raise the question of whether mature oligodendrocytes are terminally differentiated or whether they retain the capacity either to generate new myelin-forming cells or to form new myelin sheaths themselves. Unfortunately, the available data do not provide clear-cut answers. Aranella & Herndon (1984) reported that following a lysolecithin lesion in the adult mouse CNS, 'dense' oligodendrocytes take up tritiated thymidine and thus are presumably able to synthesize DNA and undergo cell division. Unfortunately no labelling of mature markers, e.g. galactocerebroside or myelin basic protein, was carried out, and the identification of the labelled 'dense' cells as 'mature' oligodendrocytes is uncertain. Similarly, Ludwin & Bakker (1988) reported that after a traumatic injury to the CNS, it is possible to find glia, apparently attached to myelin sheaths, that have taken up tritiated thymidine. Unfortunately, the curvature of the slender processes extending from these cells to the adjacent myelin sheaths is too great to permit unambiguous resolution of the membrane continuity necessary to establish that they are indeed connected. Again the developmental stage of the cells that took up tritiated thymidine is uncertain.

Studies of spontaneous remyelination in adult animals after cuprizone intoxication (Blakemore, 1974), lysolecithin injection (Blakemore *et al.*, 1977) or viral infections (Godfraind *et al.*, 1989) show that the capacity to remyelinate is present under some conditions. But what is the source of the remyelinating cells? In contrast to the earlier reports of tritiated thymidine uptake by mature oligodendrocytes, Godfraind *et al.* (1989) found no evidence of proliferation of differentiated oligodendrocytes 'as a substantial component of the response to injury'. Rather, the newly formed oligodendrocytes appeared to arise from precursors. Similarly, a study of normal development by Reynolds & Wilkin (1991) concluded that oligodendrocyte precursors take up tritiated thymidine but mature oligdendrocytes do not.

Even if mature oligodendrocytes were unable to multiply, they might, nevertheless, be able to form myelin. Indeed, a recent paper by Duncan *et al.* (1992) reported success in myelin formation in 6 of 8 hosts using mixed glia derived from adult donors and in 3 of 7 hosts using purified adult glia that had been sorted using the O1 (antigalactocerebroside) label. Whether the myelin seen was formed by oligodendrocytes that were mature at the time of transplantation is uncertain, however, in view of the possibility of some precursor cells among the mixed glia, even after cell sorting, and the even larger number of precursors found after culture of the sorted cells. Although the proportion of precursor cells may be small, even a 2% contaminant population of dividing cells, would require only six divisions to outnumber the original 98% non-dividing cells. Thus, precursor cells could still account for the modest myelin formation illustrated in that study.

The interesting possibility that adult glia cultured in the presence of fetal serum and FGF may de-differentiate and regain their ability to form myelin, was suggested by earlier *in vitro* studies (Wood & Bunge, 1991). In addition, there is the intriguing further possibility that antibody labelling of adult glia, in preparation for cell sorting, could activate signalling pathways leading to de-differentiation as well.

The conclusion that transplanted precursors are far more suitable than tranplanted adult glia in forming myelin has been supported by a quantitative study by Warrington *et al.* (1993) who injected measured numbers of highly purified neonatal glia at specific developmental stages. They showed a ~5-fold advantage of early precursors over either late precursors or mature glia with respect to *number* of myelin patches formed by the donor glia within the host brain. Moreover, the *size* of the patches formed by the early precursors was larger as well, suggesting an even larger disparity.

In our experience with transplantation of glia derived from adult animals, and not treated with growth factors, we have found myelin formation only rarely, including one case in which the donor cells had been treated with O1 followed by fluorescein isothiocyanate (FITC)-labelled anti-IgM. Moreover, the myelin formed by adult transplants was small in amount and showed no significant spread rostrally or caudally beyond the spinal segment in which it was found (Fig. 12.5).

Thus, the ultimate failure of remyelination in MS, despite the presence of large numbers of mature oligodendrocytes immediately around MS lesions, could reflect the inability, or 'limited ability' (Prineas, 1985), of mature cells to reform myelin and the ultimate depletion of the 'immature oligodendrocytes' associated with fresh MS plaques (Prineas *et al.*, 1989).

Direct evidence is not yet available on the stage of development of the

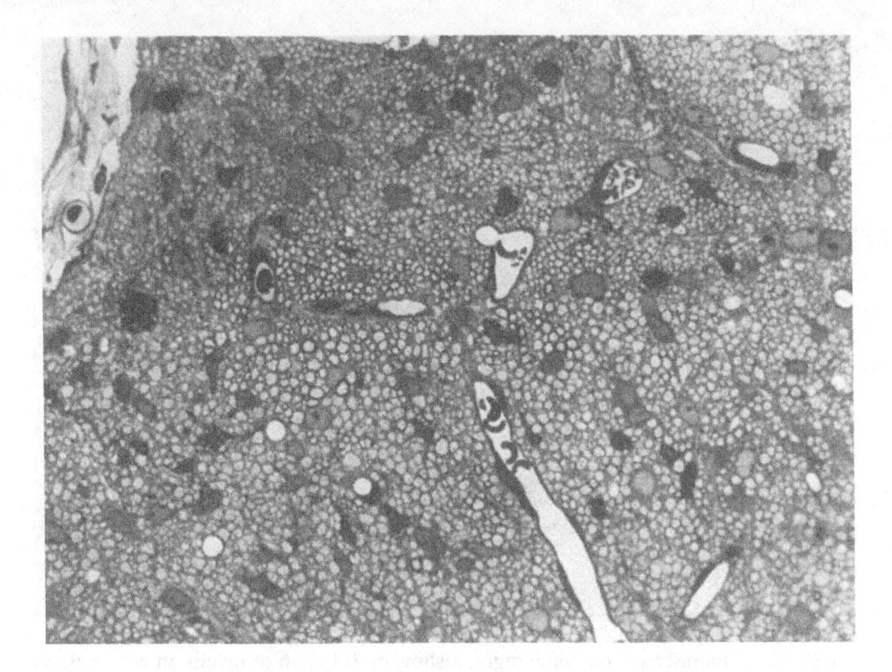

Fig. 12.5. Small cluster of myelinated fibres in md rat dorsal columns formed after transplantation of immunolabelled glia from adult rat.

glial cells that initiate myelin formation. It remains possible that precursor cells may bear adhesion molecules or other surface recognition components essential for the initial interaction with axons during myelin formation, but that these components are no longer expressed in adult oligodendrocytes (cf. Prineas *et al.*, 1989).

In summary, it is still not clear whether differentiated oligodendrocytes *in vivo* are able to recapitulate their previous behaviour and re-initiate the process of myelin formation, whether they can be induced to de-differentiate to a more primitive form that *is* capable of initiating myelin formation, or whether they are terminally differentiated and are incapable of either dividing or reforming myelin.

There is a clear precedent for the last possibility. In the case of skeletal muscle, excision, homogenization and reimplantation of a muscle is followed by its regeneration. This phenomenon provoked a controversy, similar to that currently surrounding remyelination, as to whether regenerated muscle is formed by mature myocytes, de-differentiated myocytes or precursor cells. In that case, it was ultimately shown that regeneration is accomplished not by the differentiated myocytes present in great abundance, but rather by a small, inconspicuous population of 'satellite' precursor cells that multiply and then differentiate to form the new muscle (Snow, 1977a). These cells

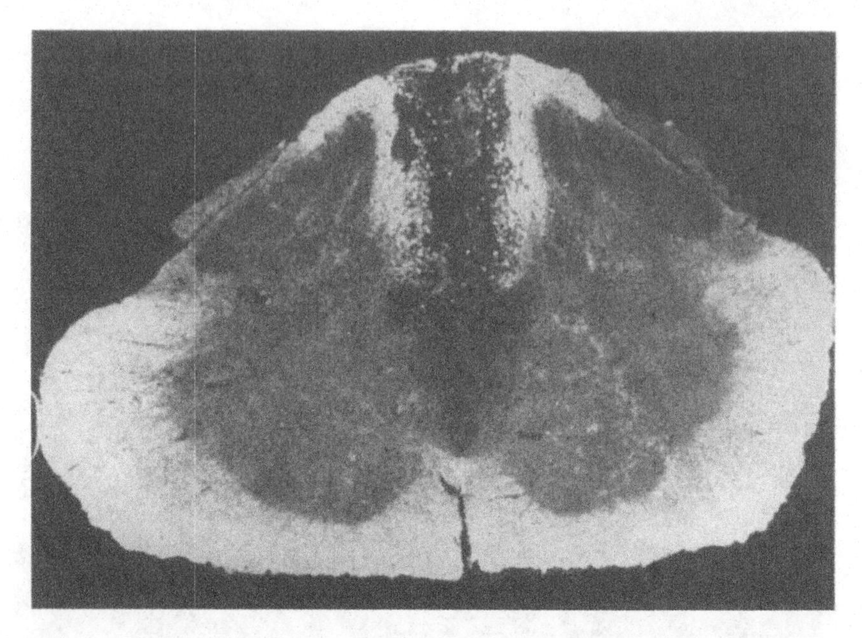

Fig. 12.6. Fluorescence photomicrograph showing O1 stain of myelin in a 4 week rat spinal cord 1 week after transplantation of anti-galactocerebroside hybridoma into dorsal columns. Cortical white matter is strongly fluorescent, including the dorsal columns, save for a midline defect representing the location of the tumour cells. From Rosenbluth *et al.* (1992).

represent only ~4% of the total number of muscle cell nuclei (Schmalbruch & Hellhammer, 1976) and diminish further in number with age (Schmalbruch & Hellhammer, 1976; Snow, 1977b).

Undifferentiated cells in the subventricular zone of the adult mammalian brain constitute a reservoir of progenitor or stem cells capable of developing into both neurones and glia (Lois & Alvarex-Buylla, 1993), and may represent a source of myelin-forming oligodendrocytes. This finding opens up the possibility that stereotactic methods could be used to obtain such cells from patients, that the cells obtained could be induced to multiply and differentiate into oligodendrocyte precursors *in vitro* and that the precursors could then be reintroduced into the same patients at sites of CNS demyelination. Such an approach would have the great virtue of providing a source of cells for autografts that would avoid all the potential problems associated with allo- or xenograft rejection.

Can host trophic or inhibitory factors affect success of myelin formation after glial transplantation?

Although there is evidence for some remyelinatiion of MS lesions by endogenous glia, the process is clearly ineffectual in many, if not most, cases.

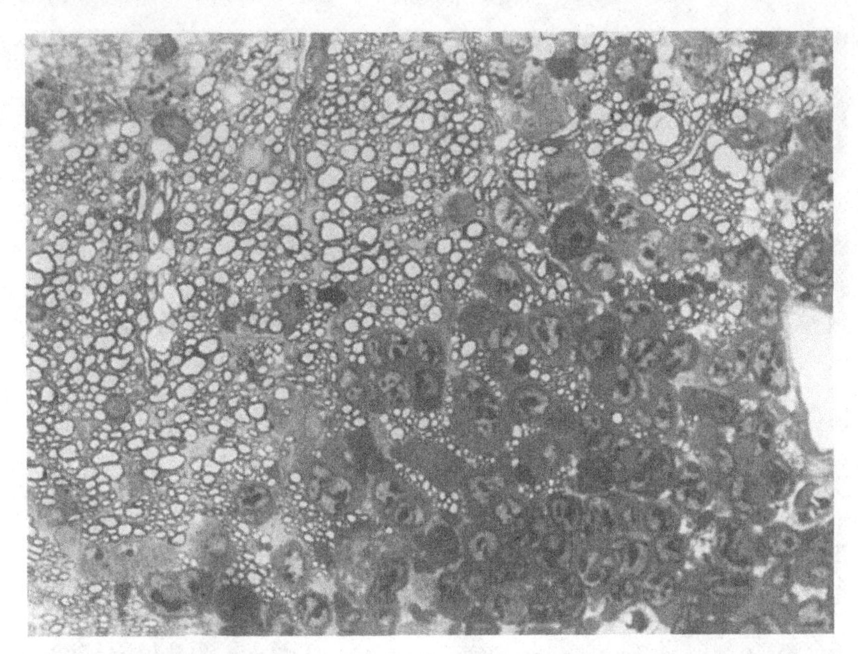

Fig. 12.7. Photomicrograph showing myelinated nerve fibres among tumour cells in dorsal columns (same preparation as in Fig. 12.6). There is little sign of myelin damage or demyelination. From Rosenbluth *et al.* (1992).

This could reflect the inability, or limited ability, of adult oligodendrocytes to remyelinate, as discussed above, or it could reflect the presence of inhibitory factors that prevent myelin formation.

It is well established that MS lesions are associated with immunoglobulin-secreting cells, and that immunoglobulins are present in the spinal fluid of MS patients (Prineas, 1985). Previous tissue culture studies have shown that culture medium containing either experimental allergic encephalomyelitis (EAE) serum (Raine *et al.*, 1978) or a monoclonal anti-galactocerebroside (Ranscht, Wood & Bunge, 1987) do indeed inhibit the formation of myelin *in vitro*.

This premise has recently been tested *in vivo* as well (Rosenbluth *et al.*, 1992, 1994). Hybridoma cells that secrete a monoclonal IgG3 anti-galactocerebroside were implanted directly into the spinal cords of either 9 day or 16 day immunosuppressed rats, and the animals examined after 1 week (Figs. 12.6–12.11). The older group, in which myelin had already formed before implantation of the hybridoma, showed numerous normal-looking myelinated fibres interspersed among the hybridoma cells and often in direct contact with them (Figs. 12.6, 12.7).

Results in the younger group, in which many dorsal column axons were still unmyelinated at the time of hybridoma implantation, were quite different, however. The dorsal columns showed a virtual absence of galactocerebroside

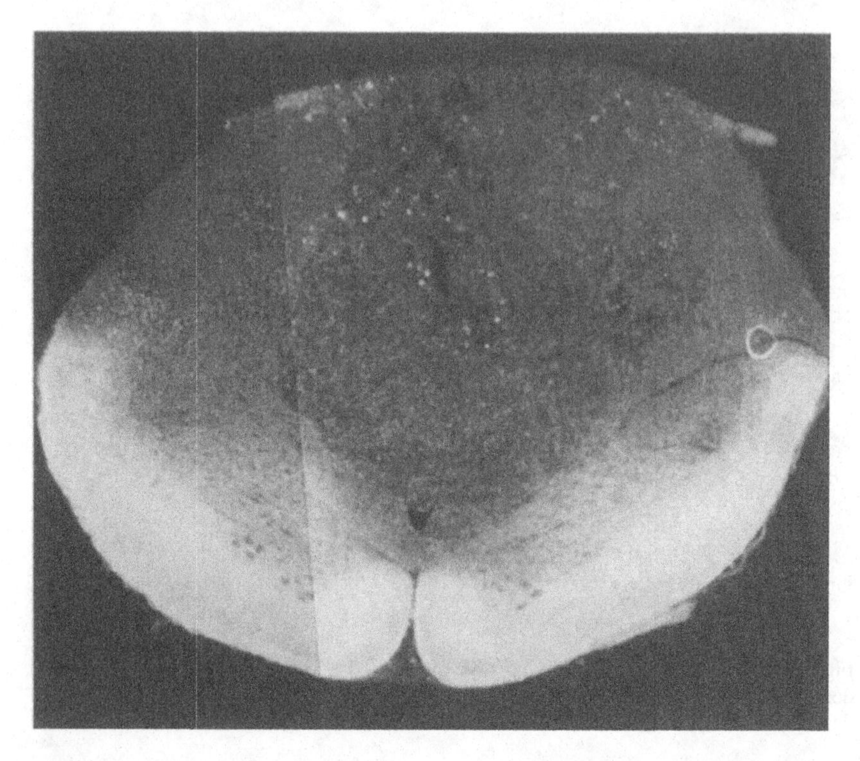

Fig. 12.8. Fluorescence photomicrograph of O1 (anti-galactocerebroside)-stained spinal cord from 16 day immunosuppressed rat. Anti-galactocerebroside hybridoma had been implanted into dorsal columns 1 week earlier. The O1-fluorescence is strong ventrally but grossly diminished dorsally. From Rosenbluth *et al.* (1992).

by immunocytochemical staining after 1 week (Fig. 12.8) in contrast to controls (Fig. 12.9), and large clusters of dorsal column axons remained persistently unmyelinated (Figs. 12.10, 12.11).

In short, the presence of the hybridoma and its secretions had no obvious effect on already-formed myelin, but prevented myelin formation around axons that were not yet myelinated at the time of implantation. This could represent inhibition of myelinogenesis by oligodendrocyte-lineage cells or a toxic effect on those cells causing their death or preventing their maturation. Implantation of control hybridomas secreting an irrelevant immunoglobulin had no such effects. Thus, in an MS patient, the presence of relevant Ig-producing cells could, in the same way, prevent the reformation of myelin by the host's own oligodendrocyte-lineage cells or, indeed, by transplanted cells.

The converse possibility, that host trophic factors are important in myelin formation by either transplanted or endogenous oligodendrocytes, has not been tested. Clearly, transplantation into neonates, where growth factors are presumably abundant, may signify little about success in adult donors, where

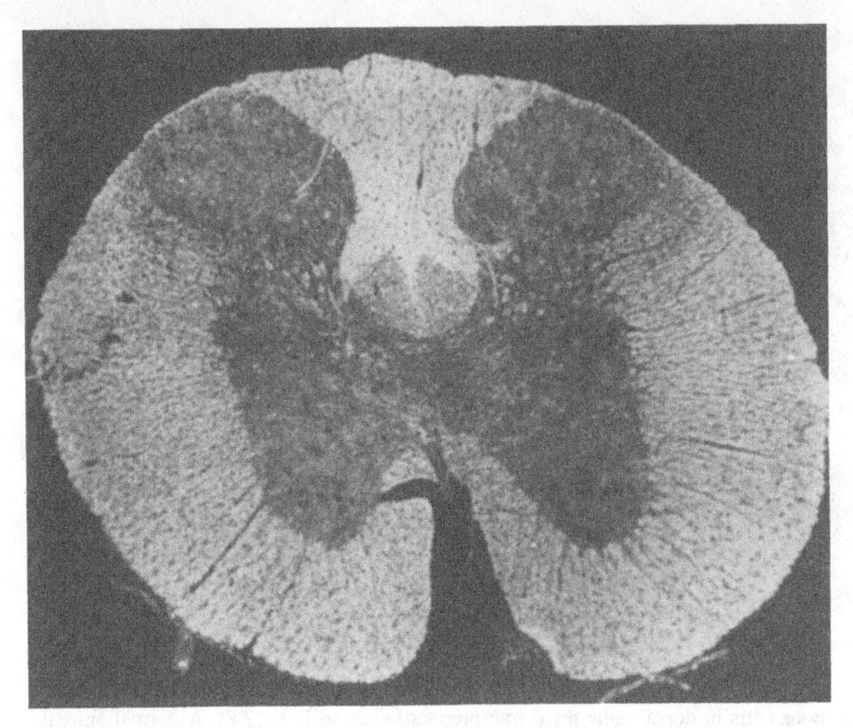

Fig. 12.9. Fluorescence photomicrograph of normal 16 day rat spinal cord immunostained with O1 (anti-galactocerebroside) antibody. Cortical white matter in this control spinal cord is strongly fluorescent ventrally and dorsally. From Rosenbluth *et al.* (1992).

growth factors may be absent from the milieu and where axons may no longer signal myelin formation. Under what conditions adult host tissue can be induced to re-express the appropriate growth factors and myelin formation signals, or whether these are in fact necessary, is unclear.

It may be significant in this regard that transplantation of fetal mouse glia into adult *shiverer* mice has been notably unsuccessful in our hands, despite the chronic dysmyelination in these animals. Yet, lysolecithin lesions in *shiverer* mouse CNS (Gout *et al.*, 1988) or ethidium bromide lesions in normal adult rat CNS (Blakemore & Crang, 1988) have been remyelinated by transplanted glia. Perhaps an acute inflammatory reaction following demyelination, or the presence of myelin breakdown products, results in the reappearance of growth factors that are otherwise absent from the adult CNS and thereby promotes myelin formation. Perhaps complete removal of axonal sheaths, whether formed by abnormal oligodendrocytes or reactive astrocytes, is essential to permit interaction with competent myelin-forming glia or to induce axons to signal myelin formation. Thus, it may be necessary to 'prime' chronic, 'burned-out' lesions in adult hosts by similar means, though less

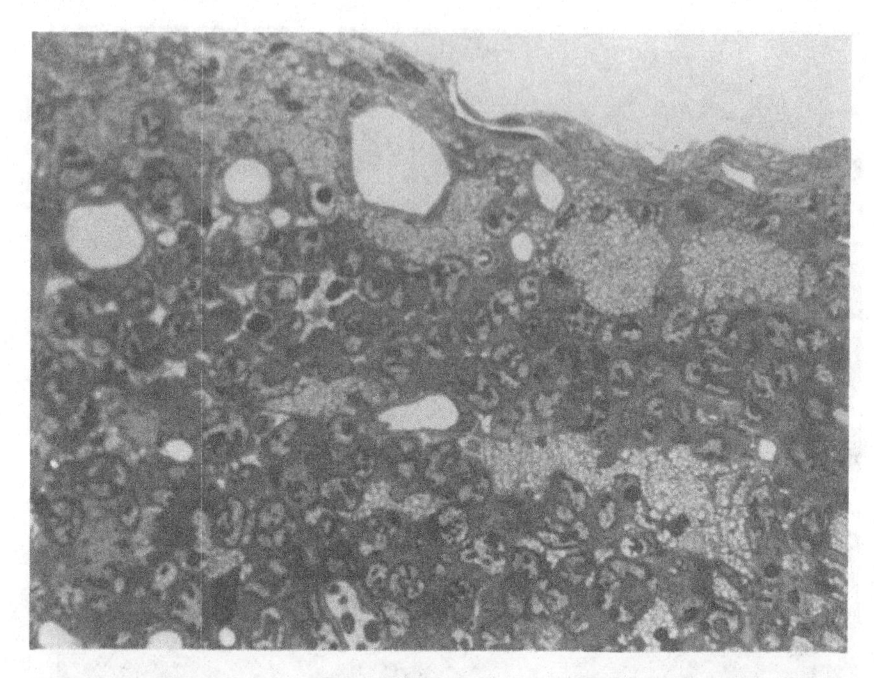

Fig. 12.10. Photomicrograph showing fascicles of amyelinated nerve fibres among tumour cells in dorsal columns (same preparation as in Fig. 12.8). A normal animal would show extensive myelination of the dorsal columns at this time (cf. Fig. 12.9). From Rosenbluth *et al.* (1992)

violent than lysolecithin or ethidium bromide injection, before glial transplantation will succeed.

Future prospects

The data reviewed above suggest: (1) that transplantation of oligodendrocyte-lineage cells is a feasible approach to overcoming human myelin loss or deficiency and potentially repairing associated functional defects, (2) that focal lesions are the ones most likely to be repaired successfully, as opposed to widespread deficiency, and (3) that fetal oligodendroglial precursors are far more likely to be effective in accomplishing this goal than unmanipulated mature oligodendrocytes derived from adult donors.

Several cautions must be kept in mind, however.

1. Long-term immunological problems may develop. Rejection of CNS allografts may occur over a time course longer than has been explored thus far in experimental animals. As pointed out above, however, the success of cyclosporine in overcoming xenograft rejection suggests that this problem may be manageable. Use of such methods as enclosing grafted cells within semipermeable

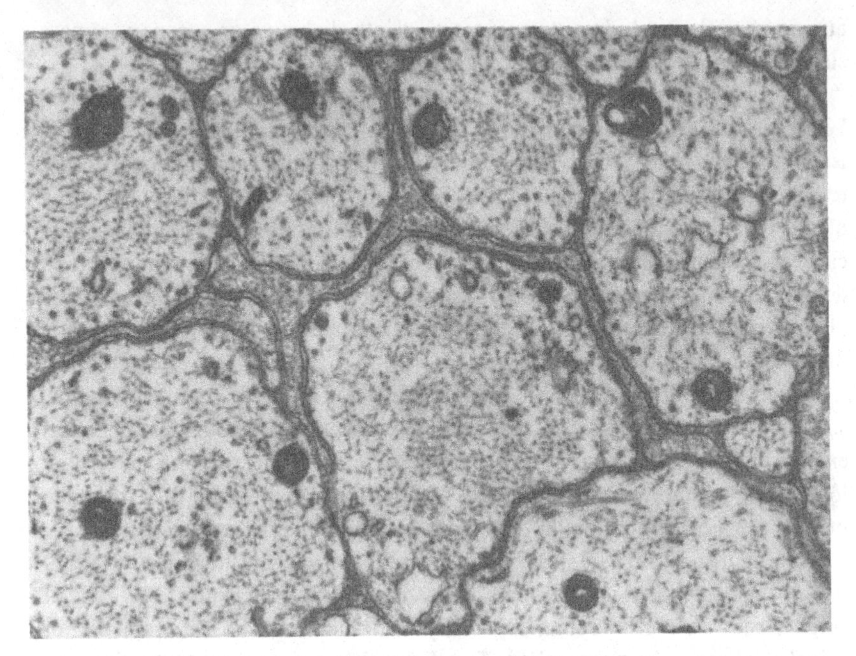

Fig. 12.11. Electron micrograph showing amyelinated fibres illustrated in Fig. 12.10. Most of the axons are bare. A few slender cell processes, probably derived from astrocytes, are present. From Rosenbluth *et al.* (1992)

chambers to isolate them from host T cells is clearly irrelevant in this case, since the goal is to have the grafted cells form myelin around individual host axons rather than secrete a soluble product into the circulation.

A further risk is that allografted cells may sensitize the host in such a way as to induce an immune response directed not only against the grafted glia but against the host's own oligodendrocytes and myelin as well. Thus, there is a risk of worsening the disease rather than improving it. Autografts of stem cells could circumvent these potential immunological problems.

2. Transplanted cells might survive but form no myelin in the absence of appropriate environmental growth factors and/or stimuli from the axons.

3. Transplanted cells might survive and form myelin around host axons, but the new myelin may then be subject to the same disease processes that damaged the host myelin in the first place. T cells or immunoglobulins directed against myelin basic protein or proteolipid protein, for example, would undoubtedly attack graft and host myelin equally, unless the isoforms of these proteins in the graft were significantly different. In this regard, the use of xenografts could be advantageous in comparison with allografts, as described above.

If the disease attacked myelin randomly, then repair of a lesion in a crucial region might be worthwhile even if there were some risk that the repair might be damaged at some time in the future. The issue of temporary benefit arises also in considering kidney transplantation into diabetics or coronary bypass surgery in hypercholesterolaemic patients.

4. Remyelination by glial grafts may not improve function. The correlation between MS lesions, demonstrated radiographically, and neurological signs is notoriously poor. The possibility exists that significant functional losses result from minor damage that would never be detected by MRI. Loosening of the paranodal junction adjacent to nodes of Ranvier, for example, could cause sufficient shunt of nodal current to block conduction even without frank demyelination. Conversely, fully demyelinated fibres within identifiable plaques may be able to conduct adequately, if imperfectly, and remyelination may confer little advantage.

In addition, although MS is characterized by relative sparing of axons, some axonal damage nevertheless occurs, perhaps as a 'bystander' phenomenon. The degree of axonal loss can be considerable. Clearly, once axons are lost, transplantation of glia is pointless, and efforts would have to be directed instead towards finding methods for promoting axonal regeneration.

5. In addition to these specific considerations, relevant to glial transplantation, there are general problems that apply to all transplants. The risk of surgery may be considerable, especially if crucial areas of the brainstem or spinal cord are to be approached. The donor tissue may carry other diseases, as yet unsuspected. The cost of transplantation surgery to society is large, and criteria would need to be formulated to distribute this service equitably. The source of donor tissue may be problematic; in particular, the use of fetal tissue for this purpose remains a significant ethical and political issue.

In conclusion, glial transplantation is a potentially useful therapeutic approach to the repair of myelin loss, especially in treating focal lesions of MS. However, significant constraints exist that may limit the usefulness of this approach. Methods to improve remyelination by endogenous glia, through the use of cytokines or transfection strategies, for example, would be preferable but are not yet in sight. Transplantation of exogenous glia into human subjects, in contrast, may soon be achievable. To what extent any of these approaches would achieve functional improvement in MS or other demyelinating diseases remains to be seen.

Acknowledgement

This work was supported by grants from the National Institutes of Health and National Multiple Sclerosis Society (US).

References

Aranella, L.S. & Herndon, R.M. (1984). Mature oligodendrocytes: division following experimental demyelination in adult animals. *Archives of Neurology*, **41**, 1162–1165.

Archer, D.R., Leven, S. & Duncan, I.D. (1994). Myelination by cryopreserved xenografts and allografts in the myelin deficient rat. *Experimental Neurology*, **125**, 268–277.
Baulac, M., Lachapelle, F., Gout, O., Berger, B., Baumann, N. & Gumpel, M. (1987). Transplantation of oligodendrocytes in the newborn mouse brain. Extension of myelination by transplanted cells: anatomical studies. *Brain Research*, **420**, 39–47.
Bjorklund, A. (1991). Neural transplantation: an experimental tool with clinical possibilities. *Trends in Neurosciences*, **14**, 319–322.
Blakemore, W.F. (1974). Remyelination of the superior cerebellar peduncle in old mice following demyelination induced by cuprizone. *Journal of the Neurological Sciences*, **22**, 121–126.
Blakemore, W & Crang, A.J. (1988). Extensive oligodendrocyte remyelination following injection of cultured central nervous system cells into demyelinating lesions in adult central nervous system. *Developmental Neuroscience*, **10**, 1–11.
Blakemore, W.F., Eames, R.A., Smith, K.J. & McDonald, W.I. (1977). Remyelination in the spinal cord of the cat following intraspinal injections of lysolecithin. *Journal of the Neurological Sciences*, **33**, 31–43.
Blakemore, W.F. & Franklin, R.J.M. (1991). Transplantation of glial cells into the CNS. *Trends in Neurosciences*, **14**, 323–327.
Bornstein, M. & Raine, C.S. (1976). The initial structural lesion in serum-induced demyelination *in vitro*. *Laboratory Investigation*, **35**, 391–401.
Bostock, H. & Sears, T.A. (1978). The internodal axon membrane: electrical excitability and continuous conduction in segmental demyelination. *Journal of Physiology (London)*, **280**, 273–301.
Chiu, S.Y. & Schwarz, W. (1987). Sodium and potassium currents in acutely demyelinated internodes of rabbit sciatic nerves. *Journal of Physiology (London)*, **391**, 631–649.
Crang, A.J. & Blakemore, W.F. (1991). Remyelination of demyelinated rat axons by transplantation of mouse oligodendrocytes. *Glia* **4**, 305–313.
Duncan, I.D., Hammang, J.P., Jackson, K.F., Wood, P.M., Bunge, R.P. & Langford, L. (1988). Transplantation of oligodendrocytes and Schwann cells into the spinal cord of the myelin deficient rat. *Journal of Neurocytology*, **17**, 351–360.
Duncan, I.D., Paino, C., Archer, D.R. & Wood, P.M. (1992). Functional capacities of transplanted cell-sorted adult oligodendrocytes. *Developmental Neuroscience*, **14**, 114–122.
Friedman, E., Nilaver, G., Carmel, P., Perlow, M., Spatz, L. & Latov, N. (1986). Myelination by transplanted fetal and neonatal oligodendrocytes in a dysmyelinating mutant. *Brain Research*, **378**, 197–207.
Friedrich, V.L. Jr & Lazzarini, R.A. (1993). Restricted migration of transplanted oligodendrocytes or their progenitors, revealed by transgenic marker MBP. *Journal of Neural Transplantation and Plasticity*, **4**, 139–146.
Gansmuller, A., Lachapelle, F., Baron-Van Evercooren, A., Hauw, J.J., Baumann, N. & Gumpel, M. (1986). Transplantation of newborn CNS fragments into the brain of *shiverer* mutant mice: extensive myelination by transplanted oligodendrocytes. II. Electron microscopic study. *Developmental Neuroscience*, **8**, 197–207.
Godfraind, C., Friedrich, V.L. Jr, Holmes, K.V. & Dubois-Dalq, M. (1989). *In vivo* analysis of glial phenotypes during a viral demyelinating disease in mice. *Journal of Cell Biology*, **109**, 2405–2416.
Gout, O., Gansmuller, A., Baumann, N. & Gumpel, M. (1988). Remyelination by transplanted oligodendrocytes of a demyelinated lesion in the spinal cord of the adult *shiverer* mouse. *Neuroscience Letters*, **87**, 195–199.

Grissmer, S. (1986). Properties of potassium and sodium channels in frog internode. *Journal of Physiology (London)*, **381**, 119–134.

Gumpel, M., Baumann, N., Raoul, M. & Jacque, C. (1983). Survival and differentiation of oligodendrocytes from neural tissue transplanted into newborn mouse brain. *Neuroscience Letters*, **37**, 307–311.

Gumpel, M., Lachapelle, F., Gansmuller A., Baulac, M., Baron-Van Evercooren, A. & Baumann, N. (1987). Transplantation of human embryonic oligodendrocytes into *shiverer* brain. *Annals of the New York Academy of Sciences*, **495**, 71–85.

Kohsaka, S., Yoshida, A., Inoue, Y., Shinozaki, T., Takamaya, H., Inoue, M., Mikoshiba, K., Takamatsu, K., Otani, M., Toya, S. & Tsukada, Y. (1986). Transplantation of bulk-separated oligodendrocytes into the brains of shiverer mutant mice: immunohistochemical and electron-microscopic studies on the myelination. *Brain Research*, **372**, 137–142.

Lachapelle, F., Duhamel-Clerin, E., Gansmuller, A., Baron-Van Evercooren, A., Villarroya, H. & Gumpel, M. (1994). Transplanted, transgenically marked oligodendrocytes survive, migrate and myelinate in the normal mouse brain as they do in the *shiverer* mouse. *European Journal of Neuroscience*, **6**, 814–824.

Lachapelle, F., Gumpel, M., Baulac, M., Jacque, C., Due, P. & Baumann, N. (1984). Transplantation of CNS fragments into the brain of *shiverer* mice: extensive myelination by implanted oligodendrocytes. I. Immunohistochemical studies. *Developmental Neuroscience*, **6**, 325–334.

Lachapelle, F., Lapie, P. & Gumpel, M. (1992). Oligodendrocytes from *jimpy* and normal mature tissue can be 'activated' when transplanted in a newborn environment. *Developmental Neuroscience*, **14**, 105–113.

Lois, C. & Alvarez-Buylla, A. (1993). Proliferating subventricular zone cells in the adult mammalian forebrain can differentiate into neurons and glia. *Proceedings of the National Academy of Sciences, USA*, **90**, 2074–2080.

Lubetzki, C., Gansmuller, A., Lachapelle, F., Lombrail, P. & Gumpel, M. (1988). Myelination by oligodendrocytes isolated from 4–6-week-old rat CNS and transplanted into newborn *shiverer* brain. *Journal of the Neurologial Sciences*, **88**, 161–175.

Ludwin, S.K. & Bakker, D.A. (1988). Can oligodendrocytes attached to myelin proliferate? *Journal of Neuroscience*, **8**, 1239–1244.

Noebels, J.L., Marcon, P.K. & Jalilian-Tehrani, M.H. (1991). Sodium channel density in hypomyelinated brain increased by myelin basic protein gene deletion. *Nature*, **352**, 431–434.

Oaklander, A.L., Pellegrino, R.G. & Ritchie, J.M. (1984). Saxitoxin binding to central and peripheral nervous tissue of the myelin deficiency (*md*) mutant rat. *Brain Research*, **307**, 393–397.

Ostermann, P.O. & Westerberg, C.E. (1975). Paroxysmal attacks in multiple sclerosis. *Brain*, **98**, 189–202.

Prineas, J.W. (1985). The neuropathology of multiple sclerosis. In *Handbook of Clinical Neurology*, vol. 3, *Demyelinating Diseases*, ed. J.C. Koester, pp. 213–257. Amsterdam: Elsevier.

Prineas, J., Kwon, E., Goldenberg, P., Ilyas, A., Quarles, R., Benjamins, J. & Sprinkle, T. (1989). Multiple sclerosis: oligodendrocyte proliferation and differentiation in fresh lesions. *Laboratory Investigation*, **61**, 489–503.

Raine, C.S., Diaz, M., Pakingan, M. & Bornstein, M.B. (1978). Antiserum-induced dissociation of myelinogenesis *in vitro*: an ultrastructural study. *Laboratory Investigation*, **38**, 397–403.

Ranscht, B., Wood, P.L. & Bunge, R.P. (1987). Inhibition of *in vitro* peripheral myelin formation by monoclonal antigalactocerebroside. *Journal of the Neurological Sciences*, **7**, 2936–2947.

Reynolds, R. & Wilkin, G.P. (1991). Oligodendroglial progenitor cells but not oligodendroglia divide during normal development of the rat cerebellum. *Journal of Neurocytology*, **20**, 216–224.

Rosenbluth, J. (1981). Freeze-fracture approaches to ionophore localization in normal and myelin-deficient nerves. *Advances in Neurology*, **31**, 391–418.

Rosenbluth, J. (1989). Role of Schwann cells in differentiation of the axolemma: consequences of myelin deficiency in spinal roots of the dystrophic mouse mutant. In *Peripheral Nerve Development and Regeneration: Recent Advances and Clinical Applications*, ed. E. Scarpini, M.G. Fiore, D. Pleasure & G. Scarlato, pp. 39–53. Padua: Liviana Press/Springer.

Rosenbluth, J. (1990). Axolemmal abnormalities in myelin mutants. *Annals of the New York Academy of Sciences*, **605**, 194–214.

Rosenbluth, J., Hasegawa, M. & Schiff, R. (1989). Myelin formation in myelin-deficient rat spinal cord following transplantation of normal fetal spinal cord. *Neuroscience Letters*, **97**, 35–40.

Rosenbluth, J., Hasegawa, M., Shirasaki, N., Rosen, C.L. & Liu, Z. (1990). Myelin formation following transplantation of normal fetal glia into myelin-deficient rat spinal cord. *Journal of Neurocytology*, **19**, 718–730.

Rosenbluth, J., Liu, Z., Guo, D. & Schiff, R. (1992). Effect of anti-GalC on CNS myelin formation. *Society for Neuroscience Abstracts*, **18**, 1449.

Rosenbluth, J., Liu, Z., Guo, D. & Schiff, R. (1993). Myelin formation by mouse glia in myelin-deficient rats treated with cyclosporine. *Journal of Neurocytology* **22**, 967–977.

Rosenbluth, J., Liu, Z., Guo, D. & Schiff, R. (1994). Inhibition of CNS myelin development *in vivo* by implantation of anti-GalC hybridoma cells. *Journal of Neurocytology*, **23**, 699–707.

Rosenbluth, J., Schiff, R., Liang, W.-L., Menna, G. Young, W. (1995). Survival, migration and differentiation of transgenic oligodendrocyte-lineage glia transplanted into spinal cord injured adult rats. *Journal of Neurotrauma* (in press).

Schmalbruch, H. & Hellhammer, U. (1976). The number of satellite cells in normal human muscle. *Anatomical Record*, **185**, 279–288.

Snow, M.H. (1977a). Myogenic cell formation in regenerating rat skeletal muscle injured by mincing. *Anatomical Record*, **188**, 201–218.

Snow, M.H. (1977b). The effects of aging on satellite cells in skeletal muscles of mice and rats. *Cell and Tissue Research*, **185**, 399–408.

Starzl, T.E., Fung, J., Tzakis, A., Todo, S., Demetris, A.J., Marino, I.R., Doyle, H., Zeevi, A., Warty, V., Michaels, M., Kusne, S., Rudert, W.A. & Trucco, M. (1993). Baboon-to-human liver transplantation. *Lancet*, **341**, 65–71.

Tontsch, U., Archer, D.R., Dubois-Dalcq, M. & Duncan, I.D. (1994). Transplantation of oligodendrocyte cell line leading to extensive myelination. *Proceedings of the National Academy of Sciences, USA*, **91**, 11616–11620.

Utzschneider, D.A., Archer, D.R., Kocsis, J.D., Waxman, S.G. & Duncan, I.D. (1994). Transplantation of glial cells enhances action potential conduction of amyelinated spinal cord axons in the myelin-deficient rat. *Proceedings of the National Academy of Sciences, USA*, **91**, 53–57.

Warrington, A.E., Barbarese, E. & Pfeiffer, S.E. (1993). Differential myelinogenic capacity of specific developmental stages of the oligodendrocyte lineage upon transplantation into hypomyelinating hosts. *Journal of Neuroscience Research*, **34**, 1–13.

Waxman, S.G., Black, J.A., Kocsis, J.D. & Ritchie, J.M. (1989). Low density of sodium channels supports action potential conduction in axons of neonatal rat optic nerve. *Proceedings of the National Academy of Sciences, USA*, **86**, 1406–1410.

Wolf, M.K., Schwing, G.B., Adcock, L.H. & Billings-Gagliardi, S. (1981). Hypomyelinated mutant mice. III. Increased myelination in mutant cerebellum co-cultured with normal optic nerve. *Brain Research*, **206**, 193–197.

Wood, P.M. & Bunge, R.P. (1991). The origin of remyelinating cells in the adult central nervous system: the role of the mature oligodendrocyte. *Glia*, **4**, 225–232.

Young, W., Rosenbluth J., Wojak, C., Sakatini, K. & Kim, H. (1989). Extracellular potassium activity and axonal conduction in spinal cord of the myelin-deficient mutant rat. *Experimental Neurology*, **106**, 41–51.

Part II
Pain

Pain is a frequent cause for people to seek medical advice. The need to have a secure understanding of the cellular basis of pain has resulted in extensive physiological, pharmacological and psychological research into its various aspects. Research, however, has often been hampered by the practical issues of identifying and recording signals in small-diameter nerves, and defining the relationship between nerve impulses recorded in animal preparations and the human experience of 'pain'. The disparate nature of pain is reflected in the varied aspects of the work presented in this section. Pain can arise not only from mechanical injury and exogenously applied noxious substances, but can also be related to muscle activity (see chapter by Westgaard) and local inflammation (chapter by Iggo). The identity of peripheral afferents whose activities gives rise to pain, together with their pathologies, is extensively reviewed by Ochoa. However, these primary afferent fibres are only the first stage of the sensory pathway and other parts of the nervous system are involved in pain syndromes. As pointed out in the chapter by Willis, while acute pain is generally obvious and easy to treat, it also produces a longer-term legacy of abnormal sensation which is less readily identifiable. Willis argues that the existence of such long-term consequences suggests a possible mechanism in the spinal cord for 'remembering' pain.

The ability to control pharmacologically both acute and chronic pain by reducing our awareness of it has a long history in both Eastern and Western cultures. The advent of increased cross-cultural dialogue in recent decades has brought with it a renewed interest in the ancient technique of acupuncture for symptomatic treatment of pain. This has prompted world-wide systematic research specifically aimed at identifying and characterizing the underlying physiological principles of the technique. The chapter by Shen presents an intriguing account of acupuncture from a physiological viewpoint.

13

Human nociceptors in health and disease

JOSÉ OCHOA

Good Samaritan Hospital and Medical Center, Oregon Health Sciences University, Portland, Oregon, USA.

Introduction

In the mid 1960s, a rare collection of normal human peripheral nerves obtained from healthy volunteers came from Chile to the Institute of Neurology in London. The material awaited further examination and quantitation by electron microscopy, which was eventually accomplished under the direction and unfailing support of W.G.P. Mair (Ochoa & Mair, 1969a, b). At that time, all we had at hand were reproductions of human unmyelinated fibres originally drawn by Remak, and micrographs of silver stains of unmyelinated axons obtained by Ranson and later by Gasser. We learned that, in man, unmyelinated fibres exhibit a specific fine structure and are 4 times more abundant than myelinated fibres. Unmyelinated fibres drop out with age and in disease. Under those circumstances the surviving axons produce tiny sprouts which make their calibre spectrum bimodal.

Basic electrophysiology

Zotterman, the impetuous Swedish scientist who first recorded impulse activity in unmyelinated (C) fibres in animals, often credited Hallin and Torebjörk with being the pioneer investigators who first recorded propagated impulse activity in C-fibres in humans (Hallin & Torebjörk, 1970; Torebjörk & Hallin, 1970). The equally outstanding work from Belgium, by Van Hees & Gybels (1972), came at about the same time. Through the use of the microneurographic technique of Hagbarth and Vallbo (Hagbarth & Vallbo 1967; Vallbo & Hagbarth, 1968), Swedes and Belgians managed to obtain single-unit recordings of C-fibre activity from undissected nerves of awake human subjects, and described receptor–response properties of C polymodal nociceptors supplying human skin. In addition the Swedes gave an account of C sympathetic neural activity in human nerve fascicles supplying both skin and muscle (Hagbarth *et al.*, 1972).

Characteristically, the cutaneous receptors of C afferent units supplying the human skin respond with high threshold to mechanical stimuli, attain relatively moderate firing rates and display intermediate rates of adaptation. These receptors respond to several different kinds of adequate stimuli, including mechanical, thermal and chemical energies. Typically, these units may show after-discharges. Torebjörk and Hallin reported that the conduction velocities for C afferent units from human skin range between 0.4 and 1.8 m/s, as measured between intradermal stimulating electrode and intraneural recording electrode (Torebjörk & Hallin, 1976). There is little written about the sensitivity of C polymodal nociceptors to low temperature as an adequate stimulus, although in an early report by Torebjörk (1974) a few polymodal nociceptors were said to respond to such stimuli with a proportionate discharge. In a prospective study on this aspect of human C-nociceptor function, Campero has found that a proportion of polymodal nociceptors with C-fibres respond consistently to low temperature, in the range of 15–0 °C. A smaller proportion of these polymodal nociceptors are responsive to mechanical and low-temperature stimuli, but not to heat (M. Campero and J. Ochoa, unpublished data).

It has been postulated that a C polymodal receptor might reach threshold cumulatively in response to different adequate energies acting upon specific transducers in its excitable membrane. Quotas of mechanical and thermal or chemical stimuli, which on their own would be subthreshold, might bring the receptor to threshold when acting additively in various combinations. This concept, which was derived from psychophysical observations on a diseased state, calls for rigorous assessment under physiological circumstances in normal controls. The basic observation, termed *cross modality threshold modulation* (XTM) (Ochoa, 1986; Ochoa *et al.*, 1987a; Culp *et al.*, 1989), was as follows: the abnormal painful sensitivity to *mechanical* stimuli of skin irritated with agents capable of sensitizing nociceptors is worsened by concomitant application of *heat*, and dramatically normalized by concomitant application of *low-temperature* stimuli.

If one accepts that the excitable receptor membrane of C polymodal nociceptors must consist of a mosaic of different transducers specifically sensitive to particular energies, and that any one population of such transducers would be capable of bringing the membrane to threshold through inward Na^+ currents, then it becomes reasonable to hypothesize on XTM. In the hyperalgesic state determined by sensitization of nociceptors by capsaicin it would be specifically the heat transducers that would be leaky and thus bring the membrane to, or near, threshold. Under those circumstances, a normally subthreshold contribution of mechanical energy acting on mechanical transmembrane (stretch) transducers would be enough to bring the membrane to threshold for firing, thus evoking a painful sensation. In that state, the application of low temperature, the antithesis of a thermal stimulus, would subtract thermal (heat) energy that would otherwise further open the leaky heat transducers. Thus one energy (temperature) modulates the threshold for another energy acting adequately on the same excitable membrane. Beyond a point, excessive low temperature might by itself activate low-temperature transducers in the C polymodal nociceptor membrane and thus enhance the overall nociceptor response. A possible clinical correlate for this prediction is found in the farsighted statement from Sir Thomas Lewis: 'Cooling abolishes the pain unless a low temperature is reached . . . that gives pain indistinguishable from that induced by heat' (1936).

Neurosecretory function of C-nociceptors

Over and above the neurophysiological function of nerve impulse generation and transmission, human C polymodal nociceptor units have the capacity to release neurotransmitters at their terminals. One consequence is antidromic vasodilatation which is confined to the cutaneous territory of the bundles of nerve fibres excited by intraneural microstimulation (INMS) (Ochoa *et al.*, 1987b). There is satisfactory evidence that this phenomenon is mediated by release of vasoactive substance P (Rosell *et al.*, 1981). The expression of C-nociceptor-mediated antidromic vasodilation is modulated by reciprocal interactions involving neural sympathetic vasoconstrictor output (Ochoa *et al.*, 1993). When both systems are co-activated during strong, painful INMS of human peripheral nerves, while monitoring cutaneous thermal emission profiles in the fields of projected cutaneous sensation, the skin temperature decreases. After termination of the stimulus, cutaneous temperature increases, eventually to decrease again upon renewed stimulation.

One concern about this experimental paradigm is that, alternatively, the regional warming might be due to sympathetic vasoconstrictor paralysis from stimulus-dependent block during INMS. However, if this were the case, the regional warming would be expected to develop during stimulation rather than after its termination, as seen regularly. Moreover, from animal experiments it is known that antidromic vasodilatation is abolished by inhibitors of substance P (Rosell *et al.*, 1981). The overall evidence indicates that vasoconstriction mediated by the sympathetic system overrides antidromic vasodilatation induced by excitation of C-nociceptor nerve fibres. Therefore, in the clinical context, fluctuations in skin temperature seen in patients suffering from 'reflex sympathetic dystrophy–causalgia–sympathetically maintained pain' (RSD-SMP) might in part be explained by reflex vasoconstriction intermittently overshadowing pre-existing antidromic vasodilatation.

Application of certain chemicals to cutaneous terminals of C polymodal nociceptors causes them to become temporarily sensitized and induces a prominent flare response, more or less visible to the naked eye depending on intrinsic characteristics of the tested skin. Whereas there is agreement on C polymodal nociceptor dependence of the flare response to chemical, thermal or mechanical injury to the skin, there is disagreement regarding the extent of the area of induced flare relative to the territories of induced hyperalgesia resulting from concomitant nociceptor hyperexcitability. It has been recently shown that the area of hyperalgesia surrounding the site of capsaicin injection (secondary hyperalgesia) is not selective to mechanical stimuli but also involves heat stimuli. Moreover, the area of capsaicin-induced vasodilatation flare matches the areas of mechanical and heat hyperalgesia, suggesting a common mechanism for all three phenomena (Serra, Campero & Ochoa, 1993).

C nociceptors and somatosensory coding and decoding

Zotterman produced persuasive neurophysiological evidence that small-calibre fibres signal pain when he recorded C action potentials from the lingual nerve of animals after heating their tongue (Zotterman, 1936). Through the ensuing half century, significant anecdotes accrued from human studies to support the concept that sensory units with small-calibre fibres mediate pain in humans. For example, in 1960, Collins *et al.* electrically stimulated exposed sural nerves in 11 awake subjects, who were asked to describe the resulting sensory quality, while monitoring compound nerve action potentials pre- and post-cordotomy. The subjects experienced pain only when slowly conducting fibre populations were recruited. Subsequently, patients with morphological and electrophysiological evidence of loss of small myelinated and unmyelinated fibres and of the Aδ and C elevations in their nerves, were reported to express loss of pain, cold and warmth sensations (Dyck & Lambert, 1969).

Comparable correlations have since been established for some forms of congenital absence of pain associated with anhidrosis (Bischoff, 1979). Torebjörk & Hallin (1973) sharpened the correlation between C afferent activity and second 'dull' pain sensation when they documented physiologically what Zotterman had observed clinically in the 1920s (see Zotterman, 1971): that is, preservation of 'second' pain sensation evoked by noxious activation of skin nerves at a time when selective A-fibre block had abolished nerve conduction, except in the slowest conducting fibres.

In the 1980s, the combined application of microneurographic recordings and INMS (Torebjörk & Ochoa, 1980) yielded strong evidence to support the concept that when primary sensory units of a particular type, as defined by stimulus–response characteristics and conduction velocity, are activated, the sensation evoked, if any, is (1) projected with remarkable accuracy to the corresponding receptive field, (2) is specific in its subjective quality, and (3) in the absence of spatial summation has a magnitude that is a function of stimulus frequency. When INMS is delivered at an endoneurial site from where unitary C-fibre activity can be recorded during stimulation of the receptive field of cutaneous C polymodal nociceptors, the sensation is also projected accurately and its subjective quality is dull burning pain. Such pain has a long reaction time and does not disappear during selective A-fibre block (Ochoa & Torebjörk, 1989). The pain is identical in its quality to that experienced from noxious stimulation of the skin during selective A-fibre block; it is also identical to a kind of causalgic pain that may emanate from nerve injury (Ochoa, 1990).

Aδ-nociceptors connected with thin myelinated fibres are much more difficult to access through microneurography than C-nociceptors (Torebjörk & Ochoa, 1990). Microstimulation of Aδ-nociceptors also evokes a first sensation that accurately overlaps their receptive field. Such sensation is a sharp pain, has a short reaction time and does disappear during A-fibre block (H.E. Torebjörk & J. Ochoa, unpublished data).

So, despite educated but not infallible scepticism (Wall, 1984), there are no good current reasons to doubt that human A and C nociceptors signal pain, and that they do so according to the law of *specificity*. Moreover, the brain 'knows the address' of every nociceptor, at least for the skin of the hand (Ochoa & Torebjörk, 1983, 1989).

The sensation evoked by INMS delivered at an intraneural C-nociceptor recording site is sometimes itch rather than dull pain. The receptor characteristics of these 'itch units' resemble those of typical polymodal C human cutaneous nociceptors (Torebjörk & Ochoa, 1981). Itch and dull pain do not transform into one another just by shifting INMS frequency. Whether these two specific qualities of sensation are subserved by separate specific channels that use a common kind of primary C unit instrument, or whether they result from different kinds of spatial interactions at central levels, remains to be established. Scratching abolishes itch sensation through central inhibition, rather than through receptor fatigue, since scratching remains effective while C nociceptors continue to be able to discharge.

Muscle nociceptors

Until recently, attempts at characterizing muscle nociceptors by receptor characteristics and conduction velocity were frustrated. It had been reported that INMS in muscle fascicles evokes a cramp-like pain that projects over the muscle belly (Ochoa & Torebjörk, 1981), but excitation of their deep group III and IV receptors, as guided by the projected sensation, was achieved only recently (Marchettini *et al.*, 1991; Simone *et al.*, 1991, 1994). In our 1981 study it had been found that the quality of muscle pain induced by INMS remains unaltered during A-fibre block, thus incriminating unmyelinated (group IV) afferents. Nevertheless, in that preliminary study it was also found that during compression-ischaemia block the reaction times of INMS-induced muscle pain shift from A to C conduction values (Ochoa & Torebjörk, 1981). Those findings suggested that both group III and IV muscle afferents are concerned with conduction of impulses eventually decoded centrally as muscle pain, but in contrast with Aδ (sharp) and C (dull) types of pain from skin, a single quality of pain is discriminated by the brain from muscle, presumably due to central convergence.

Pathophysiology of human nociceptors

The repertoire of microscopically recognizable pathological states of unmyelinated fibres is limited to axonal degeneration and sprouting. Following the experimental claim that immature sprouts from small-caliber fibres may misbehave as ectopic impulse generators (Wall & Gutnick, 1974) it has been repeatedly hypothesized that pathological sprouting of sick nociceptors might be a source of discharges leading to neuropathic pains (Brown, Martin & Asbury 1976). It is accepted that immature nociceptor sprouts may be unduly mechanosensitive, thus explaining the painful sign of Tinel (Brown & Iggo, 1963). It also remains possible that such sprouts might be the sources of continuing ectopic discharges responsible for the symptom of chronic pain.

Whatever the actual role of immature axon sprouts might be in the context of chronic pain, it seems certain that the symptom of stimulus-induced neuropathic pain (hyperalgesia, allodynia) calls for a different kind of explanation. Two kinds of abnormal nociceptor mechanisms, initially discovered through animal experimental work, have been proposed with reasonable foundation to operate in neuropathic painful conditions in human patients. One is sensitization of nociceptors and the other is central release of nociceptor input due to removal of inhibitory co-activated input. Both these syndromes are characterized by chronic pain, mechanical hyperalgesia, thermal hyperalgesia and abnormal skin temperature. However, the thermal events are reversed: one syndrome expresses hot skin and heat hyperalgesia and the other expresses cold skin and cold hyperalgesia.

'ABC syndrome'

In the syndrome of sensitization of peripheral C-nociceptors, described by Sir Thomas Lewis under the term 'erythralgia', the patients have 'a redness of the skin, associated with tenderness, the gentlest manipulation elicits pain, pain is also provoked by warming, cooling abolishes the pain . . . and the pain burns' (Lewis, 1936). When it was explicitly documented by microneurography that in this syndrome there exist sensitized nociceptors accounting for both the sensory and vasomotor antidromic phenomena, it was called the 'ABC syndrome' (Angry Backfiring C nociceptor syndrome) (Ochoa, 1986; Ochoa *et al.*, 1987a; Cline, Ochoa & Torebjörk, 1989). The syndrome is clinically identical to the acute experimental state induced in human volunteers by application of capsaicin to the skin (Culp *et al.*, 1989). Indeed, under those circumstances there is dose-dependent rubor, spontaneous pain and hyperalgesia. Both in volunteers and in patients, passive cooling abolishes spontaneous burning pain and mechanical hyperalgesia (XTM). This beneficial effect of cooling persists after block of myelinated fibres and therefore cannot be explained through central gating exerted by cold-specific input; the low temperature must directly rectify the excitability of sensitized peripheral receptors. The ABC syndrome is not a specific disease, related to a specific aetiology; it is pathophysiology-related; that is what a syndrome is. We have described it in patients with mononeuropathy, radiculopathy and a variety of polyneuropathies.

'Triple cold syndrome'

In the reverse state, that is the 'triple cold syndrome' (Ochoa & Yarnitsky, 1990; 1994; Ochoa, 1992), the skin is cold and pale and pain is provoked by cooling. Warming may improve the pain. The fact that the pain is burning again suggests that C nociceptors are involved. If 'hot' patients theoretically have leaky heat transducers in their polymodal membrane, then 'cold' patients might have leaky transducers for low temperature, because in animal cornea C-nociceptors may indeed become specifically sensitized to low temperature (P.H. Reeh, personal communication). However, for reasons discussed elsewhere (Ochoa & Yarnitsky, 1993), the alternative interpretation of central disinhibition is more appealing. Indeed, in addition to cold *hyper*algesia these patients express cold *hypo*esthesia in their 'burning' limbs. As it turns out, blocking cold sensation in normal volunteers disinhibits pain induced by low temperature: as cold sensation becomes blunted, pain induced by low temperature becomes hyperalgesic and its quality switches from cold pain to burning pain (Yarnitsky & Ochoa, 1990). This is interpreted as a release phenomenon *à la* Henry Head. The chronically cold skin would thus provide a built-in stimulus for disinhibited pain. Why is the skin cold in these 'triple cold syndrome' patients? We believe this is because the small-fibre neuropathy has caused partial degeneration of postganglionic sympathetic fibres with ensuing arteriolar denervation, supersensitivity and chronic vasoconstriction. For this reason, local somatic nerve blocks, which normally warm up a territory of skin, cause a sluggish response in these patients (Ochoa & Yarnitsky, 1993).

'Reflex sympathetic dystrophy'

There is a fashionable line of thinking that calls for stringent scrutiny. It supposes that, following mild peripheral injury, C-nociceptors from the region might develop subtle and clinically undetectable hyperexcitability. This would determine chronic nociceptor input to the spinal cord, that would in turn cause chronic hyperexcitability of wide dynamic range neurones and 'secondary' brush-induced hyperalgesia (Torebjörk, Lundberg & LaMotte, 1992; Torebjörk, 1993). There are strong reasons to believe that secondary central changes of the kind hypothesized for RSD-SMP patients, which can definitely be replicated acutely in animal experiments following overt nerve injury, cannot be extrapolated to explain chronic human painful syndromes in the absence of evidence of nerve dysfunction (M. Campero & J. Ochoa, unpublished data). Such pseudoneuropathic syndromes are traditionally rationalized under the purely descriptive and deceptive term 'reflex sympathetic dystrophy' (M. Campero & J. Ochoa, unpublished data; Ochoa, 1993; Ochoa & Verdugo, 1993a, b). In our opinion, such thinking amounts to the *post hoc* fallacy opportunely denounced by Sir Wilfred Trotter half a century ago (1943).

Acknowledgement

It was fortunate, at the start of the work reviewed here, that at the Institute of Neurology there were people curious about what the fine structure of human unmyelinated axons might be. Amongst them were David Landon, P.K. Thomas, Roger Gilliat, Ian McDonald and, of course, Tom Sears who had written seminal physiological papers on C-fibres in collaboration with Peter Nathan. The help and encouragement from all of these is now belatedly acknowledged.

References

Bischoff, A. (1979). Congenital insensitivity to pain anhidrosis: a morphometric study of sural nerve and cutaneous receptors in the human prepuce. In *Advances in Pain Research and Therapy*, vol. 3, ed. J.J. Bonica, J.C. Liebeskind & D.G. Albe-Fessard, pp. 63–65. New York: Raven Press.

Brown, A.G. & Iggo, A. (1963). The structure and function of cutaneous 'touch corpuscles' after nerve crush. *Journal of Physiology (London)*, **165**, 28–29.

Brown, M. , Martin, J.R. & Asbury, A.K. (1976). Painful diabetic neuropathy: a morphometric study. *Archives of Neurology*, **33**, 164–171.

Cline, M.A., Ochoa, J.L. & Torebjörk, H.E. (1989). Chronic hyperalgesia and skin warming caused by sensitized C nociceptors. *Brain*, **112**, 621–647.

Collins, W.F., Nulsen, F.E. & Randt, C.T. (1960). Relation of peripheral nerve fibre size and sensation in man. *Archives of Neurology*, **3**, 381–385.

Culp, W.J., Ochoa, J., Cline, M. & Dotson, R. (1989). Heat and mechanical hyperalgesia induced by capsaicin. *Brain*, **112**, 1317–1331.

Dyck, P.J. & Lambert, E.H. (1969). Dissociated sensation in amyloidosis: compound action potential, quantitative histologic and teased-fibre, and electron microscopic studies of sural nerve biopsies. *Archives of Neurology*, **20**, 490–507.

Hagbarth, K.-E., Hallin, R.G., Hongell, A., Torebjörk, H.E. & Wallin, B.G. (1972). General characteristics of sympathetic activity in human skin nerves. *Acta Physiologica Scandinavica*, **84**, 164–176.

Hagbarth, K.-E. & Vallbo, A.B. (1967). Mechanoreceptor activity recorded percutaneously with semi-microelectrodes in human peripheral nerves. *Acta Physiologica Scandinavica*, **69**, 121–122.

Hallin, R.G & Torebjörk, H.E. (1970). Afferent and efferent C units recorded from human skin nerves *in situ*. *Acta Societatis Medicorum Upsaliensis*, **75**, 277–281.

Lewis, T. (1936). *Vascular Disorders of the Limbs Described for Practitioners and Student*, pp. 93–106. London: Macmillan.

Marchettini, P., Simone, D., Mense, S. & Ochoa, J.L. (1991). Identification of group III muscle nociceptors in human nerves. *Neurology*, **41** (Suppl. 1), 1645.

Ochoa, J.L. (1986). The newly recognized painful ABC syndrome: thermographic aspects. *Thermology*, **2**, 65–107.

Ochoa, J.L. (1990). Neuropathic pains from within: personal experiences, experiments, and reflection on mythology. In *Recent Achievements in Restorative Neurology: Altered Sensation and Pain*, ed. M.R. Dimitrijevic, P.D. Wall & O. Lindholm. Basel: Karger.

Ochoa, J.L. (1992). Thermal hyperalgesia as a clinical symptom. In *Hyperalgesia and Allodynia*, Second Bristol Myers Squibb Symposium on Pain, ed. W. Willis, pp. 151–165. New York: Raven Press.

Ochoa, J.L. (1993). Editorial. Essence, investigation and management of 'neuropathic' pains: hopes from acknowledgment of chaos. *Muscle and Nerve*, **16**, 997–1008.

Ochoa, J.L., Cline, M., Comstock, W., Culp, W.J., Dotson, R., Marchettini, P. & Torebjörk, H.E. (1987a). Painful syndrome newly recognized: hyperalgesia with cross modality threshold modulation and rubor. *Society for Neuroscience Abstracts*, **13**, 189.

Ochoa, J.L., Comstock, W.J., Marchettini, P. & Nizamuddin, G. (1987b). Intrafascicular nerve stimulation elicits regional skin warming that matches the projected field of evoked pain. In *Fine Afferent Nerve Fibers and Pain*, vol. 44, ed. R.F. Schmidt, H.-G. Schaible & C. Vahle-Hinz, pp. 475–479. Weinheim: VCH

Ochoa, J.L. & Mair, W.G. P. (1969a). The normal sural nerve in man. I. Ultrastructure and numbers of fibres and cells. *Acta Neuropathologica*, **13**, 197–216.

Ochoa, J.L. & Mair, W.G. P. (1969b). The normal sural nerve in man. II. Changes in axons and Schwann cells due to age. *Acta Neuropathologica*, **13**, 217–239.

Ochoa, J.L. & Torebjörk, H.E. (1981). Pain from skin and muscle. *Pain* (Supplement) **1**, 87.

Ochoa, J.L. & Torebjörk, H.E. (1983). Sensations evoked by intraneural microstimulation of single mechano-receptor units innervating the human hands. *Journal of Physiology (London)*, **342**, 633–654.

Ochoa, J.L. & Torebjörk, H.E. (1989). Sensations evoked by intraneural microstimulation of C nociceptor fibres in human skin nerves. *Journal of Physiology (London)*, **415**, 583–599.

Ochoa, J.L. & Verdugo, R. (1993a). Reflex sympathetic dystrophy: definition and history of the ideas. A critical review of human studies. In *The Evaluation and Management of Clinical Autonomic Disorders*, ed. P.A. Low, pp. 473–492. Boston: Little, Brown.

Ochoa, J.L., & Verdugo, R. (1993b). The mythology of reflex sympathetic dystrophy and sympathetically maintained pains. *Physical Medicine and Rehabilitation Clinics of North America*, **4**, 151–163.

Ochoa, J.L. & Yarnitsky, D. (1990). Triple cold ('CCC') painful syndrome. *Pain*, Supplement **5**, S278.

Ochoa, J.L. & Yarnitsky, D. (1994). The triple cold painful syndrome: cold hyperalgesia, cold hypoesthesia and cold skin in peripheral nerve disease. *Brain*, **117**, 185–197.

Ochoa, J.L., Yarnitsky, D., Marchettini, P., Dotson, R. & Cline, M. (1993). Interactions between sympathetic vasoconstrictor outflow and C nociceptor-induced antidromic vasodilatation. *Pain*, **54**, 191–196.

Rosell, J., Olgart, L., Gazelius, B., *et al.* (1981). Inhibition of antidromic and substance-P induced vasodilatation by a substance P agonist. *Acta Physiologica Scandinavica*, **111**, 381–382.

Serra, J., Campero, M. & Ochoa, J.L. (1993). 'Secondary' hyperalgesia (capsaicin) mediated by C-nociceptors. *Society for Neuroscience Abstracts*, **19**, 965.

Simone, D., Caputi, G., Marchettini, P. & Ochoa, J.L. (1991). Slowly conducting, high threshold mechanoreceptors microrecorded from human striated muscle: receptor responses, axonal conduction velocity and projected pain. *Society for Neuroscience Abstracts*, **17**, 1368.

Simone, D.A., Marchettini, P., Caputi, G. & Ochoa, J.L. (1994). Identification of muscle afferents subserving sensation of deep pain in humans. *Journal of Neurophysiology*, **72**, 883–889.

Torebjörk, H.E. (1974). Single unit activity in afferent and sympathetic C fibres recorded from intact human skin nerves. Ph.D. thesis, University of Uppsala.

Torebjörk, H.E. (1993). Human microstimulation and intraneural microstimulation in the study of neuropathic pain. *Muscle and Nerve*, **16**, 1063–1065.

Torebjörk, H.E. & Hallin, R.G. (1970). C-fibre units recorded from human sensory nerve fascicles *in situ*. *Acta Societatis Medicorum Upsaliensis*, **75**, 81–84.

Torebjörk, H.E. & Hallin, R.G. (1973). Perceptual changes accompanying controlled preferential blocking of A and C fibre responses in intact human skin nerves. *Experimental Brain Research*, **16**, 321.

Torebjörk, H.E. & Hallin, R.G. (1974*a*). Identification of afferent C units in intact human skin nerves. *Brain Research*, **67**, 387–403.

Torebjörk, H.E. & Hallin, R.G. (1974*b*). Responses in human cutaneous A and C fibres to repeated electrical intradermal stimulation. *Journal of Neurology, Neurosurgery and Psychiatry*, **37**, 653–664.

Torebjörk, H.E. & Hallin, R.G. (1976). Skin receptors supplied by unmyelinated C fibres in man. In *Sensory Functions of the Skin in Primates, with Special Reference to Man. Wenner-Gren International Symposium 27*, ed. Y. Zotterman, pp. 475–487. Oxford: Pergamon Press.

Torebjörk, H.E., Lundberg, L.E.R. & LaMotte, R.H. (1992). Central changes in processing of mechanoreceptive input in capsaicin-induced secondary hyperalgesia in humans. *Journal of Physiology (London)*, **448**, 765–780.

Torebjörk, H.E. & Ochoa, J.L. (1980). Specific sensations evoked by activity in single identified sensory units in man. *Acta Physiologica Scandinavica*, **110**, 445–447.

Torebjörk, H.E. & Ochoa, J.L. (1981). Pain and itch from C-fibre stimulation. *Society for Neuroscience Abstracts*, **7**, 228.

Torebjörk, H.E. & Ochoa, J.L. (1990). New method to identify nociceptor units innervating glabrous skin of the human hand. *Experimental Brain Research*, **81**, 509–514.

Trotter, W. (1941). Observations and experiment and their use in the medical sciences. In *The Collected Papers of Wilfred Trotter, FRS*. Oxford: Oxford University Press.

Vallbo, A.B. & Hagbarth, K.-E. (1968). Activity from skin mechanoreceptors recorded percutaneously in awake human subjects. *Experimental Neurology*, **21**, 270–289.

Van Hees, J. & Gybels, J.M. (1972). Pain related to single afferent C fibres from human skin. *Brain Research*, **48**, 397–400.

Wall, P.D. (1984). The hyperpathic syndrome: a challenge to specificity theory. In *Somatosensory Mechanisms. Wenner-Gren International Symposium 41*, ed. C. Von Euler, O. Franzen, U. Lindblom & D. Ottoson, pp. 327–337. London: Macmillan Press.

Wall, P.D. & Gutnick, M. (1974). Properties of afferent nerve impulses originating from a neuroma. *Nature*, **248**, 740.

Yarnitsky, D. & Ochoa, J.L. (1990). Release of cold-induced burning pain by block of cold-specific afferent input. *Brain*, **113**, 893–902.

Zotterman, Y. (1936). Specific action potential in the lingual nerve of cat. *Scandinavian Archives of Physiology*, **75**, 105.

Zotterman, Y. (1971) *Touch, Tickle and Pain*, part 2. New York: Pergamon Press.

14

Sensory consequences of inflammation

A. IGGO

Department of Preclinical Veterinary Sciences, University of Edinburgh, Scotland, UK

Inflammation alters the properties of sensory receptors, in particular those with small-calibre afferent fibres, the Aδ and C groups, that include among their numbers the nociceptors. Early attempts to follow the lead of Keele & Armstrong (1964) in their classical studies of algogens acting on the cantharadin blister base, used electrophysiological methods to explore the recently accessible non-myelinated afferent units (Fjallbrant & Iggo, 1961). In these studies on healthy normal animals the algogens tested had relatively weak actions, and in some cases appeared to excite non-nociceptors as effectively as they did nociceptors. More recent work in other laboratories (Franz & Mense, 1975; Beck & Handwerker, 1976; Perl, 1976) extended the search to include, among other algogens, bradykinin (BK) and the prostanoids. Interactions of such agents with a potent non-steroidal anti-inflammatory drug (NSAID), acetylsalicyclic acid (ASA), were also relatively disappointing in the sense that they had little action in interfering with the action of administered BK or prostanoids, such as prostaglandins E_1 or E_2 (PGE_1, PGE_2) (Mense, 1982).

The development of several animal models of inflammation has now provided an opportunity to examine with more precision the nature of the inflammatory response and, in the present context, the behavioural (de Castro *et al.*, 1981), physiological (Guilbaud, Iggo & Tegner, 1985) and pharmacological (e.g. Guilbaud & Iggo, 1985) consequences of inflammation. In my own laboratory we have been using a model based on the complete Freund's adjuvant (CFA) arthritic rat (Pearson & Wood, 1959), developed for electrophysiological study of joint afferent fibres, initially with Guilbaud and Tegner (Guilbaud *et al.*, 1985) and subsequently modified to give a monoarthritis restricted to one joint (Grubb *et al.*, 1991). This particular model is proposed as an example of chronic pain where the inflammatory condition lasts for several weeks, allowing time for the development of any consequential changes in the processing of sensory information in the ascending sensory pathways. An alternative model, of short-lasting inflammation, has been used by Schmidt and colleagues, where the inflammatory agent was carrageenan or kaolin acting on the knee joint of the cat.

Joint capsule sensory receptors in arthritic rats

A characteristic feature of CFA arthritic rats is an enhanced sensitivity to gentle manipulation of the inflamed ankle joint, a procedure that is hardly noticed by normal rats (Kayser & Guilbaud, 1984). Electrophysiological recording from joint capsule afferent fibres also reveals an enhanced sensitivity to quantitatively controlled mechanical stimuli: the thresholds with von-Frey hairs were 13–26 mN in normality, compared with 0.4–1.7 mN in arthritic animals. In addition, sensory receptors in inflamed joints became spontaneously active and discharged sensory impulses in the anaesthetized animals, in striking contrast to the normals, where the joint capsule mechanonociceptors are normally silent. Indeed, in normal animals it was often difficult to find sensory units when mechanically stimulating the joint capsule. The enhancement of sensitivity did not otherwise alter the discharge characteristics: the sensory receptors were slowly adapting and had single spot-like receptive fields. In those experiments where it was possible to test the effects of flexion and extension of the inflamed ankle joint, there was a vigorous discharge from the arthritic receptors, in contrast to normals.

Pharmacology of sensitized joint capsule receptors

As mentioned above, there is considerable interest in identifying the algogens that are effective in enhancing the sensitivity of the joint nociceptors, partly with the objective of seeking pharmacological means to minimize pain and discomfort in arthritic patients. In our laboratories using the FCA rat model and close-arterial or intravenous injection of BK, substance P (SP), neurokinins A and B (NKA, NKB) and prostaglandin F_2 (PGF_2), we have not detected a significant rise in the responses of the sensory receptors in arthritic rats. Tests have further established that selective antagonists are equally effective against injected drugs in the normal and in the arthritic animal. The enhanced background and mechanical sensitivity of FCA joint capsule nociceptors is apparently unaffected, or at least is not reversed when the animals are treated with selective agonists for the algogens including BK, 5-hydroxytryptamine (5HT) and SP, suggesting that their enhanced responsiveness is not due to the single action of any of the algogens tested in this way.

One exception to this litany of failure has been the action of NSAID, such as lysine acetylsalicylic acid (L-ASA). The lack of effect of ASA on sensory receptors in normal animals (Mense, 1982) has already been mentioned. In the FCA rat, in contrast, ASA can be very potent in reducing both background activity and the response to mechanical stimuli (Fig. 14.1).

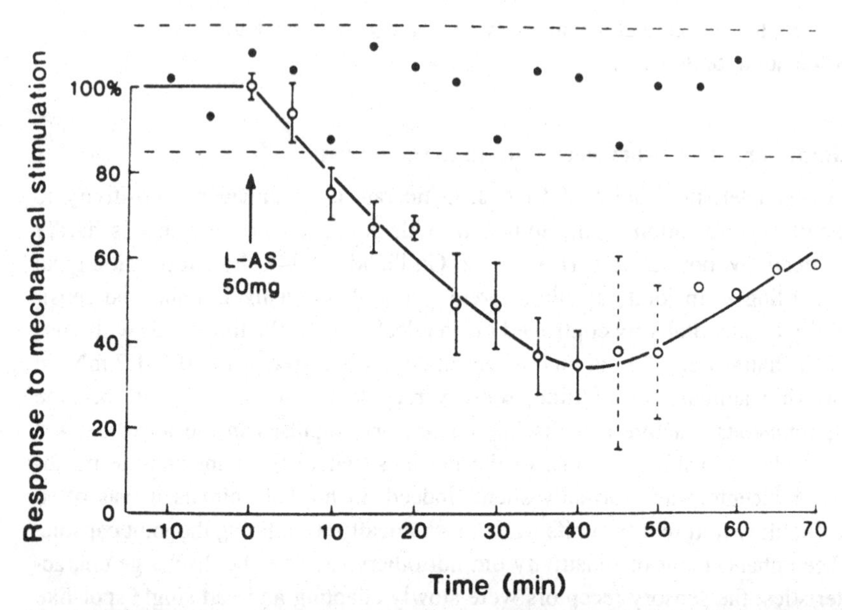

Fig. 14.1. The effect of intravenous lysine acetylsalicylate (L-AS) (50 mg ASA-equivalent/kg), administered at the arrow, on the responses to mechanical stimulation of joint capsule mechanoreceptors in normal (filled circles) and arthritic (open circles) rats. The dashed horizontal lines mark the 95% confidence limits for the normal receptors (the results from three receptors were pooled). The continuous line is drawn through the data points (mean ± SE) for pooled results from eight arthritic rats. Data were available from all animals up to 40 minutes (continuous error bars), for three of them for 50 minutes (dashed error bars), and for two of them up to 70 minutes. There is much greater individual variability in the rates of recovery than in the decline in sensitivity. Lysine acetylsalicylate had no effect on the normal receptors, but clearly depressed the responses in the arthritic animals. The pooled points in the graphs are calculated means and the vertical lines the standard errors of the means. From Guilbaud *et al.* (1985).

As can be seen from the figure, L-ASA had no effect on the response of joint capsule receptors to mechanical stimulation in normal animals, but within a few minutes there was a reduction in the arthritic animals that reached a maximum after about 40 minutes. The response declined with a half-life of 13–25 minutes, a value consistent wth a view that a prostaglandin less stable than PGE_2 was responsible. This result led us to test other prostanoids with the result that PGI_2 and its stable analogue, cicaprost, were found to be effective excitants in both normal (Birrell *et al.*, 1991) and arthritic preparations, where PGI_2 was more effective than PGE_2 in reversing the depression of response caused by L-AS (Birrell *et al.*, 1991). The absence of an available selective antagonist to the prostaglandins has so far prevented us from testing whether the enhanced activity of the arthritic sensory receptors can be accounted for by one or more of the prostaglandins. The prostaglandins were also able to enhance the excitatory action of BK.

Other putative chemical excitants continue to appear. Currently, there is active investigation of the possibility that opioid receptors, present in greater quantity in the paw tissue of FCA rats (Hassan *et al.*, 1993) may play a role in anti-nociception. Yet another new candidate is nitric oxide (NO), although testing of its role in affecting the sensory receptors in the arthritic rat is only at an early stage.

Central consequences of peripheral inflammation

Studies of the central consequences of peripheral inflammation in the CFA arthritic rat preceded systematic electrophysiological investigation of the peripheral sensory receptors. An early result was reported by Menetrey & Besson (1982) who found a high level of sustained discharge in the dorsal horn in a region that normally was silent in the absence of noxious peripheral stimulation. This superficial region of the dorsal horn (lamina I–II) is populated by nociceptor-specific (NS) neurones that do not carry a resting discharge and require noxious peripheral sensory stimuli to be excited. The striking change in neuronal activity in the FCA rats can now be accounted for by the presence of continuous background activity in the sensitized joint capsule receptors. Menetry & Besson could further enhance the neuronal activity in the dorsal horn by mild pressure, flexion, or extension of the joints, a result consistent with the view that the exaggerated activity was driven by the sensory receptors. A conclusive test would have come from experiments that blocked the incoming afferent activity, but this has not proved possible.

At higher levels in the sensory pathway there is also evidence for altered neuronal acticity in FCA rats. Gautron & Guilbaud (1982) reported that neurones in the ventrobasal thalamus exhibited high levels of background activity that could be further enhanced by light pressure on inflamed joints and by gentle flexion and extension of the joints. Here, too, there was evidence for abnormal neuronal response that could be further enhanced by what in the normal animal would be innocuous stimuli. An added test to tie the responses back to the periphery was the administration of L-ASA. In doses similar to those used subsequently in the peripheral sensory experiments (Guilbaud & Iggo, 1985), there was a reduction both in background discharge and in response to light pressure and joint flexion (Guilbaud *et al.*, 1982). Not only was the degree of depression similar to that found for the sensory receptors but also the time course of the action was similar (Fig. 14.2), further evidence that the altered central neuronal responses were driven to at least some degree by the antecedent activity in the sensory receptors.

The mechanisms in the spinal cord that underlie the high level of activity of dorsal horn neurones and their ready excitation by normally innocuous peripheral stimuli can be considered from several aspects. A contributing

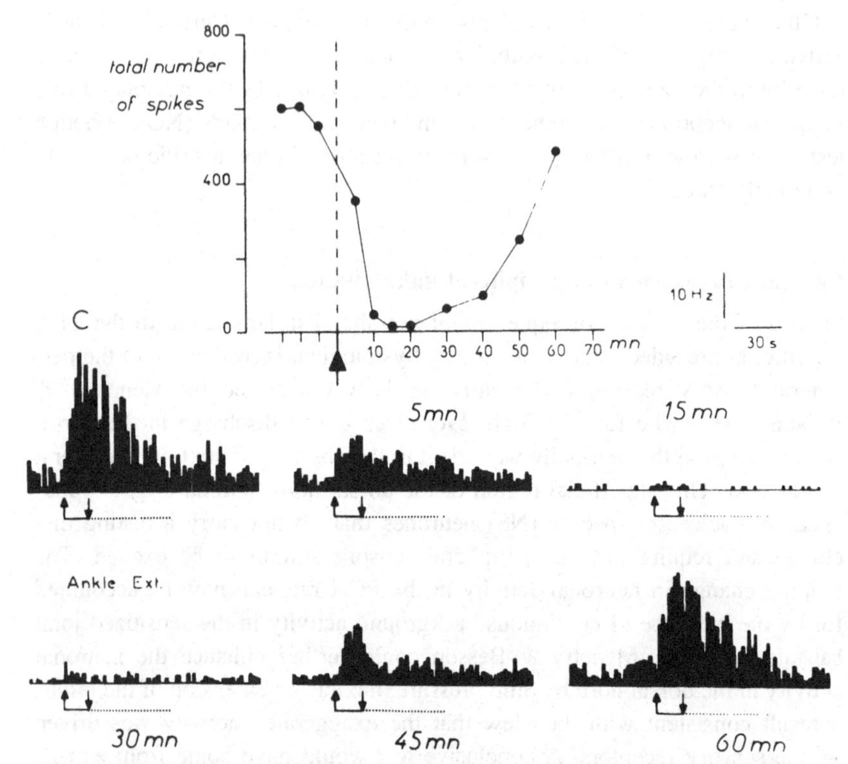

Fig. 14.2. An individual example illustrating a particular strong depressive effect of L-ASA (50 mg/kg intravenously, administered at large arrow) upon responses of a VB neurone initially activated by an extension of the contralateral ankle (C, control response). The stimulation is applied during 15 s between the two small arrows: the duration considered for the calculation of the total number of spikes in the response included the duration of the initial post-effect which is denoted by the dotted line. In this case the spontaneous firing rate was slightly decreased by L-ASA. mn, minutes. From Guilbaud *et al.* (1982).

factor to the high level of activity in neurones driven by non-myelinated nociceptors is 'wind-up' (Mendell, 1966), the phenomenon of a progressive increase in activity of NS and other neurones by a sustained input from C-nociceptors. This normal process by itself could account for some of the heightened dorsal horn discharge, since there is a continuous abnormal inflow of impulses from C-nociceptors in the inflamed tissues. Another question relates to the effect of afferent inflow on the release of transmitters and modulators from afferent fibre synapses. Using an elegant new technique, the antibody microprobe, Schaible *et al.* (1990) assessed SP release in the dorsal horn. SP increases in amount in the sensory nerves in CFA rats and much of it appears to be transported to the periphery despite the apparent lack of effect of SP on the nociceptor responses to mechanical stimulation. Schaible *et al.*

(1990) have shown in the cat that SP is released in the superficial dorsal horn by noxious stimulation of the digits, for example. A substantially greater amount of SP is released, presumably in the substantia gelatinosa and in lamina I, when light pressure is applied to the inflamed knee joint of a cat, indicating at least the high level of activity in the C-nociceptors. Such massive release of SP, and presumably other transmitters, may contribute to the hypersensitivity of central neurones (Ren *et al.*, 1992; Schaible *et al.*, 1993).

Finally, the question has to be asked whether the plastic changes detectable at spinal levels of the sensory pathway are dependent and consequential on the peripheral sensory hypersensitivity, or whether they become self-sustaining leading to a state of central hyperalgesia that persists after the peripheral changes have been resolved. The question remains unanswered.

References

Beck, P.W. & Handwerker, H.O. (1976). Bradykinin and serotinin effects on various types of cutaneous nerve fibres. *Pflügers Archiv*, **347**, 209–222.

Birrell, G.J., McQueen, D.S., Iggo, A., Coleman, R.A. & Grubb, B.D. (1991). PGI_2-induced activation and sensitization of articular mechanonociceptors. *Neuroscience Letters*, **124**, 5–8.

deCastro Costa, M., DeSutter, P., Gybels, J. & Van Hees, I. (1981). Adjuvant-induced arthritis in rats: a possible animal model of chronic pain. *Pain*, **10**, 173–185.

Fjallbrant, N. & Iggo, A. (1961). The effect of histamine, 5-hydroxytryptamine and acetylcholine on cutaneous afferent fibres. *Journal of Physiology (London)*, **156**, 578–590.

Franz, M. & Mense, S. (1975). Muscle receptors with group IV afferent fibres responding to application of bradykinin. *Brain Research*, **92**, 369–383.

Gautron, M. & Guilbaud, G. (1982). Somatic responses of ventrobasal thalamic neurones in polyarthritic rats. *Brain Research*, **237**, 459–471.

Grubb, B.D., Birrell, G.J., McQueen, D.S. & Iggo, A. (1991). The role of PGE_2 in the sensitization of mechanoreceptors in normal and inflamed ankle joints of the rat. *Experimental Brain Research*, **84**, 383–392.

Guilbaud, G., Benoist, J.M., Gautron, M. & Kayser, V. (1982). Aspirin clearly depresses responses of ventrobasal thalamus neurones to joint stimuli in arthritic rats. *Pain*, **13**, 153–163.

Guilbaud, G. & Iggo, A. (1985). The effect of lysine acetylsalicylate on joint capsule mechanoreceptors in rats with polyarthritis. *Experimental Brain Research*, **61**, 164–168.

Guilbaud, G., Iggo, A. & Tegner, R. (1985). Sensory receptors in ankle joint capsules of normal and arthritic rats. *Experimental Brain Research*, **58**, 29–40.

Hassan, A.H.S., Ableitner, A., Stein, C. & Herz, A. (1993). Inflammation of the rat paw enhances axonal transport of opioid receptors in the sciatic nerve and increases their density in the inflamed tissue. *Neuroscience*, **55**, 185–195.

Kayser, V. & Guilbaud, G. (1984). Further evidence for changes in the responsiveness of somatosensory neurones in arthritic rats: a study of the posterior intralaminar region of the thalamus. *Brain Research*, **323**, 144–147.

Keele, C.A. & Armstrong, D. (1964). *Substances Producing Pain and Itch*. London: Edward Arnold.

Lamour, Y., Guilbaud, G. & Willer, J.C. (1983). Altered properties and laminar distribution of neuronal responses to peripheral stimulation in the SmI cortex of arthritic rats. *Brain Research*, **273**, 183–187.

Mendell, L.M. (1966). Physiological properties of unmyelinated fibre projections to the spinal cord. *Experimental Neurology*, **16**, 316–332.

Menetrey, D. & Besson, J.-M. (1982). Electrophysiological characteristics of dorsal horn cells in rats with cutaneous inflammation resulting from chronic arthritis. *Pain*, **13**, 343–364.

Mense, S. (1982). The reduction of the bradykinin-induced activation of feline group II and IV muscle receptors by acetylsalicylic acid. *Journal of Physiology (London)*, **326**, 269–283.

Pearson, C.M. & Wood, F.D. (1959). Studies of arthritis and other lesions induced in rats by injection of mycobacterial adjuvant. I. General clinical and pathological characteristics and some modifying factors. *Arthritis and Rheumatism*, **2**, 440–459.

Perl, E.R. (1976). Sensitization of nociceptors and its relation to sensation. *Advances in Pain Research and Therapy*, **1**, 17–28.

Ren, K., Hylden, J.L.K., Williams, G.M., Ruda, M.A. & Dubner, R. (1992). The effects of a non-competitive NMDA receptor antagonist MK-801 on behavioural hyperalgesia and dorsal horn neuronal activity in rats with unilateral inflammation. *Pain*, **50**, 331–344.

Schaible, H.-G., Grubb, B.D., Neugeaeur, V. & Oppmann, M. (1993). The effects of NMDA antagonists on neuronal activity in cat spinal cord evoked by acute inflammation of the knee joint. *European Journal of Neuroscience*, **3**, 981–991.

Schaible, H.-G., Jarrott, B., Hope, P.J. & Duggan, A.W. (1990). Release of immunoreactive substance P in the spinal cord during development of acute arthritis in the knee joint of the cat: a study with antibody microprobes. *Brain Research*, **529**, 214–223.

15

Non-voluntary muscle activity and myofascial pain syndromes

R.H. WESTGAARD

Division of Organization and Work Science, University of Trondheim, Trondheim, Norway

Myofascial pain syndromes

Pain syndromes of suspected myofascial origin are one of the most prevalent problems in clinical neurology today. These syndromes include tension headache, temporomandibular pain, myofascial pain relating to hyperextension injury ('whiplash') or located to muscles such as sternocleidomastoid or trapezius, as well as subgroups of chronic low back pain. Many of these conditions are located to cervical or suboccipital muscles and constitute a separate group in the classification scheme for chronic pain published in *Pain* (Merskey, 1986). The pain appears to originate in muscles or connective tissue associated with muscle. The aetiology is not known, but parameters indicating a negative emotional state (depression, anxiety), stressful conditions at home or at work (time pressure, emotional conflicts, perceived poor psychosocial work environment) or mechanical stress (posture, repetitive work tasks) are often shown to correlate with the pain condition (Holm *et al.*, 1986; Ellertsen & Kløve, 1987; Kamwendo, Linton & Moritz, 1991; Winkel & Westgaard, 1992).

Myofascial pain syndrome in the trapezius muscle is often caused by mechanical stressors at the work place, such as maintained postures or repetitive work tasks. The trapezius muscle is particularly vulnerable in its function as a stabilizer of the scapula and lifter of the shoulder girdle. In this research area work tasks demanding a high level of muscle force, causing muscle fatigue with a feeling of discomfort and presumably causing muscle pain in the longer term, were initially studied. The remedy was thought to be a reduction in muscle activity level, but continuing research has shown that this was not always helpful. In fact, the prevalence rate for musculoskeletal pain in the shoulder can be just as high in sedentary work with stressful working conditions as it is in heavy manual work.

Our own studies of presumed work-related pain syndromes in the shoulder–neck region have shown that muscle activity in the trapezius can vary more than 10-fold, from 0.5% to 10% of the electromyographic (EMG) signal at maximal voluntary force (%EMG_{max}), for workers performing

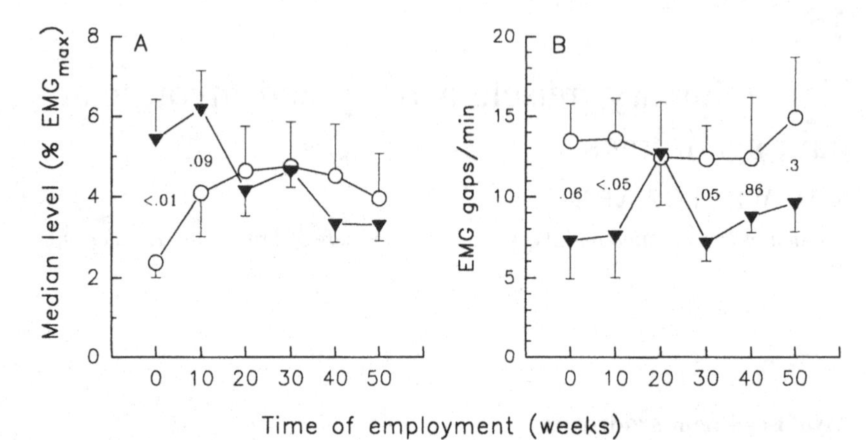

Fig. 15.1. The mean value of the median contraction level (A) and the total number of EMG gaps (B) for prospective patients (filled symbols) and non-patients (open symbols) performing the same, light repetitive work tasks plotted as a function of time of employment. Hazard ratio with confidence interval (Cox regression with time-dependent covariates) for median contraction level and total number of EMG gaps when excluding the first recording was 1.41 (CI 0.93–2.13) and 0.94 (CI 0.89–0.99), respectively. From Veiersted, *et al.* (1993) with permission.

similar sedentary and light manual work (Jensen *et al.*, 1993). Within this range of muscle activity levels there is a poor or no correlation between the pain symptoms reported and the level of muscle activity recorded in cross-sectional studies. A longitudinal study of healthy young females employed in light industry showed that the median trapezius activity level (group mean 4 %EMG_{max}) did not distinguish those who developed trapezius myalgia from those who remained symptom free (Veiersted, Westgaard & Andersen, 1993; Fig. 15.1A).

Epidemiological evaluation of risk factors brought up negative emotional factors (depression, perceived poor psychosocial work environment) and previous experience of similar pain symptoms, which may indicate reduced tolerance for muscle loads, as the only factors consistently correlated with pain symptoms in the shoulder–neck region during the preceding 12 months (Westgaard, Jensen & Hansen, 1993). These risk factors are similar to those reported for other myofascial pain syndromes. Workers reporting these risk factors did not show an elevated muscle activity level when performing their normal work tasks. For subjects not reporting the two risk factors there was a clear increase in pain symptoms with increasing muscle load, but with a considerable scatter around the regression line. The effect of psychosocial stress and sensitivity to muscle load, as indicated by previous pain symptoms, was largest in the occupational group with the lowest mechanical load.

Previous pain symptoms, but not psychosocial stress, were also associated with pain in the lower trunk and the extremities in these studies. Psychosocial stress is a prominent risk factor for cervical myofascial pain syndromes and many of the patients with symptoms in the shoulder region also reported symptoms of tension headache.

Non-voluntary muscle activity in the laboratory and at the workplace

Many experimental studies have set out to demonstrate an association between myofascial pain syndromes and muscle activity, with mixed results. This literature was reviewed in depth by Flor & Turk (1989) who concluded on the basis of 'methodologically sound' studies that there is evidence for a symptom-specific, stress-related elevation of muscle activity for patients with tension headache, but with considerable inter-individual variation in EMG responses. The EMG levels observed in these studies are generally low, typically around 10 μV root mean square, corresponding to 5–10 $\%EMG_{max}$ for the frontalis muscle and about 1 $\%EMG_{max}$ for the trapezius muscle.

Some of the studies reviewed by Flor & Turk (1989) concern non-voluntary muscle activity, first reported by Jacobson (1930). He showed an EMG response in arm muscles when the subject was asked to imagine bending the arm. Later he and others demonstrated non-voluntary muscle activity in situations relating to mental imagery, problem solving, anxiety or frustration and high mental loads. Non-voluntary muscle activity can thus be initiated by stressors in the environment, stressors that are risk factors for myofascial pain syndromes.

We have become interested in non-voluntary muscle activity as a possible contributing factor to occupational trapezius myalgia. Methods to study this phenomenon in the laboratory and to detect evidence of such activity at the workplace have been developed, based on a complex two-choice reaction time test presented on a VDU screen (Westgaard & Bjørklund, 1987). In these studies care is taken to minimize motor movement, by indicating responses with slight finger movement. Fig. 15.2 (upper panel) shows a typical surface EMG recording of non-voluntary activity from the upper trapezius muscle. This activity is characterized by low second-to-second variability, except for sudden jumps or more gradual increase or decrease in tension level ('gradients'). Similar activity patterns can be observed in widely separate muscles (Westgaard & Bjørklund, 1987), and parallel recordings from several muscles have demonstrated that this phenomenon is much more common in facial, neck and shoulder muscles than in muscles of the lower trunk and the extremities (M. Waersted & R.H. Westgaard, unpublished data).

Recent experiments with simultaneous recording of surface EMG and single motor units have shown that single motor units can be continuously active throughout a period of non-voluntary muscle activity with a duration of 10 minutes (Fig. 15.2, lower panel). Variation in the activity level of the surface EMG was commonly mirrored by similar oscillations in the

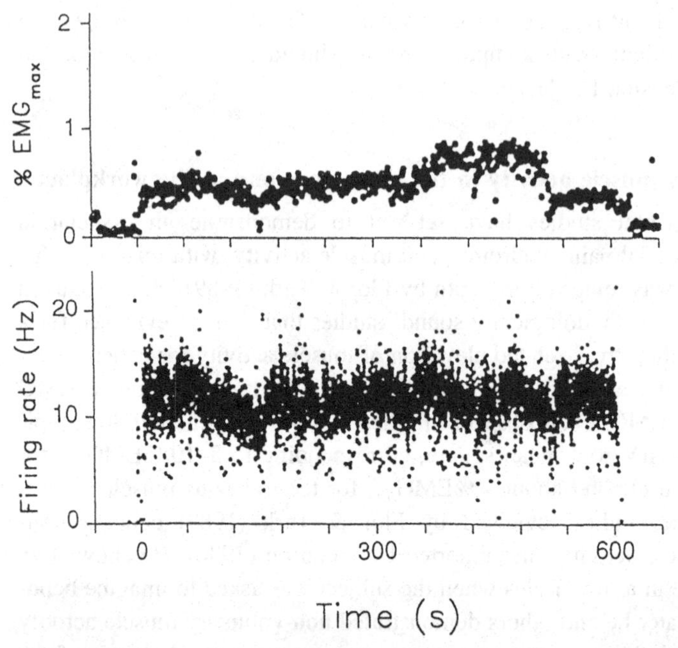

Fig. 15.2. Simultaneous time plots at 1 s resolution of integrated surface EMG (upper panel) and instantaneous firing frequency of a single motor unit (lower panel) from the upper trapezius muscle. The reaction time task starts at time 0 s and ends at time 600 s. From Waersted, Eken & Westgaard (1993).

instantaneous firing rate of single motor units, suggesting synchronous modulation of presynaptic input to the active motor units. Sudden changes in the surface EMG level were paralleled either by recruitment or derecruitment of single motor units (e.g. time 0 s, Fig. 15.2) or by a transient reduction in firing rate (e.g. time 530 s, Fig. 15.2). In other experiments showing a period of continuously increasing or decreasing surface EMG level, new motor units were recruited or derecruited throughout this period, but when recruited they retained a stable firing rate.

The continuous activation of single motor units is consistent with one hypothesized mechanism for the development of myofascial pain syndromes, namely overexertion of a small number of motor units in an otherwise quiescent muscle. Epidemiological studies of the development of trapezius myalgia at the workplace show that a considerable risk of long-term health effects may exist at activity levels down to 1–2% maximum voluntary contraction, or even below this level. The lack of a dose–effect relation at such low activity levels can be understood if the pain symptoms are caused by activation of a subset of motor units, rendering measurements of total muscle function a poor pointer to pathophysiological processes.

We have attempted to look for evidence of non-voluntary muscle activity in recordings of occupational muscle activity with low EMG levels. An analysis was introduced to quantify periods with quiescent EMG activity ('EMG gaps'), defined as periods of at least 0.2 s below 0.5 $\%EMG_{max}$ (Veiersted, Westgaard & Andersen, 1990). The previously mentioned longitudinal study of healthy females performing manual work tasks, where prospective patients were not distinguished on the basis of median muscle load, showed that the prospective patients with diagnosed trapezius myalgia had a significantly lower occurrence of EMG gaps than did those remaining healthy (Fig. 15.1B).

The hypothesis of overexertion of low-threshold motor units is supported by findings of increased occurrence of type I fibres with mitochondrial deficiencies ('ragged-red fibres') in the trapezius of patients with shoulder myalgia (Larsson *et al.*, 1988, 1990). The finding of unusually large type I muscle fibres in a similar patient material can also be interpreted in support of a sustained activation of a subpopulation of muscle fibres (Lindman *et al.*, 1991).

The neurophysiology of non-voluntary muscle activity

The most interesting presynaptic inputs to the motoneurones in relation to non-voluntary muscle activity are the serotonergic raphe-spinal system and the descending noradrenergic projections from nucleus coeruleus and subcoeruleus. These monoaminergic systems from the brainstem show a profuse innervation of motor nuclei throughout the spinal cord with individual axons sending collaterals to many spinal segments, and are predominantly monosynaptic excitatory to motoneurones (Aghajanian & Vandermaelen 1986; Holstege & Kuypers, 1987). Coeruleus cells in conscious animals are activated by various challenging or stressful stimuli (Jacobs, 1986). Medial pathways from the vestibular nuclei, the medullary reticular and the pontine reticular formations project predominantly to motoneurones of axial and proximal muscles with some pathways not extending beyond the upper thoracic level (Kuypers, 1981). This distribution is consistent with the observed distribution of muscle activation in non-voluntary muscle activity.

Fibres in the raphe magnus nuclei project to the superficial zone of the dorsal horn of the spinal cord and depress afferent nociceptive input. Electrical stimulation of the nuclei locus coeruleus has an inhibitory effect on pain transmission in the dorsal horn (Mokha, McMillan & Iggo, 1986). Thus, activation of the monoaminergic systems in the brainstem reticular formation has both a general facilitating effect on spinal motoneurones and an inhibitory effect on the pain transmission pathways. Interestingly, in long-term experiments with non-voluntary muscle activity, we have observed that some patients with chronic tension headache reported reduced pain as the experiment progressed, despite a high level of non-voluntary muscle activity (unpublished observations).

Abnormal activation of brainstem nuclei has recently been implicated as a pathophysiological mechanism in tension headache, following the demonstration that the second temporalis exteroceptive suppression period (ES2) is reduced in patients with tension headache (Schoenen *et al.*, 1987). This reflex inhibits voluntary masseter and temporalis activity and is part of the jaw opening reflex, elicited by painful peri- or intraoral stimulation. Serotonin-dependent pathways are likely to take part in the modulation of this reflex as ES2 is increased by serotonin agonists and reduced by a serotonin reuptake blocker (Schoenen, 1993). ES2 is also influenced by the psychological and physical state of well-being (Wallasch *et al.* 1993).

Peripheral or central mechanisms in myofascial pain syndromes?

The evidence cited above points to central mechanisms in the modulation of tension-type myofascial pain syndromes, and serves as part of the basis for the current debate regarding peripheral or central mechanisms in tension headache. Clinical signs of trigger points, palpatable bands in the muscle, pressure-sensitive tendons, etc., point to peripheral mechanisms. Other evidence is an increased level of stress-related non-voluntary muscle activity in tension headache patients. Occupational myofascial pain syndromes have so far been assumed to have a peripheral aetiology, secondary to muscle fatigue. Accepting that some forms of occupational myofascial syndromes develop without gross manifestation of muscle fatigue, the same debate regarding peripheral versus central mechanisms can be raised for occupational pain syndromes.

Hypotheses regarding peripheral mechanisms in the development of myofascial pain are frequently related to sustained elevation of intracellular calcium or release of potassium. Nociceptors in muscle are sensitive to potassium, but this response habituates relatively fast. A sustained high level of free calcium in the muscle cells can have several toxic effects including formation of free radicals, increased phospholipase activity and mitochondrial calcium overload, the last consistent with the reported increase of ragged-red fibres in occupational myalgia. These mechanisms operate on the individual muscle fibres independently of the activation level for the muscle.

Pain-eliciting mechanisms released by subsets of motor units are a prerequisite for explaining the marginal dose–effect relation between occupational muscle load exposure, measured by surface EMG, and health effects. The activation of single motor units in non-voluntary muscle activity shows that situations with overexertion of a small number of motor units can exist. Indicators of remaining, low tension levels in EMG recordings differentiate between patients and healthy controls in a prospective and some cross-sectional studies. This evidence can be interpreted in support of peripheral mechanisms in the generation of muscle pain syndromes. However, pain mechanisms initiated by peripheral factors are likely to be heavily modulated by central mechanisms. In particular, the possibility exists that the same central mechanism, i.e. descending input from the limbic system, both initiates myofascial pain symptoms through prolonged activation of muscle fibres and modulates these through gating effects at the medullary level. This hypothesis is speculative at present, but serves as a basis for future investigations.

References

Aghajanian, G.K. & Vandermaelen C.P. (1986). Specific systems of the reticular core: serotonin. In *Handbook of Physiology*, section 1, *The Nervous System*, ed. V.B. Mountcastle, pp. 237–256. Oxford: Oxford University Press.

Ellertsen, B. & Kløve, H. (1987). MMPI patterns in chronic muscle pain, tension headache, and migraine. *Cephalalgia*, **7**, 65–71.

Flor, H. & Turk, D.C. (1989). Psychophysiology of chronic pain: do chronic pain patients exhibit symptom-specific psychological responses? *Psychological Bulletin*, **105**, 215–259.

Holm, J.E., Holroyd, K.A., Hursey, K.G. & Penzien, D.B. (1986). The role of stress in recurrent tension headache. *Headache*, **26**, 160–167.

Holstege, J.C. & Kuypers, H.G.J.M. (1987). Brainstem projections to spinal motoneurones: an update. *Neuroscience*, **23**, 809–821.

Jacobs, B.L. (1986). Single unit activity of locus coeruleus neurons in behaving animals. *Progress in Neurobiology*, **27**, 183–194.

Jacobson, E. (1930). Electrical measurements of neuromuscular states during mental activities. I. Imagination of movement involving skeletal muscle. *American Journal of Physiology*, **91**, 567–608.

Jensen, C., Nilsen, K., Hansen, K. & Westgaard, R.H. (1993). Trapezius muscle load as a risk indicator for occupational shoulder–neck complaints. *International Archives of Occupational Environmental Health*, **64**, 415–423.

Kamwendo, K., Linton, S.J. & Moritz, U. (1991). Neck and shoulder disorders in medical secretaries. I. Pain prevalence and risk factors. *Scandinavian Journal of Rehabilitation Medicine*, **23**, 127–133.

Kuypers, H.G.J.M. (1981). Anatomy of the descending pathways. In *Handbook of Physiology, The Nervous System*, vol. 2, *Motor Control*, part I, ed. J.M. Brookhart, V.B. Mountcastle, V.B. Brooks & S.R. Geiger, pp. 597–666. Bethesda, MD: American Physiological Society.

Larsson, S.-E., Bengtsson, A., Bodegård, L., Henriksson, K.G. & Larsson, J. (1988). Muscle changes in work related chronic myalgia. *Acta Orthopaedica Scandinavica*, **59**, 552–556.

Larsson, S.-E., Bodegård, L., Henriksson, K.G. & Öberg, P.Å. (1990). Chronic trapezius myalgia: morphology and blood flow studied in 17 patients. *Acta Orthopaedica Scandinavica*, **61**, 394–398.

Lindman, R., Hagberg, M., Ängqvist, K.-A., Söderlund, K., Hultman, E. & Thornell, L.-E. (1991). Changes in muscle morphology in chronic trapezius myalgia. *Scandinavian Journal of Work Environment and Heath*, **17**, 347–355.

Merskey, H. (ed.) (1986). Classification of chronic pain: descriptions of chronic pain syndromes and definitions of pain terms. *Pain* (Supplement 3), 226pp.

Mokha, S.S., McMillan J.A. & Iggo, A. (1986). Pathways mediating descending control of spinal nociceptive transmission from the nuclei locus coeruleus (LC) and raphe magnus (NRM) in the cat. *Experimental Brain Research*, **61**, 597–606.

Schoenen, J. (1993). Exteroceptive suppression of temporalis muscle activity in patients with chronic headache and in normal volunteers: methodology, clinical and pathophysiological relevance. *Headache*, **33**, 3–17.

Schoenen, J., Jamart, B., Gerard, P., Lenarduzzi, P. & Delwaide, P.J. (1987). Exteroceptive suppression of temporalis muscle activity in chronic headache. *Neurology*, **37**, 1834–1836.

Veiersted, K.B., Westgaard, R.H. & Andersen, P. (1990). Pattern of muscle activity during stereotyped work and its relation to muscle pain. *International Archives of Occupational Environmental Health*, **62**, 31–41.

Veiersted, K.B., Westgaard, R.H. & Andersen, P. (1993). Electromyographic evaluation of muscular work pattern as a predictor of trapezius myalgia. *Scandinavian Journal of Work Environment and Health*, **19**, 284–290.

Waersted M., Eken, T. & Westgaard, R.H. (1993). Psychogenic motor unit activity: a possible muscle injury mechanism studied in a healthy subject. *Journal of Musculoskeletal Pain*, **1**, 185–190.

Wallasch, T.-M., Niemann, U., Kropp, P. & Weinschütz, T. (1993). Exteroceptive silent periods of temporalis muscle activity: correlation with neuropsychological findings. *Headache*, **33**, 121–124.

Westgaard, R.H. & Bjørklund, R. (1987). Generation of muscle tension additional to postural muscle load. *Ergonomics*, **30**, 911–923.

Westgaard, R.H., Jensen, C. & Hansen, K. (1993). Individual and work-related risk factors associated with symptoms of musculoskeletal complaints. *International Archives of Occupational Environmental Health*, **64**, 405–413.

Winkel, J. & Westgaard, R.H. (1992). Occupational and individual risk factors for shoulder–neck complaints. II. The scientific basis (literature review) for the guide. *International Journal of Industrial Ergonomics*, **10**, 85–104.

16

Is there a mechanism for the spinal cord to remember pain?

W.D. WILLIS, JR

Department of Anatomy & Neurosciences and Marine Biomedical Institute, University of Texas Medical Branch, Galveston, Texas, USA

Introduction

Pain is often an immediate consequence of injury of somatic tissue or of nerves. In addition, longer-term sensory events may follow such injuries, including primary and secondary hyperalgesia, allodynia, and the development of a chronic pain state (Wall & Melzack, 1989; Bonica, 1990). Acute pain is generally easy to recognize and to treat. However, the longer-term sensory changes are more difficult to recognize, are poorly understood, and are difficult to manage. Clearly, a theoretical foundation is needed on which to base clinical investigations of subacute and chronic pain states.

In the past several years, a number of experimental approaches have been used to improve our understanding of the neural events that follow acute injury. Evidence about sensory events has come from behavioural studies, and evidence about neural events has been derived from electrophysiological recordings and pharmacological manipulations. The studies can generally be subdivided into those stressing peripheral mechanisms, as exemplified by the chapters by Ochoa, and by Iggo (this volume), and those emphasizing central mechanisms, as reviewed in this chapter and that by Shen (this volume).

A model of secondary hyperalgesia

When the skin is damaged, pain often occurs immediately, and then primary and secondary hyperalgesia may develop (Lewis, 1942; Hardy, Wolff & Goodell, 1952). The secondary hyperalgesia may be accompanied by mechanical allodynia (LaMotte *et al.*, 1991, 1992; Torebjörk, Lundberg & LaMotte, 1992). Primary hyperalgesia can last for days, and it is attributable to the sensitization of nociceptors (Bessou & Perl, 1969; Fitzgerald & Lynn, 1977; Meyer & Campbell, 1981; LaMotte, Thalhammer & Robinson, 1983). Secondary hyperalgesia begins within minutes after injury and lasts hours (Hardy *et al.*, 1952); it may be accompanied by mechanical allodynia (LaMotte *et al.*, 1991, 1992). Secondary hyperalgesia and allodynia cannot be ascribed to long-term changes in nociceptors, although they result from the activation of nociceptors (Baumann *et al.*, 1991, LaMotte *et al.*, 1991, 1992). Instead, these sensory phenomena are caused by changes in the excitability of neurones in the central nervous system (LaMotte *et al.*, 1991, 1992; Torebjörk *et al.*, 1992).

The mechanisms involved in the central changes that result in secondary hyperalgesia and allodynia have been examined in experiments on the primate spinal cord, using the discharges of spinothalamic tract (STT) cells to model sensory changes (Simone *et al.*, 1991, Dougherty & Willis, 1992). Nociceptors in the skin are activated by intradermal injection of capsaicin. In human subjects, this causes pain, primary and secondary hyperalgesia and mechanical allodynia (LaMotte *et al.*, 1991, 1992; Torebjörk *et al.*, 1992). In anaesthetized monkeys, intradermal injection of capsaicin causes a powerful activation of STT cells that lasts for about 15–20 minutes, followed by an enhanced responsiveness of the neurones to innocuous and noxious mechanical stimuli applied both to skin near the injection site (not illustrated) and to skin in a wide surrounding area (Fig. 16.1). The increased responses to mechanical stimuli develop within 15 minutes of the capsaicin injection, and they last for at least $1\frac{1}{2}$ hours (Fig. 16.2). The STT cells undergo an increased excitability to iontophoretically applied excitatory amino acids with the same time course as the changes in responsiveness to mechanical stimulation (Fig. 16.2).

The injection of capsaicin into the skin activates chiefly C-nociceptors (but also some Aδ-nociceptors: Baumann *et al.*, 1991). C-nociceptors are likely to release both fast- and slow-acting neurotransmitters in the spinal cord, including excitatory amino acids, such as glutamate (GLU), and peptides, such as substance P (SP) (Battaglia & Rustioni, 1988; DeBiasi & Rustioni, 1988). Iontophoretic co-application of *N*-methyl-D-aspartic acid (NMDA) or of quisqualic acid (QUIS), but not α-amino-3-hydroxy-5-methylisoxazole-4-proprionic acid (AMPA), and SP onto STT cells often causes a long-lasting increase in responses to repeated applications of the same excitatory amino acid and to mechanical stimulation of the skin (Fig. 16.3; Dougherty & Willis, 1991; Dougherty *et al.*, 1993). Curiously, when responses to NMDA are increased, those to QUIS or AMPA may be reduced, and when responses to QUIS are increased, those to NMDA may be reduced (Dougherty *et al.*, 1993).

To determine whether neurotransmitters, such as excitatory amino acids and/or SP, are actually involved in the sensitization of STT cells following intradermal injection of capsaicin, selective antagonists were administered into the dorsal horn by microdialysis to block the action of these agents on their receptors. The non-NMDA receptor antagonist, CNQX, essentially eliminates the responses of STT cells to mechanical, thermal and chemical stimuli (Fig. 16.4). Evidently, synapses using an excitatory amino acid acting at non-NMDA receptors are a vital link in afferent pathways that activate STT cells (Dougherty

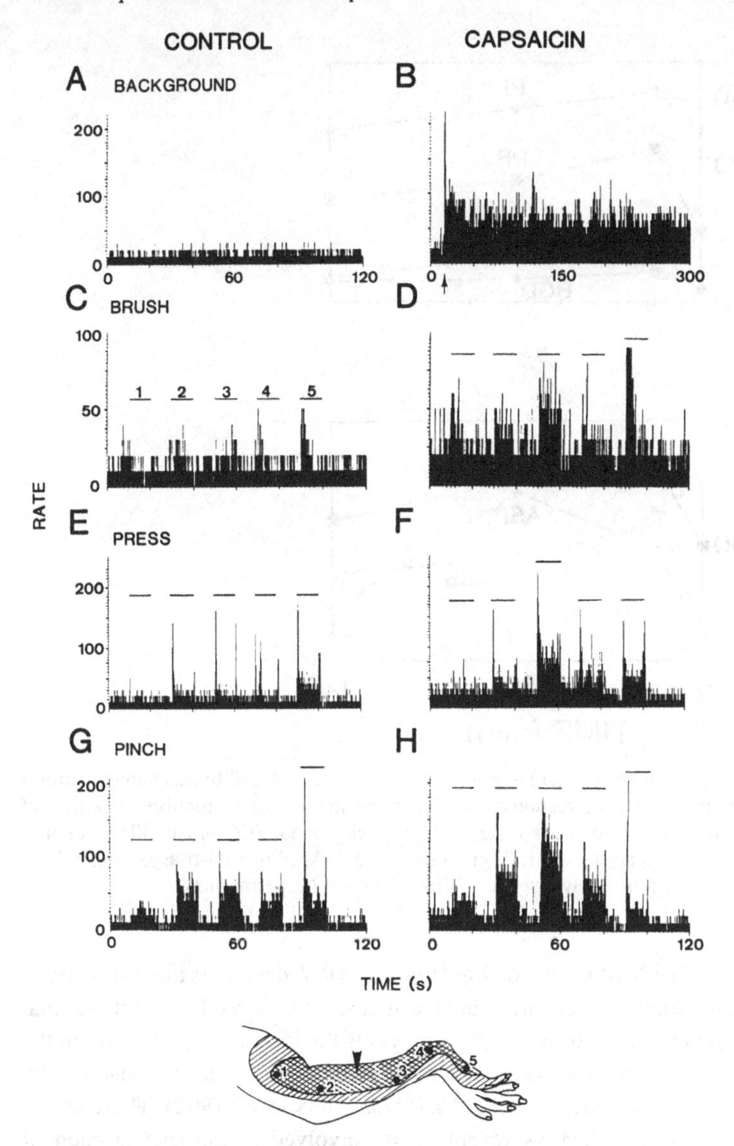

Fig. 16.1. Increases in the responses of a spinothalamic tract (STT) cell to mechanical stimuli following intradermal injection of capsaicin. The peristimulus time histograms show the background activity (A) and the responses of an STT cell to capsaicin (B) and to innocuous (BRUSH) and noxious (PRESS, PINCH) mechanical stimuli (C-H) applied at five different sites across the receptive field (see drawing at bottom). The control responses to mechanical stimulation are in (C), (E) and (G), and those following capsaicin are in (D), (F) and (H). The capsaicin, which was injected at the site marked by the arrowhead in the drawing, caused the receptive field to increase in size (cf. doubly and singly hatched areas). From Dougherty & Willis (1992) with permission.

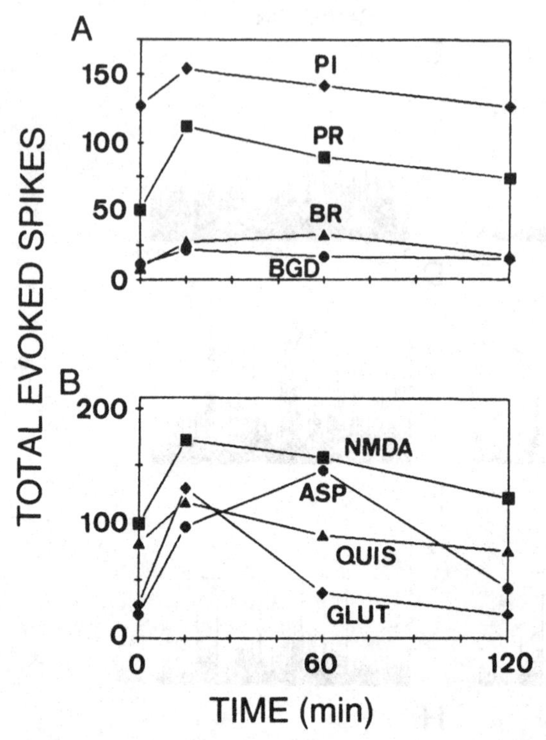

Fig. 16.2. Time courses of increases in responses of an STT cell to mechanical stimuli are shown in (A) and in responses of the same neuron to iontophoretic pulses of several excitatory amino acids in (B) BGD, background; BR, brush; PR, pressure; PI, pinch; ASP, aspartate; GLUT, glutamate; NMDA, *N*-methyl-D-aspartate; QUIS, quisqualic acid. From Dougherty & Willis, (1992) with permission.

et al., 1992). The NMDA receptor antagonist, AP-7, decreases the responses of STT cells to noxious mechanical and thermal stimuli, as well as to intradermal injection of capsaicin; furthermore, it prevents the long-lasting increase in the responses of these neurones to mechanical stimulation of the skin that usually follows intradermal capsaicin (Fig. 16.4; Dougherty *et al.*, 1992). These observations suggest that NMDA receptors are involved in the transmission of nociceptive information to STT cells and, in addition, are mediators of the sensitization of STT cells that occurs following intradermal injection of capsaicin. The NK1 receptor antagonist, CP-96,345, also prevents the sensitization of STT cells to mechanical stimuli, suggesting that SP is also involved in the process (Willis *et al.*, 1992). Since CP-96,345 has an additional action on Ca^{2+} channels, the effects of CP-96,344 were also examined. This agent has a similar action on Ca^{2+} channels, but does not antagonize NK-1 receptors. CP-96,344 had no effect on the sensitization of STT cells. Thus, SP is likely to contribute to the sensitization of STT cells.

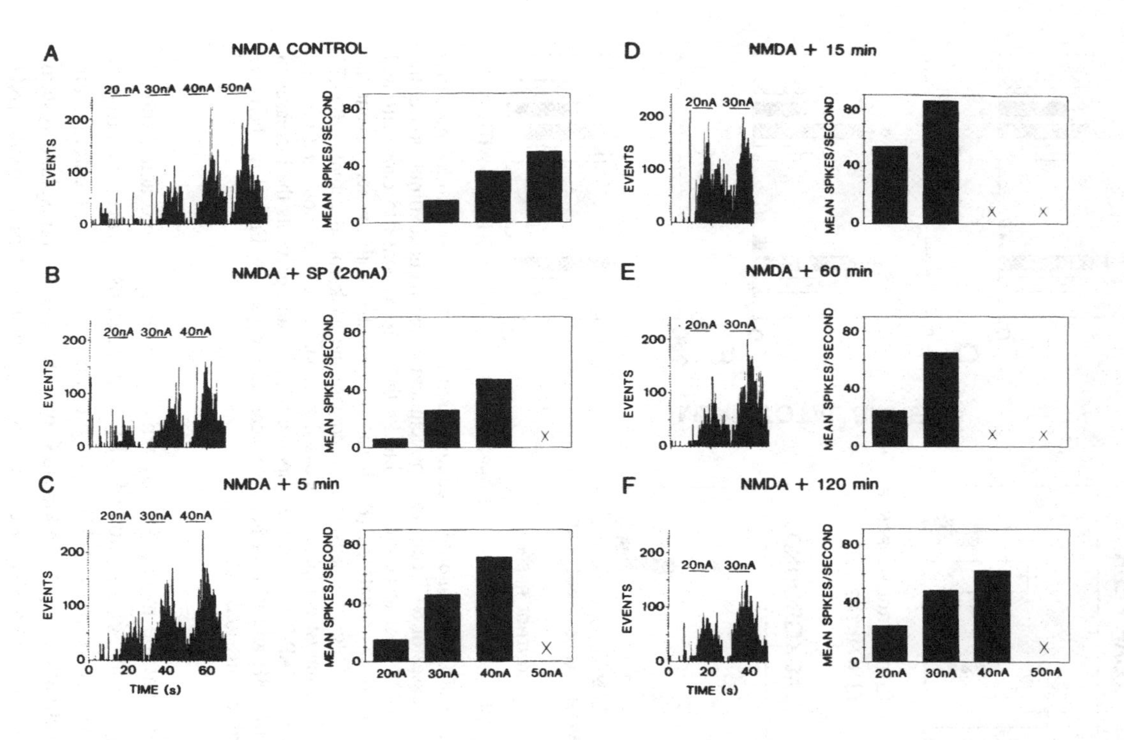

Fig. 16.3. Long-lasting increase in the responses of an STT cell to iontophoretic applications of NMDA following co-application of NMDA and SP. The peristimulus time histograms show the responses to graded doses of NMDA; the bars show the sizes of these responses after subtraction of the background activity. Control responses are in (A), and the responses during co-application of SP are in (B). (C)–(F) show the responses to NMDA at the times indicated following cessation of the application of SP. From Dougherty & Willis, (1991) with permission.

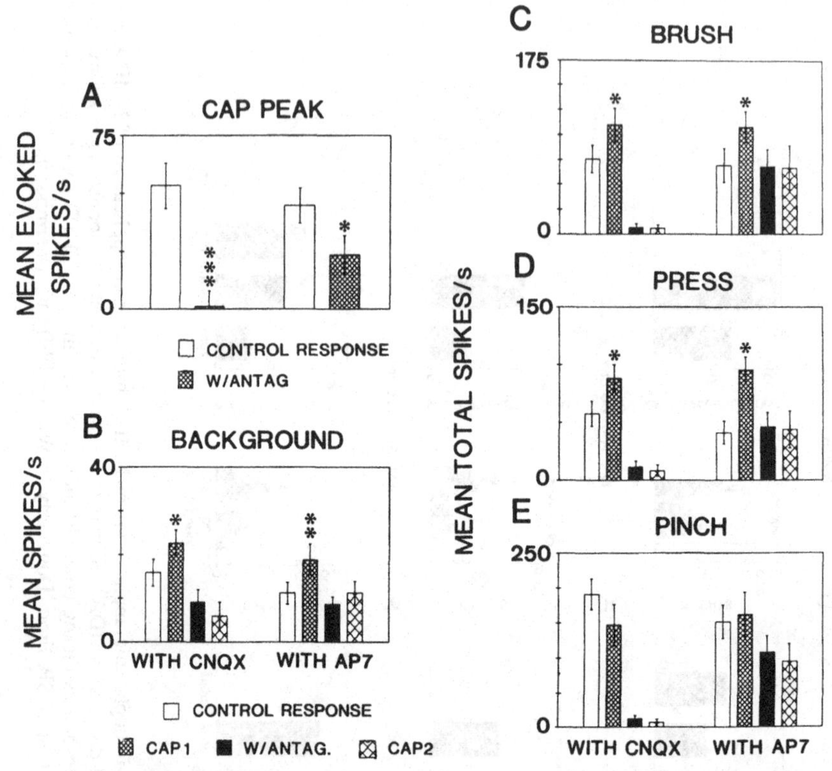

Fig. 16.4. Effect of CNQX and AP-7 administered into the dorsal horn by microdialysis on the responses of STT cells to capsaicin and to mechanical stimulation of the skin. In (A), CNQX almost completely blocks the peak response to capsaicin, and AP-7 reduces it to about half. In (B), it can be seen that both agents prevent the ability of the second dose of capsaicin to increase the background activity. (C)–(E) show that CNQX almost completely blocks the responses to innocuous and noxious mechanical stimuli. AP-7 reduces the response to pinch and prevents the increases in responses to BRUSH and PRESS normally seen after capsaicin. From Dougherty, *et al.* (1992) with permission.

We believe that the mechanism by which excitatory amino acids and SP cause a long-lasting sensitization of STT cells depends on the activation of second messenger systems. NMDA might cause an influx of calcium and subsequent activation of protein kinase C or of the inositol phosphate system. Recently, we have obtained evidence that the release of nitric oxide (NO) may also be involved, since the nitric oxide synthase blocker, L-nitroarginine methyl ester, but not the inactive isomer *N*-nitroarginine methyl ester, prevents the sensitization (Paleček, Palečková & Willis, 1993). Thus, NO is one of several second messengers that may be involved in the responses to activation of NMDA receptors. Other excitatory amino acids, such as QUIS, presumably interact with metabotrophic glutamate receptors, which in turn engage second messengers. SP is thought to cause the release of calcium from intracellular stores, among other actions.

Models of neuropathic pain

Bennett & Xie (1988) described a rat model of neuropathic pain in which four loose ligatures were tied around the sciatic nerve. Within a few days, the animals develop a characteristic syndrome in which they guard the ipsilateral hindlimb and demonstrate a shortened latency to noxious heat stimulation of the paw; they also show signs of allodynia (Bennett & Xie, 1988; Attal *et al.*, 1990). The apparent hyperalgesia lasts for weeks or even months. The sciatic nerve is severely damaged, with degeneration of most of the myelinated and a large fraction of the unmyelinated axons distal to the site of the ligatures (Carlton *et al.*, 1991). Evidence has been obtained that the behavioural signs in this syndrome are due to ectopic spike generators that develop in the damaged primary afferent fibres (Kajander & Bennett, 1992). Recordings from STT cells in neuropathic rats reveal a number of changes in their activity, including an increase in the background activity and frequent after-discharges following noxious stimulation (Paleček *et al.*, 1992b). Curiously, the threshold for responses to noxious heat stimuli are unchanged from normal.

A comparable neuropathic pain syndrome is produced by tight ligation of one-third to one-half of the sciatic nerve (Seltzer, Dubner & Shir, 1990). More recently, another tight ligation model has been introduced by Kim & Chung (1992). In this model, one or two spinal nerves are ligated just distal to the dorsal root ganglion. In rats, L5 and L6 are suitable segments; L4 is avoided since ligation of this spinal nerve results in a substantial motor deficit. It could be shown that injury to nerve fibres distal to the dorsal root ganglion is critical to the development of behavioural signs of mechanical allodynia and heat hyperalgesia in the neuropathic animals (Sheen & Chung, 1993). The Chung model is of particular interest, since clear evidence has been obtained indicating that the behavioural signs are dependent on the integrity of the sympathetic nervous system (Kim & Chung, 1991; Kim *et al.*, 1993).

Although an attempt has been made to produce a neuropathic pain model in monkeys using the loose ligature technique, it was not successful (Paleček *et al.*, 1992a). However, it has been possible to produce what appears to be a neuropathic pain syndrome in monkeys using a tight ligature of the L7 spinal nerve (Carlton *et al.*, 1994). The animals behave as though they have thermal hyperalgesia, as well as mechanical and cold allodynia. Recordings from STT cells in the L6 segment (Paleček *et al.*, 1992a) reveal that the responses to innocuous and noxious mechanical and thermal stimuli are enhanced compared with those in normal animals and also with the responses of STT neurones on the contralateral side of the spinal cord (Fig. 16.5). However, the responses of

STT cells in L7 are generally depressed. It is not clear as yet whether or not the neuropathic state in monkeys is sympathetically maintained.

The neuropathic pain state in rats and monkeys lasts for weeks. It is noteworthy that a number of anatomical changes can be observed in the dorsal horn during this timeframe. For example, in rats subjected to loose ligatures of the sciatic nerve, there are alterations in the peptide content of the dorsal horn, and there is an increase in immunohistochemical staining for lectins, such as soybean agglutinin and RL-29 (Cameron *et al.*, 1991; Garrison, Dougherty & Carlton, 1993). There is also an increase in immunostaining for GAP-43, a protein that is found in the growth cones of regenerating neurones. In addition, some large neurones in the dorsal horn on the neuropathic side become immunoreactive for antibodies to RL-29 (Cameron *et al.*, 1993). Thus, a number of morphological changes occur in the spinal cord in association with the neuropathic pain syndrome. In addition, many of the neurones of the dorsal horn and in areas of the brain receiving somatosensory information become metabolically hyperactive, as shown by increased 2-deoxyglucose uptake (Mao *et al.*, 1992a, 1993a). Many of these changes are likely to be attributable to the activation of second messenger systems, leading to functional changes in the central nervous system, and also to the induction of gene expression, which can result in structural changes that may last weeks or even indefinitely.

Clearly, experimental models of neuropathic pain should lead to a better understanding of the mechanisms underlying some forms of chronic pain. In addition, the availability of such models provides an opportunity for testing the efficacy of drugs or electrical stimulation procedures in relieving this type of pain (e.g. Mao *et al.*, 1992b, c, 1993b; Yamamoto & Yaksh, 1992; Kuypers & Gybels, 1993; Sotgui, 1993).

Conclusions

On the basis of experimental evidence of the kind reviewed here, it is clear that there are both short- and long-term mechanisms by which the spinal cord can 'remember' pain. The short-term mechanisms seem to involve the activation of second messenger systems by release of excitatory amino acids and peptides, such as SP, into the spinal cord dorsal horn. The long-term mechanisms may in addition involve morphological changes, undoubtedly secondary to the induction of gene expression. Similar changes are likely to operate at the level of the brain.

Acknowledgements

The work done in the author's laboratory was supported by NIH grants NS 09743 and NS 11255.

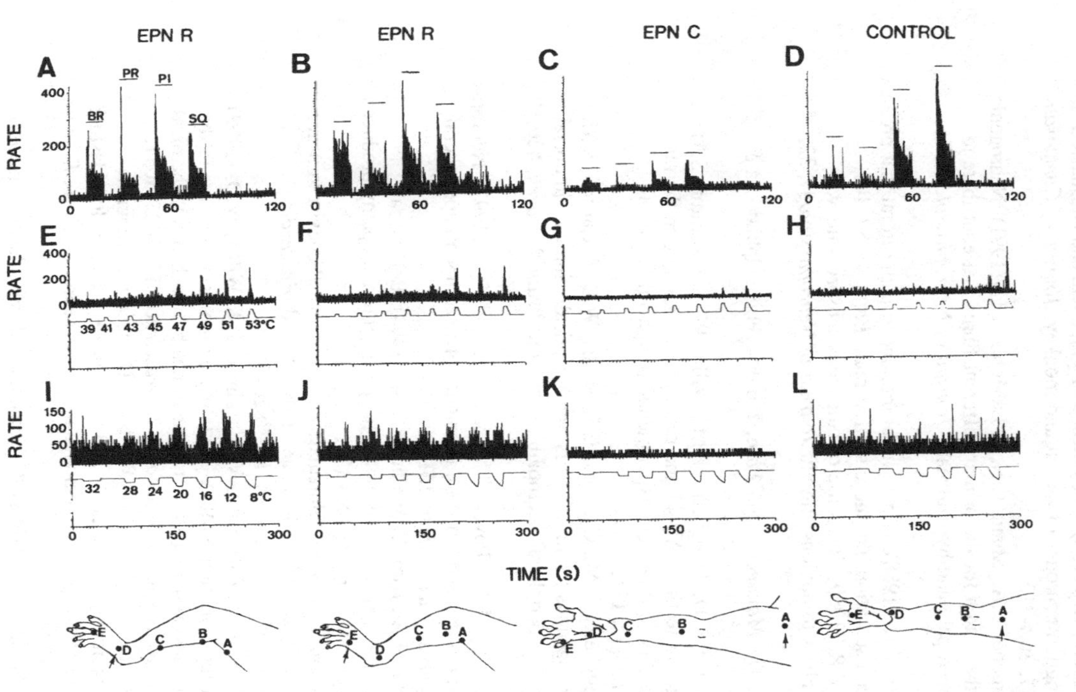

Fig. 16.5. Increased responses of STT cells to mechanical and thermal stimuli in an animal that was made neuropathic by ligation of the L7 spinal nerve 14 days previously. The peristimulus time histograms in the first two columns were recorded from STT cells in the L6 segment; those in the third column were from an STT cell in L7; and those in the fourth column were from an STT cell on the contralateral side. The responses to mechanical stimulation in (A) and (B) and the thermal responses in (E), (F), (I) and (J) were abnormally large (compared with those of STT cells in a large reference population), whereas the responses in (C), (G) and (K) were small, and those in (D), (H) and (L) were no different from normal. The drawings at the bottom show the sites used for mechanical stimulation in the experiment; those indicated by arrows are the sites stimulated for the responses that are illustrated. From Paleček, *et al.* (1992a) with permission.

References

Attal, N., Jazat, F., Kayser, V. & Guilbaud, G. (1990). Further evidence for 'pain-related' behaviours in a model of unilateral peripheral mononeuropathy. *Pain*, **41**, 235–251.

Battaglia, G. & Rustioni, A. (1988). Coexistence of glutamate and substance P in dorsal root ganglion neurones of the rat and monkey. *Journal of Comparative Neurology*, **277**, 302–312.

Baumann, T.K., Simone, D.A., Shain, C.N. & LaMotte, R.H. (1991). Neurogenic hyperalgesia: the search for the primary afferent fibres that contribute to capsaicin-induced pain and hyperalgesia. *Journal of Neurophysiology*, **66**, 212–227.

Bennett, G.J. & Xie, Y.K. (1988). A peripheral mononeuropathy in rat that produces disorders of pain sensation like those seen in man. *Pain*, **33**, 87–107.

Bessou, P. & Perl, E.R. (1969). Response of cutaneous sensory units with unmyelinated fibres to noxious stimuli. *Journal of Neurophysiology*, **32**, 1025–1043.

Bonica, J.J. (1990). *The Management of Pain*, 2nd edn. Philadelphia: Lea & Febiger.

Cameron, A.A., Cliffer, K.D., Dougherty, P.M., Willis, W.D. & Carlton, S.M. (1991). Changes in lectin, GAP-43 and neuropeptide staining in the rat superficial dorsal horn following experimental peripheral neuropathy. *Neuroscience Letters*, **131**, 249–252.

Cameron, A.A., Dougherty, P.M., Garrison, C.J., Willis, W.D. & Carlton, S.M. (1993). The endogenous lectin RL-29 is transynaptically induced in dorsal horn neurones following peripheral neuropathy in the rat. *Brain Research*, **620**, 64–71.

Carlton, S.M., Dougherty, P.M., Pover, C.M. & Coggeshall, R.E. (1991). Neuroma formation and numbers of axons in a rat model of experimental peripheral neuropathy. *Neuroscience Letters*, **131**, 89–92.

Carlton, S.M., Lekan, H.A., Kim, S. & Chung, J.M. (1994). Behavioral manifestations of an experimental model for peripheral neuropathy produced by spinal nerve ligation. *Pain*, **56**, 155–156.

De Biasi, S. & Rustioni, A. (1988). Glutamate and substance P coexist in primary afferent terminals in superficial laminae of spinal cord. *Proceedings of the National Academy of Sciences, USA*, **85**, 7820–7824.

Dougherty, P.M., Paleček, J., Palečková, V., Sorkin, L.S. & Willis, W.D. (1992). The role of NMDA and non-NMDA excitatory amino acid receptors in the excitation of primate spinothalamic tract neurones by mechanical, chemical, thermal, and electrical stimuli. *Journal of Neuroscience*, **12**, 3025–3041.

Dougherty, P.M., Paleček, J., Zorn, S. & Willis, W.D. (1993). Combined application of excitatory amino acids and substance P produces long-lasting changes in responses of primate spinothalamic tract neurones. *Brain Research Reviews*, **18**, 227–246.

Dougherty, P.M. & Willis, W.D. (1991). Enhancement of spinothalamic neuron responses to chemical and mechanical stimuli following combined microiontophoretic application of *N*-methyl-D-aspartic acid and substance P. *Pain*, **47**, 85–93.

Dougherty, P.M. & Willis, W.D. (1992). Enhanced responses of spinothalamic tract neurons to excitatory amino acids accompany capsaicin-induced sensitization in the monkey. *Journal of Neuroscience*, **12**, 883–894.

Fitzgerald, M. & Lynn, B. (1977). The sensitization of high threshold mechanoreceptors with myelinated axons by repeated heating. *Journal of Physiology (London)*, **265**, 549–563.

Garrison, C.J., Dougherty, P.M. & Carlton, S.M. (1993). Quantitative analysis of substance P and calcitonin gene-related peptide immunohistochemical staining in the dorsal horn of neuropathic MK-801-treated rats. *Brain Research*, **607**, 205–214.

Hardy, J.D., Wolff, H.G. & Goodell, H. (1952) *Pain Sensation and Reactions*. New York: Williams & Wilkins. (Reprinted New York: Hafner, 1967.)

Kajander, K.C. & Bennett, G.J. (1992). Onset of a painful peripheral neuropathy in rat: a partial and differential deafferentation and spontaneous discharge in Aβ and Aδ primary afferent neurones. *Journal of Neurophysiology*, **68**, 734–744.

Kim, S.H. & Chung, J.M. (1991). Sympathectomy alleviates mechanical allodynia in an experimental animal model for neuropathy in the rat. *Neuroscience Letters*, **134**, 131–134.

Kim, S.H. & Chung, J.M. (1992). An experimental model for peripheral neuropathy produced by segmental spinal nerve ligation in the rat. *Pain*, **50**, 355–363.

Kim, S.H., Na, H.S., Sheen, K. & Chung, J.M. (1993). Effects of sympathectomy on a rat model of peripheral neuropathy. *Pain*, **55**, 85–92.

Kupers, R.C. & Gybels, J.M. (1993). Electrical stimulation of the ventroposterolateral thalamic nucleus (VPL) reduces mechanical allodynia in a rat model of neuropathic pain. *Neuroscience Letters*, **150**, 95–98.

LaMotte, R.H., Lundberg, L.E.R. & Torebjörk, H.E. (1992). Pain, hyperalgesia and activity in nociceptive C units in humans after intradermal injection of capsaicin. *Journal of Physiology (London)*, **448**, 749–764.

LaMotte, R.H., Shain, C.N., Simone, D.A. & Tsai, E.F.P. (1991). Neurogenic hyperalgesia: psychophysical studies of underlying mechanisms. *Journal of Neurophysiology*, **66**, 190–211.

LaMotte, R.H., Thalhammer, J.G. & Robinson, C.J. (1983). Peripheral neural correlates of magnitude of cutaneous pain and hyperalgesia: a comparison of neural events in monkey with sensory judgments in human. *Journal of Neurophysiology*, **50**, 1–26.

Lewis, T. (1942). *Pain*. New York: Macmillan.

Mao, J., Mayer, D.J. & Price, D.D. (1993a). Patterns of increased brain activity indicative of pain in a rat model of peripheral mononeuropathy. *Journal of Neuroscience*, **13**, 2689–2702.

Mao, J., Price, D.D., Coghill, R.C., Mayer, D.J. & Hayes, R.L. (1992a). Spatial patterns of spinal cord [^{14}C]-2-deoxyglucose metabolic activity in a rat model of painful peripheral mononeuropathy. *Pain*, **50**, 89–100.

Mao, J., Price, D.D., Hayes, R.L., Lu, J. & Mayer, D.J. (1992b). Intrathecal GM1 ganglioside and local nerve anesthesia reduce nociceptive behaviors in rats with experimental peripheral mononeuropathy. *Brain Research*, **584**, 28–35.

Mao, J., Price, D.D., Hayes, R.L., Lu, J., Mayer, D.J. & Frenk, H. (1993b). Intrathecal treatment with dextrorphan or ketamine potently reduces pain-related behaviors in a rat model of peripheral mononeuropathy. *Brain Research*, **605**, 164–168.

Mao, J., Price, D.D., Mayer, D.J., Lu, J. & Hayes, R.L. (1992c). Intrathecal MK-801 and local nerve anesthesia synergistically reduce nociceptive behaviors in rats with experimental peripheral mononeuropathy. *Brain Research*, **576**, 254–262.

Meyer, R.A. & Campbell, J.N. (1981). Myelinated nociceptive afferents account for the hyperalgesia that follows a burn to the hand. *Science*, **213**, 1527–1529.

Paleček, J., Dougherty, P.M., Kim, S.H., Palečková, V., Lekan, H., Chung, J.M., Carlton, S.M. & Willis, W.D. (1992a). Responses of spinothalamic tract neurons to mechanical and thermal stimuli in an experimental model of peripheral neuropathy in primates. *Journal of Neurophysiology*, **68**, 1951–1966.

Paleček, J., Palečková, V., Dougherty, P.M., Carlton, S.M. & Willis, W.D. (1992b). Responses of spinothalamic tract cells to mechanical and thermal stimulation of skin in rats with experimental peripheral neuropathy. *Journal of Neurophysiology*, **67**, 1562–1573.

Paleček, J., Palečková, V. & Willis, W.D. (1993). The role of nitric oxide in hyperalgesia induced by capsaicin injection in primates. *Abstracts of 7th World Congress on Pain*, pp. 226–227. Seattle: IASP Publications.

Seltzer, Z., Dubner, R. & Shir, Y. (1990). A novel behavioral model of neuropathic pain disorders produced in rats by partial sciatic nerve injury. *Pain*, **43**, 205–218.

Sheen, K. & Chung, J.M. (1993). Signs of neuropathic pain depend on signals from injured nerve fibres in a rat model. *Brain Research*, **610**, 62–68.

Simone, D.A., Sorkin, L.S., Oh, U., Chung, J.M., Owens, C., LaMotte, R.H. & Willis, W.D. (1991). Neurogenic hyperalgesia: central neural correlates in responses of spinothalamic tract neurones. *Journal of Neurophysiology*, **66**, 228–246.

Sotgui, M.L. (1993). Descending influence on dorsal horn neuronal hyperactivity in a rat model of neuropathic pain. *NeuroReport*, **4**, 21–24.

Torebjörk, H.E., Lundberg, L.E.R. & LaMotte, R.H. (1992). Central changes in processing of mechanoreceptive input in capsaicin-induced secondary hyperalgesia in humans. *Journal of Physiology (London)*, **448**, 765–780.

Wall, P.D. & Melzack, R. (eds.) (1989). *Textbook of Pain*, 2nd edn. London: Churchill Livingstone.

Willis, W.D., Paleček, J., Palečková, V., Ragland, J. & Dougherty, P.M. (1992). Neurokinin receptor antagonists modify the responses of primate STT neurones to cutaneous stimuli. *Neuroscience Abstracts*, **18**, 1023.

Yamamoto, T. & Yaksh, T.L. (1992). Studies on the spinal interaction of morphine and the NMDA antagonist MK-801 on the hyperesthesia observed in a rat model of sciatic mononeuropathy. *Neuroscience Letters*, **135**, 67–70.

17

The neurophysiological basis of pain relief by acupuncture

EH SHEN

Shanghai Brain Research Institute, Shanghai, China.

Introduction

Acupuncture has been a medical practice in China for several thousand years. In ancient times, pointed stone slivers were used. With time metal probes were developed, but much coarser than the fine needles presently used. Acupuncture has been used to treat various diseases, including appendicitis, bacillary dysentery, gastric ulcer, cholelithiasis, hypertension, poliomyelitis, insomnia, malaria, and even epilepsy. Although good success rates have been claimed for treatment of these and other diseases, owing to natural remission of some diseases during the long course of treatment some results are questionable. For instance, little or no significant difference has been observed in the success rates of acupuncture versus placebo treatment (administration of vitamin B) for malaria. In contrast, there is little doubt of the analgesic effect of acupuncture. Pain is relieved immediately after needling of the acupuncture points in many cases, particularly cases of pain caused reflexly by a certain injury. Pain thresholds have been observed to be elevated by acupuncture in normal human subjects (Chiang *et al.*, 1973). Acupuncture has also been employed instead of anaesthesia for surgical operations, though with restrictions.

The physiological basis of acupuncture analgesia has been extensively studied in China, but only some of these results have been introduced to the Western world (Nathan, 1978; Han & Terenius, 1982). The present chapter summarizes work related to mechanisms of acupuncture analgesia from the neurophysiological point of view, with particular emphasis on the descending inhibitory systems involved.

Tender spots of remote referred pain

Besides using the points on the traditional meridian system, acupuncturists often use the so called *Ash'i* points. *Ash'i*, which means 'isn't it?', is often asked when the acupuncturist is trying to find tender spots of referred pain. Needling such tender spots usually produces better results for relieving pain.

In addition to segmental referred pain, which has been well documented in text-books, there is a kind of remote referred pain. For example, gastric ulcer may cause tender spots in specific regions of the ear auricle. Needling these spots was able to stop stomach pain. Acupuncturists even drew a map of the pinna showing regions referred to various visceral organs and body regions.

The map was shown to be validated to some degree when we examined it among patients using a double-masked method. The phenomenon was then tested on monkeys (Wu, Du & Shen, 1962). The monkeys were trained to receive daily examination of tender spots on the pinna with a round-tipped rod fixed on a steel spring in a holder which was pressed with a known pressure. After control mapping for a few days, the monkey was anaesthetized and given noxious irritant such as application of capsicum extract on one hindlimb. Tender spots appeared on both pinnae after wakening, and increased in number gradually, then disappeared after 7–10 days. Injection of 0.1 ml of 2% procaine solution into the third ventricle eliminated all tender spots, which would reappear after 13–60 minutes. Intravenous injection of chemicals meant to suppress activity of the reticular formation, such as chlorpromazine (0.2–0.4 mg/kg) or nembutal (0.7 mg/kg), augmented the reaction to the pressure test and increased the number of spots. In contrast, chemicals such as LSD_{25} (30–40 μg/kg), amphetamine (1.6 mg/kg) or adrenaline (0.05–0.1 mg/kg), which are thought to increase the excitability of the reticular formation, suppressed the reaction to the pressure test and decreased the number of spots. It is suggested that central mechanisms underlying such remote referred pain may involve the medial portion of the rostral brainstem, and may be subject to influence from the reticular formation.

Muscle contraction surrounding the needle

During manoeuvring of acupuncture needles the patient feels a kind of soreness, numbness, heaviness and swelling; these are called acupuncture sensations. In addition, the acupuncturist feels something sucking the needle, which is called the manual feeling of *Dech'i*. Both the acupuncture sensations and the manual feeling are thought to be essential for ensuring the curative effect. We assumed that the sucking force originates from muscle contraction in deep tissues of the punctured region. To study this assumption, experiments were performed on healthy human subjects (Acupuncture Group of Shanghai Institute of Physiology, 1973). An experienced acupuncturist was employed to do the needling manoeuvres. Using an acupuncture needle, insulated except at the tip, to serve both for puncture and for recording electromyogram (EMG), we found that the sucking force sensed by the acupuncturist, the amplitude of the EMG and the intensity of the acupuncture sensations felt by the subject showed a positive correlation. The local muscular activity was found to be a reflex, since both the manual feeling and the EMG activity disappeared after lumbar anaesthesia when the point for needling was on a lower extremity. The reflexly induced muscular contraction is supposed to serve as positive feedback to recruit more deep receptors into action and thus strengthen the acupuncture effect.

Nerve fibres activated by acupuncture

The acupuncture sensations must involve the activities of deep receptors in the punctured region. Recordings of afferent impulses from the fine filaments dissected from the deep peroneal nerve of the cat showed firing of both myelinated and unmyelinated fibres during needling of the region innervated by the nerve (Wei *et al.*, 1973; Wei, Zhang & Feng, 1978). Most activity originated from pressure receptors and stretch receptors. Manipulation of the acupuncture needle activated the stretch receptors and low-threshold pressure receptors first. After a longer period of manoeuvring the high-threshold pressure receptors became involved. Some units continued firing after cessation of manipulation or withdrawal of the needle. Mapping the receptive field of the afferent fibres of pressure receptors showed concentrations of pressure receptors in the commonly used acupuncture points of *Zusanli*, *Shangjushu*, etc.

Clinicians trained in Western medicine usually pass electric current pulses through the needle (electroacupuncture, EA) instead of using manual manipulation. To visualize the nerve fibres mediating the acupuncture sensation, a metal recording electrode was inserted close to the median nerve of human volunteers to monitor the averaged action potentials (Lin *et al.*, 1986). Long periods of vascular occlusion of the arm blocked the EA feeling together with blockade of type I and II fibres, but the sensation produced by manual needling could be completely blocked only after blockade of type III fibres.

In animal experiments workers have stimulated nerve trunks in attempts to obtain effects that resemble the effects of acupuncture. In this way the action potentials can be monitored from the nerve being stimulated. It was found that activation of large-diameter fibres (types I, II) was sufficient for inhibition of the jaw movement reflex (Lu *et al.*, 1979), viscero-somatic reflex (Shen, Tsai & Lan, 1975) and wide dynamic range units in the spinocervical tract (Wu *et al.*, 1974); however, when small-diameter fibres were activated, inhibition was more profound (Shen *et al.*, 1975; Chen *et al.*, 1986).

Modulation of nociceptive transmission at the spinal level

Intracellular recording of wide dynamic range dorsal horn cells at L7–S1 of the spinal cat showed that EA at the hindlimb evoked inhibitory postsynaptic potentials (IPSPs) or hyperpolarization (Wu *et al.*, 1986). The responses to noxious thermal stimuli were inhibited during the whole period of 5 Hz EA.

However, owing to the polysynaptic nature of the IPSPs, only very short-lasting hyperpolarization and inhibition of less than a minute were observed if the frequency of EA was increased to 100 Hz.

Role of descending inhibition

In an animal model for reproducing the acupuncture effect for relieving visceral pain, it was noticed that supraspinal structure must play a key role in the inhibitory effect of acupuncture on a viscero-somatic reflex (VSR) (Shen *et al.*, 1975). Reflex discharges were recorded from an intercostal nerve in response to a single pulse delivered to the splanchnic nerve of cats. The VSR could be elicited only when the stimulus intensity reached the threshold of Aδ fibres and was eliminated by a small dosage of the morphine derivative, fentanyl (8 μg/kg, intravenously). The VSR was augmented, and the inhibitory effect of EA on VSR no longer persisted, after high spinal transection, even if the stimulation intensity was decreased to reduce the VSR to the size before transection.

The inhibitory effect of acupuncture on wide dynamic range dorsal horn cells was also shown to be more prominent in the intact animal than in the spinal animal (Wu *et al.*, 1986). Inhibition in the intact animal was attenuated after topical application of iced saline to the thoracic cord to block the descending inhibitory pathway. The hyperpolarization elicited by 100 Hz EA lasted longer in the intact animal than that in the spinal animal.

The descending inhibitory pathway

To explore the descending pathways from the supraspinal structures exerting the inhibitory effect, various lesions were placed in the cord at T1–3 (Shen *et al.*, 1975). Complete section of the dorsal column and ventral funiculi did not affect the inhibition of the VSR by EA. Bilateral lesions placed in the dorsolateral funiculi (DLF), adjacent to the dorsal horns, produced the most effective release of the reflex and reduced the inhibitory effect of EA. It should be mentioned that the diffuse noxious inhibitory controls (DNIC) revealed by the French group (Le Bars *et al.*, 1979a, b) are very similar to the phenomenon described here, and the descending pathway mediating DNIC is also the same (Villanueva, Chitour & Le Bars, 1986b).

Interestingly, the VSR already eliminated by fentanyl reappeared immediately after spinal transection or section of the DLF, implying that the action of fentanyl on this spinal reflex also depends on the supraspinal structures and is mediated by the same descending pathways (Shen, *et al.*, 1975). These findings are similar to later results describing the descending pathway in the analgesic action of morphine and brain stimulation on the pain reaction to pinch of the hindlimb of rats (Basbaum *et al.*, 1977).

The ascending pathway for acupuncture messages

In the course of exploration of the descending pathways, lesions placed in the ventrolateral funiculi (VLF) did not release the VSR from tonic descending inhibition, yet acupuncture at the hindlimb could no longer inhibit the VSR. It is thus reasonable to consider the VLF as the ascending pathway for acupuncture message. This point was verified in a simple way. A needle was inserted into a point in the forelimb, from where the afferent impulses entered the cervical cord, above the level of VLF lesions. Current passing through this needle in the forelimb inhibited the VSR completely while acupuncture at the hindlimb was ineffective, showing that lesions in the VLF blocked the acupuncture message ascending from the hindlimb (Shen *et al.*, 1975).

A commonly used animal model for acupuncture analgesia is stimulating the snout of rabbit with radiant heat to measure the threshold of the pain reaction, which would be elevated by acupuncture. In this model the ascending pathway for the afferent impulses from the acupuncture point was also identified to be in the VLF (Chiang *et al.*, 1975). Again, the ascending pathway mediating the DNIC was also found in the ventrolateral quadrant (Villanueva *et al.*, 1986a).

The supraspinal structures

The failure of intercollicular transection to interrupt the descending inhibitory effect of acupuncture suggests that the supraspinal structures involved must be located in the brainstem caudal to the superior colliculus (Shen *et al.*, 1975). By means of various lesions and sections placed in the brainstem, the central structure was localized to the medial region of the medulla including the nucleus raphe magnus (NRM) (Du & Chao, 1976). Microinjection of 5,6-dihydroxytryptamine (5,6-DHT) into the NRM, to selectively destroy serotonergic cells, diminished the acupuncture effect on the VSR in the cat (Du *et al.*, 1978) and on the tail-flick reflex in the rat (Chiang *et al.*, 1979). Sustained EA (20–30 minutes) significantly increased the radioactivity of the cerebroventricular perfusate in rabbits pretreated with labelled serotonin ([^{3}H]5HT) (Yi *et al.*, 1977), and enhanced the fluorescence of serotonergic cells in the NRM and the nucleus raphe dorsalis (NRD) shown with the Falck–Hillarp method (Tung, Chiang & Fu, 1978; Tung & Chiang, 1981). These findings imply that both release and synthesis of 5HT are increased by acupuncture.

Descending noradrenergic (NA) axons appear to act synergically with the serotonergic pathway to produce the acupuncture effect (Ye, Feng & Shen, 1983). Using 6-hydroxydopamine (6-OHDA) to destroy the descending NA pathway partially diminished the acupuncture analgesic effect on the tail-flick reflex of rats. If 5,6-DHT and 6-OHDA are combined to destroy both the

5HT and NA descending fibres, the acupuncture effect was blocked almost completely. The descending NA neurones in A_1 of the medulla were demonstrated to be activated during EA (Xie, 1981).

Effects on sensory versus motor pathway

The descending inhibitory effect of acupuncture on a reflex or on the dorsal horn cells does not necessarily mean that it blocks ascending nociceptive transmission, since it could be inhibiting the motor part of the reflex or the interneurones in a reflex arc. No work has been done on spinothalamic tract neurones. The work on the spinocervical tract fibres was done only in spinal animals (Wu, Chao & Wei, 1974). Experiments have been designed to solve this problem (Shen, Ma & Lan, 1978). Evoked potentials from the orbital cortex of cats were recorded following splanchnic nerve stimulation at intensities that stimulated Aδ fibres. The inhibitory effect of acupuncture on these evoked potentials did not persist after section of the descending inhibitory path DLF at high thoracic levels or after a lesion was placed in the NRM. Evidently, the diminution of the cortical potentials during acupuncture is in fact inhibition at the spinal level. Acupuncture activates descending inhibitory centres which thereafter send impulses via DLF to block ascending nociceptive transmission from the spinal cord to the cortex.

How does acupuncture affect the motor pathway? Stimulation of the motor cortex of the cat elicited discharges of the intercostal nerve from which VSR was also recorded. The motor cortex–intercostal response (MIR) and the VSR could occlude or facilitate each other, suggesting that they share a common final pathway. Acupuncture inhibited the VSR but not the MIR, or even facilitated the latter. This indicates that acupuncture acts to inhibit the sensory but not the motor part of the reflex arc (Shen *et al.*, 1982).

Descending presynaptic inhibition

Presynaptic inhibition by way of the VLF–NRM–DLF recurrent circuit has been studied (Shen, Ma & Lan, 1980). Stimulation of the DLF at T3 evoked a dorsal root potential (DRP) at a lower level, e.g. L6, with a latency of 3–5 ms. Stimulation of the NRM also evoked a DRP recorded at the same dorsal root with a longer latency (36 ± 3 ms). If we stimulated an isolated strand of VLF which was connected with the spinal cord only at the rostral end, a DRP was evoked at L6 with an even longer latency (48 ± 6 ms). The time course of the inhibition of the VSR or a flexor reflex, conditioned by stimulation of the DLF, NRM or isolated VLF, showed a correlation with the DRP. Measurement of the length constant of the fibres involved in the DRP showed that they fell into the group of fine myelinated fibres. Destruction of the NRM abolished the DRP evoked by stimulation of the isolated

VLF. Section of the DLF eliminated the DRP evoked by stimulation of the NRM or isolated VLF. The DRP is thus considered to be evoked through a VLF–NRM–DLF recurrent circuit. Electrical stimulation of the forelimb nerve evoked a DRP at L6 through this circuit. However, stimulation of the hindlimb nerve evoked a depolarization of the L6 dorsal root, dominated by a short-latency segmentally evoked DRP, which masked the long-latency component of the DRP mediated through the recurrent circuit.

Interaction at the thalamic level

In the thalamus of rats and rabbits neurones responding with specific long sustained discharges to noxious stimuli were found in nucleus parafasciculus (Pf), nucleus centralis lateralis (CL) and the area around the posterior commissure (Chang, 1973). The response could be attenuated by EA or pinch of Achilles tendon, and totally eliminated by intravenous injection of morphine (3 mg/kg). Stimulation of the nucleus centrum medianum (Lo *et al.*, 1978), NRD (Chang *et al.*, 1979) or bulbar reticular formation (Shen & Ho, 1976) inhibited nociceptive discharges in Pf or CL. After the dorsal half of the spinal cord of rabbit was cut at T3–4, the inhibitory effect of forelimb acupuncture on the pain sensitive neurones in Pf was reduced (Chang *et al.*, 1979). Stimulation of the caudal stump of the isolated dorsal spinal cord inhibited the pain-sensitive neurones in Pf of rabbit (Chang, Xie & Qiao, 1982). It seems plausible that descending inhibition also participates in the acupuncture effect on thalamic nociceptive neurones.

Endorphins and acupuncture

The studies on the role played by endorphins in acupuncture analgesia have been thoroughly reviewed by Han & Terenius (1982). Activation of the endogenous opioid system by acupuncture explains why acupuncture and morphine share common neural substrates. However, as Han and Terenius stated, the acupuncture effect consists of something more than the action of morphine: 'Acupuncture activates multiple afferent pathways, leading to altered activity in numerous CNS systems.' It has long been known that somatic stimulation can change autonomic imbalance to balance (Gellhorn, 1957). Acupuncture is, in essence, a special kind of somatic sensory input. Normalization in addition to the analgesic effect makes acupuncture an important practice in the clinics.

References

Acupuncture Group of Shanghai Institute of Physiology (1973). Electromyographic activity produced locally by acupuncture manipulation. *Chinese Medical Journal*, **53**, 532–535.

Basbaum, A.I., Marley, N.J.E., O'Keefe, J. & Clanton, C.H. (1977). Reversal of morphine and stimulus-produced analgesia by subtotal spinal cord lesions. *Pain*, **3**, 43–56.

Chang, H.T. (1973). Integrative action of thalamus in the process of acupuncture for analgesia. *Scientia Sinica*, **16**, 25–60.

Chang, H.Y., Xie, Y.F. & Qiao, J.T. (1982). Effect of stimulation of caudal stump of isolated spinal cord on the nociceptive responses of units in nucleus parafascicularis. *Acta Physiologica Sinica*, **34**, 78–83.

Chang, K.L., *et al.* (1979). Effect of stimulation of midbrain raphe nuclei on the discharge of pain sensitive cells in nucleus parafasciculus of thalamus and its significance in acupuncture analgesia. *Acta Physiologica Sinica*, **31**, 209–218.

Chen, L.S., *et al.* (1986). Analysis of afferent fibres for acupuncture analgesia. In *Acupuncture, Moxibustion and Acupuncture Anaesthesia*, ed. H.T. Chang, pp. 348–354. Beijing: Science Press.

Chiang, C.Y., Chang, C.T., Chu, H.L. & Yang, L.F. (1973). Peripheral afferent pathway for acupuncture analgesia. *Scientia Sinica*, **16**, 210–217.

Chiang, C.Y., Liu, J.Y., Chu, T.H., Pai, Y.H. & Chang, S.C. (1975). Studies on spinal ascending pathway for effect of acupuncture analgesia in rabbits. *Scientia Sinica*, **18**, 651–658.

Chiang, C.Y., Tu, H.C., Chao, Y.F., Pai, Y.H., Ku, H.K., Cheng, J.K., Shan, H.Y. & Yang, F.Y. (1979). Effects of electrolytic lesions or intracerebral injection of 5,6-dihydroxytryptamine in raphe nuclei on acupuncture analgesia in rats. *Chinese Medical Journal*, **92**, 129–136.

Du, H.J. & Chao, Y.F. (1976). Localization of central structures involved in descending inhibitory effect of acupuncture on viscerosomatic reflex discharges. *Scientia Sinica*, **19**, 137–148.

Du, H.J., Shen, E., Dong, X.W., Jiang, Z.H., Ma, W.X., Fu, L.W., Jin, G.Z., Zhang, Z.D., Han, Y.F., Yu, L.P. & Feng, J. (1978). Effects on acupuncture analgesia by injection of 5,6-dihydroxytryptamine in cat: a neurophysiological, neurochemical and fluorescence histochemical study. *Acta Zoologica Sinica*, **24**, 11–20.

Gellhorn, E. (1957). *Autonomic Imbalance and the Hypothalamus*. Minneapolis: University of Minnesota Press.

Han, J.S. & Terenius, L. (1982). Neurochemical basis of acupuncture analgesia. *Annual Reviews of Pharmacology and Toxicology*, **22**, 193–220.

Le Bars, D., Dickenson, A.H. & Besson, J.M. (1979a). Diffuse noxious inhibitory controls (DNIC). I. Effect on dorsal horn convergent neurones in the rat. *Pain* **6**, 283–304.

Le Bars, D., Dickenson, A.H. & Besson, J.M. (1979b). Diffuse noxious inhibitory controls (DNIC). II. Lack of effect on nonconvergent neurones, supraspinal involvement and theoretical implication. *Pain*, **6**, 305–327.

Lin, W.Z., *et al.* (1986). Observations on the afferent fibres mediating the acupuncture sensation. In *Studies on Acupuncture, Moxibustion and Acupuncture Anaesthesia*, ed. H.T. Chang, pp. 323–330. Beijing: Science Press.

Lo, F.S., *et al.* (1978). Inhibition of nociceptive discharges of parafascicular neurones by direct electrical stimulation of nucleus centrum medianum. *Scientia Sinica*, **21**, 533

Lu, G.W., *et al.* (1979). Role of peripheral afferent nerve fibre in acupuncture analgesia elicited by needling point zusanli. *Scientia Sinica*, **22**, 680–692.

Nathan, P.W. (1978). Acupuncture analgesia. *Trends in Neurosciences*, **1**, 21–23.

Shen, E., Ma, W.H. & Lan, C. (1978). Involvement of descending inhibition in the effect of acupuncture on the splanchnically evoked potential in the orbital cortex of cat. *Scientia Sinica*, **21**, 677–85.

Shen, E., Ma, W.H. & Lan, C. (1980). Dorsal root potential evoked by a spino-bulbo-spinal recurrent circuit. *Acta Physiologica Sinica*, **32**, 61–67.
Shen, E., Ouyang, S., Ma, W.X. & Lan, Q. (1982). Influence of acupuncture, raphe stimulation or morphine on intercostal discharges elicited by stimulation of motor cortex. *Kexue Tongbao*, **27**, 1235–1240.
Shen, E., Tsai, T. T. & Lan, C. (1975). Supraspinal participation in the inhibitory effect of acupuncture on viscero-somatic reflex discharges. *Chinese Medical Journal*, **1**, 431–440.
Shen, K.F. & Ho, S.F. (1976). Effect of stimulation of bulbar reticular formation on the long-latency evoked responses in the region of centrolateral nucleus of thalamus. *Kexue Tongbao*, **21**, 507–510.
Tung, H.W. & Chiang, C.H. (1981). Changes in 5-hydroxytryptamine and noradrenaline fluorescence intensity in the nucleus raphe magnus during acupuncture analgesia in rat. *Acta Physiologica Sinica*, **33**, 24–29.
Tung, H.W., Chiang, C.H. & Fu, L.W. (1978). Changes in monoamine fluorescence intensity in the rat's midbrain raphe nuclei and locus coeruleus in the process of acupuncture analgesia. *Acta Biochimica et Biophysica Sinica*, **10**, 119–125.
Villanueva, L., Peschanski, M., Calvins, B. & Le Bars, D. (1986a). Ascending pathway in the spinal cord involved in triggering of diffuse noxious inhibitory controls in the rat. *Journal of Neurophysiology*, **55**, 34–55.
Villanueva, L., Chitour, D. & Le Bars, D. (1986b). Involvement of the dorsolateral funiculus in the descending spinal projections responsible for diffuse noxious inhibitory controls in the rat. *Journal of Neurophysiology*, **56**, 1185–1195.
Wei, J.Y., Feng, C.C., Chu, T.H. & Chang, S.C. (1973). Observations on activity of some deep receptors in cat hind limb during acupuncture. *Kexue Tongbao*, **18**, 184–186.
Wei, J.Y., Zhang, S.J. & Feng, J.Z. (1978). Activation of unmyelinated muscle afferents by acupuncture or pressure exerted on muscle. *Acta Zoologica Sinica*, **24**, 21–28.
Wu, C.P., Chao, C.C. & Wei, J.Y. (1974). Inhibitory effect produced by stimulation of afferent nerves on responses of cat dorsolateral fasciculus fibres to noxious stimulus. *Scientia Sinica*, **17**, 688–697.
Wu, C.P., Xing, J.H., Xing, B.R., Lu, J.G. & Chen, Y.C. (1986). Inhibitory effect of electric acupuncture on dorsal horn neurones of the spinal cord. In *Studies on Acupuncture, Moxibustion and Acupuncture Anaesthesia*, ed. H.T. Chang *et al.*, pp. 30–36. Beijing: Science Press.
Wu, W.Y., Du, H.J. & Shen, E. (1962). Experimental referred tender spots on the pinna of monkey. *Acta Physiologica Sinica*, **25**, 78–86.
Xie, Y.K. (1981). The effect of monoamines and acupuncture on neuronal activity in the A1 nuclear region of the medulla in cats. *Acta Physiologica Sinica*, **33**, 30–35.
Ye, W.L., Feng, X.C. & Shen, E. (1983). Evaluation of the role played by different monoaminergic descending pathways in acupuncture analgesic effect in rats. *Kexue Tongbao*, **28**, 1555–1559.
Yi, C.C., Lu, T.H., Wu, S.H. & Tsou, K. (1977). A study on the release of ^{3}H-5-hydroxytryptamine from brain during acupuncture and morphine analgesia. *Scientia Sinica*, **20**, 113–124.

Part III
Control of central nervous system output

Most of the chapters in this section could have been included under the heading of 'motor control', but we have chosen the present heading because we wished to include some areas which, although they might normally be considered outside motor control, may nevertheless be particularly relevant in a pathological situation. Thus we start with Crill's review of channels in neocortical neurones, knowledge of which is not only fundamental to an understanding of upper motor neurone physiology, but is also relevant to many other types of neurone. Similarly, the hippocampus may not normally be considered part of motor systems, yet the way in which it displays seizure behaviour (see chapter by Jefferys) may well be a model for disorders of other cortices. Accompanying this, the chapter from Andersen contains important data vital for any experimenter wishing to extrapolate from *in vivo* experiments in this field to equivalent results in behaving animals.

The chapters by Taylor & Durbaba, Aminoff & Goodin and Prochazka *et al.* cover areas more normally considered under motor control. That by Taylor & Durbaba on the physiological significance of fusimotor drives is particularly relevant to this volume because it draws comparisons between the control of limb muscles and the control of respiratory muscles. This latter is an area covered in a further nine chapters in this section. Respiratory movements are some of the most automatic, yet the respiratory muscles are used for one of the most voluntary actions, speech, as well as many other more or less automatic actions (see chapter by Grélot *et al.*). Respiratory movements and the physiology of the neurones controlling them can thus be considered exemplary for motor control.

We have included chapters ranging from the membrane and synaptic properties of the central respiratory neurones and the ways in which they are organized in order to produce a patterned rhythmic output both in health and disease (chapters by Berger *et al.*, Hilaire *et al.*, Grélot *et al.*, Duron, Seers), through considerations of reflexes and the operation of the different muscles (chapters by Romaniuk *et al.*, Goldman), to very direct consideration of pathological conditions, particularly those of neurological origin (chapters by

Howard and by Mills & Green). The chapter by Seers also emphasizes that, as well as being related to voluntary movement, the vital function of respiration is also tied in with the control of the heart. The circle is completed by the chapter by Nisimaru, which shows how cardiovascular control is itself directly related to control of other movements via the cerebellum.

Finally we have also included two chapters on the pathophysiology of two more traditional motor areas of the brain: the inferior olive (Burke) and the human motor cortex (Freund).

18

Synaptic transduction in neocortical neurones

WAYNE E. CRILL

Department of Physiology and Biophysics, University of Washington, Seattle, Washington, USA.

Introduction

Neurotransmitters mediate information transfer between neurones and in most cell types synaptic activity evokes slow potentials that are transduced into all-or-nothing action potentials. The primary site for integration of neuronal activity is on the soma and dendrites where input from many different sources converge and generate spike trains in neurones by summed synaptic currents. The molecular events during synaptic communication also provide many sites for modulation. Here I consider only the last step of the postsynaptic transduction operation where slow synaptic potentials are transduced into nerve impulses. This process is controlled by the type and density of voltage and chemically gated ion channels in the postsynaptic neurone. The model system discussed here is the pyramidal neurones in layer V of rat and cat somatosensory cortex. The pattern of evoked repetitive firing in pyramidal neurones has functional significance. Lemon & Mantel (1989) found that even a single spike in a corticospinal neurone produced a measurable facilitation of the target muscle electromyogram.

Types of channels in neocortical neurones

Central neurones uniformly have many types of ion channels. No one type occurs only in a single class of neurones. The collection of ion channel types provides great diversity in the voltage dependence, kinetics and modulation by neurotransmitters. In a specific neurone type modest changes in density, location or modulation of ion channels can markedly alter neurone behaviour. The paper physiologist can easily generate models with many different response patterns. We are just beginning to understand the role of ion channel types and subtypes in the transduction of synaptic currents into spike trains. This chapter briefly summarizes our laboratory's contribution to this problem. Table 18.1 lists a dozen ionic currents identified by physiological and pharmacological tests in regular firing cat and rat layer V Betz cells. The currents

Table 18.1. *Identified channel types in regular firing neocortical pyramidal neurones*

Current	Blockers	Primary function
Depolarizing currents		
I_{Na}	TTX	Action potential depolarization
I_{NaP}	TTX	Constant threshold; adds to synaptic currents
I_h	Caesium	Keeps E_m near threshold; burst firing in some cell types
$I_{Ca(T)}$	Nickel	? Burst firing
$I_{Ca(L)}$	Dihydropyridines	Intracellular signalling; activates K^+ currents
$I_{Ca(N)}$	ω-conotoxin	Intracellular signalling; activates K^+ currents
$I_{Ca(other\ ?P)}$	ω-Aga-IVA toxin	Intracellular signalling; activates K^+ currents
Hyperpolarizing currents		
I_{KTf}	TEA	Action potential repolarization
I_{KTs}	4-AP	Action potential repolarization
$I_{K(Ca)}$	Apamin	Medium AHP
$I_{K(Ca)}$	Muscarinic and β-adrenergic agonists	Slow AHP; adaptation
$I_{K(Na)}$	Muscarinic and β-adrenergic agonists	Slow AHP; adaptation

TTX, tetrodotoxin; TEA, tetraethylammonium; 4-AP, 4-aminopyridine; AHP, after-hyperpolarization.

carried by Na^+, Ca^{2+} or a mixture of K^+ and Na^+ ions are depolarizing and the K^+ currents are repolarizing.

All neurones of the mammalian central nervous system that generate all-or-nothing action potentials have a classical Hodgkin & Huxley (1952) voltage-dependent and tetrodotoxin (TTX)-sensitive sodium conductance (I_{Na}) with activation and inactivation gating properties. In regular spiking Betz cells four pharmacologically different Ca^{2+} channels are present but the flow of Ca^{2+} currents contributes little to the electrogenic response of the neurones. Blocking Ca^{2+} currents has no measurable effect on action potential shape (but see Ca^{2+}-mediated K^+ currents below). The low voltage activated T type Ca^{2+} current is prominent in thalamic and inferior olive neurones and contributes to bursting behaviour. Since the layer V neurones studied in our laboratory show only regular repetitive firing a relatively small T current was expected and demonstrated by Sayer, Schwindt & Crill (1990). The high

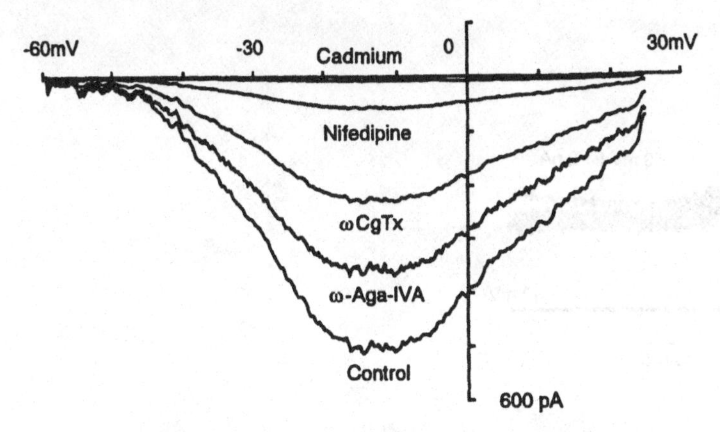

Fig. 18.1. Current–voltage relationship for high voltage activated (HVA) Ca^{2+} currents using a ramp voltage command (1 mV/ms) in the presence of sequential application of ω-Aga-IVA, ω-conotoxin GVIA (ω-CgTx), nifedipine and cadmium. Whole-cell voltage-clamp traces measured with a patch electrode from an acutely isolated rat neocortical pyramidal neurone are shown. From Brown *et al.* (1994).

voltage activated (HVA) Ca^{2+} current (Fig. 18.1) is composed of three nearly equal pharmacological identified components: an L type blocked by dihydropyridines; an N type blocked by ω-conotoxin GVIA; and a P type channel blocked by ω-Aga-IVA toxin (Brown *et al.*, 1994). In neocortical cells the major electrogenic function of HVA Ca^{2+} entry is to activate repolarizing K^+ currents. The Catterall Laboratory has found a high concentration of N Ca^{2+} channels on the dendrites of neurones (Westenbroek *et al.*, 1992). This spatial separation may reflect the location of synaptic terminals that release neurotransmitters with selective effects on different channel types. For example, in rat cortical neurones activation of metabotrophic glutamate receptors selectively decreases the L-type Ca^{2+} channels (Sayer, Schwindt & Crill, 1992).

Two voltage-dependent K^+ channels (I_{KTf} and I_{KTs}) are present in regular spiking neurones (Fig. 18.2). Both K^+ currents are activated rapidly and therefore contribute to action potential repolarization (Spain, Schwindt & Crill, 1991a). Blocking either I_{KTf} with 1 mM tetraethylammonium ions or I_{KTs} with 4-aminopyridine increases the duration of an evoked action potential. I_{KTf} and I_{KTs} show voltage-dependent inactivation like the A channels originally described in invertebrates (Connor & Stevens, 1971). I_{KTf} is inactivated within 20 ms whereas inactivation of I_{KTs} takes several hundred milliseconds. Spain, Schwindt & Crill (1991b) have shown that the transient K^+ currents inhibit subsequent firing after a conditioning hyperpolarization (which removes inactivation of the transient K^+ currents) in a select group of cortical pyramidal neurones.

The after-hyperpolarization following one or more action potentials reflects the time course of different K^+ conductance mechanisms. Kinetic properties

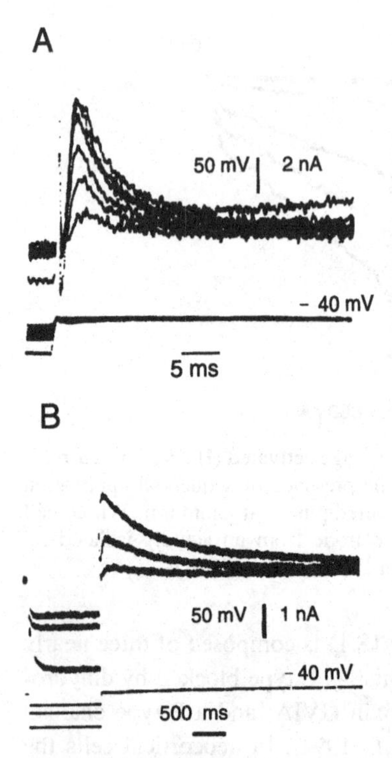

Fig. 18.2. Voltage-dependent transient K^+ currents recorded from layer V pyramidal cells from neocortical slices from cat using single electrode voltage-clamp. (A) and (B) show superimposed current traces (top) evoked by voltage-clamp commands from different conditioning potentials to –40 mV. (A) I_{KTf} at a fast time base; (B) I_{KTs} at a slower time base. From Spain, Schwindt & Crill (1991a).

permit the various K^+ currents to play different roles in the pattern of evoked activity. A single action potential has a fast after-hyperpolarization and a medium duration after-hyperpolarization (fAHP and mAHP). Repetitive firing induces an additional longer slow after-hyperpolarization (sAHP) (Schwindt *et al.*, 1988*a*). The fAHP is caused by the activation of the two voltage-dependent K^+ currents described above (I_{KTf} and I_{KTs}). An apamin-sensitive Ca^{2+}-dependent K^+ current hyperpolarizes the neurone during the mAHP. Selective block of this conductance mechanism markedly increases the instantaneous repetitive firing rate without any effect upon slow adaptation. A mixture of a Na^+-dependent and a Ca^{2+}-dependent K^+ current causes the sAHP. Activation of noradrenaline β receptors blocks the sAHP and the slow phase of adaptation in repetitive firing (Foehring, Schwindt & Crill, 1989). The sAHP and consequent adaptation to repetitive firing are the primary electric effects of Ca^{2+} entry during action potentials.

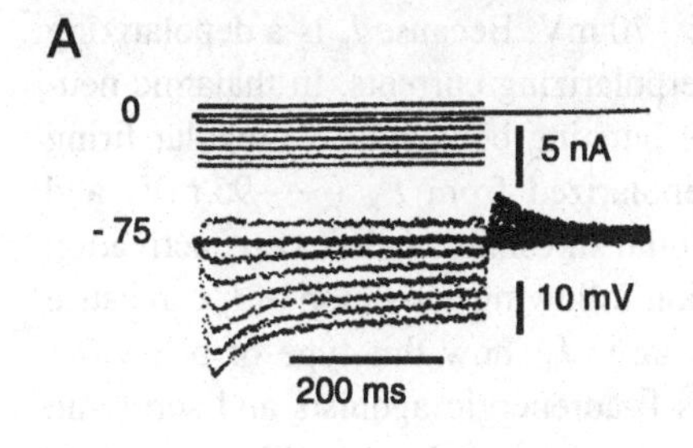

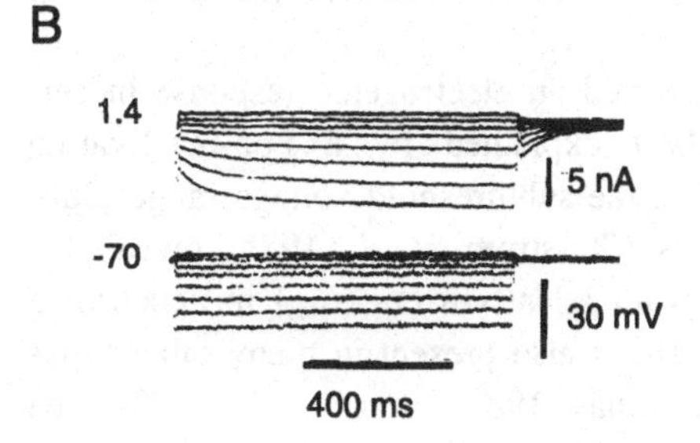

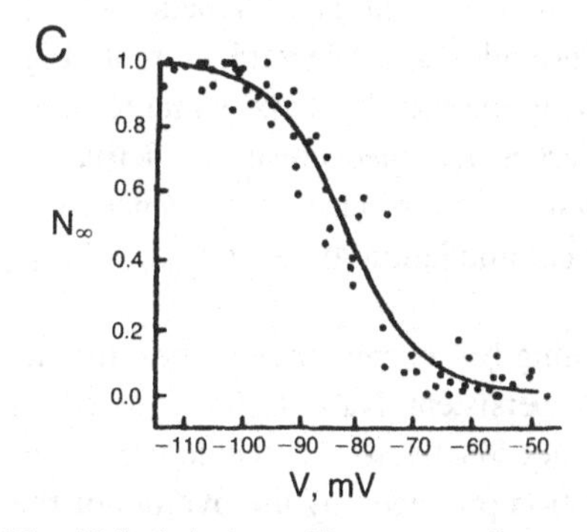

Fig. 18.3. I_h in layer V neurones from cat neocortical slice preparation. (A) Voltage responses (bottom) to a series of hyperpolarizing constant current pulses. (B) Membrane current (top) measured with sharp electrode single electrode voltage clamp to a series of hyperpolarizing voltage commands (bottom). (C) Plot of relative inward rectifier conductance (N_∞) as function of membrane potential for five neurones. From Spain, Schwindt & Crill (1987).

Subthreshold depolarizing currents

Two depolarizing currents are activated at membrane potentials negative to values where the transient Na^+ conductance is first detected. The ubiquitous I_h (Fig. 18.3) is activated by hyperpolarization and is a non-specific cation current that reverses at about −30 mV (Spain, Schwindt & Crill, 1987). The steady-state activation shown in Fig. 18.3C illustrates that I_h is partially activated at resting potentials of −65 to −70 mV. Because I_h is a depolarizing current it will resist the effects of hyperpolarizing currents. In thalamic neurones this function allows I_h to control bursting behaviour; in regular firing cortical neurones I_h keeps the cell depolarized from E_K (~ −95 mV) and within striking distance of action potential threshold. I_h has slow activation and deactivation kinetics; a depolarization following activation of I_h can cause rebound firing. Only Betz cells with a large I_h show this type of behaviour (Spain *et al.*, 1991*b*). In thalamic cells β-adrenergic agonists and serotonin increase I_h by shifting its voltage dependence towards zero (McCormick & Pape, 1990).

Llinás & Sugimori (1980) first reported an electrogenic response in cerebellar Purkinje cells that was best explained by a non-inactivating TTX-sensitive Na^+ current activated in the subthreshold voltage range. Subsequently, voltage-clamp experiments (Stafstrom *et al.*, 1985) revealed a subthreshold persistent or non-inactivating Na^+ current (I_{Nap}) in neocortical neurones. This conductance mechanism is also present in many other types of mammalian neurones (Alonso & Llinás, 1989; French *et al.*, 1990; see review by Llinás, 1988). The conductance is activated rapidly so the current can sum with the briefest synaptic input. In neocortical layer V neurones I_{NaP} is small compared with I_{Na} but it still provides a net inward current (Fig. 18.4) and therefore sets threshold by driving the membrane potential to more depolarized values where the much larger I_{Na} mechanism is activated. Because threshold is set by a non-inactivating inward current, I_{NaP} , neocortical pyramidal neurones show minimal accommodation to a slowly depolarizing inward current.

Since the discovery of the non-inactivating Na^+ current in neurones investigators have debated the mechanism. A persistent Na^+ conductance could reflect a subset of Na^+ channels that do not inactivate – a 'window conductance' in the Hodgkin & Huxley formulation produced by the overlap of the steady-state activation and inactivation curves for the transient Na^+ conductance. The concept of a window current applies only to the classical axon model where activation and inactivation gating are independent processes. Neurones often have a more rapid *voltage-independent* inactivation process (Aldrich, Corey & Stevens, 1983). With this model of Na^+ channel gating no window current is present.

Recently we have recorded Na^+ channel currents from soma on-cell patches and inside-out excised patches in slice and acutely isolated neocortical neurones (Alzheimer, Schwindt & Crill, 1993). A consistent but small percentage of depolarizations revealed single Na^+ channels with either brief late openings or sustained openings (Fig 18.4) . Although no patch contained a single channel and it is possible the non-inactivating responses were from a seperate set of channels, we believe our observations are best explained by assuming that Na^+ channels have two modes of

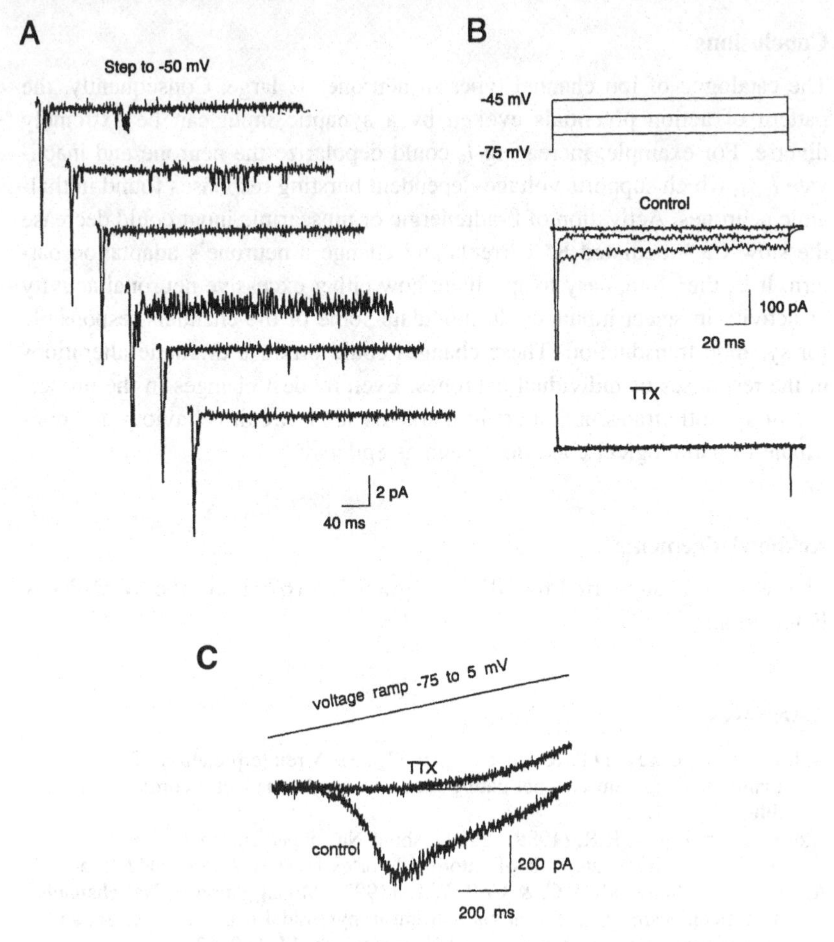

Fig. 18.4. Persistent Na^+ current recorded from rat neocortical pyramidal neurones. (A) Multi-channel recording showing the opening of Na^+ channels in response to a step potential change from −100 mV to −50 mV. Channels show early transient opening; late brief openings (traces 1 and 2); and persistent openings (trace 4). (B) Whole-cell Na^+ current in another neurone in response to step voltage commands. Initial transient current is truncated. (C) Whole-cell current in another neurone to a ramp voltage command.

behaviour: the inactivating response to depolarization and a much less frequent non-inactivating mode. The biophysical properties of the non-inactiving openings were identical to those of the transient Na^+ channels. Supporting this idea is the observation that cloned RIII and I Na^+ channel subtypes occasionally fail to inactivate and show infrequent prolonged reopenings (Moorman *et al.*, 1990; Zhou *et al.*, 1991). Calculations reveal that the channels in the non-inactivating gating mode can support the recorded whole-cell persistent current.

Conclusions

The catalogue of ion channel types in neurones is large. Consequently, the pattern of action potentials evoked by a synaptic input can be extremely diverse. For example, increasing I_h could depolarize the neurone and inactivate $I_{Ca(T)}$ which supports voltage-dependent bursting responses found in thalamic neurones. Activation of β-adrenergic or muscarinic input could decrease the slow Ca^{2+}-mediated K^+ currents and change a neurone's adaptation pattern. It is, therefore, easy to speculate how either excessive neuronal activity or activity in select inputs could modulate some of the channels responsible for synaptic transduction. These changes could produce dramatic alterations in the responses of individual neurones. Even modest changes in the properties of synaptic transduction could markedly alter circuit behaviour and contribute to pathological conditions such as epilepsy.

Acknowledgements

This work was supported by NINCDS grant NS 16792 and the W.M. Keck Foundation.

References

Aldrich, R.W., Corey, D.P. & Stevens, C.F. (1983). A reinterpretation of mammalian sodium channel gating based on single channel recording. *Nature*, **306**, 436–441.

Alonzo, A. & Llinás, R.R. (1989). Subthreshold Na^+-dependent theta-like rhythmicity in stellate cells of entorhinal cortex layer II. *Nature*, **342**, 175–177.

Alzheimer, C., Schwindt, P.C. & Crill, W.E. (1993). Modal gating of Na^+ channels as a mechanism of persistent Na^+ current in pyramidal neurones from rat and cat sensorimotor cortex. *Journal of Neuroscience*, **13**, 660–673.

Brown, A.M., Sayer, R.J., Schwindt, P.C. & Crill, W.E. (1994). P-type calcium channels in rat neocortical neurones. *Journal of Physiology (London)*, **475**, 197–205.

Connor, J.A. & Stevens, C.F. (1971). Voltage clamp studies of a transient outward membrane current in gastropod neural somata. *Journal of Physiology (London)*, **213**, 21–30.

Foehring, R.C., Schwindt, P.C. & Crill, W.E. (1989). Norepinephrine selectively reduces slow Ca^{2+}- and Na^+-mediated K^+ currents in cat neocortical neurons. *Journal of Neurophysiology*, **61**, 245–256.

French, C.R., Sah, P., Buckett, K.J. & Gage, P.W. (1990). A voltage-dependent persistent sodium current in mammalian hippocampal neurons. *Journal of General Physiology*, **95**, 1139–1157.

Gage, P.W., Lamb, G.D. & Wakefield, B.T. (1989). Transient and persistent sodium currents in normal and denervated mammalian skeletal muscle. *Journal of Physiology (London)*, **418**, 427–439.

Hodgkin, A.L. & Huxley, A.F. (1952). A quantitative description of membrane current and its application to conduction and excitation in nerve. *Journal of Physiology (London)*, **117**, 500–544.

Lemon, R.N. & Mantel, G.W.H. (1989). The influence of changes in discharge frequency of corticospinal neurones on hand muscles in the monkey. *Journal of Physiology (London)*, **413**, 351–378.

Llinás, R.R. (1988). The intrinsic electrophysiological properties of mammalian neurons: insights into central nervous system function. *Science*, **242**, 1654–1664.

Llinás, R. & Sugimori, M. (1980). Electrophysiological properties of *in vitro* Purkinje cell somata in mammalian cerebellar slices. *Journal of Physiology (London)*, **305**, 171–195.

McCormick, D. & Pape, H.-C. (1990). Noradrenergic and serotonergic modulation of a hyperpolarization-activated cation current in thalamic relay neurones. *Journal of Physiology (London)*, **431**, 319–342.

McCormick, D. & Prince, D.A. (1987). Post-natal development of electrophysiological properties of rat cerebral cortical pyramidal neurones. *Journal of Physiology (London)*, **393**, 743–762.

Moorman, J.R., Kirsch, G.E., VanDongen, A.M.J., Joho, R.H. & Brown, A.M. (1990). Fast and slow gating of sodium channels encoded by a single mRNA. *Neuron*, **4**, 243–252.

Sayer, R.J., Brown, A.M., Schwindt, P.C. & Crill, W.E. (1993). Calcium currents in acutely isolated human neocortical neurons. *Journal of Neurophysiology*, **69**, 1596–1606.

Sayer, R.J., Schwindt, P.C. & Crill, W.E. (1990). High- and low-threshold calcium currents in neurons acutely isolated from rat sensorimotor cortex. *Neuroscience Letters*, **120**, 175–178.

Sayer, R.J., Schwindt, P.C. & Crill, W.E. (1992). Metabotropic glutamate receptor modulates high-threshold calcium current of neurons acutely isolated from rat neocortex. *Journal of Neurophysiology*, **68**, 833–842.

Schwindt, P.C., Spain, W.J., Foehring, R.C., Stafstrom, C.E., Chubb, M.C. & Crill, W.E. (1988a). Multiple potassium conductances and their functions in neurons from cat sensorimotor cortex *in vitro*. *Journal of Neurophysiology*, **59**, 424–449.

Schwindt, P.C., Spain, W.J., Foehring, R.C., Chubb, M.C. & Crill, W.E. (1988b). Slow conductances in neurons from cat sensorimotor cortex *in vitro*: their modulation by neurotransmitters and their role in slow excitability changes. *Journal of Neurophysiology*, **59**, 450–467.

Spain, W.J., Schwindt, P.C. & Crill, W.E. (1987). Anomalous rectification in neurones from cat sensorimotor cortex *in vitro*. *Journal of Neurophysiology*, **57**, 1555–1576.

Spain, W.J., Schwindt, P.C., & Crill, W.E. (1991a) Two transient potassium currents in layer V pyramidal neurones from cat sensorimotor cortex. *Journal of Physiology (London)*, **434**, 591–607.

Spain, W.J., Schwindt, P.C. & Crill, W.E. (1991b). Post-inhibitory excitation and inhibition in layer V pyramidal neurones from cat sensorimotor cortex. *Journal of Physiology (London)*, **434**, 609–626.

Stafstrom, C.E., Schwindt, P.C., Chubb, M.C. & Crill, W.E. (1985). Properties of persistent sodium conductance and calcium conductance of layer V neurons from cat sensorimotor cortex *in vitro*. *Journal of Neurophysiology*, **53**, 152–170.

Westenbroek, R.E., Hell, J.W., Warner, C., Dubel, S.J., Snutch, T.P. & Catterall, W.P. (1992). Biochemical properties and subcellular distribution of an N-type calcium channel 1 subunit. *Neuron*, **9**, 1099–1115.

Zhou, J., Potts, J.F., Trimmer, J.S., Agnew, W.S. & Sigworth, F.J. (1991). Multiple gating modes and the effect of modulating factors on the I sodium channel. *Neuron*, **7**, 775–785.

19

Cortical circuits, synchronization and seizures

JOHN G.R. JEFFERYS

Department of Physiology and Biophysics, St Mary's Hospital Medical School, Imperial College of Science, Technology and Medicine, London, UK

The epilepsies as a group are amongst the most common of neurological disorders, so that about 1% of the population experience more than one seizure during their lifetime. They provide a particularly fertile area for the interaction of basic neurophysiology and clinical neurology. Epileptic activity can be considered a pathological extreme of normal neuronal synchronization. Studies of experimental epilepsies both help us understand the fundamentals of the underlying disease processes and provide the impetus for basic research on brain function. They do indeed provide a 'Window to Brain Mechanisms' to borrow the title of an earlier monograph (Lockard & Ward, 1980).

A key feature of epileptic foci is the abrupt transition from apparently normal neuronal activity to the intense and synchronous discharge of many, if not all, of the neurones in the focus. The extent and duration of this synchrony varies. For present purposes we can divide discharges from epileptic foci into: brief and localized 'interictal' bursts, which last <200 ms; 'polyspikes' or 'after-discharges', which last several seconds but which also tend to remain localized at the focus; and 'seizures', which last tens of seconds up to 2 minutes, and which normally involve large areas of the brain, and may generalize into most of the neocortex, resulting in motor fits (Fig. 19.1). Normally epileptic seizures are self-limiting, but when these constraints fail status epilepticus results, which is a life-threatening medical emergency, and which is likely to cause substantial brain damage.

Considerable progress has been made in understanding the cellular and network properties responsible for brief epileptic discharges in rat and guinea pig hippocampus and neocortex under a variety of conditions. The modern era of this kind of research started with the identification *in vivo* of an essential cellular correlate of the interictal discharge in the 'paroxysmal depolarization shift', which is large depolarization of neuronal membranes synchronous with the epileptic discharge (Matsumoto & Ajmone-Marsan, 1964; Prince, 1968). We now know that it results from the summation of excitatory postsynaptic potentials (EPSPs) which occur simultaneously at many of the synapses made on each cell (Johnston & Brown, 1981; Traub &

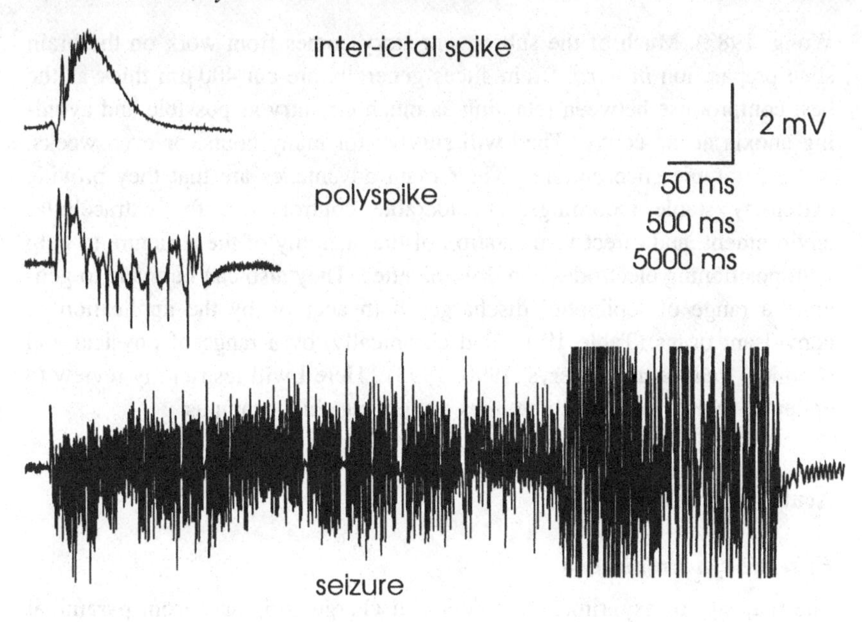

Fig. 19.1. Classes of epileptic activity. Discharges from epileptic foci can be classed by their duration into: interictal events, polyspike and seizures (note the 10-fold changes in the timescales of the successive traces). These records are from bipolar electrodes chronically implanted in the hippocampal CA3 region of a rat which had received a minute dose of tetanus toxin into the hippocampus 14 days previously. The seizure generalized to motor areas during the final 15 s, resulting in large movement artifacts. From G.T. Finnerty (1993) PhD thesis, University of London.

Table 19.1. *Cellular actions of convulsant drugs*

Convulsant drug	Cellular action	Reference
Bicuculline, penicillin, picrotoxin, tetanus toxin (direct effect)	Blocks $GABA_A$-mediated inhibition	Dingledine & Gjerstad (1979); Schwartzkroin & Prince (1978); Wong, Traub & Miles (1986)
high-$[K^+]_o$ (8 mM)	Increases cellular excitability; impairs inhibition	Rutecki, Lebeda & Johnston (1985); Chamberlin, Traub & Dingledine (1990); Traub & Dingledine (1990)
4-Aminopyridine	Blocks K^+ channels, resulting in increased transmitter release	Buckle & Haas (1982); Ives & Jefferys (1990)
low-$[Mg^{2+}]_o$ (<0.2 mM)	Unblocks NMDA receptors at excitatory synapses; reduces inhibition by <20%	Mody, Lambert & Heinemann (1987); Traub, Jefferys & Whittington (1994)

Wong, 1982). Much of the subsequent story comes from work on the brain slice preparation *in vitro*. Brain slices generally are cut 400 μm thick as the best compromise between retaining as much circuitry as possible and avoiding anoxia at the centre. They will survive for many hours, or even weeks, in 'organotypic slice culture'. Their main advantages are that they provide extremely stable recordings, considerable control over the extracellular environment, and direct visualization of the anatomy of the structure to help with positioning electrodes and drug pipettes. They also can be made to generate a range of 'epileptic' discharges both acutely, by the application of convulsant drugs (Table 19.1), and chronically, by a range of physical and chemical treatments (Jefferys, 1990, 1993). Here I will restrict my review to epileptic foci in rat (and sometimes guinea pig) hippocampus.

Acute electrophysiology

Epileptic synchronization

The majority of experimental epileptic discharges originate from pyramidal cells in the hippocampal region known as CA3. We now recognize that several factors contribute to its low seizure threshold. CA3 pyramidal cells are large, have complex intrinsic properties (notably calcium spikes in the dendrites which lead to bursts of sodium spike discharges at the soma). They also are interconnected by a network of excitatory synaptic connections. The essential feature for epileptogenesis is this network of glutamatergic synaptic connections between the CA3 pyramidal cells (Traub & Wong, 1982; Wong, Traub & Miles, 1986). It provides the fast synchronization of the onset of the discharge, primarily through the non-NMDA (*N*-methyl-D-aspartic acid) or AMPA (α-amino-3-hydroxy-5-methylisoxazole-4-propionic acid) subclass of glutamate receptor. These recurrent excitatory synapses are relatively powerful. Paired intracellular recordings have shown that single CA3 pyramidal cells can depolarize postsynaptic pyramidal cells by ~1 mV, 10-fold greater than many other central synapses. However, monosynaptic EPSPs of this size are found between only about 1–2% of pairs of CA3 pyramidal cells (Miles & Wong, 1987); there probably are many more structural synapses, which either are silent or weaker than the sensitivity of the sharp electrode recordings used in these studies. Normally this recurrent, or mutual excitation does not cause synchronous discharges, but it will do so, given a minimum number ('aggregate') of pyramidal cells (a few thousand: Miles, Wong & Traub, 1984), under a variety of experimental conditions, for instance: blocked synaptic (GABAergic) inhibition, enhanced synaptic excitation and increased neuronal excitability (Fig. 19.2).

In a series of computer simulations of networks of realistic CA3 pyramidal neurones performed in close association with experiments (reviewed in

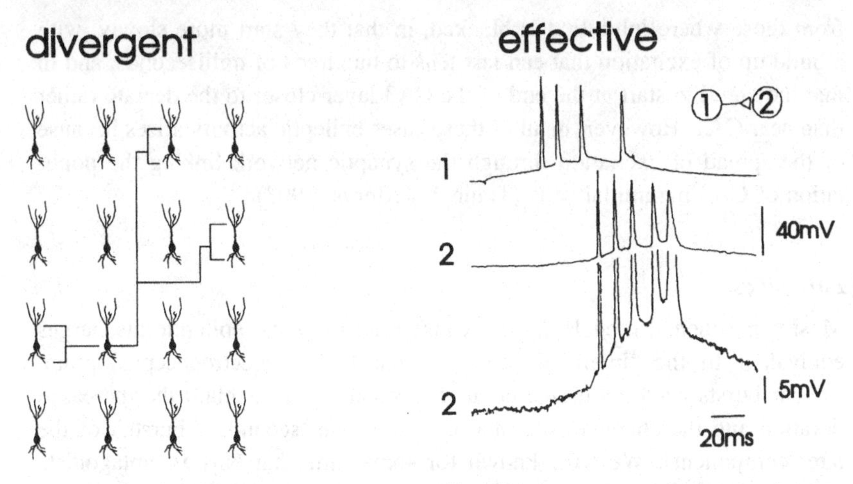

Fig. 19.2. Key features of the CA3 region necessary for epileptic activity in a range of models. The mutually excitatory connections between CA3 pyramidal cells must (i) be divergent and (ii) be powerful. Most convulsant drugs enhance the probability of successful excitation of the postsynaptic cell, by reducing inhibition, increasing excitation, etc. (see Table 19.1). From Wong *et al.* (1986), with permission.

Traub & Miles, 1991), Traub *et al.* have shown that the synaptic connectivity found in CA3 is sufficient for the synchronization of the population into epileptic discharges if two conditions are met. Firstly the connections must be divergent, i.e. each pyramidal cell must, on average, synapse onto more than one other pyramidal cell. Secondly, these synapses must be powerful enough to discharge the postsynaptic cell. In normal tissue the synapses are not powerful enough (the probability of discharge is <5%). However, if the presynaptic cell fires a complex spike (a train of 5–7 sodium action potentials driven by a dendritic calcium spike) then the probability of discharge increases to <90% (Wong *et al.*, 1986; Traub & Miles, 1991). Convulsant drugs act by increasing the probability of discharging the postsynaptic cell in various ways (Table 19.1); mostly they have the effect of promoting complex spikes (intrinsic bursts), with the notable exception of 4-aminopyridine (Traub & Jefferys, 1993). It is possible, though still not proven, that chronic models and clinical foci may depend on changes in the connectivity of the network due to the sprouting of new synapses which is known to occur in a number of models (Tauck & Nadler, 1985; Ben-Ari & Represa, 1990). Low-$[Mg^{2+}]_o$ bursts differ from many others in that they persist in the presence of drugs that block the AMPA (or non-NMDA) class of excitatory receptor, which shows both that NMDA receptors are located at the recurrent synapses between CA3 pyramidal cells, and that they are capable of recruiting the population into an epileptic discharge, although normally this is done by the AMPA receptors. Those models which retain inhibition differ in detail from those where inhibition is blocked, in that they start more slowly, with a build up of excitation that can last tens to hundreds of milliseconds, and in that they tend to start at the end of the CA3 layer closer to the dentate rather than near CA1. However, in all of these cases epileptic activity arises because of the spread of excitation through the synaptic network linking the population of CA3 pyramidal cells (Traub & Jefferys, 1993).

Polyspikes

Most experimental models *in vitro* cause brief (100 ms) epileptic discharges equivalent to the 'interictal spike' of the human electroencephalogram. Longer bursts need additional cellular mechanisms to explain the increased duration and the characteristic phasic quality, the 'secondary bursts', of the later components. We have known for some time that NMDA antagonists block longer bursts but have little effect on short, interictal bursts or on the initial part, the primary burst, of polyspikes. This suggested to us that the prolongation of epileptic discharges into polyspikes is due to the prolonged kinetics of depolarizing current from the NMDA class of glutamate receptor. The best estimates we have been able to make for its kinetics in intact mammalian tissue at physiological temperatures is a 5 ms rising phase and a time constant of 150 ms for the decay, compared with 2 ms for both these measures of the AMPA current (Traub, Miles & Jefferys, 1993). This does not, however, explain the organization of the secondary bursts. We have recently proposed a model in which the current through NMDA channels sustains synaptic excitation for longer than the processes (mainly outward K^+ membrane currents) that otherwise terminate the burst (Fig. 19.3). Dendritic Ca^{2+} currents are activated during the primary burst. In the mature hippocampus they are mainly of a high-threshold variety, unlike in the thalamus where low-threshold Ca^{2+} currents provide the tendency to oscillate, providing the rebound excitation in the model of synchronous oscillations in the thalamus first proposed by Andersen & Sears (1964). In the hippocampal model the Ca^{2+} spikes are terminated by outward K^+ currents which are activated by $Ca^{2+}{}_i$. However, as the outward K^+ current wanes, it ultimately gives way to sustained inward dendritic current through NMDA receptors. This in turn retriggers Ca^{2+} spikes and initiates the next secondary burst. Essentially the dendrites constitute an oscillator which is driven by the tonic inward current through NMDA receptors. The secondary bursts in individual cells are synchronized by the discharge of the first few cells causing the release of more glutamate at synapses, which causes an additional inward current through the AMPA receptor that recruits the rest of the population and also recharges the NMDA receptor. Steady current injection into CA3 pyramidal cell dendrites does indeed cause an oscillating response (Traub *et al.*, 1993). This model provides many predictions that can, and will, be tested experimentally. The

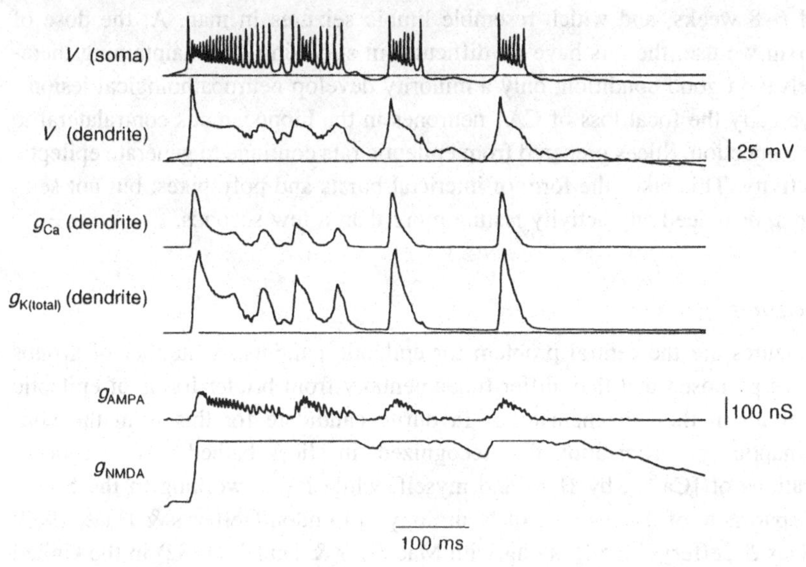

Fig. 19.3. Polyspike activity in the soma and an apical dendritic compartment 0.6 length constants away of a single simulated cell from a 100 cell network (top two traces). This discharge pattern is synchronized by current through the AMPA class of glutamatergic synaptic receptor, and sustained by the slower current through the NMDA class of receptor. The secondary bursts of the second half of the discharge are shaped by the intrinsic oscillation of the Ca^{2+} and K^{+} currents of the dendrites. The slow after-hyperpolarization appears as a tail on the total K^{+} conductance, and plays a key role, with the NMDA conductance, in determining the interval between secondary bursts. From Traub *et al.* (1993) with permission.

polyspike ultimately terminates with the build up of one or more of: slow $GABA_B$-mediated inhibition, slow after-hyperpolarizations, or NMDA receptor desensitization.

This model shows at the network level how NMDA antagonists have their anticonvulsant actions, which were identified many years ago (Coutinho-Netto *et al.*, 1981; Croucher, Collins & Meldrum, 1982). Unfortunately NMDA receptors have other roles, and no selective antagonists have so far had a good enough therapeutic index to reach clinical practice. We have proposed on the basis of our model that drugs targeting the repetitive use of high-threshold voltage-sensitive Ca^{2+} channels might be a better bet.

Chronic models

Single seizures do not qualify as epilepsy. Chronic models provide an experimental means of addressing the consequences of recurrent seizures. Our own interests centre on epilepsy induced in rats by intrahippocampal tetanus toxin. The rats experience intermittent epileptic seizures, which recur over a period of 6–8 weeks, and which resemble limbic seizures in man. At the dose of toxin we use, the rats have no difficulty in surviving and maintaining themselves in good condition; only a minority develop neuropathological lesions, typically the focal loss of CA1 neurones in the hippocampus contralateral to the injection. Slices prepared from epileptic rats continue to generate epileptic activity. This takes the form of interictal bursts and polyspikes, but not seizures, or indeed any activity lasting more than a few seconds.

Seizures

Seizures are the central problem for epileptic patients. A number of groups have proposed that they differ fundamentally from briefer forms of epileptic activity in their mechanism. A favourite candidate for this is in the non-synaptic synchronization first recognized, in slices bathed in low concentrations of $[Ca^{2+}]_o$, by Haas and myself, while I was working in the Sobell Department of the Institute of Neurology in London (Jefferys & Haas, 1982; Haas & Jefferys, 1984), along with MacVicar & Dudek (1982) in the United States. Essentially these bursts are synchronized by electrical interactions or 'field bursts', and by increased $[K^+]_o$ (which could be termed spreading excitation), and they can last tens of seconds (Konnerth, Heinemann & Yaari, 1986; Jensen & Yaari, 1988; Traynelis & Dingledine, 1989). However, our work on the tetanus toxin focus *in vivo* shows that seizures do not arise in this manner (Fig. 19.1). If they did then the phase lag between CA3 (the initiation site for interictal discharges) and CA1 (where field bursts start) should reverse. In fact CA3 always leads CA1 during tetanus toxin seizures *in vivo* (Finnerty & Jefferys, 1993). We believe that the substrate for seizures does differ from those for interictal spikes and polyspikes, in extent, not in location. Thus we know that both hippocampi always are involved, and suspect that structures such as the entorhinal cortex and cingulate cortex play critical roles (Hawkins & Mellanby, 1987; Finnerty & Jefferys, 1993). The notion that the seizure focus may differ from the interictal foci in being distributed may have important implications for epilepsy surgery; it may be sufficient to cut the long-range connections between the separate nodes of the distributed focus, rather than to remove all the grey matter within the extended focus.

Cellular mechanisms of chronic epileptic foci

The immediate action of the tetanus toxin is to block inhibitory synapses, but this reverses after about 4 weeks, following (due to?) increased expression of the mRNA for the synthetic enzyme for GABA (Najlerahim *et al.*, 1992). However, the tissue remains epileptogenic for some time after the recovery of synaptic inhibition (i.e. of GABA release or monosynaptic IPSCs). In the hippocampus this period lasts a few weeks (Empson & Jefferys, 1993; Whittington & Jefferys, 1993), while in the neocortex it can last for many months (Empson *et al.*, 1993). We have evidence, for the hippocampal focus, that while the inhibitory neurones are still present, they may not be adequately excited; this is similar to the 'dormant basket cell' hypothesis proposed by R. Sloviter in another context (Sloviter, 1991). This has been implicated in another chronic model of epilepsy, the relapse into recurrent seizures some weeks after electrically induced status epilepticus (Bekenstein & Lothman, 1993). Other factors have been suggested from other chronic experimental models such as kindling, for which there is some evidence in the tetanus toxin model, in particular the 'sprouting' of new excitatory connections and, perhaps related, the increased use of NMDA receptors at excitatory synapses (Tauck & Nadler, 1985; Ben-Ari & Represa, 1990; Simpson, Wheal & Williamson, 1991; Masukawa *et al.*, 1991; Cronin *et al.*, 1992).

Conclusion

The detailed exploration of the features of neuronal networks that cause neuronal synchronization, and epileptic activity of various kinds, will continue to contribute to our understanding of normal brain function, while identifying the unique features of seizure activity will, we believe, lead to more rational approaches to treatment.

Acknowledgements

The author is a Wellcome Senior Lecturer, and thanks: the Wellcome Trust for its generous support; Simon Colling, Ruth Empson, Gerald Finnerty, Alison Watts and Miles Whittington for valuable discussions and for performing much of the work described here; and last, but by no means least, Tom Sears for providing the opportunity and encouragement to start down this path.

References

Andersen, P. & Sears, T.A. (1964). The role of inhibition in the phasing of spontaneous thalamocortical discharge. *Journal of Physiology (London)*, **173**, 459–480.

Bekenstein, J.W. & Lothman, E.W. (1993). Dormancy of inhibitory interneurones in a model of temporal lobe epilepsy. *Science*, **259**, 97–100.

Ben-Ari, Y. & Represa, A. (1990). Brief seizure episodes induce long-term potentiation and mossy fibre sprouting in the hippocampus. *Trends in Neurosciences*, **13**, 312–318.

Buckle, P.J. & Haas, H.L. (1982). Enhancement of synaptic transmission by

4-aminopyridine in hippocampal slices of the rat. *Journal of Physiology (London)*, **326**, 109–122.

Chamberlin, N.L., Traub, R.D. & Dingledine, R. (1990). Role of EPSPs in initiation of spontaneous synchronized burst firing in rat hippocampal neurones bathed in high potassium. *Journal of Neurophysiology*, **64**, 1000–1008.

Coutinho-Netto, J., Abdul-Ghani, A.S., Collins, J.F. & Bradford, H.F. (1981). Is glutamate a trigger factor in epileptic hyperactivity? *Epilepsia*, **22**, 289–296.

Cronin, J., Obenaus, A., Houser, C.R. & Dudek, F.E. (1992). Electrophysiology of dentate granule cells after kainate-induced synaptic reorganization of the mossy fibres. *Brain Research*, **573**, 305–310.

Croucher, M.J., Collins, J.F. & Meldrum, B.S. (1982). Anticonvulsant action of excitatory amino acid antagonists. *Science*, **216**, 899–901.

Dingledine, R. & Gjerstad, L. (1979). Penicillin blocks hippocampal IPSPs unmasking prolonged EPSPs. *Brain Research*, **168**, 205–209.

Empson, R.M., Amitai, Y., Jefferys, J.G.R. & Gutnick, M.J. (1993). Injection of tetanus toxin into the neocortex elicits persistent epileptiform activity but only transient impairment of GABA release. *Neuroscience*, **57**, 235–239.

Empson, R.M. & Jefferys, J.G.R. (1993). Inhibitory synaptic function in primary and secondary chronic epileptic foci induced by intrahippocampal tetanus toxin in the rat. *Journal of Physiology (London)*, **465**, 595–614.

Finnerty, G.T. & Jefferys, J.G.R. (1993). Spontaneous epileptiform activity in freely moving rats with epileptic foci induced by tetanus toxin. *Journal of Physiology (London)*, **467**, 319P

Haas, H.L. & Jefferys, J.G.R. (1984). Low-calcium field burst discharges of CA1 pyramidal neurones in rat hippocampal slices. *Journal of Physiology (London)*, **354**, 185–201.

Hawkins, C.A. & Mellanby, J.H. (1987). Limbic epilepsy induced by tetanus toxin: a longitudinal electroencephalographic study. *Epilepsia*, **28**, 431–444.

Ives, A.E. & Jefferys, J.G.R. (1990). Synchronization of epileptiform bursts induced by 4-aminopyridine in the *in vitro* hippocampal slice preparation. *Neuroscience Letters*, **112**, 239–245.

Jefferys, J.G.R. (1990). Basic mechanisms of focal epilepsies. *Experimental Physiology*, **75**, 127–162.

Jefferys, J.G.R. (1993). The pathophysiology of epilepsies. In *The Epilepsies*, ed. J. Laidlaw, A. Richens & D.W. Chadwick, pp. 241–276. Edinburgh: Churchill Livingstone.

Jefferys, J.G.R. & Haas, H.L. (1982). Synchronized bursting of CA1 pyramidal cells in the absence of synaptic transmission. *Nature*, **300**, 448–450.

Jensen, M.S. & Yaari, Y. (1988). The relationship between interictal and ictal paroxysms in an *in vitro* model of focal hippocampal epilepsy. *Annals of Neurology*, **24**, 591–598.

Johnston, D. & Brown, T.H. (1981). Giant synaptic potential hypothesis for epileptiform activity. *Science*, **211**, 294–297.

Konnerth, A., Heinemann, U. & Yaari, Y. (1986). Nonsynaptic epileptogenesis in the mammalian hippocampus *in vitro*. I. Development of seizure-like activity in low extracellular calcium. *Journal of Neurophysiology*, **56**, 409–423.

Lockard, J.S. & Ward, A.A., Jr (1980). *Epilepsy: A Window to Brain Mechanisms*. New York: Raven Press.

MacVicar, B.A. & Dudek, F.E. (1982). Electronic coupling between granule cells of the rat dentate gyrus: physiological and anatomical evidence. *Journal of Neurophysiology*, **47**, 579–592.

Masukawa, L.M., Higashima, M., Hart, G.J., Spencer, D.D. & O'Connor, M.J. (1991). NMDA receptor activation during epileptiform responses in the dentate gyrus of epileptic patients. *Brain Research*, **562**, 176–180.

Matsumoto, H. & Ajmone-Marsan, C. (1964). Cortical cellular phenomena in experimental epilepsy: interictal manifestations. *Experimental Neurology*, **9**, 286–304.

Miles, R. & Wong, R.K.S. (1987). Latent pathways revealed after tetanic stimulation in the hippocampus. *Nature*, **329**, 724–726.

Miles, R., Wong, R.K.S. & Traub, R.D. (1984). Synchronized afterdischarges in the hippocampus: contribution of local synaptic interactions. *Neuroscience*, **12**, 1179–1189.

Mody, I., Lambert, J.D. & Heinemann, U. (1987). Low extracellular magnesium induces epileptiform activity and spreading depression in rat hippocampal slices. *Journal of Neurophysiology*, **57**, 869–888.

Najlerahim, A., Williams, S.F., Pearson, R.C.A. & Jefferys, J.G.R. (1992). Increased expression of GAD mRNA during the chronic epileptic syndrome due to intrahippocampal tetanus toxin. *Experimental Brain Research*, **90**, 332–342.

Prince, D.A. (1968). The depolarization shift in 'epileptic' neurones. *Experimental Neurology*, **21**, 467–485.

Rutecki, P.A., Lebeda, F.J. & Johnston, D. (1985). Epileptiform activity induced by changes in extracellular potassium in hippocampus. *Journal of Neurophysiology*, **54**, 1363–1374.

Schwartzkroin, P.A. & Prince, D.A. (1978). Cellular and field potential properties of epileptogenic hippocampal slices. *Brain Research*, **147**, 117–130.

Simpson, L.H., Wheal, H.V. & Williamson, R. (1991). The contribution of non-NMDA and NMDA receptors to graded bursting activity in the CA1 region of the hippocampus in a chronic model of epilepsy. *Canadian Journal of Physiology and Pharmacology*, **69**, 1091–1098.

Sloviter, R.S. (1991). Permanently altered hippocampal structure, excitability, and inhibition after experimental status epilepticus in the rat: the 'dormant basket cell' hypothesis and its possible relevance to temporal lobe epilepsy. *Hippocampus*, **1**, 41–66.

Tauck, D.L. & Nadler, J.V. (1985). Evidence of functional mossy fibre sprouting in hippocampal formation of kainic acid-treated rats. *Journal of Neuroscience*, **5**, 1016–1022.

Taylor, C.P. & Dudek, F.E. (1982). Synchronous neural afterdischarges in rat hippocampal slices without active chemical synapses. *Science*, **218**, 810–812.

Taylor, C.P. & Dudek, F.E. (1984). Synchronization without active chemical synapses during hippocampal afterdischarges. *Journal of Neurophysiology*, **52**, 143–155.

Traub, R.D. & Dingledine, R. (1990). Model of synchronized epileptiform bursts induced by high potassium in CA3 region of rat hippocampal slice: role of spontaneous EPSPs in initiation. *Journal of Neurophysiology*, **64**, 1009–1018.

Traub, R.D. & Jefferys, J.G.R. (1994). Are there unifying principles underlying the generation of epileptic discharges *in vitro*? *Progress in Brain Research*, **102**, 383–394.

Traub, R.D., Jefferys, J.G.R. & Whittington, M.A. (1994). Enhanced NMDA conductance can account for epileptiform activity induced by low Mg^{++} in the rat hippocampal slice. *Journal of Physiology (London)*, **478**, 379–393.

Traub, R.D. & Miles, R. (1991). *Neuronal Networks in the Hippocampus*. Cambridge: Cambridge University Press.

Traub, R.D., Miles, R. & Jefferys, J.G.R. (1993). Synaptic and intrinsic conductances shape picrotoxin-induced synchronized after-discharges in the guinea-pig hippocampal slice. *Journal of Physiology (London)*, **461**, 525–547.

Traub, R.D. & Wong, R.K.S. (1982). Cellular mechanism of neuronal synchronization in epilepsy. *Science*, **216**, 745–747.

Traynelis, S.F. & Dingledine, R. (1989). Modification of potassium-induced interictal bursts and electrographic seizures by divalent cations. *Neuroscience Letters*, **98**, 194–199.

Whittington, M.A. & Jefferys, J.G.R. (1993). Epileptic activity outlasts disinhibition in area CA3 and the dentate gyrus *in vitro* during the tetanus toxin model of chronic epilepsy in the rat. *Journal of Physiology (London)*, **473**, 165P.

Wong, R.K.S., Traub, R.D. & Miles, R. (1986). Cellular basis of neuronal synchrony in epilepsy. *Advances in Neurology*, **44**, 583–592.

20

Physiologically induced changes of brain temperature and their effect on extracellular field potentials

P. ANDERSEN, E. MOSER AND V. JENSEN

Department of Neurophysiology, Institute of Basic Medical Sciences, University of Oslo, Norway

Temperature is an important factor for nervous activity. In physiological conditions we expect the temperature-regulating mechanisms of the homeothermic animal to prevent large temperature changes. However, in addition to the well-documented changes in brain temperature during feeding and sleeping (Abrams & Hammel, 1964), we have recently observed other, surprisingly large variations of brain temperature in behaving rats (Moser, Mathiesen & Andersen, 1993). There are also indications that brain temperature may change appreciably in man, both during exercise and after cooling of the facial skin (Nielsen, 1988). Furthermore, drugs and anaesthesia may cause the temperature of the brain to drop appreciably.

In the present short survey, we review the temperature changes that can be recorded from the brain of freely moving rats and the consequences such changes have on field and unitary potentials.

Field potentials and brain temperature in freely moving rats

In rats swimming in a Morris water maze (Morris, 1984) at 18 °C (Moser *et al.*, 1993), we noticed quite large and fast changes of simultaneously recorded hippocampal field potentials (Fig. 20.1A, B). There was an increased latency of both the field excitatory postsynaptic potential (f-EPSP) and the population spike. Surprisingly, the latter increased greatly in amplitude. The slope of the f-EPSP, measured at its maximum, was reduced. All field potential changes depended on the temperature of the water, but were still quite large at the usually employed temperature of 25 °C. While rats swam for 5 minutes, implanted thermistors showed that their brain temperatures fell from nearly 38 °C to less than 30 °C (Fig. 20.1C). This temperature represents the threshold for hypothermically induced impairment of spatial learning (Panakhova, Buresova & Bures, 1984; Rauch, Welch & Gallego, 1989; Vanderwolf, 1991). When rats swam in water of different temperatures, the field potential changes were less developed the warmer the water. In fact, in water above 38 °C, all parameters change in the opposite direction from those described

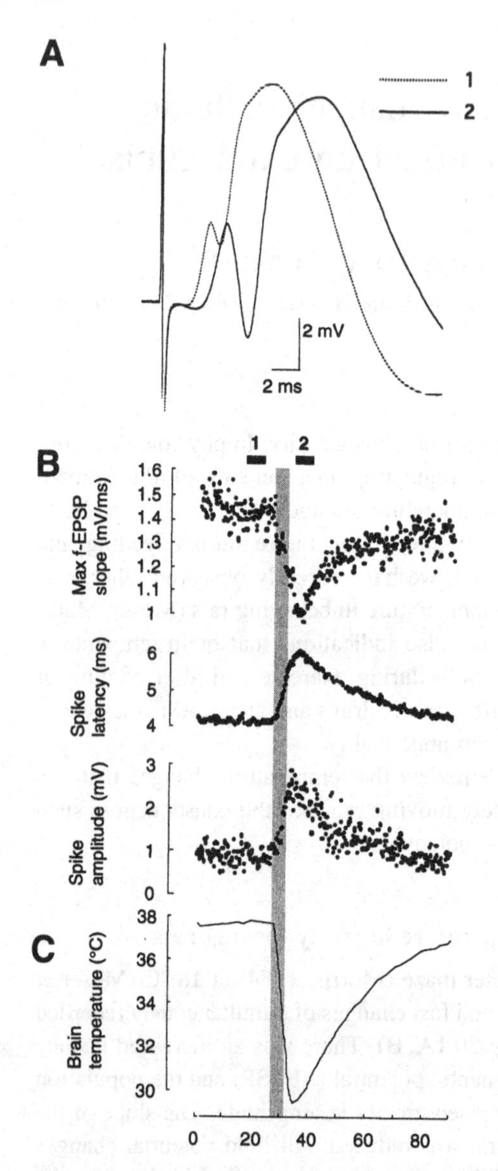

Fig. 20.1. Field potential changes caused by cooling of the brain. Field potentials of the dentate gyrus in response to stimulation of the perforant path before, during and after swimming in a water maze at 18 °C. (A) Superimposed records taken at average brain temperatures of 37.6 °C (trace 1) and 29.5 °C (trace 2) before and after swimming for 5 minutes, respectively. (B) Plots of the maximal slope of the field excitatory postsynaptic potential (EPSP), the population spike amplitude and latency during the same experiment. (C) Brain temperature, recorded from the contralateral brain half, at a depth just above the hippocampus. The vertical grey stripe indicates the swimming.

above. These results strongly suggest that swimming in water colder than the body temperature cooled the brain, which resulted in the alterations of the field potentials (Andersen, 1960). This was verified by recording field potentials and brain temperature in resting animals during gentle warming produced either by a heating lamp or by blowing warm air into the cage.

More surprisingly, when rats explored a new environment on their own drive and at a leisurely pace, the field potentials changed in the opposite direction from those shown in Fig. 20.1. Because the f-EPSP to a constant stimulus increased, such data were initially taken as a sign of enhanced synaptic transmission and named short-term exploratory modulation (STEM: Sharp, McNaughton & Barnes, 1989; Green, McNaughton & Barnes, 1990). When controlled with implanted thermistors, it turned out that the brain temperature increased appreciably, even with such moderate motor activity (Moser *et al.*, 1993).

Other changes of field potentials could also be due to temperature changes in the brain. The alteration of hippocampal field potentials during the sleep–wake cycle, which has been ascribed to state-dependent excitability changes (Winson & Abzug, 1977), could to a certain extent be due to simultaneous temperature changes (Kovalzon, 1973; Obàl *et al.*, 1985). Likewise, the *N*-methyl-D-aspartic acid (NMDA)-receptor channel blocker MK-801 produces hypothermia, while 3–3(2-arboxypiperazin-4-yl)-propyl-1-phosphonate, another NMDA receptor blocker, did not (Corbett *et al.*, 1990). These findings open the possibility that some of the effects of 2-amino-5-phosphone-pentanoate on learning (Morris *et al.*, 1986; Davis, Butcher & Morris, 1992) could be due to a change in brain temperature.

Activity-dependent brain temperature changes

Parallel changes of the field potentials and brain temperature were also seen during other types of self-paced, spontaneous behaviour. When a rat was left to itself in semi-darkness in its home cage, the brain temperature usually declined slowly in parallel with a relaxation of the animal. In Fig. 20.2A, the rat was immobile and went to sleep during the first 2 hours. During two periods of REM sleep (arrows) the temperature increased in parallel with the occurrence of muscle jerks. After about 2 hours the animal became active and showed within the next 5 hours four periods of spontaneous activity (asterisks), during which the brain temperature increased by between 0.8 °C and 1.2 °C. In parallel with the brain temperature changes, the field potential of the dentate fascia in response to stimulation of the perforant path also changed (Fig. 20.2B, C). The most characteristic changes were the latency shifts of the field EPSP, but the reduction of the population spike amplitude is also clear (Fig. 20.2C). During such periods we noted a significant increase

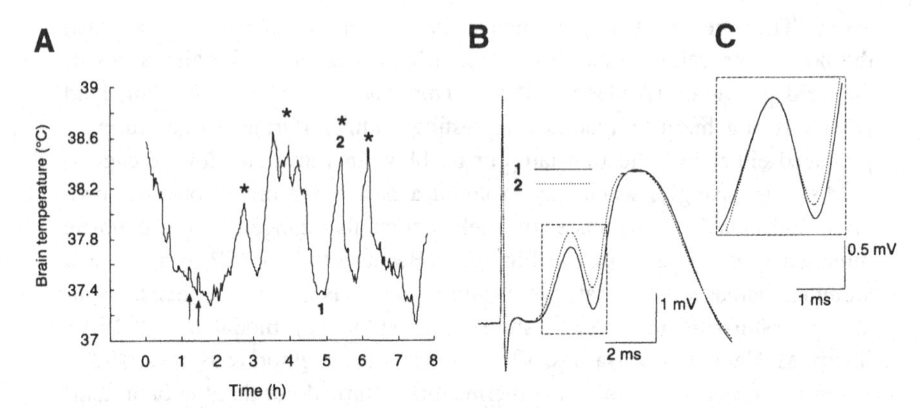

Fig. 20.2. Spontaneous changes of brain temperature. (A) Brain temperature recorded from a rat during 8 hours of spontaneous behaviour. The arrows indicate two periods of REM sleep, while the asterisks signal four periods of exploratory activity. (B) Superimposed field potential record taken at the times indicated by the numbers 1 and 2 in (A). The records in the inset are enlarged in (C), and aligned upon the initial, positive peak to illustrate the amplitude change of the spike.

in the temperature, even during exploratory activity of modest intensity. On average, the brain temperature increased by 0.11 °C per minute during forced exploration, a warming which could measure as much as 3 °C when the animals were highly excited and continuously moving around in the environment.

Gradient of brain temperature

By recording from different depths, we noted a temperature gradient in the ventro-dorsal direction in both freely moving and anaesthetized rats. In the resting condition, the difference between the base and the dorsal part of the brain usually measured about 1 °C. This gradient was maintained at various brain temperatures, both when the animal increased its brain temperature by moving around and when the temperature was reduced during anaesthesia. Thermistors placed at the same depth in the transverse or rostro-caudal plane did not record any significant temperature difference. Thus the rat brain can be seen as a relatively flat sheet of tissue with a vertically directed temperature gradient. For measurement purposes, therefore, the thermistor or thermocouple should be placed at a level which corresponds to the depth of the tissue which generates the signal of interest. The close association between movement and increased brain temperature, and the consistently warmer base compared with the top of the brain are in accord with the work of Abrams and Hammel, who found that arterial blood is the main source of brain warming in behaving animals (Abrams & Hammel, 1964, 1965; Abrams, Stolwijk & Graichen, 1965).

Cause of the temperature-induced field potential changes

The explanation for the temperature-induced changes of the field potential rests on simple biophysical facts. During cooling the increased latency of the f-EPSP is in part due to a reduced conduction velocity. The conduction velocity has a temperature coefficient of 1.6, similar to that of the upstroke of the action potential (Hodgkin & Katz, 1949). The latency of the f-EPSP also incorporates the release of the transmitter at the synapses involved (Schiff & Somjen, 1985). The much higher temperature coefficient of transmitter release (about 3.0; Katz & Miledi, 1965) explains the delayed onset and the slower rate of rise of the f-EPSP. The most surprising finding is that of an enhanced population spike at reduced brain temperature (Fig. 20.1A, B). The reason for this apparent paradox rests on four points. Firstly, each individual action potential has a larger amplitude in a cooled than in a warm condition (Schoepfle & Erlanger, 1941). The reason is that the repolarizing K^+ currents are more temperature-sensitive than the Na^+ currents (Hodgkin & Katz, 1949). Consequently, the onset of the repolarization comes somewhat later in the cool condition, allowing the action potential to reach closer to the equilibrium potential for sodium and thus to achieve a larger amplitude in a warm condition. Secondly, the repolariziation of the action potential proceeds more slowly in the cold situation, giving the action potential a longer duration. Thirdly, cooling gives a slight depolarization of the membrane. This depolarization is so small, however, that it has little effect on the size of the action potential, but it will tend to give a slightly reduced threshold for activation of the cell (Hodgkin & Katz, 1949). The most important factor is the fourth, which is the enhanced efficiency of summation of individual action potentials to compound action potentials (Ritchie & Straub, 1956). Cooling causes much greater broadening of each individual action potential than dispersion of their onset times because of dissimilar temperature coefficients for the two processes.

Model of summation effect

The simple model in Fig. 20.3 illustrates this phenomenon. To the left a single action potential is drawn at a cool and a warm temperature, showing the difference in amplitude and time course. Below, to the left are drawn ten superimposed action potentials, slightly dispersed to imitate somewhat different conduction velocities. In the upper right of the figure are the same potentials drawn on a smaller vertical scale. Below a simple summation of the individual action potentials to a compound action potential is shown, and how the cool situation is associated with a larger amplitude of the compound potential. Under extracellular recording conditions, the signal will be differentiated and show a di- or triphasic form, but the principle of increased

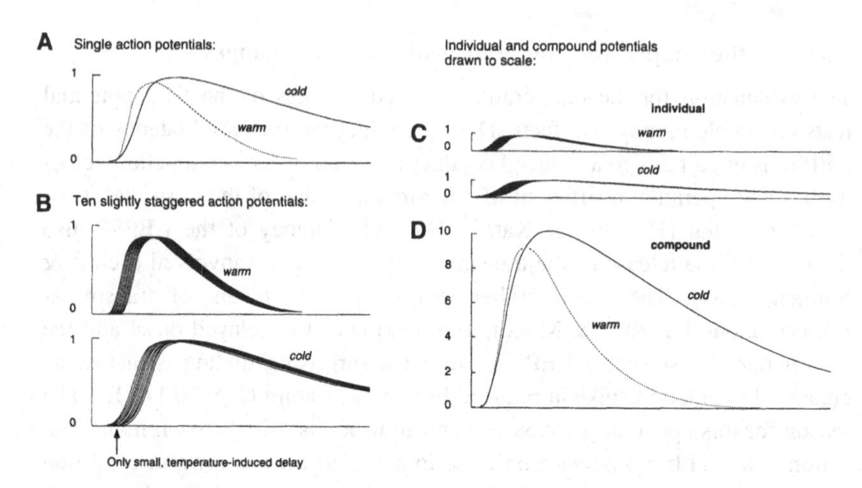

Fig. 20.3. Model illustrating the increased summed signals during cooling. (A) An action potential as seen in warm conditions superimposed on one as seen after cooling. (B) Ten superimposed action potentials, identical to those in (A), but with slightly different latencies, simulating different conduction velocities of their parent axons. (C) As (B), but drawn with smaller gain. (D) Algebraic summation of the individual traces in (C) to compound potentials.

efficiency of summation will give the same type of result as the monophasically drawn potentials.

Does the enhanced population potential in the cooled brain mean a larger number of discharging cells?

Under normal conditions, we interpret a changed size of the population spike as an altered number of cells discharging. In their original work on the hippocampus, Andersen, Bliss & Skrede (1971) measured a number of individual action potentials and compared their occurrence with the envelope of the population spike. For both antidromic and orthodromically induced signals, the population spike was, in fact, a simple summation of individual action potentials caused by neuronal discharges. Consequently the area under the population spike gives a good index of the number of discharging cells. Because the spread of activation is usually quite limited, the size of the population spike can also be used as an acceptable index of the number of discharging cells.

However, during a situation with altered brain temperature this rule no longer holds. In spite of an enhanced population spike during cooling, the available evidence suggests that either a constant number or a slightly reduced number of cells takes part in the discharge. For example, when cooling the dorsal column nuclei, all components of the signals induced by stimulation of peripheral nerves are enhanced (Andersen *et al.*, 1972). The N-wave,

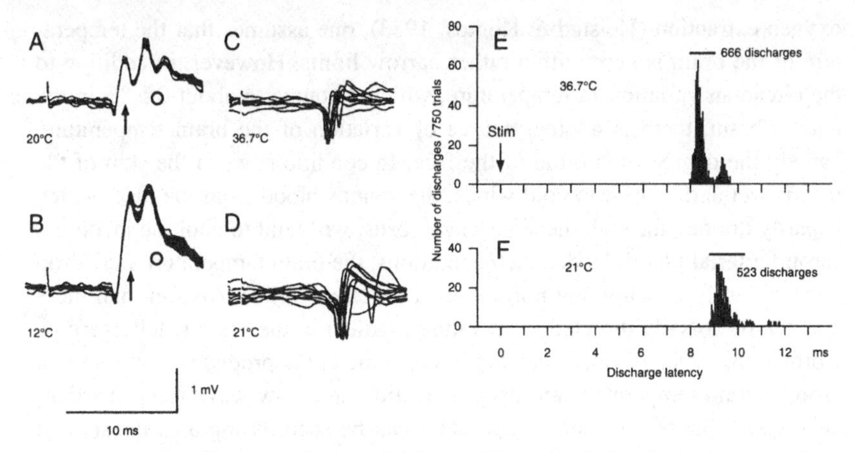

Fig. 20.4. Cooling-induced enhanced field potentials and cellular discharge properties. (A), (B) Field potentials from the cuneate nucleus in response to stimulation of the superficial radial nerve, taken during cooling to 20 °C and 12 °C, respectively. The arrow labels the afferent fibre volley and the circle signals the N-wave. (C), (D) Cuneate cell responding to radial nerve volleys at 10 Hz at cuneate surface temperatures of 36.7 °C and 21 °C, respectively. (E), (F) Latency histogram of the same cell in response to 750 trials. The horizontal bars label monosynaptic responses.

which is a blend of field synaptic potentials and discharges of cuneate neurones, is greatly enhanced by cooling (Fig. 20.4A, B). However, simultaneous recording from individual cuneate units showed an increased latency and reduced probability of discharge (Fig. 20.4C-F). A moderately reduced discharge probability was seen already on cooling by 5 °C. Larger changes occurred progressively with greater degrees of cooling (Fig. 20.4F). The sensitivity to changes of the discharge probability depends upon the regularity and size of the EPSP, particularly at high frequencies. The EPSP again depends upon the number of quanta usually delivered at the synapse in question. In this regard, the dorsal column nuclei are probably quite robust, showing faithful transmitter release, even at relatively high frequencies of the afferent impulses. Cortical synapses, with fewer quanta and much lower probabilities of release, are likely to be more temperature-sensitive. Thus, without proper recording from individual neurones in the brain during temperature alterations, it is not possible to be certain what a population spike change involves. However, on the basis of the cuneate data we may predict that each trial probably gives a reduced number of discharges in spite of an enhanced population spike.

Brain temperature changes in man

It is tacitly assumed that the brain temperature of man is constant. Knowing the efficient autoregulation of the brain circulation and the nearly constant oxygen extraction (Heistad & Kontos, 1983), one assumes that the temperature of the brain is kept within rather narrow limits. However, in addition to the circadian variation in temperature, which amounts to about 0.5 °C in the human brain, there is another source of variation of the brain temperature, namely the degree of cooling of the face. In conditions when the skin of the face is wet and subject to cold winds, the venous blood from the face, which is partly drained through the sinus cavernosus, will tend to cool the incoming carotid arterial blood. Under such conditions, the brain temperature may drop as much as 1 °C below the normal level (Nielsen, 1988). Also in man there is a ventro-dorsally directed temperature gradient in the brain (Mellergård & Nordström, 1990, 1991). Other conditions which may produce a considerable drop in brain temperature are deep relaxation and slow wave sleep. Particularly large changes in brain temperature can be seen during anaesthesia and accidental or intended hypothermia (Fay & Smith, 1941). Since appreciable changes in the evoked field potential may occur with temperature fluctuations as small as 1 °C, one should be aware of these changes in assessing the detailed nature of evoked potentials.

Conclusion

Increased or decreased brain temperature will affect all electrical signals, irrespective of their nature or site of generation. Thus, temperature effects can be observed in all types of evoked potentials and spontaneous signals, whether they are generated in the brainstem or in the cortex. However, due to the inflow of warm arterial blood at the base of the brain, the most dramatic changes will take place at the other end of the gradient, that is to say along the dorsal surface. Consequently, in humans the signals generated in the dorsalmost parts of the cerebral cortex are most sensitive to physiological or externally induced changes in brain temperature.

References

Abrams, R. & Hammel, H.T. (1964). Hypothalamic temperature in unanesthetized albino rats during feeding and sleeping. *American Journal of Physiology*, **206**, 641–646.

Abrams, R. & Hammel, H.T. (1965). Cyclic variations in hypothalamic temperature in unanesthetized rats. *American Journal of Physiology*, **208**, 698–702.

Abrams, R.M., Stolwijk, J.A.J. & Graichen, H. (1965). Brain temperature and blood flow in unanesthetized albino rats. *Life Sciences*, **4**, 2399–2410.

Andersen, P. (1960). Interhippocampal impulses. III. Basal dendritic activation of CA3 neurons. *Acta Physiologica Scandinavica*, **48**, 209–230.

Andersen, P., Bliss, T.V.P. & Skrede, K.K. (1971). Unit analysis of hippocampal population spikes. *Experimental Brain Research*, **13**, 208–221.

Andersen, P., Gjerstad, L. & Pasztor, E. (1972). Effect of cooling on synaptic transmission through the cuneate nucleus. *Acta Physiologica Scandinavica*, **84**, 433–447.

Corbett, D., Evans, S., Thomas, C., Wang, D. & Jonas, R. (1990). MK-801 reduced cerebral ischemic injury by inducing hypothermia. *Brain Research*, **514**, 300–304.

Davis, S., Butcher, S.P. & Morris, R.G.M. (1992). The NMDA receptor antagonist D-2-amino-5-phosphonopentanoate (D-AP5) impairs spatial learning and LTP *in vivo* at intracerebral concentrations comparable to those that block LTP *in vitro*. *Journal of Neuroscience*, **12**, 21–23.

Fay, T. & Smith, G.W. (1941). Observations on reflex reponses during prolonged periods of human refrigeration. *Archives of Neurology and Psychiatry*, **45**, 215–222.

Green, E.J., McNaughton, B.L. & Barnes, C.A. (1990). Exploration-dependent modulation of evoked responses in fascia dentata: dissociation of motor, EEG and sensory factors and evidence for a synaptic efficacy change. *Journal of Neuroscience*, **10**, 1445–1471.

Heistad, D.D. & Kontos, H.A. (1983). Cerebral circulation. In *Handbook of Physiology*, ed. J.T. Shepherd, F.M. Abboud & S.R. Geiger, sect. 2, vol. III, part 1, pp. 137–182. Bethesda, MD: American Physiological Society.

Hodgkin, A.L. & Katz, B. (1949). The effect of temperature on the electrical activity of the giant axon of the squid. *Journal of Physiology (London)*, **109**, 240–249.

Katz, B. & Miledi, R. (1965). The effect of temperature on the synaptic delay at the neuromuscular junction. *Journal of Physiology (London)*, **181**, 656–670.

Kovalzon, V.M. (1973). Brain temperature variations during natural sleep and arousal in white rats. *Physiology and Behaviour*, **10**, 667–670.

Mellergård, P. & Nordström, C.-H. (1990). Epidural temperature and possible intracerebral temperature gradients in man. *British Journal of Neurosurgery*, **4**, 31–38.

Mellergård, P. & Nordström, C.-H. (1991). Intracerebral temperature in neurosurgical patients. *Neurosurgery*, **28**, 709–713.

Morris, R.G.M. (1984). Development of a water-maze procedure for studying spatial learning in the rat. *Journal of Neuroscience Methods*, **11**, 47–60.

Morris, R.G.M., Anderson, E., Lynch, G.S. & Baudry, M. (1986). Selective impairment of learning and blockade of long-term potentiation by an *N*-methyl-D-aspartate receptor antagonist, AP5. *Nature*, **319**, 774–776.

Moser, E., Mathiesen, I. & Andersen, P. (1993). Association between brain temperature and dentate field potentials in exploring and swimming rats. *Science*, **259**, 1324–1326.

Nielsen, B. (1988). Natural cooling of the brain during outdoor bicycling? *Pflügers Archiv*, **411**, 456–461.

Obàl, F., Rubicsek, G., Alföldi, P., Sáry, G. & Obàl, F. (1985). Changes in the brain and core temperatures in relation to the various arousal states in rats in the light and dark periods of the day. *Pflügers Archiv*, **404**, 73–79.

Panakhova, E., Buresova, O., Bures, J. (1984). The effect of hypothermia on the rat's spatial memory in the water tank task. *Behavioral and Neural Biology*, **42**, 191–196.

Rauch, T.M., Welch, D.I. & Gallego, L. (1989). Hypothermia impairs performance in the Morris water maze. *Physiology and Behavior*, **45**, 315–320.

Ritchie, J.M. & Straub, R.W. (1956). The effect of cooling on the size of the action potential of mammalian non-medullated fibres. *Journal of Physiology (London)*, **134**, 712–717.

Sharp, P.E., McNaughton, B.L. & Barnes, C.A. (1989). Exploration-dependent modulation of evoked responses in fascia dentata: fundamental observations and time course. *Psychobiology*, **17**, 257–269.

Schiff, S.J. & Somjen, G.G. (1985). Effects of temperature on synaptic transmission in hippocampal tissue slices. *Brain Research*, **345**, 279–284.

Schoepfle, G.M. & Erlanger, J. (1941). The action of temperature on the excitability, spike height and configuration, and the refractory period observed in the responses of single medullated nerve fibres. *American Journal of Physiology*, **134**, 694–704.

Vanderwolf, C.H. (1991). Effects of water temperature and core temperature on rat's performance in a swim-to-platform test. *Behavioural Brain Research*, **44**, 105–106.

Winson, J. & Abzug, C. (1977). Gating of neuronal transmission in the hippocampus: efficacy of transmission varies with behavioural state. *Science*, **196**, 1223–1225.

21

Fusimotor control of the respiratory muscles

ANTHONY TAYLOR AND RADE DURBABA

Sherrington School of Physiology, United Medical and Dental Schools, St Thomas' Hospital Campus, London, UK

The first observations

Following the demonstration by Leksell (1945) of the fusimotor effects of γ-efferent fibres and their detailed study by Kuffler, Hunt & Quilliam (1951), their chief action was thought to be to bias spindle afferent firing. Hunt & Kuffler (1951) emphasized the usefulness of this in preventing silencing of spindles during active muscle shortening, while Merton (1953) visualized it as a way of operating a 'length follow-up servo' system. By the end of the 1950s much ground work had been done on the fusimotor system, principally in hindlimb muscles, and Granit's view had become established that the γ loop was generally co-activated with direct drive to α-motoneurones (Granit, 1955; Granit, Kellerth & Szumski, 1966).

In the early 1960s, the respiratory system was recognized as a convenient model for the organization of motor control. Tom Sears was involved in a series of studies laying the foundations for this approach. Intracellular recording demonstrated the monosynaptic connections of muscle afferents to intercostal motoneurones (Eccles, Sears & Shealy, 1962) and the inhibitory as well as excitatory effects of descending respiratory drive (Sears, 1964c). He also provided important background information on the afferent and efferent fibre diameter spectra (Sears, 1964a) and on motor unit types in intercostal muscles (Andersen & Sears, 1964) and introduced recording from natural intramuscular nerve filaments of intercostal muscles, in which the action potentials of α and γ motor fibres could be distinguished by their different amplitudes (Sears, 1962, 1963). It appeared that γ units were controlled by descending respiratory drives similar to those reaching the α units (Fig. 21.1A, B). A similar conclusion arose (Fig. 21.1C) from the parallel work by von Euler's group (Critchlow & Euler, 1963; Eklund, Euler & Rutkowski, 1964), which had shown that the rhythmically modulated fusimotor activity was more than enough to compensate for muscle shortening (Fig. 21.2).

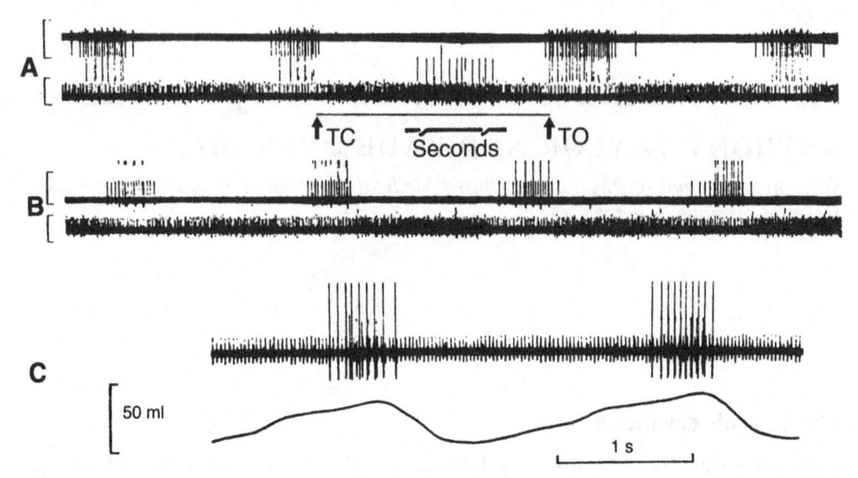

Fig. 21.1. (A) and (B) Records of α and fusimotor activity. In each case the upper and lower traces are from inspiratory and expiratory nerve filaments respectively. The expiratory record contains γ spikes only except when an α unit is recruited after tracheal closure (TC). TO, tracheal opening. From Sears (1964b). (C) The upper trace shows a tonically firing small γ unit and a larger modulated one firing at the same time as the large α spike. Tidal volume is indicated by the lower trace, with inspiration upwards. From Eklund *et al.* (1964) with permission of the authors.

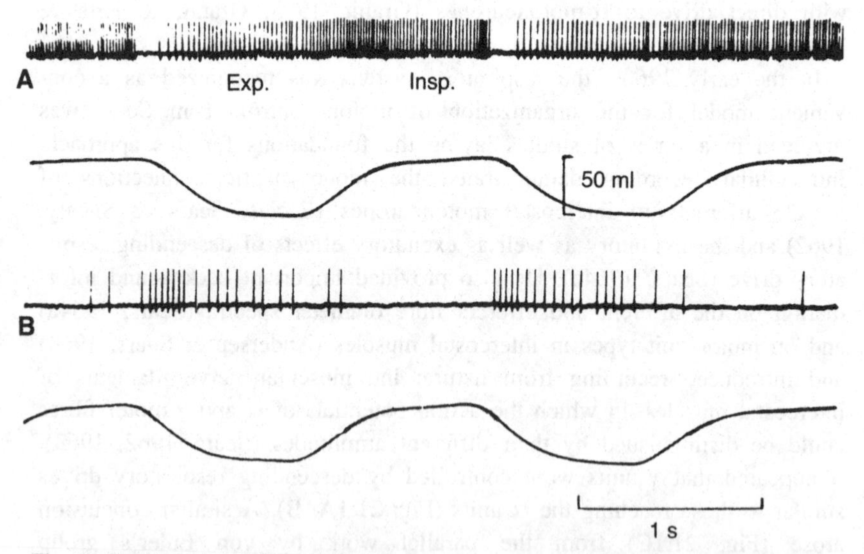

Fig. 21.2. (A) and (B) Records from the same inspiratory muscle spindle afferent, but with (B) recorded 3.5 minutes after applying lidocaine to the nerve to block γ activity. Note the reversal of phase of discharge relative to the tidal volume (lower) trace. The original intercostal space record has been removed. After Critchlow & Euler (1963).

Static and dynamic fusimotor systems

Matthews (1962) had shown that γ-efferents could be divided into static and dynamic types according to their effects on spindle responses to stretch. This began a series of attempts to deduce the patterns of activity occurring in the two different types in various natural movements. It was noted in the intercostal nerve filament preparations (e.g. Fig. 21.1) that whilst many γ-efferents discharged rhythmically with respiration, others maintained essentially constant tonic firing throughout the respiratory cycle (Eklund *et al.*, 1964; Corda, Euler & Lennerstrand, 1966). These authors did consider at that time whether the two classes of gamma activity might relate to their being static or dynamic fusimotor in function, but preferred the view that whilst modulated units were essentially respiratory, the tonic ones were expressing a postural function. Euler & Peretti (1966) also recorded from a large number of external intercostal spindle afferents in dorsal roots and designated them as belonging to

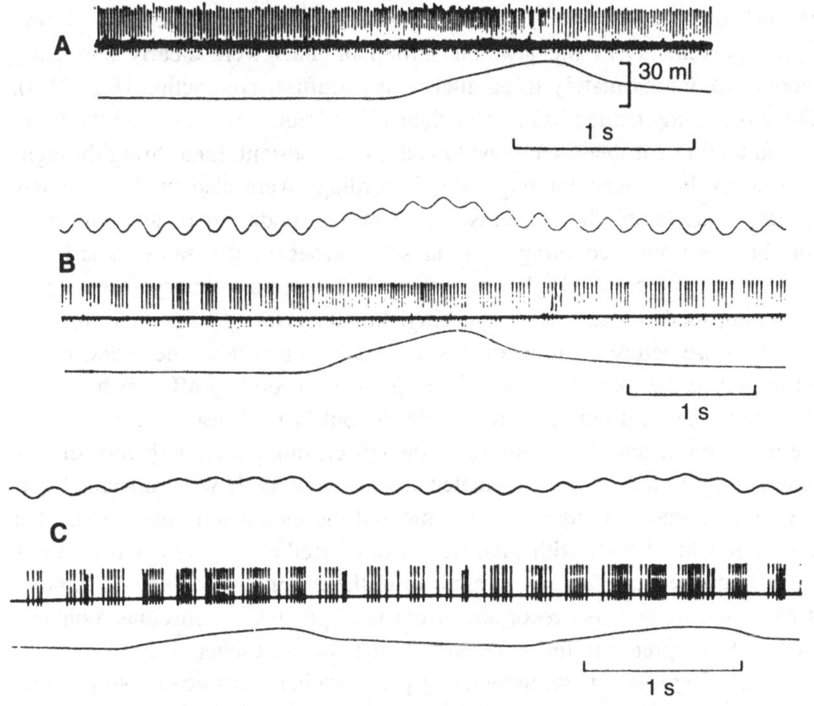

Fig. 21.3. (A) A secondary type afferent from an external intercostal shows firing increasing during active inspiratory shortening. (B) A primary type afferent showing reduced sensitivity to sinusoidal stretch during inspiration. (C) Another primary type unit showing enhanced sinusoidal response during inspiration. In (B) and (C) the top trace indicates intercostal space width; the lowest trace in each record indicates tidal volume. From Euler & Peretti (1966).

primary or secondary endings according to their ability to follow high-frequency vibration. The majority of both primary and secondary endings showed the effects of fusimotor activation in parallel with skeleto-motor contraction (Fig. 21.3). Since secondary endings are rarely affected by dynamic fusimotor activity, it might have been considered that the modulated fusimotor units were static and the tonic ones dynamic, but Euler and Peretti still preferred an interpretation based on separate postural and respiratory functions.

Fusimotor studies in other systems

Unfortunately, very little further work was done directly on intercostal fusimotor neurones for a long while, but related studies were pursued separately in jaw muscles involved in masticatory movements and in hindlimb muscles during locomotion. By recording from intramuscular nerve filaments in the masseter muscle of lightly anaesthetized cats it was possible again to distinguish two classes of behaviour amongst fusimotor neurones (Appenteng, Morimoto & Taylor, 1980; Taylor, Appenteng & Morimoto, 1981). During cyclic movements of the jaw, one group of units were clearly phasically modulated approximately in parallel with extrafusal contraction (Fig. 21.4). The other group tended to increase their firing frequency following the onset of stimulation but then maintained much more constant, tonic firing throughout the cyclic movement sequence. Recordings were also made from jaw elevator muscle spindle afferents under similar conditions of anaesthesia as for the fusimotor recordings and in some cases in the same experiment (Gottlieb & Taylor, 1983). Afferents were designated as primary or secondary according to their response to succinylcholine. Primary afferent firing was shown to rise before the movement sequence began, with a time course parallel to that of the tonic fusimotor firing pattern. Secondary afferent firing, on the other hand, did not show this build up, but instead peaked at the end of the jaw opening and the beginning of the jaw closing phase, with very similar time course to that of the modulated fusimotor units. It was concluded that the tonic fusimotor units were dynamic and the modulated ones static. This conclusion fitted well with interpretations of studies of afferent recordings made in conscious cats (and referred to in the papers quoted above), though it must be admitted that recordings from jaw spindles in conscious monkeys were not interpreted in the same way (Goodwin & Luschei, 1975).

In other experiments on locomoting pre-mamillary cats again two patterns of fusimotor unit discharge were recognized in filaments of triceps surae nerve (Murphy, Stein & J. Taylor, 1984). One was designated phasically modulated and the other tonically modulated, but it will be convenient to use the same nomenclature as above and to call them modulated and tonic respectively. The modulated units regularly had higher resting firing rates

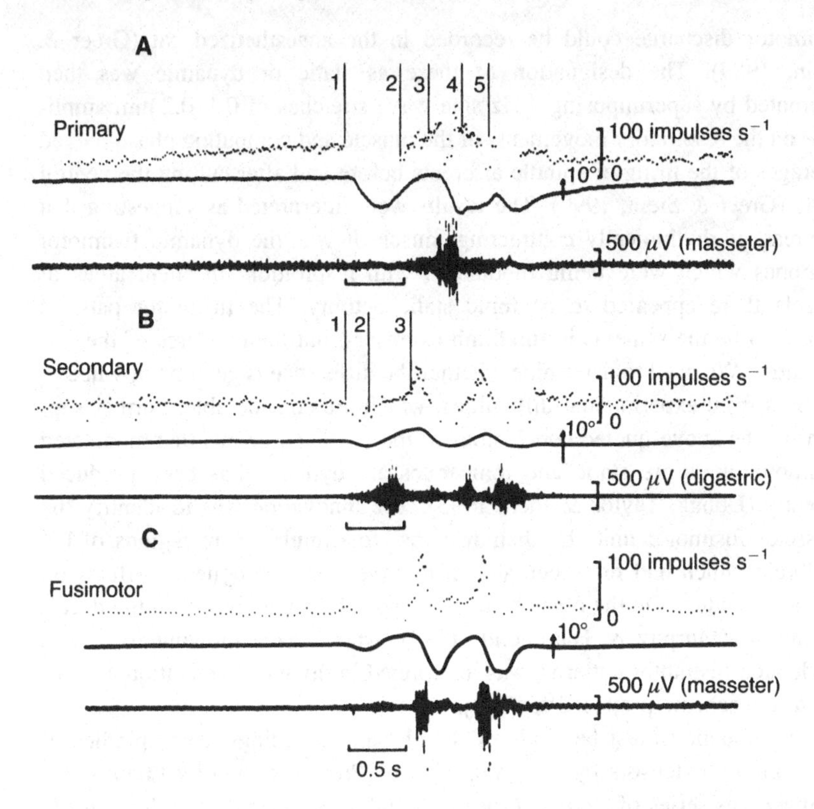

Fig. 21.4. Records of a jaw elevator muscle spindle primary (A) and secondary (B) and a masseter nerve fusimotor unit (C). Jaw movement records are shown below the instantaneous frequency plots in each case, with opening upwards. EMG records are shown for the masseter in (A) and (C) and for the digastric in (B). From Appenteng *et al.* (1980).

than the tonic ones. In separate experiments single ventral root γ axons demonstrated to have dynamic actions had high firing rates while static axons had low rates. Putting these two observations together it was concluded that in extensor muscles during locomotion the modulated γ units were dynamic whilst the tonic ones were static. This conclusion was the converse of that reached above for jaw elevators in cyclic movements, but was subsequently backed up by further experiments in which sinusoidal stretching of the triceps showed the expected changes in sensitivity of spindle primary afferents (J. Taylor, Stein & Murphy, 1985).

The return of the intercostals

More recently the same laboratory has turned attention again to the intercostal muscles. First, it was confirmed that tonic and modulated patterns of fusimotor discharge could be recorded in the anaesthetized cat (Greer & Stein, 1989). The designation of these as static or dynamic was then attempted by superimposing 4 Hz sine wave stretches of 0.1–0.2 mm amplitude on the respiratory movements of the muscle and computing phase-locked averages of the firing of spindle afferents before and after cutting the ventral roots (Greer & Stein, 1990). The results were interpreted as suggesting that in areas of rhythmically contracting muscle it was the dynamic fusimotor neurones which were being modulated with respiration. In silent areas of muscle there appeared to be tonic static activity. The fusimotor patterns appear to be the same as in hindlimb extensors, but the converse of the jaw elevators. We need to determine whether the difference is genuine or whether it arises from experimental difficulties, which are considerable. Further support for the above-quoted conclusions of the jaw experiments that modulated fusimotor units are static and tonic ones are dynamic has been produced recently (Donga, Taylor & Jüch, 1993). The innovation was to identify the masseter fusimotor units by their response to stimulation in regions of the midbrain which had first been shown to have selective dynamic effects on the jaw spindles. On the other hand, new work on ankle *flexors* in the decerebrate cat (Murphy & Hammond, 1993), while again showing tonic and modulated fusimotor patterns, was interpreted in favour of activation according to the scheme proposed for the jaw.

Some mention must be made of the chronic recordings from spindles of cat hindlimb extensors by the groups led by Prochazka and by Loeb. From an ingenious series of studies, Prochazka and his colleagues (Hulliger *et al.*, 1985) were able to deduce that for particular movements, fusimotor activity is largely tonic, either static or dynamic according to the motor task. Alpha–gamma linkage was not thought to be a general rule, but might occur in strong efforts. Thus there is a distinct difference between these data and those derived from decerebrate cats. Interestingly, it has also been the case that reflex inputs to fusimotor neurones, so clearly seen in reduced preparations, are not at all evident in conscious animals.

The future

Taking all the various lines of evidence together it still does not seem possible at the moment to come to general conclusions regarding the patterns of fusimotor activity occurring in natural movements. The intercostal muscles are potentially very good subjects for study in this connection because of their continued rhythmic activity in anaesthesia and yet even here our knowledge is still very incomplete. It would be particularly useful to have chronic recordings of spindle activity made as in the case of the jaws and the hindlimbs. Direct intercostal fusimotor recordings in reduced preparations should also be repeated and an attempt made to identify them by central stimulation.

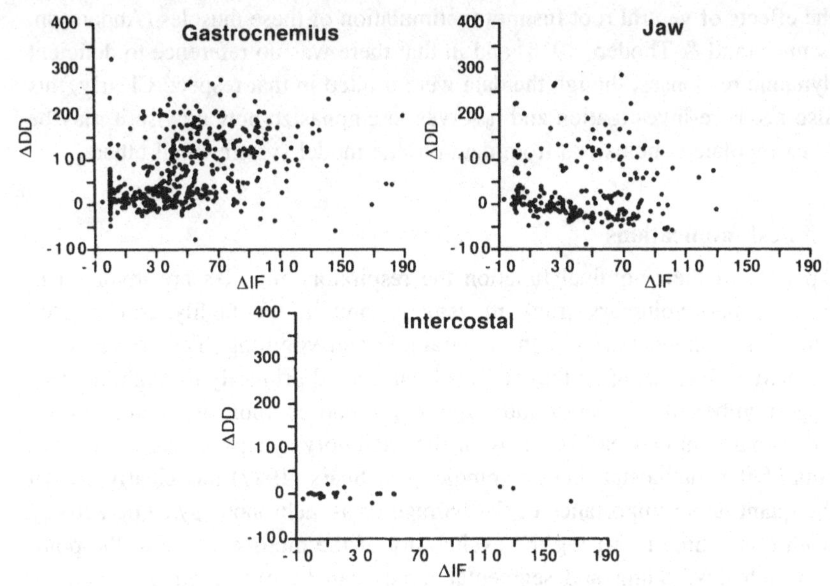

Fig. 21.5. Plots of the responses of populations of gastrocnemius, jaw elevator and intercostal spindle afferents to succinylcholine. Note that none of the 25 intercostal afferents shows significant increases in Dynamic Difference (DD). After Taylor *et al.* (1992a, b).

In preliminary work directed to this end, we have recently been making recordings from intercostal muscle spindle afferents in anaesthetized cats. A technique which we have found very useful in the study of jaw and hindlimb spindles is to classify afferents by the effect on their responses to ramp and hold stretches of a standardized intravenous dose of succinylcholine (Taylor *et al.*, 1992a, b). The intention is to assess the relative contributions to each afferent's activity of the intrafusal dynamic bag_1 and the static bag_2 muscle fibres. With cyclically repeating stretches, we measure the frequency just before each stretch as the Initial Frequency (IF) and the increment from this to the peak frequency as the Dynamic Difference (DD). The increases in these caused by succinylcholine are symbolized by ΔIF and ΔDD and seem to measure the effects of the bag_2 and bag_1 fibres respectively. Applying this test to 25 intercostal spindle afferents recorded from dorsal root filaments in the cat we have been very surprised to find signs only of bag_2 effects. It can be seen from Fig. 21.5 that while jaw elevator spindles and gastrocnemius spindles show a wide range of uncorrelated values of ΔIF and ΔDD, the intercostal muscles show increments only in IF and not in DD. The succinylcholine is obviously reaching the spindles as seen by increases in IF, presumed due to contraction of bag_2 fibres. The absence of effects on bag_1 fibres must therefore mean either that they are insensitive to the drug or that there is some peculiarity of their innervation. We know of only one study of the effects of ventral root fusimotor stimulation of these muscles (Andersson, Lennestrand & Thoden, 1968) and in that there was no reference to deficient dynamic responses, though the data were limited in that respect. Clearly, this also needs re-investigation and it serves to emphasize how unwise it may be to extrapolate conclusions from data on one model system to all others.

Clinical implications

Apart from their nominal function the respiratory muscles are involved in postural and voluntary trunk movements and in the highly co-ordinated activities of phonation, coughing, straining and vomiting. The convergence of many influences upon the muscles is subserved primarily through the very large number of excitatory and inhibitory synapses upon the motoneurones ('the final common path'). Many of the excitatory synapses belong to afferents of the intercostal muscle spindles and Sears (1977) has clearly shown the quantitative importance of the barrage of asynchronous synaptic activity from this source in raising the excitability of the motoneurones to the point at which descending and segmental inputs can be expressed by repetitive firing of motor units. The fusimotor inputs to the intercostal muscle spindles in turn largely determine the total spindle afferent firing. Thus, the patterns of the fusimotor output present at any time will be very influential in determining the postural set and the reflex response to disturbances from all sources. There seems little doubt that the mean position of the chest wall is set normally to achieve optimum mechanical efficiency in breathing. Very probably the cerebellum will be involved in plastically adapting the system to this end (Baker, Seers & Sears, 1992) and one obvious means by which the cerebellum can exert its control is via adjustments of the fusimotor outflow. A variety of observations attest to the potency of the cerebellar influence on fusimotor discharge (Eklund *et al.*, 1964). Thus, in seeking a full understanding of the way in which disturbances in respiratory mechanics arise, for example in asthma, we will need to take account of the neurophysiological considerations briefly reviewed above. The subject is still far from fully understood, but there does now exist a sound framework on which to build.

Acknowledgements

We are grateful for support from the Research Endowments Committee of St Thomas Hospital, Action Research and the Medical Research Council.

References

Andersen, P. & Sears, T.A. (1964). The mechanical properties and innervation of fast and slow motor units in the intercostal muscles of the cat. *Journal of Physiology (London)*, **173**, 114–129.

Andersson, B.F., Lennerstrand, G. & Thoden, U. (1968). Fusimotor effects on position and velocity sensitivity of spindle endings in the external intercostal muscle of the cat. *Acta Physiologica Scandinavica*, **74**, 285–300.

Appenteng, K., Morimoto, T. & Taylor A. (1980). Fusimotor activity in masseter nerve of the cat during reflex movements. *Journal of Physiology (London)*, **305**, 415–431.

Baker, S., Seers, C. & Sears, T.A. (1992). Respiratory modulation of afferent transmission to the cerebellum. In *Respiratory Control: Central and Peripheral Mechanisms*, ed. D.F. Speck, M.S. Dekin, W.R. Revelette & D.T. Frazier. Lexington: University Press of Kentucky.

Corda, M., Euler, C. von & Lennerstrand, G. (1966). Reflex and cerebellar influences on α and on 'rhythmic' and 'tonic' γ-activity in the intercostal muscle. *Journal of Physiology (London)*, **184**, 898–923.

Critchlow, V. & Euler, C. von (1963). Intercostal muscle spindle activity and its γ motor control. *Journal of Physiology (London)*, **168**, 820–847.

Donga, R., Taylor, A. & Jüch, P.J.W. (1993). The use of midbrain stimulation to identify the discharges of static and dynamic fusimotor neurones during reflex jaw movements in the anaesthetized cat. *Experimental Physiology*, **78**, 15–23.

Eccles, R.M., Sears, T.A. & Shealy, C.N. (1962). Intra-cellular recording from respiratory motoneurones of the thoracic spinal cord in the cat. *Nature*, **193**, 844–846.

Eklund, G., Euler, C. von & Rutkowski, S. (1964). Spontaneous and reflex activity of intercostal gamma motoneurones. *Journal of Physiology (London)*, **171**, 139–163.

Euler, C. von. & Peretti, G. (1966). Dynamic and static contributions to the rhythmic gamma activation of primary and secondary spindle endings in external intercostal muscle. *Journal of Physiology (London)*, **187**, 501–516.

Goodwin, G.M. & Luschei, E.S. (1975). Discharge of spindle afferents from jaw-closing muscles during chewing in alert monkeys. *Journal of Neurophysiology*, **38**, 560–571.

Gottlieb, S. & Taylor, A. (1983). Interpretation of fusimotor activity in cat masseter nerve during reflex jaw movements. *Journal of Physiology (London)*, **345**, 423–438.

Granit, R. (1955). *Receptors and Sensory Perception*. Connecticut: Yale University Press.

Granit, R., Kellerth, J.-O. & Szumski, A.J. (1966). Intracellular recording from extensor motoneurons activated across the gamma loop. *Journal of Neurophysiology*, **29**, 530–544.

Greer, J.J. & Stein, R.B. (1989). Patterns of gamma motoneurone activity in the external intercostal muscles of the cat during respiration. *Brain Research*, **477**, 369–372.

Greer, J.J. & Stein, R.B. (1990). Fusimotor control of muscle spindle sensitivity during respiration in the cat. *Journal of Physiology (London)*, **422**, 245–264.

Hulliger, M., Zangger, P., Prochazka, A. & Appenteng, K. (1985). Fusimotor 'set' vs alpha-gamma linkage in voluntary movements in cats. In *Electromyography and Evoked Potentials*, ed. A. Struppler & A. Windl, pp. 56–63. Berlin: Springer.

Hunt, C.C. & Kuffler, S.W. (1951). Stretch receptor discharges during muscle contraction. *Journal of Physiology (London)*, **113**, 298–315.

Kuffler, S.W., Hunt, C.C. & Quilliam, J.P. (1951). Function of medullated small muscle-nerve fibres in mammalian ventral roots: efferent muscle spindle innervation. *Journal of Neurophysiology*, **14**, 29–54.

Leksell, L. (1945). The action potential and excitatory effects of the small ventral root fibres to skeletal muscle. *Acta Physiologica Scandanavica*, **10**, 1–84.

Matthews, P.B.C. (1962). The differentiation of two types of fusimotor fibres by their effects on the dynamic response of muscle spindle primary endings. *Quarterly Journal of Experimental Physiology*, **47**, 324–333.

Merton, P.A. (1953). Speculations on servo control of movement. In *The Spinal Cord*, ed. G.E.W. Wolstenholme, pp. 245–255. London: Churchill.

Murphy, P.R. & Hammond, G.R. (1993). The locomotor discharge characteristics of ankle flexor γ-motoneurones in the decerebrate cat. *Journal of Physiology (London)*, **462**, 59–70.

Murphy, P.R., Stein, R.B. & Taylor, J. (1984). Phasic and tonic modulation of impulse rates in γ-motoneurons during locomotion in premammillary cats. *Journal of Neurophysiology*, **53**, 228–243.

Sears, T.A. (1962). The activity of the small motor fibre system innervating respiratory muscles of the cat. *Australian Journal of Science*, **25**, 102.

Sears, T.A. (1963). Activity of fusimotor fibres innervating muscle spindles in the intercostal muscles of the cat. *Nature*, **197**, 1013–1014.

Sears, T.A. (1964a). The fibre calibre spectra of sensory and motor fibres in the intercostal nerves of the cat. *Journal of Physiology (London)*, **172**, 150–161.

Sears, T.A. (1964b). Efferent discharges in alpha and fusimotor fibres of intercostal nerves of the cat. *Journal of Physiology (London)*, **174**, 295–315.

Sears, T.A. (1964c). The slow potentials of thoracic respiratory motoneurones and their relation to breathing. *Journal of Physiology (London)*, **175**, 404–424.

Sears, T.A. (1977). The respiratory motoneurone and apneusis. *Federation Proceedings*, **36**, 2412–2420.

Taylor, A., Appenteng, K. & Morimoto, T. (1981). Proprioceptive input from the jaw muscles and its influence on lapping, chewing, and posture. *Canadian Journal of Physiology and Pharmacology*, **59**, 636–644.

Taylor, A., Durbaba, R. & Rodgers, J.F. (1992a). The classification of afferents from muscle spindles of the jaw-closing muscles of the cat. *Journal of Physiology (London)*, **456**, 609–628.

Taylor, A., Rodgers, J.F., Fowle, A.J. & Durbaba, R. (1992b). The effect of succinylcholine on cat gastrocnemius muscle spindle afferents of different types. *Journal of Physiology (London)*, **456**, 629–644.

Taylor, J., Stein, R.B. & Murphy, R.B. (1985). Impulse rates and sensitivity to stretch of soleus muscle spindle afferent fibres during locomotion in premammillary cats. *Journal of Neurophysiology*, **53**, 341–360.

22

Cerebral accompaniments and functional significance of the long-latency stretch reflexes in human forearm muscles

MICHAEL J. AMINOFF AND DOUGLAS S. GOODIN

Department of Neurology, University of California, San Francisco, California, USA

A series of electromyographic (EMG) discharges can be recorded following the sudden stretch of a muscle that is either isometrically contracting or in motion. The first discharge (M1) is generally agreed to reflect activity of the monosynaptic stretch reflex. The nature and physiological basis of the later EMG activity is controversial (Chan, 1983; Marsden, Rothwell & Day, 1983), but there appears to be a reflex component (the so-called M2 response) followed by an overlapping component (M3) that is also of reflex origin but is influenced to a greater degree by voluntary mechanisms in some circumstances (Hammond, 1956; Lee & Tatton, 1978; Rothwell, Traub & Marsden, 1980; Jaeger, Gottlieb & Agarwal, 1982). Late EMG activity of similar appearance can be elicited either by interrupting a limb in motion or by perturbing a limb held isometrically against a constant force.

To account for this late EMG activity, some authors have favoured purely segmental mechanisms such as (1) grouped discharges in group Ia primary afferents (Hagbarth *et al.*, 1980, 1981); (2) separate discharges in fast- and slow-conducting afferent fibres following the stretch stimulus (Matthews, 1984); or (3) slowly conducting polysynaptic pathways within the cord (Hultborn & Wigstrom, 1980). Others, by contrast, have postulated that a long-loop, possibly transcortical, reflex pathway is involved (Lee & Tatton, 1978; Rothwell *et al.*, 1980; Marsden *et al.*, 1983; Abbruzzese *et al.*, 1985; Matthews, Farmer & Ingram, 1990). Even lesion experiments in animals have led to conflicting results. Thus, Tatton *et al.* (1975) found that postcentral lesions eliminated the late response, whereas Miller & Brooks (1981) found no such effect. Some authors have studied the relationship between somatosensory evoked cerebral potentials (SEPs) recorded over the scalp in response to muscle stretch and the various components of the late EMG response in order to support the concept of a long-loop reflex pathway subserving these EMG responses (Conrad, Dressler & Benecke, 1984; Abbruzzese *et al.*, 1985). These studies, however, provide only indirect evidence for the existence of a long-loop pathway because mechanical somatosensory stimulation will elicit SEPs regardless of whether any late EMG activity is produced.

There is also controversy about the degree of *volitional modulation* of the late EMG response to a sudden perturbation that stretches a muscle. Certain authors have reported some degree of voluntary control (Hammond, 1956; Thomas, Brown & Lucier, 1977; Lee & Tatton, 1978; Jaeger *et al.*, 1982), but more recent work (Rothwell *et al.*, 1980; Rothwell, Day & Marsden, 1986) has suggested that the apparent volitional modulation is related to an overlap of the reflex responses with subsequent voluntary activity, especially under circumstances in which the timing of the perturbation can be predicted by the subject. In earlier studies in our laboratory, we recorded these late EMG responses following the interruption of a ballistic movement (Goodin, Aminoff & Shih, 1990). We found that the ability of a subject to predict the onset of a perturbing stimulus markedly attenuated the amplitude of the late EMG responses. Thus, when the onset of the perturbation could be accurately predicted by the subject, both the M2 and M3 components of the late EMG response were attenuated regardless of whether the subject was required to oppose the perturbation (Fig. 22.1). Such a dependency of the M2 and M3 components of late EMG activity on expectancy indicates the importance of suprasegmental mechanisms in the regulation of these reflex components. Indeed, only *phasic* supraspinal influences can account for the effects of expectancy, and this in turn suggests the operation of long-loop, possibly transcortical, reflexes.

The functional role of this activity is similarly uncertain. Some authors regard this late EMG activity as part of a 'servo-mechanism' which helps to restore a limb to an intended or previous position (Marsden, Merton & Morton, 1976; Marsden, Rothwell & Day, 1983; Chan, 1983; Bedingham & Tatton, 1984), whereas others have argued that it is involved in the *control of limb stiffness* (Houk, 1976; Kwan, Murphy & Repeck, 1979; Lee, Murphy & Tatton, 1983). Our findings of late EMG activity in both agonist and antagonist muscles supports the latter hypothesis (Goodin *et al.*, 1990). Indeed, others have correlated the changes in the late reflex responses with the occurrence and severity of rigidity in patients with Parkinson's disease (Tatton & Lee, 1975; Marsden *et al.*, 1978; Mortimer & Webster, 1979; Berardelli, Sabra & Hallett, 1983; Rothwell *et al.*, 1983; Cody *et al.*, 1986).

Our observation that the late EMG activity is expectancy dependent (Goodin *et al.*, 1990) suggested to us a means of eliminating the mechanically elicited SEP, so that we could examine more directly whether there were any cerebral accompaniments of the late EMG activity. In particular, we subtracted from the cerebral and muscle responses to unpredictable stretch of a muscle the corresponding but attenuated responses to a predictable stretch. The mechanical stimulus was the same in both circumstances, and the subtraction technique therefore eliminated the contribution of the mechanically elicited SEP to the response recorded over the scalp.

By this means, we were able to record cerebral potentials that persisted

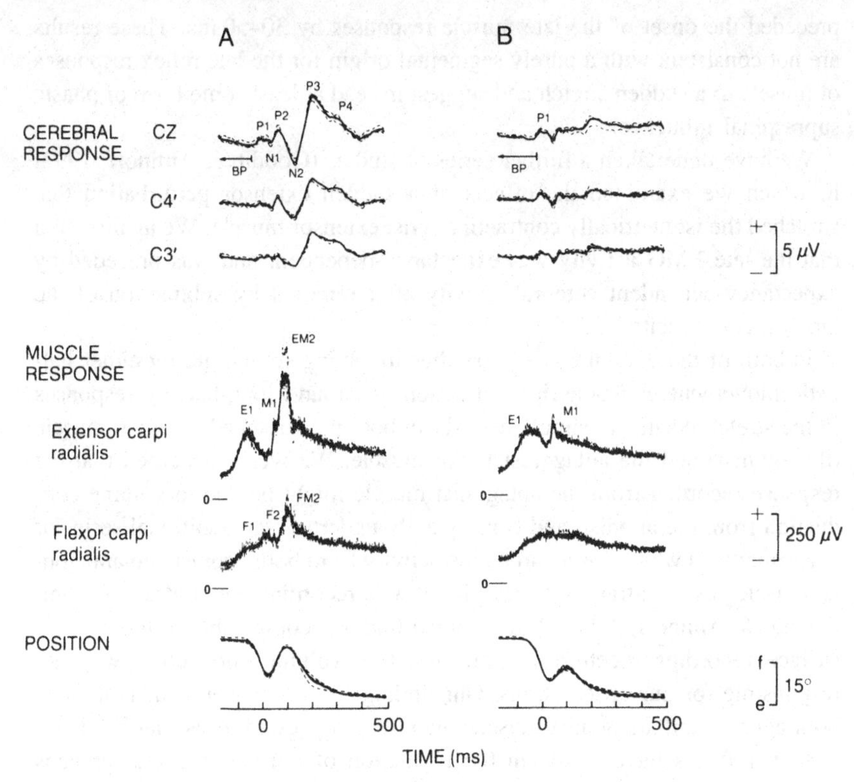

Fig. 22.1. Grand average responses for an extensor movement against a constant 2.3 N force that is unexpectedly interrupted (A) or interrupted predictably on every trial (B) by an increase in the force to 18.4 N. Subjects were required to compensate for, i.e. to oppose, the increased torque. The three panels in each condition show the cerebral responses (top), the rectified EMG responses (middle), and the position trace (bottom). Horizontal bars indicate the zero point for the rectified EMG traces. The calibration of the responses is shown at the bottom right of each panel. The 'f' and 'e' in the position trace calibration indicate flexion and extension, respectively. In each case, three preliminary grand averages have been superimposed (dashed lines) and the overall grand average also superimposed (continuous line) to demonstrate the reproducibility of the findings. On the time axis, 0 ms indicates the time when the increased torque was applied. It can be seen that the late EMG activity is considerably attenuated when the interruption occurred predictably, and the cerebral responses are also markedly altered. See text for further details.

after removal by subtraction of the potentials related to the SEP, and we found that they preceded the late muscle activity induced by the sudden stretch of the wrist extensor muscle during performance of a ballistic extensor movement. These potentials were present when the late responses were present, absent when they were not present, and became progressively smaller when the late responses became smaller, suggesting a tight coupling between these events and the late EMG activity. The earliest of these cerebral events preceded the onset of the late muscle responses by 30–50 ms. These results are not consistent with a purely segmental origin for the late reflex responses of muscle to a sudden stretch and suggest instead at least some form of phasic supraspinal influence.

We have undertaken a further series of studies (Goodin & Aminoff, 1992) in which we examined the effects of a sudden extensor perturbation that stretched the isometrically contracting wrist extensor muscle. We again found that the late EMG activity was expectancy-dependent and was preceded by expectancy-dependent cerebral activity after removal by subtraction of the mechanically elicited SEP.

In both of these studies, i.e. in studies involving stretch during either ballistic movement or isometric contraction, we found long-latency responses to the stretch occurring simultaneously in both the stretched extensor muscle (the agonist) and the antagonist flexor muscle. We were concerned that the response recorded from the antagonist muscle might be due to volume conduction from the agonist, and consequently undertook an additional series of experiments in which we recorded the activity from both agonist and antagonist muscles using intramuscularly placed wire recording electrodes (Chequer, Goodin & Aminoff, 1991). We obtained findings comparable to those using surface recording electrodes, indicating that volume conduction was not responsible for our observations. Our findings of simultaneous activation of both agonist and antagonist muscles therefore suggested to us that the long-latency reflexes have a role in the regulation of muscle stiffness, perhaps serving as a base on which subsequent voluntary or corrective movements are activated.

The late reflex muscle responses to stretch, however, probably have a dual function. Although the late EMG activity is simultaneously present in both the stretched muscle and its antagonist, suggesting its involvement in the control of limb stiffness, the size of the late EMG activity is always greater in the stretched muscle than its antagonist, suggesting also a servo-role for these responses (Goodin & Aminoff, 1992). In more recent work (Chequer *et al.*, 1994), subjects were asked to maintain the limb in a fixed position and oppose any displacement that occurred. We again found that late EMG activity was present in both agonist and antagonist although the amount of integrated EMG activity for *both* the M2 and M3 components of the late response was greater in the agonist muscle (i.e. the muscle actually stretched by the perturbing torque) than in the antagonist. The ratio of agonist to antagonist activity, however, was significantly larger for the M2 than for the M3 component. These results indicate that both the M2 and M3 components of the late EMG response have dual functional role, being involved both in a servo-mechanism and in the control of limb stiffness. Each component, however, has a distinctive role. M2 seems more involved in the servo-role, whereas M3 seems more involved in the control of limb stiffness.

References

Abbruzzese, G., Berardelli, A., Rothwell, J.C., Day, B.L. & Marsden, C.D. (1985). Cerebral potentials and electromyographic responses evoked by stretch of wrist muscles in man. *Experimental Brain Research*, **58**, 544–551.

Bedingham, W. & Tatton, W.G. (1984). Dependence of EMG responses evoked by imposed wrist displacements on pre-existing activity in the stretched muscles. *Canadian Journal of Neurological Sciences*, **11**, 272–280.

Berardelli, A., Sabra, A.F. & Hallett, M. (1983). Physiological mechanisms of rigidity in Parkinson's disease. *Journal of Neurology, Neurosurgery and Psychiatry*, **46**, 45–53.

Chan, C.W.Y. (1983). Segmental versus suprasegmental contributions to long-latency stretch responses in man. In *Motor Control Mechanisms in Health and Disease*, ed. J.E. Desmedt, pp. 467–487. New York: Raven Press.

Chequer, R.S., Goodin, D.S. & Aminoff, M.J. (1991). The nature of suprasegmental influences in the late reflex activity in human forearm muscles. *Brain Research*, **566**, 284–289.

Chequer, R.S., Goodin, D.S., Aminoff, M.J. & Maeztu, C. (1994). Late electromyographic activity following stretch in human forearm muscles: physiological role. *Brain Research*, **641**, 273–278.

Cody, F.W.J., MacDermott, N., Matthews, P.B.C. & Richardson, H.C. (1986). Observations on the genesis of the stretch reflex in Parkinson's disease. *Brain*, **109**, 229–249.

Conrad, B., Dressler, D. & Benecke, R. (1984). Changes of somatosensory evoked potentials in man as correlates of transcortical reflex mediation? *Neuroscience Letters*, **46**, 97–102.

Goodin, D.S. & Aminoff, M.J. (1992). The basis and functional role of the late EMG activity in human forearm muscles following wrist displacement. *Brain Research*, **589**, 39–47.

Goodin, D.S., Aminoff, M.J. & Shih, P.Y. (1990). Evidence that the long latency stretch responses in the human wrist extensor muscles involve a transcerebral pathway. *Brain*, **113**, 1075–1091.

Hagbarth, K.-E., Hagglund, J.V., Wallin, E.U. & Young, R.R. (1981). Grouped spindle and electromyographic responses to abrupt wrist extension movements in man. *Journal of Physiology (London)*, **312**, 81–96.

Hagbarth, K.-E., Young, R.R., Hagglund, J.V. & Wallin, E.U. (1980). Segmentation of human spindle and EMG responses to sudden muscle stretch. *Neuroscience Letters*, **19**, 213–217.

Hammond, P.H. (1956). The influences of prior instruction to the subject on an apparently involuntary neuromuscular response. *Journal of Physiology (London)*, **132**, 17P-18P.

Houk, J.C. (1976). An assessment of stretch reflex function. *Progress in Brain Research*, **44**, 303–314.

Hultborn, H. & Wigstrom, H. (1980). Motor response with long latency and maintained duration evoked by activity in Ia afferents. In *Spinal and Supraspinal Mechanisms of Voluntary Motor Control and Locomotion. Progress in Clinical Neurophysiology*, vol. 8, ed. J.E. Desmedt, pp. 99–116. Basel: Karger.

Jaeger, R.J., Gottlieb, G.L. & Agarwal, G.C. (1982). Myoelectric responses at flexors and extensors of human wrist to step torque perturbations. *Journal of Neurophysiology*, **48,** 388–402.

Kwan, H.C., Murphy, J.T. & Repeck, M.W. (1979). Control of stiffness by the medium latency electromyographic response to limb perturbation. *Canadian Journal of Physiology and Pharmacology*, **57**, 277–285.

Lee, R.G., Murphy, J.T. & Tatton, W.G. (1983). Long latency reflexes in man: mechanisms, functional significance and changes in patients with Parkinson's disease or hemiplegia. In *Motor Control Mechanisms in Health and Disease*, ed. J.E. Desmedt, pp. 489–508. New York: Raven Press.

Lee, R.G & Tatton, W.G. (1978). Long loop reflexes in man: clinical applications. In *Cerebral Motor Control in Man: Long Loop Mechanisms. Progress in Clinical Neurophysiology*, vol. 4, ed. J.E. Desmedt, pp. 320–333. Basel: Karger.

Marsden, C.D., Merton, P.A. & Morton, H.B. (1976). Servo action in the human thumb. *Journal of Physiology (London)*, **257**, 1–44.

Marsden, C.D., Merton, P.A., Morton, H.B. & Adam, J. (1978). The effect of lesions of the central nervous system on long-latency stretch reflexes in the human thumb. In *Cerebral Motor Control in Man: Long Loop Mechanisms. Progress in Clinical Neurophysiology*, vol. 4, ed. J.E. Desmedt, pp. 334–341. Basel: Karger.

Marsden, C.D., Rothwell, J.C. & Day, B.L. (1983). Long-latency automatic responses to muscle stretch in man: origin and function. In *Motor Control Mechanisms in Health and Disease*, ed. J.E. Desmedt, pp. 509–539. New York: Raven Press.

Matthews, P.B.C. (1984). Evidence from the use of vibration that the human long-latency stretch reflex depends upon spindle secondary afferents. *Journal of Physiology (London)*, **348**, 383–415.

Matthews, P.B.C., Farmer, S.F. & Ingram, D.A. (1990). On the localization of the stretch reflex of intrinsic hand muscles in a patient with mirror movements. *Journal of Physiology (London)*, **428**, 561–577.

Miller, A.D. & Brooks, V.B. (1981). Late muscular responses to arm perturbations persist during supraspinal dysfunctions in monkeys. *Experimental Brain Research*, **41**, 146–158.

Mortimer, J.A. & Webster, D.D. (1979). Evidence for a quantitative association between EMG stretch responses and parkinsonian rigidity. *Brain Research*, **162**, 169–173.

Rothwell, J.C., Day, B.L. & Marsden, C.D. (1986). Habituation and conditioning of the human long-latency stretch reflex. *Experimental Brain Research*, **63**, 205–215.

Rothwell, J.C., Obeso, J.A., Traub, M.M. & Marsden, C.D. (1983). The behaviour of the long-latency stretch reflex in patients with Parkinson's disease. *Journal of Neurology, Neurosurgery and Psychiatry*, **46**, 35–44.

Rothwell, J.C., Traub, M.M. & Marsden, C.D. (1980). Influence of voluntary intent on the human long-latency stretch reflex. *Nature*, **286**, 496–498.

Tatton, W.G. & Lee, R.G. (1975). Evidence for abnormal long-loop reflexes in rigid parkinsonian patients. *Brain Research*, **100**, 671–676.

Tatton, W.G., Forner, S.D., Gerstein, G.I., Chambers W.W. & Liu C.N. (1975). The effect of postcentral cortical lesions on motor responses to sudden upper limb displacements in monkeys. *Brain Research*, **96**, 108–113.

Thomas, J.S., Brown, J. & Lucier, G.E. (1977). Influence of task set on muscular responses to arm perturbations in normal subjects and Parkinson patients. *Experimental Neurology*, **55**, 618–628.

23

The cerebellum and proprioceptive control of movement

ARTHUR PROCHAZKA, MONICA GORASSINI AND JANET TAYLOR*

Division of Neuroscience, University of Alberta, Edmonton, Alberta, Canada.

Introduction

Sherrington (1906) called the cerebellum the 'head ganglion of the proprioceptive system'. When it is damaged, movement deficits result. In humans, certain characteristic abnormalities arise, including locomotor ataxia, intention tremor and hypermetria. Hypotonia, or generalized weakness, may also result. Holmes (1922, 1939) provided a detailed and definitive description of these symptoms in his studies of soldiers wounded in World War I. He concluded that cerebellar lesions led to abnormalities in the rate, regularity and force of voluntary movement. But there were reasons to doubt that the cerebellum is the primary generator of motor commands: several months after a cerebellar lesion, the accuracy of movement in humans can sometimes return to near-normal. Shortly after complete cerebellar ablation, animals can still initiate voluntary movements and perform goal-directed activities, albeit inaccurately (Mackay & Murphy, 1979). It has therefore been posited that the cerebellum provides adjustment and co-ordination of CNS centres and pathways which *are* the primary generators of motor commands (Luciani, 1915; Holmes, 1917; Lorento de Nó, 1924; Rosenblueth, Wiener & Bigelow, 1943). MacKay & Murphy (1979) coined the term 'accessory gain adjustment' to describe this role.

Gain control of proprioception

Gamma efferent nerve fibres control muscle spindle sensitivity by activating the small intrafusal muscle fibres within spindles (review: Matthews, 1972). By the early 1950s it was clear that spindle responses to stretch could be greatly modulated by γ action. Furthermore, large fluctuations in both spindle afferent and γ efferent firing had been observed when the excitatory state of the central nervous system varied, either because of strong sensory input, electrical stimulation, or local inactivation (e.g. Granit & Kaada, 1952; Granit, Holmgren & Merton, 1955; Sears, 1964; Corda, von Euler &

* Present address: Prince of Wales Medical Research Institute, Sydney, NSW, Australia.

Lennerstrand, 1966). The electrical stimulation experiments were in fact designed to explore the role of the γ motor system and to identify the central structures which controlled it. Two areas were very effective in activating γ-motoneurones: the reticular formation and the cerebellum (Granit & Kaada, 1952; Granit *et al.*, 1955). Interestingly, reticular stimulation produced diffuse, tonic changes in γ action, often unrelated to phasic muscle contraction. It was suggested that the control of the γ system was in some way related to arousal (Granit & Kaada, 1952). Granit (1955) saw the reticular formation as the 'collecting network' for this diffuse control, because it had 'properties characteristic of γ activity, such as long duration, broad front of attack, and great range'. However, a separate mode of control had also been posited, in which the γ motoneurones were the 'ignition mechanism' to initiate movement, being co-activated with α-motoneurones, and slightly leading them (Merton, 1951; Granit, 1955). Several variations on this theme were proposed: 'follow-up servo action' (Merton, 1953), α–γ linkage (Granit, 1955), 'servo-assistance' and 'α–γ coactivation' (Matthews, 1972). The notion that γ-motoneurones were phasically turned on and off with α-motoneurones tended to overshadow the idea of tonic activation related to arousal until the early 1980s, when task-dependence of spindle firing observed in awake animals again suggested a measure of independence of α- and γ-motoneurones (Prochazka & Wand, 1981; Thach, Perry & Schieber, 1982; Schieber & Thach, 1985; Prochazka, 1989).

Cerebellar control of the γ system was further pursued by Gilman (1969a, b) in acute experiments at various times after chronic lesions in cats and monkeys. Generally speaking, the tonic firing of spindle afferents and γ-motoneurones was depressed after cerebellectomy. What is less commonly realized is that dynamic stretch-sensitivity of spindle primary endings hardly changed (Gilman, 1969b; Kornhauser, Bromber & Gilman, 1982). With hindsight, the overall effects on spindle responses were actually very modest, particularly when compared with the changes now known to occur spontaneously in the conscious animal. However, in the clinical teaching of the subject, cerebellar deficits came to be associated with fusimotor abnormalities (Brodal, 1981; Kandel, Schwartz & Jessel, 1991; Nichols, Martin & Wallace, 1992; Thach, Goodkin & Keating, 1992).

The data which implicated the cerebellum in fusimotor control were obtained in anaesthetized or decerebrate animals, incapable of willed movement and with abnormal descending drive to the spinal cord. The data were also far from consistent: after cerebellectomy the tonic firing of spindles and γ-motoneurones was reduced (Gilman, 1969a, b; Van der Meulen & Gilman, 1965), but in other experiments, *activation* of the cerebellar nuclei caused reduced spindle firing (Vitek, 1984). Though stimulation of the *medial* vermis of the cerebellar cortex reduced tonic spindle firing, stimulation of *lateral* vermis and hemispheres increased it (Granit & Kaada, 1952; Granit *et al.*, 1955).

Recently, some interesting, albeit indirect evidence emerged from chronic recordings in monkeys. The sensitivity of neurones in somatosensory cortex to proprioceptive feedback was increased during cerebellar cooling (Hore & Flament, 1988). It was suggested that the cerebellar deficit had increased, rather than reduced, fusimotor activity, particularly of the γ_d type, and this had raised spindle sensitivity to the point of destabilizing the reflex arc, causing tremor. Of course the changes in somatosensory cortex could also have been due to modulation of ascending transmission at brainstem and thalamic levels (e.g. Jiang, Lamarre & Chapman, 1990). But the possibility that major

Fig. 23.1. Schematic showing task-dependent changes in γ drive to muscle spindles. Adapted with permission from Trend (1987).

cerebellar symptoms could result from excessive γ action prompted our group to test the matter directly (Gorassini, Prochazka & Taylor, 1993).

The firing of muscle spindle afferents was recorded in intact cats before and during reversible inactivation of the cerebellar interpositus and dentate nuclei with lidocaine. Two interrelated hypotheses were tested: (1) cerebellar nuclear inactivation interferes with the control of muscle spindle sensitivity; (2) cerebellar ataxia is indirectly caused by aberrant muscle spindle sensitivity. As the control of muscle spindle sensitivity in the normal cat appears to be task- and context-dependent (Fig. 23.1), spindle sensitivity had to be compared in different situations before and during cerebellar inactivation.

Cerebellar motor deficits

Testing these hypotheses required an evaluation of the deficits caused by the cerebellar inactivation. Ataxia is a widely used, but imprecise term. It includes dysmetria, dyssynergia, decomposition of movements, tremor and drunken gait. A loss of co-ordination between segments was seen in the cats, which could be termed dyssynergia or decomposition of movement; however, this was very difficult to quantify. Tremor was not seen consistently, nor was there convincing evidence of hypotonia (the centrality of this symptom is in fact in some doubt: Hore, 1987; Diener & Dichgans, 1992). When gait abnormalities occurred, they were associated with a concomitant loss of balance and dysmetria of voluntary movement.

Fig. 23.2 illustrates some typical features of ataxic gait in cats during ipsilateral interpositus inactivation: high-stepping, premature termination of swing followed by exaggerated placement (also described by Udo *et al.*, 1980), loss of lateral stability, limb cross-over and contralateral correction resulting in abnormally wide-based stance. Postural effects included head sway, tonic flexion of the forelimb, and excessive extensor thrust of the hindlimb, often causing the foot to lose purchase and slip backwards into hyperextension. Corrective movements tended to be delayed and hypermetric. These features have much in common with the classical symptoms of human cerebellar gait (Holmes, 1922): 'the placing of the foot is frequently incorrect, most commonly it is abducted too much, but it is often brought in front of the other, so that in advancing the latter he may trip over it and fall towards the normal side'; 'the movements of the homolateral leg are irregular, the foot is often raised too high owing to excessive flexion at the hip, and may either fall to the floor inertly or be brought to it with undue force'. Interestingly, although cerebellar motor deficits were quite apparent during interpositus and fastigial inactivation, dentate injections produced no noticeable motor effects.

Fig. 23.2. Video tracings of ataxic gait during lidocaine inactivation of interpositus. Top: hyperflexion ('high-stepping') of both the ipisilateral hindlimb and forelimb. Second panel: the forelimb halts momentarily in mid-air (premature termination of swing phase), body sways to right. Third panel: left forelimb crosses over right forelimb. Bottom: balance regained by a lateral corrective movement of the right forelimb.

Spindle sensitivity during ataxic episodes

The general impression was that muscle spindles responded to length changes in much the same way during ataxia as they did before and after. Fig. 23.3 shows that the averaged firing of hamstrings spindle primary endings during step cycles was virtually unchanged during interpositus inactivation. The sensitivity was quantified by plotting mean instantaneous firing rate against the response of a first-order model of the length signal (see Gorassini *et al.*, 1993 for details).

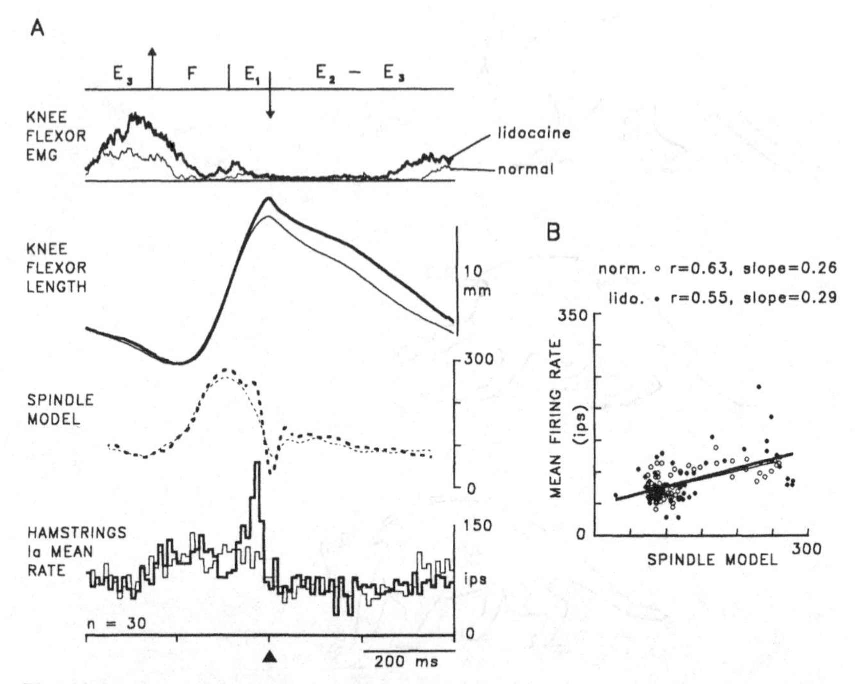

Fig. 23.3. (A) Averaged data from three knee flexor Ia afferents, each contributing 10 normal step cycles (thin lines) and 10 ataxic step cycles (thick lines), i.e. 60 steps in all. Top: step cycle phases (E_2 and E_3, stance; F and E_1, swing) arrows mark foot lift-off and contact. Individual cycles were aligned at the black triangle to peak knee flexor length and averaged. Firing rate profiles were computed by summing afferent spikes from aligned cycles in 20 ms bins. The hyperflexion during ataxic swing phases is reflected in augmented EMG and increased muscle displacement. Firing rates in ataxic steps were similar to normal steps, except for a transient increase at the end of E_1. Dashed traces show length signals after filtering with a transfer function which mimicked the response characteristics of passive spindle Ia afferents to large-amplitude length changes. (B) Scatter plots and regression lines of the firing rate profiles versus filtered length for normal steps (open circles) and ataxic steps (filled circles). The slope (spindle sensitivity) was not significantly different during ataxia. Reproduced from Gorassini *et al.* (1993), with permission.

As mentioned above, spindle sensitivities are known to change in different tasks and contexts. Fig. 23.4 shows the averaged responses of the same primary endings to imposed movements of the limb, before and during interpositus inactivation. Sensitivity was very high during the imposed movements compared with that during walking (cf. Figs. 23.3 & 23.4). This was also the case during interpositus inactivation. Apart from a small, statistically insignificant increase in spindle primary sensitivity during ataxic gait (Fig. 23.3), the only sign of unusual fusimotor action was seen when cats made corrective movements during postural instability, or during turning manoeuvres. These situations probably involved extra attention or arousal, in which increased γ

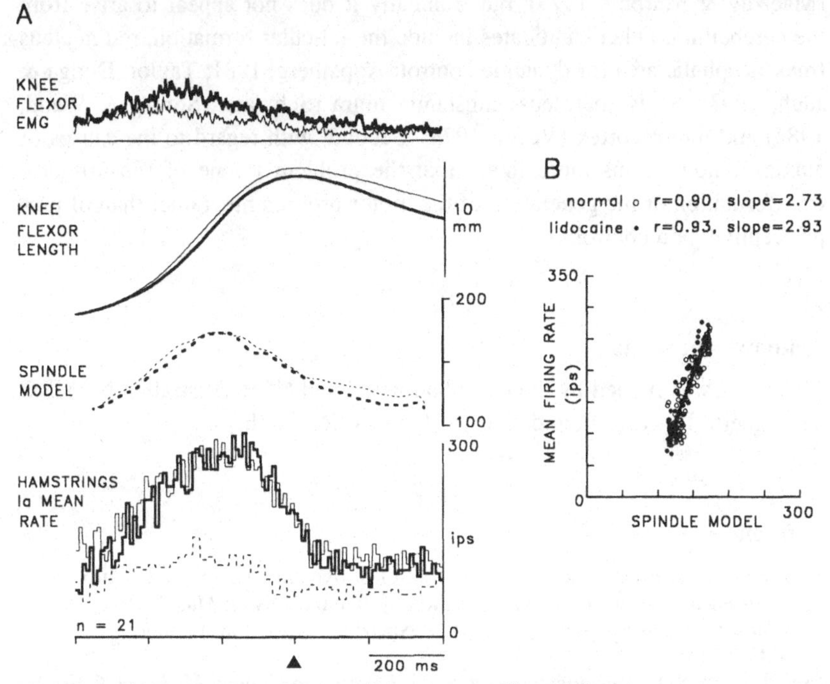

Fig. 23.4. (A) Responses of same afferents as in Fig. 23.3 to imposed movements before (thin lines) and during (thick lines) interpositus inactivation and during deep anaesthesia (dashed lines; length traces not included, but of similar profile). Note the high firing rate in response to imposed stretch in the awake animal, whether ataxic or not, compared with that during anaesthesia. This indicates strong γ_d action. (B) Scatter plots of mean firing rate versus filtered length. Note the 10-fold increase in slope of the regression lines compared with Fig. 23.3B, indicating a much higher level of γ_d action in both normal and ataxic states (i.e. the context-dependent change in γ action was still present during ataxia). Reproduced from Gorassini *et al.* (1993) with permission.

drive would be expected. Spindles recorded during dentate inactivation did not change their response properties. This is not too surprising, since dentate injections did not cause ataxia.

In summary, there was little if any change in γ control of muscle spindle sensitivity during inactivation of dentate or interpositus nuclei. Two conclusions were drawn. First, the interpositus and dentate nuclei are not the primary controllers of the fusimotor system. Second, ataxia does not result from disordered proprioceptive sensitivity. For various reasons, these conclusions are still tentative and require further confirmation (see Gorassini *et al.*, 1993). However, it is clear that the large changes in γ action which we expected to see during ataxia were simply not there and, in retrospect, perhaps we were wrong to expect them. Where does this leave us? First, γ modulation of spindle sensitivity is an excellent example of 'metasystemic control'

(MacKay & Murphy, 1979), but ironically it does not appear to arise from the cerebellum. Other candidates include the reticular formation, red nucleus (mesencephalic area for dynamic control: Appelberg, 1981; Taylor, Donga & Jüch, 1993), locus coeruleus, substantia nigra (Schwartz, Sontag & Wand, 1984) and motor cortex (Vedel, 1973). Second, with regard to the causes of ataxia, it now seems more likely that the problem is one of sensorimotor transformation or the generation of the motor programme, rather than of proprioceptive gain control.

Acknowledgements

This work was funded by the Canadian MRC and NCE, Australian NHMRC and Alberta Heritage Foundation for Medical Research.

References

Appelberg, B. (1981). Selective central control of dynamic gamma motoneurones utilised for the functional classification of gamma cells. In *Muscle Receptors and Movement*, ed. A. Taylor & A. Prochazka, pp. 97–108. London: Macmillan.

Brodal, A. (1981). *Neurological Anatomy in Relation to Clinical Medicine*. Oxford: Oxford University Press.

Corda, M., Euler, C. von & Lennerstrand, G. (1966). Reflex and cerebellar influences on α and on 'rhythmic' and 'tonic' γ-activity in the intercostal muscle. *Journal of Physiology (London)*, **184**, 898–923.

Diener, H.-C. & Dichgans, J. (1992). Pathophysiology of cerebellar ataxia. *Movement Disorders*, **7**, 95–109.

Gilman, S. (1969a). Fusimotor fibre responses in the decerebellate cat. *Brain Research*, **14**, 218–221.

Gilman, S. (1969b). The mechanism of cerebellar hypotonia: an experimental study in the monkey. *Brain*, **92**, 621–638.

Gorassini, M., Prochazka, A. & Taylor, J. (1993). Cerebellar ataxia and muscle spindle sensitivity. *Journal of Neurophysiology*, **70**, 1853–1862.

Granit, R. (1955). *Receptors and Sensory Perception*. New Haven: Yale University Press.

Granit, R., Holmgren, B. & Merton, P.A. (1955). The two routes for excitation of muscle and their subservience to the cerebellum. *Journal of Physiology (London)*, **130**, 213–224.

Granit, R. & Kaada, B.R. (1952). Influence of stimulation of central nervous structures on muscle spindles in cat. *Acta Physiologica Scandinavica*, **27**, 130–160.

Holmes, G. (1917). The symptoms of acute cerebellar injuries due to gunshot injuries. *Brain*, **40**, 461–535.

Holmes, G. (1922). Clinical symptoms of cerebellar disease and their interpretation. The Croonian Lectures, III. *Lancet*, ii, 59–65.

Holmes, G. (1939). The cerebellum of man. Tenth Hughlings Jackson Memorial Lecture. *Brain*, **62**, 1–30.

Hore, J. (1987). Loss of set-dependent reactions during cerebellar dysfunction causes limb instability. In *Higher Brain Functions: Recent Explorations of the Brain's Emergent Properties*, ed S.P. Wise, pp. 101–111. New York: Wiley

Hore, J. & Flament, D. (1988). Changes in motor cortex neural discharge associated with the development of cerebellar ataxia. *Journal of Neurophysiology*, **60**, 1285–1302.

Jiang, W., Lamarre, Y. & Chapman, C.E. (1990). Modulation of cutaneous cortical evoked potentials during isometric and isotonic contractions in the monkey. *Brain Research*, **536**, 69–78.

Kandel, E.R., Schwartz, J.H. & Jessell, T.M. (1991). *Principles of Neural Science*. New York: Elsevier.

Kornhauser, D., Bromberg, M.B. & Gilman, S. (1982). Effects of lesions of fastigial nucleus on static and dynamic responses of muscle spindle primary afferents in the cat. *Journal of Neurophysiology*, **47**, 977–986.

Lorento de Nó, R. (1924). Etudes sur le cerveau postérieur. III. Sur les connexions cerebelleuses des fascicules afferents au cervelet, et sur la fonction de cet organe. *Travaux du Laboratoire de Recherches Biologique de l'Université de Madrid*, **22**, 51–65.

Luciani, L. (1915). *The Hindbrain: Human Physiology, Muscular and Nervous System*, Vol 3, trans. F.A. Welby, London: Macmillan.

Mackay, W.A. & Murphy, J.T. (1979). Cerebellar modulation of reflex gain. *Progress in Neurobiology*, **13**, 361–417.

Matthews, P.B.C. (1972). *Mammalian Muscle Receptors and Their Central Actions*, pp. 195–262. London: Edward Arnold.

Merton, P.A. (1951). The silent period in a muscle of the human hand. *Journal of Physiology (London)*, **114**, 183–198.

Merton, P.A. (1953). Speculations on the servo-control of movement. In *CIBA Foundation Symposium. The Spinal Cord*, ed. G.E.W. Wolstenholme, pp. 247–255. London: Churchill.

Nichols, J.G., Martin, A.R. & Wallace, B.G. (1992). *From Neuron to Brain*, 3rd edn. Sunderland, MA: Sinauer Associates.

Prochazka, A. (1989). Sensorimotor gain control: a basic strategy of motor systems? *Progress in Neurobiology*, **33**, 281–307.

Prochazka, A. & Wand, P. (1981). Independence of fusimotor and skeletomotor systems during voluntary movement. In *Muscle Receptors and Movement*, ed. A. Taylor, & A. Prochazka, pp. 229–243. London: Macmillan.

Rosenblueth, A., Wiener, N. & Bigelow, J. (1943). Behavior, purpose and teleology. *Philosophy of Science*, **10**, 18–24.

Schieber, M.H. & Thach, W.T. (1985). Trained slow tracking. II. Bidirectional discharge patterns of cerebellar nuclear, motor cortex, and spindle afferent neurones. *Journal of Neurophysiology*, **54**, 1228–1270.

Schwartz, M., Sontag, K.-H. & Wand, P. (1984). Sensory-motor processing in substantia nigra pars reticulata in conscious cats. *Journal of Physiology (London)*, **347**, 129–147.

Sears, T.A. (1964). Efferent discharges in alpha and fusimotor fibres of intercostal nerves of the cat. *Journal of Physiology (London)*, **174**, 295–315.

Sherrington, C.S. (1906). *The Integrative Action of the Nervous System*. New Haven: Yale University Press.

Taylor, A., Donga, R. & Jüch, P.J.W. (1993). Fusimotor effects of midbrain stimulation on jaw muscle spindles of the anaesthetized cat. *Experimental Brain Research*, **93**, 37–45.

Thach, W.T., Goodkin, H.P. & Keating, J.G. (1992). The cerebellum and the adaptive coordination of movement. *Annual Reviews of Neuroscience*, **15**, 403–442.

Thach, W.T., Perry, J.G. & Schieber, M.H. (1982). Cerebellar output: body images and muscle spindles. *Experimental Brain Research, Supplement*, **6**, 440–454.
Trend, P.St.J. (1987). Gain control in proprioceptive reflex pathways. PhD. thesis, University of London.
Udo, M., Matsukawa, K., Kamei, H. & Oda, Y. (1980). Cerebellar control of locomotion: effects of cooling cerebellar intermediate cortex in high decerebrate and awake walking cats. *Journal of Neurophysiology*, **44**, 119–134.
Van der Meulen, J.P. & Gilman, S. (1965). Recovery of muscle spindle activity in cats after cerebellar ablation. *Journal of Neurophysiology*, **28**, 943–957.
Vedel, J.P. (1973). Cortical control of dynamic and static gamma motoneurone activity. In *New Developments in Electromyography and Clinical Neurophysiology*, ed. J.E. Desmedt, vol. 3, pp. 126–135. Basel: Karger.
Vitek J.L. (1984). The effects of dentate nuclear output via pathways not involving the sensorimotor cortex on the amplitude and organization of the stretch reflex in decerebrate cats. PhD thesis, University of Minnesota.

24

Roles of the lateral nodulus and uvula of the cerebellum in cardiovascular control

NAOKO NISIMARU

Department of Physiology, Oita Medical University, Oita, Japan

The cerebellum is involved in the control of not only motor but also autonomic functions (Moruzzi, 1950; Ito, 1984). Blood pressure, heart rate and respiration have been shown to be affected by electrical stimulation of the anterior vermis and the fastigial nucleus (Moruzzi, 1950; Ban, Hilliard & Sawyer, 1960; Achari & Downman, 1970; Nisimaru & Kawaguchi, 1984). Previously, we reported that electrical stimulation of localized regions in the posterior lobe (lobules VII, VIII and X) produced inhibition of renal sympathetic nerve activity and a fall in blood pressure in anaesthetized rabbits (Nisimaru, Yamamoto & Shimoyama, 1984b; Nisimaru & Watanabe, 1985). More recently, Bradley *et al.* (1987a, b) also reported that electrical stimulation of the medial uvula (lobule IXb) induced cardiovascular responses in rabbits and cats. Interestingly, they showed that the effect of stimulation of the medial uvula was reversed when the anaesthesia wore off in decerebrate rabbits. However, little attention has been paid to the involvement of the lateral portion of the nodulus–uvula in cardiovascular function, on which I shall focus in this chapter. As an index of cardiovascular effects of cerebellar stimulation, we recorded, integrated and averaged efferent discharges from renal sympathetic nerves, which were affected by cerebellar stimulations with trains of only a few pulses. For stimulus mapping, a platinum–iridium needle electrode (diameter 200 μm, excluding the layer of insulation) was inserted into the cerebellum.

Stimulation of the lateral nodulus–uvula in anaesthetized rabbits.

In rabbits anaesthetized with α-chloralose plus urethane (30 and 600 mg/kg), lateral parts of the dorsal nodulus and the most ventral uvula were stimulated with a train of ten brief current pulses no stronger than 100 μA (200 Hz). Depression of renal sympathetic nerve activity was induced from those sites which fell within a small region extending over about 1 mm longitudinally through the dorsal nodulus encroaching the border to the ventral uvula, at 2.7–3.7 mm lateral to the midline (Fig. 24.1) (Nisimaru & Watanabe, 1985).

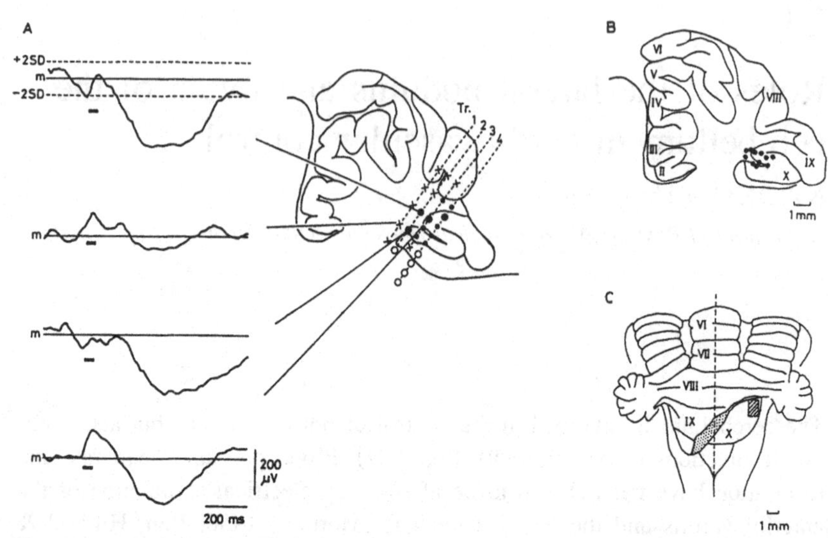

Fig. 24.1. Effects of stimulation of the nodulus and uvula upon renal sympathetic nerve activity. (A) Parasagittal section of the cerebellum at 3.3 mm to the right of the midline (for identification of details see (B)). Tr. 1–4, electrode tracks. Filled circles, sites where an appreciable depression was induced with currents of less than 100 μA; sizes of the filled circles vary according to the amplitude of the depression (small dots, 12–30% of the mean; large dots, over 30%). Open circles, sites where a facilitation of the renal sympathetic nerve activity was induced with currents of 100 μA. Crosses, ineffective sites with currents of 200 μA. Traces, averaged renal sympathetic nerve activity under stimulation at the sites indicated in the diagram of the cerebellum; horizontal bars in these traces indicate periods of stimulation with a train of ten pulses (current intensity 100 μA). Both the mean value (m) and the standard deviation (SD) of spontaneous nerve activity were calculated from 170 successively recorded traces (Nisimaru *et al.*, 1984b). (B) Parasagittal section of the cerebellum as in (A). II–X, lobule numbers of the cerebellar vermis. (C) Dorsal view of the cerebellum. (B) and (C) summarize the results from 11 rabbits. Note in (C) that a part of the uvula was removed in order to illustrate the effective stimulating region over the dorsal nodulus (cut face is stippled). Filled circles in (B) indicate the most effective sites, one in each rabbit, for inducing depression of renal sympathetic nerve activity with currents of less than 100 μA. The obliquely hatched rectangular area in (C) encloses the effective stimulating area, extending between 2.9 and 3.7 mm from the midline and over 1 mm longitudinally. From Nisimaru & Watanabe (1985) with permission.

When the stimulating trains were sustained over several seconds, the blood pressure was also depressed (Fig. 24.2A). This fall in blood pressure was associated with a transient increase followed by a slight decrease in the renal arterial blood flow and with an increase in the femoral arterial blood flow (Fig. 24.2A).

Paton & Spyer (1992) reported that the middle region of lobule IX, specifically sublobule b, was the only region of the posterior vermis from which alterations in the cardiovascular and respiratory system were evoked

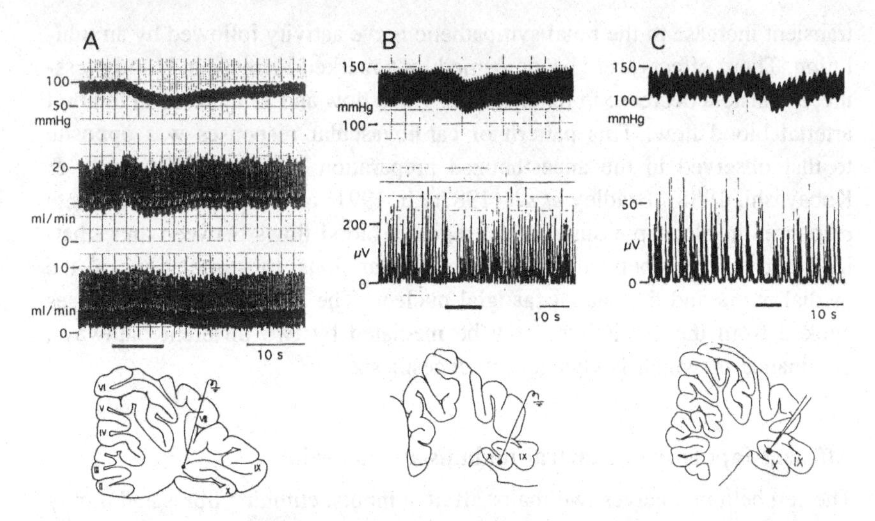

Fig. 24.2. Effects of electrical or chemical stimulation of the lateral nodulus–uvula upon sympathetic nerve activity, blood pressure and regional blood flows in anaesthetized rabbits. (A) The effects of electrical stimulation (current intensity, 50 µA; pulse width, 200 µs; frequency, 200 Hz). Top trace: blood pressure. Middle trace: renal arterial blood flow. Bottom trace: femoral arterial blood flow. (B) Similar stimulating effects to those shown in (A) but in another rabbit (current intensity, 200 µA). Upper trace: blood pressure. Lower trace: integrated renal sympathetic nerve activity. (C) The effects of chemical stimulation with an injection of 0.2 M L-glutamate (100 nl). Upper trace: systemic blood pressure. Lower trace: sympathetic nerve activity. Below the traces in (A), (B) and (C) the parasagittal section of the cerebellum at 2.8, 4.2 and 3.2 mm, respectively, is shown to indicate the stimulation site and type of stimulating electrode. The horizontal bars show when the stimulus was applied.

consistently during electrical or chemical activation in the rabbit. In order to exclude the possibility that our results, obtained by stimulating the lateral nodulus–uvula, were due to current spread or activation of passing fibres, we injected L-glutamate (0.2 M, 100 nl) into the lateral nodulus–uvula as a chemical stimulant. The results from glutamate injection were identical to those from electrical stimulation (Fig. 24.2B, C), confirming that the cardiovascular effect seen in Figs. 24.1 and 24.2A arises from activation of cerebellar neurones in the lateral nodulus–uvula. It should be noted that Bradley, Gherladucci & Spyer (1991) demonstrated that stimulation of lobule X in some cats induced some cardiovascular responses.

Stimulation of the lateral nodulus–uvula in unanaesthetized decerebrate rabbits

We repeated similar experiments in unanaesthetized, decerebrate rabbits. Electrical stimulation of the lateral nodulus–uvula with repetitive pulses lasting for several seconds (current intensity less than 100 µA) induced a large transient increase in the renal sympathetic nerve activity followed by an inhibition. These effects were accompanied by a marked increase in blood pressure, a transient decrease in renal arterial blood flow and an increase in femoral arterial blood flow. This pattern of cardiovascular responses was opposite to that observed in the anaesthetized preparation (Nisimaru, Katayama & Kobayashi, 1988). Bradley *et al.* (1987a, b, 1991) also reported the opposite responses in blood pressure and in regional blood flows between anaesthetized and unanaesthetized decerebrate preparations to stimulation of the medial uvula and the caudal fastigial nucleus. The cardiovascular responses evoked from the cerebellum may be mediated by two different pathways, dominance of which is changed by anaesthesia.

Afferent inputs to the lateral nodulus–uvula region

The cerebellum receives two major afferent inputs: climbing fibres and mossy fibres. Localized injection of horseradish peroxidase (HRP) into rabbit's nodulus revealed six longitudinal zones of the nodulus receiving climbing fibre projections from five different subdivisions of the inferior olive (Katayama & Nisimaru, 1988). It is suggested that the small lateral area of the nodulus–uvula involved in cardiovascular control receives climbing fibres from the rostrolateral part of the medial accessory olive and/or the dorsomedial cell column in the contralateral inferior olive. Injection of HRP into the lateral nodulus–uvula region in rabbits also revealed that this region bilaterally receives mossy fibre inputs from vestibular nuclei, prepositus hypoglossi nucleus, Roller nucleus, intercalatus nucleus and the medial fasciculus longitudinalis (Katayama & Nisimaru, 1987).

We examined the field potentials evoked in the lateral nodulus–uvula following stimulation of cardiovascular afferent nerves. Electrical stimulation of the left aortic and vagal afferent nerves produced prominent field potentials in the dorsal region of right lateral nodulus with a latency of 7–10 ms, positive in the molecular layer and negative in the granular layer as shown in Fig. 24.3 (Nisimaru, Aramaki & Watanabe, 1984a; Nisimaru, Katayama & Watanabe, 1986). These potentials were significantly attenuated during high-frequency stimulation of the afferent nerves (Fig. 24.3). The laminal profile and frequency response characteristics are identical to those described earlier for climbing fibre responses (Maekawa & Takeda, 1975).

We also observed the mossy fibre responses in the lateral nodulus–uvula by stimulation of aortic and vagal afferent nerves (Nisimaru *et al.*, 1984a; Nisimaru *et al.*, 1986), as reported previously for mossy fibre responses (Maekawa & Takeda, 1975). An electrophysiological study has shown that vestibular fibres project to the vestibulocerebellum (nodulus and uvula) through both mossy and climbing fibre pathways (Precht, Simpson & Llinás,

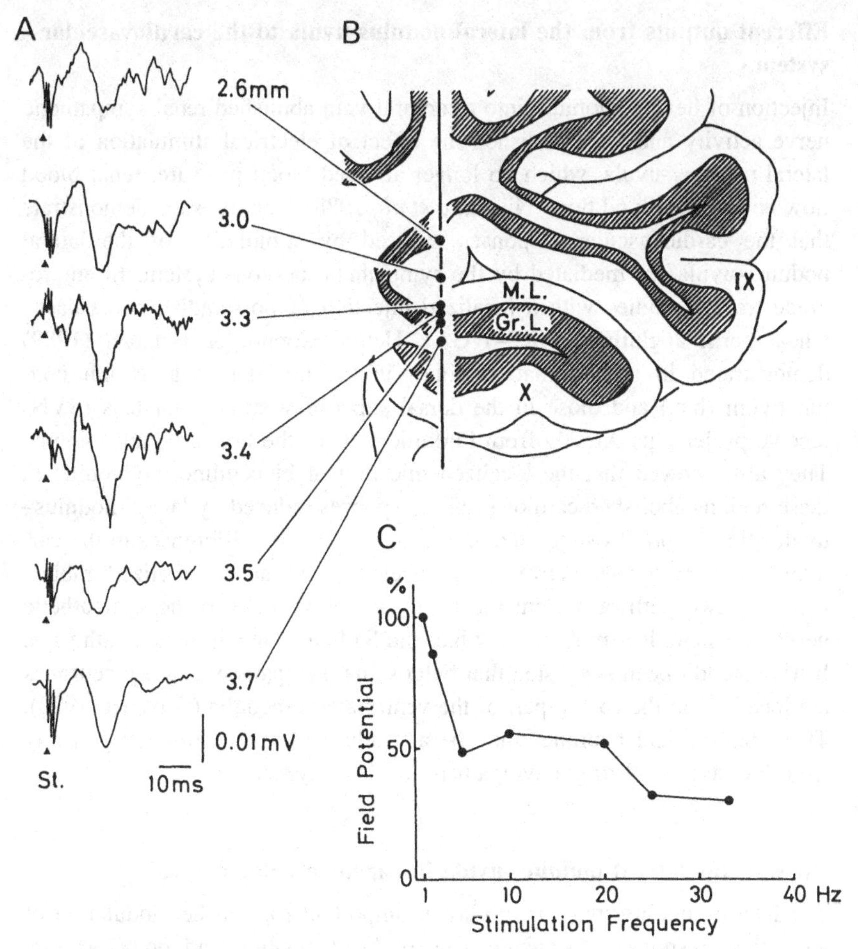

Fig. 24.3. Climbing fibre responses evoked in the lateral nodulus–uvula by stimulation of vagal afferent fibres at the neck area. (A) Depth profiles of the field potentials evoked in the right lateral nodulus following stimulation of the left vagal afferent nerves. Recording sites are shown in (B). Distances are shown from the surface of lobule VIIIb. (B) Part of the parasagittal section of the cerebellum at 3.4 mm lateral to the midline (lobules IX and X). (C) Graphical representation of sizes of field potentials versus stimulation frequency. Sizes of potentials at each stimulation frequency are represented as a percentage of those evoked by stimulation at 1/s. St., time of the stimulation (3 pulses; interval, 1 ms; intensity, 800 μA); M.L., molecular layer; Gr.L., granular layer.

1976). Our results, together with this observation, suggest that Purkinje cells in the lateral nodulus–uvula receive cardiovascular and vestibular information through both the mossy and climbing fibre systems and then integrate this information.

Efferent outputs from the lateral nodulus-uvula to the cardiovascular system

Injection of hexamethonium into a femoral vein abolished renal sympathetic nerve activity and also abolished the effect of electrical stimulation of the lateral nodulus–uvula, which no longer affected blood pressure, renal blood flow or femoral blood flow (Nisimaru *et al.*, 1988). These results demonstrate that the cardiovascular responses induced by stimulation of the lateral nodulus–uvula are mediated by the sympathetic nervous system. In anterograde tracing studies with a localized injection of horseradish peroxidase–wheat germ agglutinin (HRP–WGA), Henry, Connor & Balaban (1989) demonstrated in rabbits that neurones in the nucleus of brachium conjunctivum (b.c.) and those in the dorsal superior vestibular nucleus (SVN) receive projections directly from Purkinje cells of the lateral nodulus–uvula. They also showed that the localized injection of bicuculline methiodide in these regions abolished cardiovascular responses induced by lateral nodulus–uvula stimulation. As suggested above, in view of the differences in the cardiovascular responses between anaesthetized and unanaesthetized rabbits there are two pathways from the lateral nodulus–uvula to the sympathetic nervous system. It is probable that b.c. and SVN mediate these two pathways. It has recently been suggested that bulbospinal sympathoexcitatory neurones are localized in the rostral part of the ventrolateral medulla (Guyenet, 1990). These bulbospinal neurones may be a candidate for the common pathway from b.c. and SVN to the sympathetic nervous system.

Roles of the lateral nodulus–uvula in cardiovascular control

The lateral nodulus–uvula may play an important role in the modulation of autonomic responses during changes of head position and body posture, because, as shown above, stimulation of the lateral nodulus–uvula induces cardiovascular effects and this region receives cardiovascular and vestibular information.

We examined changes of renal sympathetic nerve activity and blood pressure during postural change in both anaesthetized and awake intact rabbits. In anaesthetized rabbits, a 30° head-up tilt caused a transient inhibition of the renal sympathetic nerve activity and a large decrease in blood pressure, both of which slowly recovered to the control level within 10–20 s (Nisimaru, Yanai & Matuzaki, 1993). These responses resemble those previously reported in anaesthetized cats, where electrical stimulation of the primary vestibular afferent nerve induced an inhibitory effect on spontaneous renal nerve discharges and caused a fall in blood pressure (Uchino *et al.*, 1970).

It is likely that under anaesthesia head-up tilt stimulates vestibular afferent nerves, which in turn inhibit renal sympathetic nerve activity reflexly, leading to the fall in blood pressure. The later gradual recovery of the renal sympathetic nerve activity and the blood pressure could be due to baroreceptors. Thus, the responses in sympathetic nerve activity and blood pressure induced by head-up tilt can be explained by a combination of vestibulo-sympathetic and baroreceptor reflexes. Neurones in the caudal paramedian reticular formation receive inputs from both the vestibular afferent nerve and the carotid sinus nerve (Yates & Yamagata, 1990) and, therefore, constitute a candidate for a brainstem pathway mediating these reflexes. In the anaesthetized condition, the cerebellum, cerebrum and other higher brainstem regions might not be involved in these responses.

In awake control rabbits, by contrast, renal sympathetic nerve activity showed a transient increase and was maintained at a slightly lower level, while blood pressure initially decreased and recovered to the control level in a few seconds following a head-up tilt (Nisimaru *et al.*, 1993). This result suggests that in awake intact rabbits the vestibulo-sympathetic and baroreceptor reflexes of sympathetic nerve activity and blood pressure induced by a head-up tilt are adaptively inhibited by the cerebellum, cerebrum or other higher brain areas.

We conducted similar experiments 5–7 days after bilateral lesioning of the lateral nodulus–uvula with local injection of kainic acid (0.1%). After bilateral destruction of the lateral nodulus–uvula, a head-up tilt caused an immediate large increase in renal sympathetic nerve activity, which was sustained at a higher level. Blood pressure also increased transiently, but then decreased and remained lower than the control level (Nisimaru *et al.*, 1993). Therefore, after bilateral destruction of the lateral nodulus–uvula, the adaptive condition could no longer be induced, but the vestibulo-sympathetic and baroreceptor reflexes following a head-up tilt could be inhibited by other higher brain areas. It is likely that adaptation of sympathetic nerve activity and blood pressure during postural changes may be immediately induced by Purkinje cells in the lateral nodulus–uvula.

Conclusion and clinical implications

The results of our experiments have indicated that cardiovascular responses apparently are mediated by two different pathways from the Purkinje cell in the lateral nodulus–uvula to preganglionic sympathetic neurones. Climbing and mossy fibres to this cerebellar region mediate relevant inputs from the aortic and vagal afferent nerves. However, vestibular and/or cerebellar influences on circulation appear to be complex (Uchino *et al.*, 1970; Doba & Reis, 1974). The discrepancy among experiments may have been caused partly by different anaesthetic conditions, and therefore by different pathways having been stimulated. Changes of inputs from the labyrinth during exessive motion and space flight may induce orthostatic intolerance (Money, 1970; Yates,

1992). The dysfunction of vestibulo-sympathetic reflexes could also produce an idiopathic orthostatic hypotension (Thomas *et al.*, 1981). The dysfunction of vestibulocerebellum and/or other higher brain areas might also be involved in the circulatory problems in these conditions.

The cerebellum, particularly the fastigial nucleus and the uvula in the posterior vermis, might play an important role in mediating vestibulo-sympathetic reflexes (Doba & Reis 1974; Bradley *et al.*, 1991). Paton & Spyer (1992) have suggested also that the medial region of uvula is important in the co-ordination of the motor and autonomic changes that occur during alerting or orientating. As described above, our studies on the effect of posture in awake rabbits indicate that Purkinje cells in the lateral nodulus–uvula inhibit the vestibulo-sympathetic and baroreceptor reflexes, and adaptively control cardiovascular responses. These studies may suggest that the nodulus and uvula in the cerebellum perform important functions during the body motions of everyday life.

In summary, the cardiovascular problems in motion and space sicknesses may be caused by a failure of control mechanisms associated with the nodulus and/or uvula. Further research concerning vestibular and cerebellar influences on the autonomic nervous system is needed.

References

Achari, N.K. & Downman, C.B.B. (1970). Autonomic effector responses to stimulation of nucleus fastigius. *Journal of Physiology (London)*, **210**, 637–650.

Ban, T., Hilliard, J. & Sawyer, C.H. (1960). Autonomic and electroencephalographic responses to stimulation of the rabbit cerebellum. *Anatomical Record*, **136**, 309.

Bradley, D.J., Ghelarducci, B., Paton, J.F.R. & Spyer, K.M. (1987a). The cardiovascular responses elicited from the posterior cerebellar cortex in the anaesthetized and decerebrate rabbit. *Journal of Physiology (London)*, **383**, 537–550.

Bradley, D.J., Ghelarducci, B. & Spyer, K.M. (1991). The role of the posterior cerebellar vermis in cardiovascular control. *Neuroscience Research*, **12**, 45–56.

Bradley, D.J., Pascoe, J.P., Paton, J.F.R. & Spyer K.M. (1987b). Cardiovascular and respiratory responses evoked from the posterior cerebellar cortex and fastigial nucleus in the cat. *Journal of Physiology (London)*, **393**, 107–121.

Doba, N. & Reis, D.J. (1974). Role of the cerebellum and vestibular apparatus in regulation of orthostatic reflexes in the cat. *Circulation Research*, **34**, 9–18.

Guyenet, P.G. (1990). Role of the ventral medulla oblongata in blood pressure regulation. In *Central Regulation of Automatic Functions*, ed. A.D. Loewy & K.M. Spyer, pp. 145–167. New York: Oxford University Press.

Henry, R.T., Connor, J.D. & Balaban, C.D. (1989). Nodulus-uvula depressor response: central GABA-mediated inhibition of α-adrenergic outflow. *American Journal of Physiology*, **256**, H1601-H1608.

Ito, M. (1984). *The Cerebellum and Neuronal Control*. New York: Raven Press.

Katayama, S. & Nisimaru, N. (1987). The origins of the climbing and the mossy fiber inputs to the cerebellar nodulus in rabbits. *Journal of the Physiological Society of Japan*, **49**, 317.

Katayama, S. & Nisimaru, N. (1988). Parasagittal zonal pattern of olivo-nodular projections in rabbit cerebellum. *Neuroscience Research*, **5**, 424–438.

Maekawa, K. & Takeda, T. (1975). Mossy fiber responses evoked in the cerebellar flocculus of rabbits by stimulation of the optic pathway. *Brain Research*, **98**, 590–595.

Money, K.E. (1970). Motion sickness. *Physiological Reviews*, **50**, 1–39.

Moruzzi, G. (1950). The cerebellar influence in the autonomic sphere. In *Problems in Cerebellar Physiology*, pp. 74–96. Springfield, IL: C.C. Thomas.

Nisimaru, N., Aramaki, H. & Watanabe, Y. (1984a). Afferent projections to the cerebellar nodulus from the cardiovascular afferents in rabbits. *Journal of the Physiological Society of Japan*, **46**, 177.

Nisimaru, N., Katayama, S. & Kobayashi, N. (1988). The cardiovascular responses evoked from the lateral nodulus-uvula of the cerebellar cortex in decerebrate and anesthetized rabbits. *Journal of the Physiological Society of Japan*, **50**, 601.

Nisimaru, N., Katayama, S. & Watanabe, Y. (1986). Mossy and climbing fiber projections to the cerebellar uvula and nodulus from the cardiovascular afferents. *Neuroscience Research, Supplement*, **3**, S74.

Nisimaru, N. & Kawaguchi, Y. (1984). Excitatory effects on renal sympathetic nerve activity induced by stimulation at two distinctive sites in the fastigial nucleus of rabbits. *Brain Research*, **304**, 372–376.

Nisimaru, N. & Watanabe, Y. (1985). A depressant area in the lateral nodulus-uvula of the cerebellum for renal sympathetic nerve activity and systemic blood pressure in the rabbit. *Neuroscience Research* **3**, 177–181.

Nisimaru, N., Yamamoto, M. & Shimoyama, I. (1984b). Inhibitory effects of cerebellar cortical stimulation on sympathetic nerve activity in rabbits. *Japanese Journal of Physiology, 34*, 539–551.

Nisimaru, N., Yanai, S. & Matuzaki, H. (1993). Role of lateral nodulus and uvula of cerebellum on orthostatic cardiovascular control in awake rabbits. In *Abstracts of the XXXII Congress of the International Union of Physiological Sciences*, 141.46/P.

Paton, J.F.R. & Spyer, K.M. (1992). Cerebellar cortical regulation of circulation. *News in Physiological Sciences*, **7**, 124–129.

Precht, W., Simpson, J.I. & Llinás, R. (1976). Responses of Purkinje cells in rabbit nodulus and uvula to natural vestibular and visual stimuli. *Pflügers Archiv*, **367**, 1–6.

Sato, Y. & Barmack, N.H. (1985). Zonal organization of olivocerebellar projections to the uvula in rabbits. *Brain Research*, **359**, 281–291.

Thomas, J.E., Schirger, A., Fealey, R.D. & Sheps, S.G. (1981). Orthostatic hypotension. *Mayo Clinic Proceedings*, **56**, 117–125.

Uchino, Y., Kudo, N., Tsuda, K. & Iwamura, Y. (1970). Vestibular inhibition of sympathetic nerve activities. *Brain Research*, **22**, 195–206.

Yates, B.J. (1992). Vestibular influences on the sympathetic nervous system. *Brain Research Reviews*, **17**, 51–59.

Yates, B.J. & Yamagata, Y. (1990). Convergence of cardiovascular and vestibular inputs on neurons in the medullary paramedian reticular formation. *Brain Research*, **513**, 166–170.

25

Central actions of curare and gallamine: implications for reticular reflex myoclonus?

W. BURKE

Departments of Anatomy & Histology, and Physiology and the Sydney Institute for Biomedical Research, (F13), University of Sydney, Sydney, NSW, Australia

Introduction

Although curare and gallamine are usually known for their neuromuscular blocking actions, it has always been appreciated that they may also have significant effects on the central nervous system (CNS). As early as 1812 (Brodie, cited in Smith *et al.*, 1947) it was believed that curare acted on the brain. There are many other references to a central action of curare in the nineteenth century. Since the initial laboratory investigations into gallamine a central action has been recognized (Salama & Wright, 1952a).

Curare

The early history of curare is interesting but not particularly informative about its central action for two reasons. Firstly, prior to the early 1940s curare extracts were a mixture of substances, of variable and uncertain composition. The introduction of the pure alkaloid, *d*-tubocurarine, meant that henceforth it was possible for the results from different laboratories to be legitimately compared. Secondly, in many experiments the drug was injected intravenously. *d*-Tubocurarine (dTC) crosses the blood–brain barrier sparingly. Under certain conditions, e.g. in certain pathological states, the blood–brain barrier may be breached and more drug may pass over. As a result, there has been much confusion and disagreement in the studies in which dTC was administered parenterally. For a proper analysis the drug must be injected into the CNS or into a brain cavity. Earlier references are summarized in McIntyre (1947) and Smith and colleagues (1947).

Injection of dTC into the subarachnoid space produced a wide range of effects. Salama & Wright (1950) injected dTC intraventricularly and intracisternally in chloralose-anaesthetized or decerebrated cats and caused excitation of the vasomotor, respiratory and cardiac systems and also elicited autonomic effects, especially on the salivary glands and bronchi. There was also heightened reflex excitability of the spinal cord and later generalized convulsions.

Intrathecal injection in an isolated spinal cord also caused excitation but much weaker than when dTC was injected into the brain. Feldberg & Fleischhauer (1963) reported tremor, myoclonic jerks, convulsions, various autonomic responses and abnormal discharges resembling the seizure discharge of epilepsy. Are these effects all due to action at a single site? It appears not. dTC has been shown to act at several sites. Chang (1953) applied dTC topically to the cerebral cortex of cats and rabbits and produced local spontaneous synchronous discharges; also the intracortical response in visual cortex to stimulation of the lateral geniculate nucleus was enhanced. Feldberg & Fleischhauer (1962, 1963) injected dTC intraventricularly in cats but by ingenious restriction of the injection to particular parts of the ventricular system were able to demonstrate that dTC acts on the hippocampus to produce seizure discharges. This observation was confirmed and extended by Lebeda, Hablitz & Johnston (1982) who applied dTC to guinea pig hippocampal slices; they demonstrated epileptiform activity in the CA3 region, both spontaneous and in response to orthodromic stimulation of the mossy fibres.

A direct action of dTC on basal ganglia is also very probable. McKenzie, Gordon & Viik (1972) injected dTC into the caudate nucleus and putamen of rats and produced choreiform activity characterized by involuntary movements of the contralateral forelimb, contralateral rotation of the head, facial grimaces and teeth chattering. Forchetti *et al.* (1980) injected dTC into the striatum of rats and produced a complex motor syndrome similar to that produced by McKenzie and colleagues. They also reported tremor fits. Al-Zamil, Bagust & Kerkut (1990), using the isolated spinal cord of the guinea pig, showed that inhibition of motor neurones produced by stimulation of an adjacent ventral root (probably mediated via Renshaw cells) and also inhibition produced by stimulation of the lateral part of the spinal cord were both antagonized by dTC; dTC also caused an increase in spontaneous activity in the ventral roots, presumed to be due to block of Renshaw cells.

Finally, electrophoretic administration of dTC into the inferior olive of rats elicited rhythmical activity (Headley, Lodge & Duggan, 1976). This is an action shared by several other substances, not all chemically similar, such as harmine, harmaline, dihydro-β-erythroidine and gallamine.

Gallamine

Gallamine also acts on several regions of the CNS. Neurones in the pericruciate cortex of cats were excited by gallamine applied electrophoretically, discharging in high-frequency bursts (Crawford & Curtis, 1966). This apparently did not affect the acetylcholine sensitivity of the cells. Administered electrophoretically to cells in the ventrobasal nucleus of the thalamus of cats, gallamine had a characteristic excitant action (Andersen & Curtis, 1964). At first the cells were excited, or the synaptic or acetylcholine-induced firing was enhanced. Later, the cells discharged in characteristic bursts of high-frequency firing in which spike amplitude was reduced; the state progressed to a complete block and refractoriness for several minutes. Similarly, cuneate neurones of cats were strongly excited by micro-iontophoretic application of gallamine, discharging in prolonged high-frequency bursts (Galindo, Krnjević & Schwartz, 1968). It was proposed that gallamine produced a relatively long-lasting depolarization. Gallamine caused a similar high-frequency discharge in medullary neurones of the cat when administered electrophoretically (Salmoiraghi & Steiner, 1963). These bursts could last up to 1 minute and ended in almost complete inactivation, followed by recovery within 1–3 minutes.

In the spinal cord of cats also, gallamine was found to be a powerful excitant of all spinal interneurones tested, producing high-frequency bursts similar to those elicited in cortical, thalamic and medullary neurones (Curtis, Ryall & Watkins, 1966). Gallamine also excited all Renshaw cells in cat spinal cord, producing high-frequency bursts and facilitating the excitant action of cholinomimetics and amino acids, later causing a depolarization block (Curtis & Ryall, 1966). Finally, gallamine has a direct action on neurones in the inferior olive as discussed in a later section.

Chemical specificity of the actions of *d*-tubocurarine and gallamine

The actions of dTC and gallamine in the peripheral nervous system are mainly, but not exclusively, nicotinic. Gallamine blocks both M1 and M2 muscarinic receptors (reviewed in Bowman, 1982), whereas dTC is considered to have only weak muscarinic affinity (Todrick, 1954). Both dTC and gallamine can block ion channels (reviewed in Bowman, 1982), including glutamate-activated ion channels in invertebrates (Yamamoto & Washio, 1979; Cull-Candy & Miledi, 1983). dTC blocks some serotonin (5-HT) receptors in visceral primary afferent neurones of the rabbit nodose ganglion (Higashi & Nishi, 1982). Gallamine blocks the delayed potassium conductance in amphibian and mammalian peripheral nerve fibres (Smith & Schauf, 1981).

In the CNS the actions of dTC and gallamine appear to be less specific and not fully characterized. Salama & Wright (1952b) injected dTC into the subarachnoid space of chloralose-anaesthetized cats to produce convulsions and then injected other drugs to see whether these would antagonize the convulsive effect of dTC. The following drugs were found to diminish the convulsions: atropine, acetylcholine, neostigmine, eserine, hexaethyltetraphosphate, tetramethylammononium iodine, nicotine, lobeline and potassium chloride. Whereas in some cases the drugs may be acting on the same receptors that react with dTC, in other cases this seems unlikely. There may be many receptors along the chain of structures between the dTC receptors and the motor neurones that are finally excited and antagonism could occur at any of these receptors. Therefore this type of experiment is not useful in characterizing the dTC receptors.

Electrophoretic application of dTC or gallamine onto single cells gives a clearer idea of which receptors are involved. For example, Curtis & Ryall (1966) found that the excitant action of gallamine on Renshaw cells of the cat was not blocked by dihydro-β-erythroidine and was also evident on other non-cholinoceptive interneurones; this clearly indicates a non-nicotinic action of gallamine. Gallamine binds to muscarinic M1 and M2 receptors on rat brain membranes (Burke, 1986). It has been suggested that gallamine may enhance acetylcholine release through blockade of M2 autoreceptors (Messer, Ellerbrock & Bohnett, 1991). These drugs have actions other than on cholinergic receptors. dTC opposes the conductance change produced by gamma-aminobutyric acid (GABA) on single cells of the isolated olfactory cortex (Scholfield, 1982) and in hippocampal neurones (Lebeda *et al.*, 1982). Gallamine potentiates the effect of glutamate on cuneate neurones (Galindo *et al.*, 1968).

Inferior olive

The remainder of this chapter will concentrate on the action of dTC and gallamine on the inferior olive (IO). This may provide an explanation for two important behavioural effects of these substances, namely myoclonic jerks and tremor. The IO is the source of climbing fibres to the cerebellar cortex, with branches to the deep cerebellar nuclei, and thus is able to initiate or modulate muscular activity. The cerebellar cortex, via the axons of the Purkinje cells, communicates with the cerebellar and vestibular nuclei, which in turn send outputs back to the IO. The various parts of the body are topographically represented in the cerebellar cortex in precise parasagittal zones. Each zone together with its associated regions of cerebellar nuclei and IO constitutes an olivo-cerebellar complex, the functional unit of the cerebellum, controlling a particular motor function (reviewed in Andersson *et al.*, 1987).

Many cells in the IO are linked to one another by gap junctions (Sotelo, Llinás & Baker, 1974; King, 1976), a feature that allows this nucleus, or parts of it, to behave as a syncytium. This feature would tend to oppose the precise topography of the cerebellum if it were not restricted or controlled in some way. Llinás (1974) proposed that this control was exerted by an input to the IO which, via inhibitory interneurone terminals, might shunt the gap junction conductance. Angaut & Sotelo (1987) in the rat and de Zeeuw *et al.* (1989) in the cat showed GABAergic terminals in close proximity to gap junctions and de Zeeuw *et al.* (1989) demonstrated that these terminals belonged to axons from the cerebellar nuclei. Such synaptic regions on the cell (or any other non-depolarizing synapse) would permit less current to flow through other gap junctions.

Some physiological evidence to support this idea has been provided by Llinás & Sasaki (1989). It is well known that certain drugs, notably harmaline, act on the IO to produce rhythmical activity in the cells at a frequency identical with the tremor which the drug also produces (Llinás & Volkind, 1973). Llinás & Sasaki (1989) recorded simultaneously from a large number of Purkinje cells and showed that when a few IO neurones were excited the subsequent discharge of Purkinje cells had a rostrocaudal organization and a very tight synchrony. Previous work had shown that this synchrony could be ascribed to the electrotonic interaction between IO neurones. Block of the GABAergic pathways by picrotoxin or bicuculline disrupted the spatial distribution of activity in the cerebellum, leading to widespread activity.

It is important to distinguish between rhythmicity of firing and synchronization (cf. de Zeeuw *et al.*, 1989). Rhythmicity of firing implies a cycle of changes in the membrane conductance of a cell either internally generated or externally imposed. Synchronization is believed to depend on the syncytial properties of the IO cells. Harmaline can generate a rhythm, and therefore a tremor, but the synchronization may be limited and this can be greatly increased by GABA blockers (Llinás & Sasaki, 1989). A myoclonic jerk implies a strong synchronization, not necessarily involving any rhythm. This distinction may be pursued further by looking at the innervation of the IO.

Apart from the GABAergic innervation from the cerebellar nuclei, the IO receives a serotonergic innervation mainly from the raphe nuclei, especially from raphe obscurus and raphe pallidus (Compoint & Buisseret-Delmas, 1988). Some of the fibres also contain substance P and there may also be fibres with substance P alone (Paré, Descarries & Wiklund, 1987). A non-cerebellar GABAergic input to the IO may also contain 5-HT and may originate from the raphe nuclei or from the nucleus reticularis gigantocellularis (NRG) (discussed in de Zeeuw *et al.*, 1989). The GABAergic input to the somata of IO cells seems to be mainly non-cerebellar; this input, therefore, may control rhythmicity whereas the GABAergic input to the dendrites (sites of the gap junctions) may control synchronization (cf. de Zeeuw *et al.*, 1989).

Sjölund, Björklund & Wiklund (1977) showed that harmaline elicited oscillations only in regions of the IO that received a 5-HT innervation and that if this was destroyed (e.g. by 5,6-dihydroxytryptamine) harmaline lost its power. The rhythmic activity of IO cells depends on oscillations of membrane potential and these depend on the presence in the somatic membrane of a calcium conductance; hyperpolarization, such as produced by harmaline, removes inactivation of the calcium conductance leading to low-threshold calcium spikes, leading to sodium spikes and followed by a long-lasting hyperpolarization and refractoriness of the calcium conductance (Llinás &

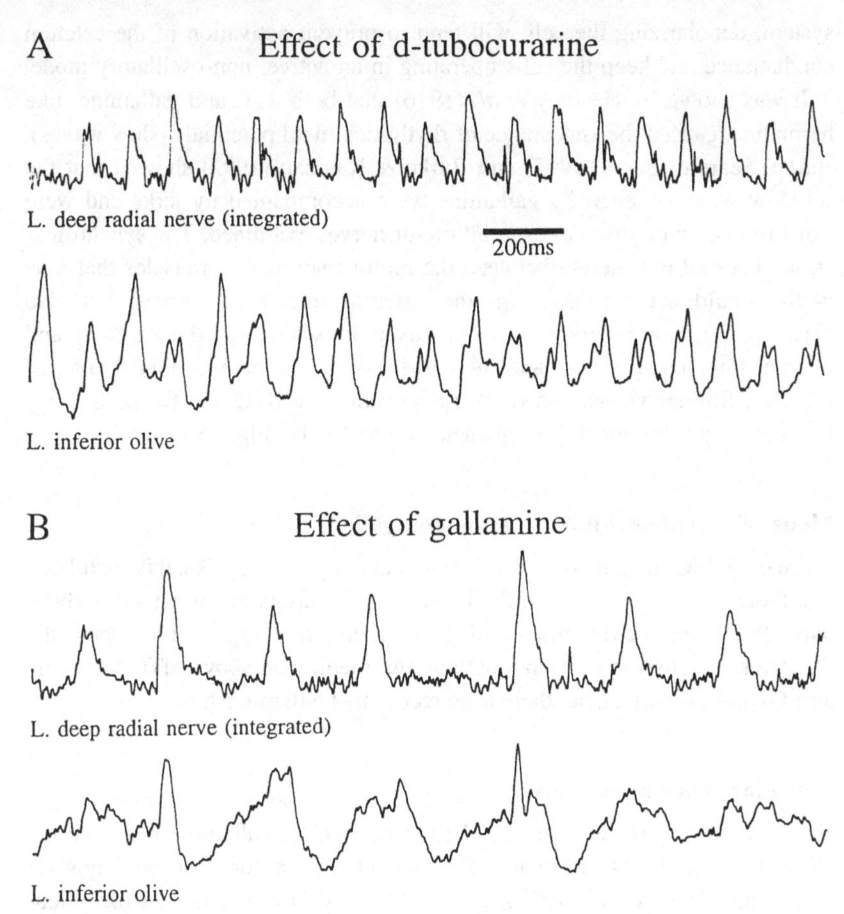

Fig. 25.1. *d*-Tubocurarine and gallamine elicit slow waves in the inferior olive and rhythmical discharges in motor nerves. (A) Two milligrams of *d*-tubocurarine were injected into the cisterna magna of a pentobarbitone-anaesthetized cat 9 hours before the records were taken. (B) Four milligrams of gallamine triethiodide were injected intracisternally about 5 hours after decerebration under a short-lasting anaesthetic; the recordings were made 30 minutes later. Time calibration applies to all records.

Yarom, 1981, 1986). Although synaptic activity can initiate or modify the rhythm, the exact way in which the 5-HT innervation acts is still not clear nor is the way in which harmaline depends on this innervation. 5-HT may hyperpolarize the membrane and there is some evidence that harmaline is acting presynaptically to increase or modify the output of the nerves (Barragan, Delhaye-Bouchaud & Laget, 1985).

The IO also receives a weak catecholaminergic innervation (Sladek & Hoffmann, 1980) and, of course, strong excitatory innervation from many regions, especially from the spino-olivary tract and from the mesodiencephalic junction (e.g. Onodera, 1984; de Zeeuw *et al.*, 1989). Any excitatory system, depolarizing the cell, will tend to prevent activation of the calcium conductance and keep the cells operating in an active, non-oscillatory mode.

It was shown by Headley *et al.* (1976) that both dTC and gallamine, like harmaline, caused the appearance of rhythmical field potentials (slow waves). Burke, Sears & Seers (1995) and Burke & Ramzan (1992) showed that the IO slow waves elicited by gallamine were accompanied by jerks and were synchronized with discharges in all motor nerves examined. The synchronization occurred not merely between the motor neurones of muscles that normally would act together, e.g. the external intercostal nerves, but also between the motor nerves of antagonistic muscles, e.g. the external and internal intercostal nerves, and the nerves to the soleus and tibialis anterior muscles. Similar waves and discharges occur with dTC and for both drugs become very rhythmical at frequencies up to 12 Hz (Fig. 25.1).

Mode of action of *d*-tubocurarine and gallamine

Since both dTC and gallamine elicit tremor and myoclonic jerks, it is postulated that there are two actions. On the IO somata the drugs might act similarly to harmaline to initiate rhythmical firing. Secondly, they might act to oppose the GABAergic action at the gap junctions. As mentioned above, dTC has some anti-GABA properties, but there is no record that gallamine has.

Reticular reflex myoclonus

The myoclonic jerks and tremor elicited by dTC or gallamine resemble the clinical condition of reticular reflex myoclonus, a form of post-hypoxic myoclonus (Hallett *et al.*, 1977), in that the jerks originate in or near the reticular formation of the brainstem and the cat, like the patient, is very stimulus-sensitive, responding with a jerk to quite mild stimuli. Gallamine can elicit slow waves in the IO, not only in the decerebrate cat but also when, in addition, the cerebellum is removed and the cat spinalized at C1 or C2 (Burke, Sears & Seers, 1995). Although dTC and gallamine can elicit rhythmical firing by electrophoretic injection into the IO (Headley *et al.*, 1976), it is possible that the myoclonic jerks might be due to an action elsewhere. It has been shown that the non-epileptic myoclonus in the baboon *Papio papio* is a reticular reflex myoclonus (Rektor *et al.*, 1993) and is cholinergic (muscarinic)-system dependent, with the NRG and the nucleus pontis oralis possibly playing a major role (Rektor *et al.*, 1984, 1986). The NRG, as noted above, may send a 5HT/GABA input to the IO. If some of this input goes to the dendrites (de Zeeuw *et al.*, 1989) then another way in which dTC and gallamine could elicit a myoclonus would be by blocking the cholinergic synapses on the cells of the NRG, thus removing a 5HT/GABA influence on the IO and permitting widespread synchronization.

Both 5-HT and GABA are implicated in various myoclonic conditions but there is evidence for both increased and decreased activity of these systems (Pranzatelli & Snodgrass, 1985). If both systems are concerned with regulating the gap junctions between IO cells it could be that a decreased activity would allow spread of the syncytium (as we are postulating is the action of dTC and gallamine) but it is also conceivable that overactivity might also lead to the same result by desensitization.

Acknowledgement

I thank Vladimir Balcar for advice.

References

Al-Zamil, Z., Bagust, J. & Kerkut, G.A. (1990). Tubocurarine and strychnine block Renshaw cell inhibition in the isolated mammalian spinal cord. *General Pharmacology*, **21**, 499–509.

Andersen, P. & Curtis, D.R. (1964). The pharmacology of the synaptic and acetylcholine-induced excitation of ventrobasal thalamic neurones. *Acta Physiologica Scandanavica*, **61**, 100–120.

Andersson, G., Ekerot, C.F., Oscarsson, O. & Schouenborg, J. (1987). Convergence of afferent paths to olivo-cerebellar complexes. In *Cerebellum and Neuronal Plasticity*, ed. M. Glickstein, C. Yeo & J. Stein, pp. 165–173. New York: Plenum Press.

Angaut, P. & Sotelo, C. (1987). The dentato-olivary projection in the rat as a presumptive GABAergic link in the olivo-cerebellar-olivary loop. An ultrastructural study. *Neuroscience Letters*, **83**, 227–231.

Barragan, L.A., Delhaye-Bouchaud, N. & Laget, P. (1985). Drug-induced activation of the inferior olivary nucleus in young rabbits. *Neuropharmacology*, **24**, 645–654.

Bowman, W. (1982). Non-relaxant properties of neuromuscular blocking drugs. *British Journal of Anaesthetics*, **54**, 147–160.

Burke, R.E. (1986). Gallamine binding to muscarinic M1 and M2 receptors, studied by inhibition of [^{3}H]pirenzepine and [^{3}H]quinuclidinylbenzilate binding to rat brain membranes. *Molecular Pharmacology*, **30**, 58–68.

Burke, W. & Ramzan, I. (1992). Myoclonus in the decerebrate cat produced by gallamine. *Brain Research*, **580**, 189–196.

Burke, W., Sears, T.A. & Seers, C. (1995). Spontaneous synchronized neural activity in decerebrate gallamine-paralysed cats. *Neuroscience*, **68**, 943–953.

Chang, H.-T. (1953). Similarity in action between curare and strychnine on cortical neurons. *Journal of Neurophysiology*, **16**, 221–233.

Compoint, C. & Buisseret-Delmas, C. (1988). Origin, distribution and organization of the serotoninergic innervation in the inferior olivary complex of the rat. *Archives of Italian Biology*, **126**, 99–110.

Crawford, J.M. & Curtis, D.R. (1966). Pharmacological studies on feline Betz cells. *Journal of Physiology (London)*, **186**, 121–138.

Cull-Candy, S.G. & Miledi, R. (1983). Block of glutamate-activated synaptic channels by curare and gallamine. *Proceedings of the Royal Society of London, Series B*, **218**, 111–118.

Curtis, D.R. & Ryall, R.W. (1966). The acetylcholine receptors of Renshaw cells. *Experimental Brain Research*, **2**, 66–80.

Curtis, D.R., Ryall, R.W. & Watkins, J.C. (1966). The action of cholinomimetics on spinal interneurones. *Experimental Brain Research*, **2**, 97–106.
de Zeeuw, C.I., Holstege, J.C., Ruigrok, T.J.H. & Voogd, J. (1989). Ultrastructural study of the GABAergic, cerebellar, and mesodiencephalic innervation of the cat medial accessory olive: anterograde tracing combined with immunocytochemistry. *Journal of Comparative Neurology*, **284**, 12–35.
Feldberg, W. & Fleischhauer, K. (1962). The site of origin of the seizure discharge produced by tubocurarine acting from the cerebral ventricles. *Journal of Physiology (London)*, **160**, 258–283.
Feldberg, W. & Fleischhauer, K. (1963). Site of action of tubocurarine reaching the brain via the cerebral ventricles. *Progress in Brain Research*, **3**, 1–19.
Forchetti, C., Scarnati, E., Pacitti, C. & Agnoli, A. (1980). Striatal cholinergic receptors and dyskinetic motor activity in the rat. *Neuroscience Letters*, **20**, 363–367.
Galindo, A., Krnjević, K. & Schwartz, S. (1968). Patterns of firing in cuneate neurones and some effects of Flaxedil. *Experimental Brain Research*, **5**, 87–101.
Hallett, M., Chadwick, D., Adam, J. & Marsden, C.D. (1977). Reticular reflex myoclonus: a physiological type of human post-hypoxic myoclonus. *Journal of Neurology, Neurosurgery and Psychiatry*, **40**, 253–264.
Headley, P.M., Lodge, D. & Duggan, A.W. (1976). Drug-induced rhythmical activity in the inferior olivary complex of the rat. *Brain Research*, **101**, 461–478.
Higashi, H. & Nishi, S. (1982). 5-Hydroxytryptamine receptors of visceral primary afferent neurones on rabbit nodose ganglia. *Journal of Physiology (London)*, **323**, 543–567.
King, J.S. (1976). The synaptic cluster (glomerulus) in the inferior olivary nucleus. *Journal of Comparative Neurology*, **165**, 387–400.
Lebeda, F.J., Hablitz, J.J. & Johnston, D. (1982). Antagonism of GABA-mediated responses by d-tubocurarine in hippocampal neurons. *Journal of Neurophysiology*, **48**, 622–632.
Llinás, R. (1974). Motor aspects of cerebellar control. Eighteenth Bowditch Lecture. *The Physiologist*, **17**, 19–46.
Llinás, R. & Sasaki, K. (1989). The functional organization of the olivo-cerebellar system as examined by multiple Purkinje cell recordings. *European Journal of Neuroscience*, **1**, 587–602.
Llinás, R. & Volkind, R.A. (1973). The olivo-cerebellar system: functional properties as revealed by harmaline-induced tremor. *Experimental Brain Research*, **18**, 69–87.
Llinás, R. & Yarom, Y. (1981). Electrophysiology of mammalian inferior olivary neurones *in vitro. Different types of voltage-dependent ionic conductances. Journal of Physiology (London)*, **315**, 549–567.
Llinás, R. & Yarom, Y. (1986). Oscillatory properties of guinea-pig inferior olivary neurones and their pharmacological modulation: an *in vitro* study. *Journal of Physiology (London)*, **376**, 163–182.
McIntyre, A.R. (1947). *Curare, Its History, Nature and Clinical Use*. Chicago: University of Chicago.
McKenzie, G.M., Gordon, R.J. & Viik, K. (1972). Some biochemical and behavioural correlates of a possible animal model of human hyperkinetic syndromes. *Brain Research*, **47**, 439–456.
Messer, W.S. Jr, Ellerbrock, B.R. & Bohnett, M. (1991). Regulation of muscarinic receptors by intrahippocampal injections of gallamine. *Brain Research*, **564**, 73–78.
Onodera, S. (1984). Olivary projections from the mesencephalic structures in the cat

studied by means of axonal transport of horseradish peroxidase and tritiated amino acids. *Journal of Comparative Neurology*, **227**, 37–49.
Paré, M., Descarries, L. & Wiklund, L. (1987). Innervation and reinnervation of rat inferior olive by neurones containing serotonin and substance P: an immunohistochemical study after 5,6-dihydroxytryptamine lesioning. *Journal of Neurocytology*, **16**, 155–167.
Pranzatelli, M.R. & Snodgrass, S.R. (1985). The pharmacology of myoclonus. *Clinical Neuropharmacology*, **8**, 99–130.
Rektor, I., Bryere, P., Valin, A., Silva-Barrat, C., Naquet, R. & Menini, C. (1984). Physostigmine antagonizes benzodiazepine-induced myoclonus in the baboon, *Papio papio*. *Neuroscience Letters*, **52**, 91–96.
Rektor, I., Bryere, P., Silva-Barrat, C. & Menini, C. (1986). Stimulus-sensitive myoclonus of the baboon *Papio papio*: pharmacological studies reveal interactions between benzodiazepines and the central cholinergic system. *Experimental Neurology*, **91**, 13–22.
Rektor, I., Švejdová, M., Silva-Barrat, C. & Menini, C. (1993). The cholinergic system-dependent myoclonus of the baboon *Papio papio* is a reticular reflex myoclonus. *Movement Disorders*, **8**, 28–32.
Salama, S. & Wright, S. (1950). Action of *d*-tubocurarine chloride on the central nervous system of the cat. *British Journal of Pharmacology*, **5**, 49–61.
Salama, S. & Wright, S. (1952a). Action on central nervous system of compounds RP 3565 and RP 3697 and of tetramethylammonium and tetraethylammonium. *British Journal of Pharmacology*, **7**, 1–13.
Salama, S. & Wright, S. (1952b). Influence of various drugs on the action of curare on the central nervous system of the cat. *British Journal of Pharmacology*, **7**, 14–24.
Salmoiraghi, G.C. & Steiner, F.A. (1963). Acetylcholine sensitivity of cat's medullary neurons. *Journal of Neurophysiology*, **26**, 581–597.
Scholfield, C.N. (1982). Antagonism of γ-aminobutyric acid and muscimol by picrotoxin, bemegride, leptazol, D-tubocurarine and theophylline in the isolated olfactory cortex. *Naunyn Schmiedebergs Archives of Pharmacology*, **318**, 274–280.
Sears, T.A., Seers, C. & Burke, W. (1987). Synchronization of neural activity in the cat following decerebration (abstract). *Neuroscience Letters, Supplement*, **27**, S121.
Sjölund, B., Björklund, A. & Wiklund, L. (1977). The indolaminergic innervation of the inferior olive. 2. Relation to harmaline induced tremor. *Brain Research*, **131**, 23–37.
Sladek, J.R. & Hoffman, G.E. (1980). Monoaminergic innervation of the mammalian inferior olivary complex. In *The Inferior Olivary Nucleus: Anatomy and Physiology*, ed. J. Courville, C. de Montigny & Y. Lamarre, pp. 145–162. New York: Raven Press.
Smith, K.J. & Schauf, C.L. (1981). Effects of gallamine triethiodide on membrane currents in amphibian and mammalian peripheral nerve. *Journal of Pharmacology and Experimental Therapeutics*, **217**, 719–726.
Smith, S.M., Brown, H.O., Tolman, J.E.P. & Goodman, L.S. (1947). The lack of cerebral effects of d-tubocurarine. *Anesthesiology*, **8**, 1–14.
Sotelo, C., Llinás, R. & Baker, R. (1974). Structural study of inferior olivary nucleus of the cat: morphological correlates of electrotonic coupling. *Journal of Neurophysiology*, **37**, 541–559.
Todrick, A. (1954). The inhibition of cholinesterases by antagonists of acetylcholine and histamine. *British Journal of Pharmacology*, **9**, 76–83.
Yamamoto, D. & Washio, H. (1979). Curare has a voltage-dependent blocking action on the glutamate synapse. *Nature*, **281**, 372–373.

26

Pathophysiology of upper motoneurone disorders

HANS-JOACHIM FREUND

Department of Neurology, University of Düsseldorf, Düsseldorf, Germany

The pyramidal syndrome

The hallmark of lesions of the motor cortex or its descending fibres in the human is the pyramidal syndrome. It is characterized by deficient force generation and impairment of selective control of the fingers. Recovery is better in proximal than in distal muscles, which almost regain their former capacity, even in cases with complete pyramidal tract (PT) transection. Within the remaining range, force control is good as is sensory guidance. There is no disturbance of movement initiation, preparation or of specifications such as movement direction, and there are no apractic phenomena.

It is an old controversy whether the increase in tone and reflexes, which is often regarded as part of the pyramidal syndrome, is due to the damage of motor cortex and its descending fibres or whether it indicates the impairment of more rostral areas. Ablation studies in monkeys and apes showed that excisions restricted to area 4 led to an initially flaccid paresis, later resolving into a state of approximately normal tonus. Hines (1929) found spasticity only after excisions involving the anterior part of area 4 and called this 'suppressor strip' or area 4s. Electrical stimulation of this strip yielded muscle relaxation. For the mediation of this inhibitory action McCulloch, Graf & Magoun (1946) described a separate non-pyramidal cortico-reticular pathway projecting from area 4s to the brainstem. Although such an inhibitory zone has not been confirmed by subsequent studies, other authors have associated spasticity with more anterior lesions centring on the superior precentral sulcus in the premotor area of the macaque cortex (Fulton & Kennard, 1934; Denny-Brown & Botterell, 1948; Woolsey *et al.*, 1952). However, this issue remained controversial because it was also observed that spasticity may occur after lesions of motor cortex in apes and humans (for review see Freund, 1987).

The major difficulty in the definition of the pyramidal syndrome is the ambiguity of the term 'pyramidal'. Clinically, pyramidal function is usually assigned to 'primary motor cortex' and thereby distinguished from extrapyramidal components. Anatomically the attribution of 'pyramidal' solely to motor cortex is not correct, because the pyramidal tract is classically defined as consisting of those fibres constituting the pyramid at the level of the medulla oblongata. Only a proportion of these fibres originates in area 4, whereas most axons come from other areas such as the premotor and parietal cortex. Thus, the term 'pyramidal syndrome' is used in a dual way, to indicate either PT damage or selective impairment of primary motor cortex and/or its descending fibres.

The ambiguities in the description of the clinical facets of the pyramidal syndrome in humans further illustrate the complexity of what is referred to as 'pyramidal'. The spectrum of functional disturbances covers a broad range: weakness, difficulty in selecting certain muscles for action, temporal deordering of muscle activation, loss of antagonistic inhibition, release of flexor reflexes, increase in muscle tone, inadequate tactile placing difficulties in postural control and impairment of manual skill.

What is primary motor cortex in the human?

The present definition of the extent of the primary motor cortex in the monkey is that it is coextensive with area 4 as outlined in the upper part of Fig. 26.1 as M1. M1 occupies a substantial part of the precentral gyrus. The dorsal (PMd) and ventral (PMv) components of premotor cortex cover the frontal part of the precentral gyrus. Both primary and premotor cortex are confined to the precentral gyrus and the bank of the central and arcuate sulci. In the human, area 4 is less represented on the gyral surface, in particular below the level of the junction of the superior frontal sulcus with the precentral sulcus (Fig. 26.1, below). In terms of somatotopic representation, this junction corresponds approximately to the region from which movements around the elbow can be elicited. The hand area and the mouth area have a much smaller representation on the gyral surface than the dorsal parts, where more rostral movements of the arm, trunk and leg are represented.

This observation, based on one brain in Brodmann's cytoarchitectural map, was further extended by a recent study of Rademacher *et al.* (1993). Their findings are shown in Fig. 26.2. The cytoarchitectonic maps of six human brain specimens illustrate that area 4 has virtually no representation on the gyral surface below the level of the junction of the superior frontal and precentral sulci. Consequently, the hand and mouth representations of area 4 are largely located in the sulcus. Their magnification factor seems surprisingly small, in contrast to the large representation shown by functional methods such as brain stimulation. When these stimulus effects are superimposed on the precentral gyrus it appears that the excitable motor cortex extends more rostrally than area 4. This raises the question whether the precentral

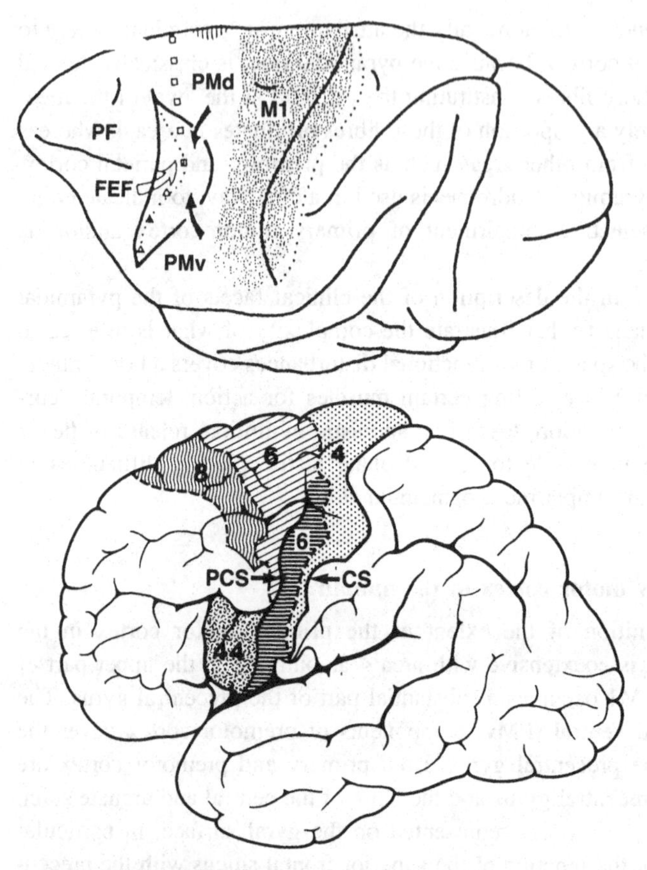

Fig. 26.1. Comparison of the motor areas in the human and monkey brain. Upper: Surface view of the monkey's cerebral cortex according to Wise *et al.* (1991). M1, primary motor cortex corresponding to area 4; PMd, dorsal premotor cortex; PMv, ventral premotor cortex; FEF, frontal eye field; PF, prefrontal cortex. Dotted lines represent the fundus of the arcuate (rostral) and central sulcus. Right: Distribution of Brodmann's agranular and dysgranular cytoarchitectonic areas 4, 6, 8 and 44 on the precentral and frontal gyri. Different shading patterns designate the different cytoarchitectronic areas and subdivide area 6 into a ventral compartment that lies on the precentral gyrus (PMv) and a dorsal part (PMd) covering the superior and middle frontal gyri. Dashed lines indicate borders between cytoarchitectronic areas. Lower part according to Freund (1991).

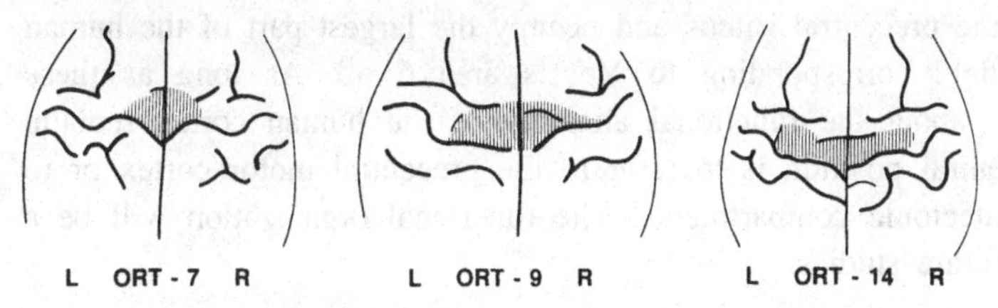

Fig. 26.2. Topography of lateral Brodmann area 4: individual variations of the surface extent of area 4 (hatching) on the lateral cerebral surface in three left (L) and three right (R) hemispheres (view from above, anterior is at the top). Modified from Rademacher *et al.* (1993) with permission.

compartment of area 6 which was called area 6 aα by Vogt is in the human part of primary motor cortex, or of premotor cortex, thus corresponding to area PMv in the monkey.

The premotor areas

The definition of premotor cortex (PM) by Dum & Strick (1989) for the macaque is that it represents that region in the frontal lobe which has direct projections to primary motor cortex (area 4) and also has direct cortico-spinal projections. For the human, Fulton (1935) delineated the premotor area as the frontal agranular cortex (area 6) rostral to motor cortex. This area shows particular enlargement in humans, where it is 5 times larger than area 4 as compared with a 1 : 1 relationship in the macaque (Bailey & von Bonin, 1951). Fig. 26.1 shows the distribution of Brodmann's frontal agranular and dysgranular cytoarchitectonic areas 6, 8 and 44 on the frontal gyri. If the human inferior precentral sulcus represents the homologue of the arcuate sulcus in the monkey, area PMv in the monkey may be the homologue of the Vogt's area 6 aα, the precentral part of area 6 of the human, as indicated in Fig. 26.1. Unfortunately there are no recent data about the connectivity of areas 4 and 6 in the human. There is also no evidence from human studies that lesions restricted to area 6 aα not affecting area 4 would have similar or different effects from lesions restricted to area 4. The use of transcranial magnetic stimulation has not clarified this issue, and, to date, there have been no positron emission tomography activation studies that could resolve the problem.

Thus it is presently unclear whether the human area 6 aα corresponds to PMv in the monkey and whether it should be considered as premotor cortex or as primary motor cortex. If the former view is taken, this would imply that the primary motor cortex would have almost no gyral representation for the hand and mouth area.

For our studies on premotor cortex we have kept to the original Fulton concept, referring to premotor cortex only for those lesions that are located anterior to the precentral sulcus and occupy the largest part of the human area 6, a field corresponding to Vogt's area 6 aβ. As long as these uncertainties about the functional anatomy of the human cortex remain, the most neutral position is to refer to the precentral motor cortex or to its cytoarchitectonic compartments. The functional organization will be a matter for future studies.

Functional recovery

The functional parcellation of motor cortex and its relation to other areas contributing to the pyramidal tract is of particular clinical relevance for the issue of functional recovery after damage to precentral cortex or to the pyramidal tract. There is firm evidence that damage to the precentral gyrus and damage to the pyramidal tract can both be almost fully compensated (Förster, 1935; Penfield & Rasmussen, 1950; Reulen *et al.*, 1992; Kunesch *et al.*, 1995; Bucy, Ladpli & Ehrlich, 1966; Nathan & Smith, 1973). As shown by cases with hemispherectomy (Müller *et al.*, 1991; Benecke, Meyer & Freund, 1990) and cordotomies (Nathan & Smith, 1973), ipsilateral pathways can mediate functional recovery. Parietal and premotor output also contribute to functional recovery. Sasaki & Gemba (1984) have shown that there is almost full functional recovery after cooling of motor cortex in the monkey. When the postcentral cortex is then cooled in addition, the formerly restored function vanishes completely without subsequent recovery.

Lesion analysis in the human shows that recovery of elementary motor functions is not related to the location and size of lesions in a simple way. Gross damage such as hemispherectomy, or transection of pyramidal tract at the level of the peduncles or the spinal cord may be compensated to an astonishing degree, whereas small lacunar infarcts with only partial damage to the pyramidal tract are followed by complete hemiplegia without any subsequent recovery even in younger patients (Freund, 1991). Consequently, the relationship between damage of the 'pyramidal system' and the ensuing functional deficit does not follow the peripheral nerve model: the more fibres or motor neurones that are damaged, the larger the deficit. Even for the precentral/pyramidal output module and the most elementary aspect of motor function, generation of force, the structure–function relationship is complex. It varies between full and no recovery after partial damage of the pyramidal tract. Therefore the pathophysiology of the upper motor neurone cannot be reduced to a consideration of the disturbed precentral output. Rather it reflects complex interactions in a distributed network with various ipsi- and contralateral components. Only when these interactions are gated properly, can the destruction of the precentral gyrus and its fibres be compensated.

References

Bailey, P.A. & von Bonin, G. (1951). *The Isocortex of Man*. Urbana: University of Illinois Press.

Benecke, R., Meyer, B.U. & Freund, H.-J. (1990). Reorganisation of descending motor pathways in patients after hemispherectomy and severe hemispheric lesions demonstrated by magnetic brain stimulation. *Experimental Brain Research*, **83**, 419–426.

Bucy, P.C., Ladpli, R. & Ehrlich, A. (1966). Destruction of the pyramidal tract in the monkey. *Journal of Neurosurgery*, **25**, 1–23.

Denny-Brown, D. & Botterell, E.H. (1948). The motor functions of the agranular frontal cortex. *Research Publications of the Association for Research into Nervous and Mental Diseases*, **27**, 235–345.

Dum, R.P. & Strick, P.L. (1991). Premotor areas: nodal points for parallel efferent systems involved in the central control of movement. In *Motor Control: Concepts and Issues*, ed. D.R. Humphrey & H.-J. Freund, pp. 383–397. Dahlem Konferenzen. New York: Wiley.

Förster, O. (1936). Motorische Felder und Bahnen. In *Handbuch der Neurologie*, vol. 6, ed. by H. Bumke & O. Förster. Berlin: Springer.

Freund, H.-J. (1987). Abnormalities of motor behavior after cortical lesions in man. In *Handbook of Physiology*, sect. 1, *The Nervous System*, ed. V.B. Mountcastle, vol. 5, *Higher Functions of the Brain*, part 2, ed. F. Plum, pp. 763–810. Baltimore: Williams & Wilkins.

Freund, H.-J. (1991). What is the evidence for multiple motor areas in the human brain? In *Motor Controls: Concepts and Issues*, ed. D.R. Humphrey & H.-J. Freund, pp. 399–411. Dahlem Konferenzen. New York: Wiley.

Fulton, J.F. (1935). A note on the definition of the 'motor' and 'premotor' areas. *Brain*, **58**, 311–316.

Fulton, J.F. & Kennard, M.A. (1934). A study of flaccid and spastic paralyses produced by lesions of the cerebral cortex in primates. *Research Publications of the Association for Research into Nervous and Mental Diseases*, **13**, 158–210.

Hines, M. (1929). On cerebral localization. *Physiological Reviews*, **9**, 462–574.

Kennard, M.A., Viets, H.R. & Fulton, J.F. (1934). The syndrome of the premotor cortex in man: impairment in skilled movements, forced grasping, spasticity and vasomotor disturbance. *Brain* **57**, 69–84.

Kunesch, E., Binkofski, F., Steinmetz, H. & Freund, H.-J. (1995). The pattern of motor deficits in relation to the site of stroke lesions. *European Neurology*, **35**, 20–26.

McCulloch, W.S., Graf, C. & Magoun, H.W. (1946). A cortico-bulbo-reticular pathway from area 4-s. *Journal of Neurophysiology*, **9**, 127–132.

Müller, F., Kunesch, E., Binkofski, F. & Freund, H.-J. (1991). Residual sensorimotor functions in a patient after right-sided hemispherectomy. *Neuropsychologia*, **29**, 125–145.

Nathan, P.W. & Smith, M.C. (1973). Effects of two unilateral cordotomies on the lower limbs. *Brain*, **96**, 471–494.

Penfield, W. & Rasmussen, T. (1950). *The Cerebral Cortex of Man*. New York: Macmillan.

Rademacher, J., Caviness, V.S., Steinmetz, H. & Galaburda, A.M. (1993). Topographical variation of the human primary cortices: implications for neuroimaging, brain mapping and neurobiology. *Cerebral Cortex*, **3**, 313–329.

Reulen, J.H., Ebling, U., Schmidt, U.D. & Ying, H. (1992). Safe surgery of lesions near the motor cortex using intra-operative mapping techniques: a report on 50 patients. *Acta Neurochirurgica*, **119**, 23–28.

Sasaki, K. & Gemba, H. (1984). Compensatory motor function of the somatosensory cortex for the motor cortex temporarily impaired by cooling in the monkey. *Experimental Brain Research*, **55**, 60–68.

Wise, S.P., Alexander, G.E., Altmann, J.S., Brooks, V.B., Freund, H.-J, Fromm, C.J., Humphrey, D.R., Sasaki, K., Strick, P.L., Tanji, J., Vogel, S. & Wiesendanger, M. (1991). What are the specific functions of the different motor areas? In *Motor Control: Concepts and Issues*, ed. D.R. Humphrey & H.-J. Freund, pp. 463–485. Dahlem Konferenzen. New York: Wiley.

Woolsey, C.N., Settlage, P.H., Meyer, D.R., Spencer, W., Hamuy, P. & Travis, A.M. (1952). Patterns of localization in precentral and 'supplementary' motor areas and their relation to the concept of a premotor area. *Research Publications of the Association for Research into Nervous and Mental Diseases*, **30**, 238–264.

27

Modulation of hypoglossal motoneurones by thyrotropin-releasing hormone and serotonin

ALBERT J. BERGER, DOUGLAS A. BAYLISS AND FÉLIX VIANA

Department of Physiology and Biophysics, University of Washington School of Medicine, Seattle, Washington, USA.

Motoneurones form part of the final common pathway in the control of the musculature. Two modes of neurotransmitter action on neurones are distinguishable: neurotransmitters can gate ion channels directly or induce alterations in ion channel behaviour indirectly by way of second-messenger systems (Hille, 1992). In the former case the action is characterized by rapid onset and brief duration (milliseconds) whereas in the latter case the effects occur over longer time spans (seconds to minutes). In recent years it has become apparent that alterations in specific ionic conductances by neurotransmitters acting via second-messenger systems can modify motoneuronal behaviour.

Two neurotransmitters that produce their action primarily through second-messenger systems are the three-amino-acid peptide thyrotropin-releasing hormone (TRH), and the indolamine serotonin (5-HT). TRH and 5-HT are of interest because the source of these transmitters to various motoneurone pools appears to be the caudal raphe nuclei in the medulla oblongata, including raphe pallidus and obscurus (Holstege & Kuypers, 1987). Along with substance P (SP) these transmitters can be found co-localized within the same raphe neurone (Johansson *et al.*, 1981). Neurones of raphe pallidus and obscurus alter their activity dramatically in different behavioural states (Jacobs & Azmitia, 1992), highest levels of activity occurring during waking and lowest levels during rapid-eye-movement (REM) sleep. It is therefore possible that TRH-, 5-HT- and SP-related inputs to motoneurones may vary with each state. An understanding of how these neurotransmitters affect motoneurones may provide new insights into changes in motor output that occur during various states (e.g. muscle atonia and reduced muscle reflexes in REM sleep).

The purpose of this chapter is to review our recent work on the actions of TRH and 5-HT on motoneurones. We focus primarily on data obtained from *in vitro* slice experiments on hypoglossal motoneurones (HMs) because they comprise an important class of respiratory motoneurones that control upper airway patency, swallowing, speech, etc. The effects of these

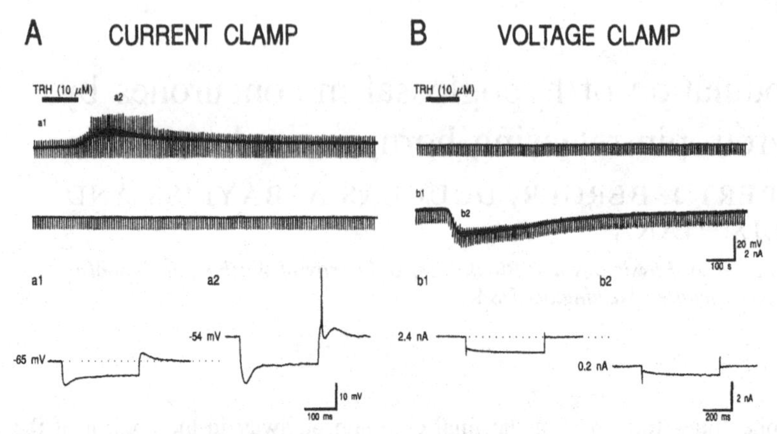

Fig. 27.1. Thyrotropin-releasing hormone (TRH) causes depolarization and/or development of an inward current in hypoglossal motoneurones (HMs). TRH (10 μM) was applied in the perfusate to two HMs. Records of membrane potential (upper panels, top traces) and current (upper panels, bottom traces) show the time course of the response to and recovery from TRH. Responses to constant-amplitude, hyperpolarizing current (A, 1.0 nA) and voltage (B, 10 mV) steps are shown on a faster time base (lower panels). (A) TRH caused a depolarization to firing threshold in this HM. Near the peak of the response, action potentials (truncated) occurred on the rebound from the hyperpolarizing pulses. The voltage response to the current pulse was smaller during the control period (a1) than at the peak of the TRH-induced depolarization (a2), indicating that TRH increased the input resistance of the cell. (B) From a holding potential of –59 mV, TRH caused the development of an inward current. Instantaneous current responses to the voltage steps were larger during the control period (b1) than at the peak of the TRH-induced inward current, indicating that TRH decreased the input conductance of the cell. From Bayliss *et al.* (1992) with permission of the American Physiological Society.

neurotransmitters on ionic mechanisms and thereby on changes in firing behaviour are of particular importance.

Effects of TRH

These experiments were performed in thick (400 μm) transverse slices taken from the adult rat medulla oblongata (Bayliss, Viana & Berger, 1992). HMs were impaled with conventional intracellular microelectrodes and studied under current- and voltage-clamp recording conditions. TRH caused depolarization and/or the development of an inward current (I_{TRH}), whether applied in the perfusate (Fig. 27.1) or by focal application. At the peak of the response to TRH there was an increase in input resistance (Fig. 27.1A), seen during voltage clamp as a decrease in input conductance (Fig. 27.1B). A possible explanation of these results is that TRH acts on HMs via reduction of a resting K^+ conductance. Several lines of experimental evidence indicated that indeed this was the case (Bayliss *et al.*, 1992). Further, TRH effects

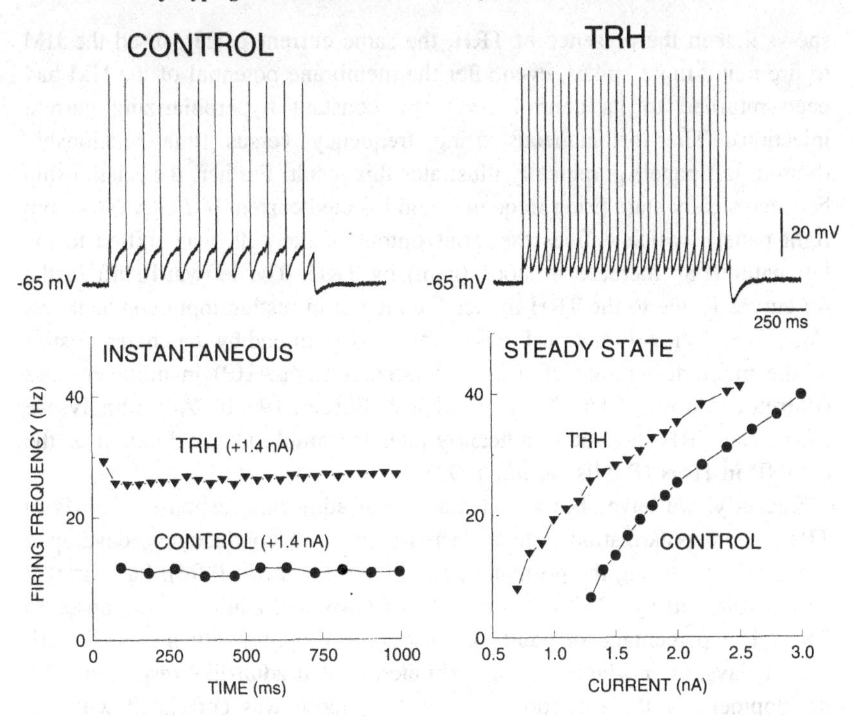

Fig. 27.2. TRH shifts the input–output relationship of HMs. The spike-firing responses to 1.4 nA depolarizing current pulses (1 s) were recorded in control conditions (top left) and then while the slice was bathed in 10 μM TRH (top right). After TRH had caused the cell to depolarize, DC current was injected to return the HM to the same membrane potential as in the control period. In the presence of TRH, the same magnitude current pulse from the same membrane potential caused the HM to discharge many more action potentials. This is illustrated graphically (bottom left) as the instantaneous firing frequency (calculated from the interspike intervals) being higher throughout the full duration of the pulse. Steady-state firing frequency (average of the instantaneous firing frequencies over the final 250 ms) was determined from a family of depolarizing current pulses in control and in TRH and plotted with respect to injected current (*f–I* curve, bottom right). The *f–I* relationship was shifted to the left by TRH, indicating that smaller current inputs were necessary to reach the threshold for minimal repetitive firing. TRH had no effect on the slope of the *f–I* curve. From Bayliss *et al.* (1992), with permission of the American Physiological Society.

were dose-dependent and reversible; long applications showed only a modest desensitization (Bayliss *et al.*, 1992). Other studies also have shown that TRH directly excites mammalian spinal and cranial motoneurones (Takahashi, 1985; Nistri, Fisher & Gurnell, 1990; Rekling, 1990; Wang & Dun, 1990b; Fisher & Nistri, 1993).

We also investigated the effects of TRH on the input–output behaviour of adult HMs (Bayliss *et al.*, 1992). We injected depolarizing current pulses (the input) and recorded the firing behaviour of HMs (the output) in control conditions and at the peak of the TRH response. Fig. 27.2 (upper panels) shows that in the presence of TRH, the same current pulse caused the HM to fire many more spikes, even after the membrane potential of the HM had been returned to its control level (by constant hyperpolarizing current injection). The instantaneous firing frequency versus time relationship (bottom left panel) graphically illustrates this result. Further, the relationship between steady-state firing frequency and injected current (*f–I* curve) (bottom right panel), which defines the input–output of the cell, was shifted to the left without an increase in slope (gain) by TRH. The leftward shift in the *f–I* curves is due to the TRH-induced reduction in resting input conductance. The slope of the *f–I* relationship is primarily determined by the characteristics of the medium-duration after-hyperpolarization (m-AHP) in motoneurones (Barrett & Barrett, 1976; Viana, Bayliss & Berger, 1993b). Accordingly, we found that TRH did not significantly alter the amplitude or duration of the m-AHP in HMs (Bayliss *et al.*, 1992).

Recently, we have shown that the typical adult-like response of HMs to TRH, i.e. depolarization with a decrease in input conductance, developed progressively during the postnatal period (Bayliss *et al.*, 1994). Early in this period (birth to age 2 days) only 25% of HMs had adult-like responses to TRH. The percentage of adult-like responses progressively increased until age 11 days, after which all cells exhibited typical adult-like responses. The development of the electrophysiological response was correlated with the increasing TRH mRNA expression in raphe neurones, and increasing density of TRH-immunoreactive fibres and TRH receptor binding in the hypoglossal nucleus.

Effects of serotonin

Serotonin directly excites spinal (Takahashi & Berger, 1990; Wang & Dun, 1990a; Lindsay & Feldman, 1993; Ziskind-Conhaim, Seebach & Gao, 1993) and cranial (facial and hypoglossal) motoneurones (Aghajanian & Rasmussen, 1989; Berger, Bayliss & Viana, 1992; Larkman & Kelly, 1992), through effects on several ionic conductances, including: enhancement of the mixed cationic (Na^+/K^+), inwardly rectifying current, I_h (Takahashi & Berger, 1990; Larkman & Kelly, 1992); enhancement of a low-voltage-activated Ca^{2+} current (Berger & Takahashi, 1990); and decrease of a resting K^+ conductance (Wang & Dun, 1990a; Larkman & Kelly, 1992).

Two *in vitro* slice preparations were used to study the effects of 5-HT on neonatal rat HMs (3–9 days old) (Berger *et al.*, 1992). In one, thin slices of the medulla (approximately 130 μm thick) were used to obtain tight-seal whole-cell recordings from visualized HMs with patch-type electrodes. In the second, thick slices of the medulla (400–500 μm) were prepared and conventional current-clamp recordings were made with sharp microelectrodes.

5-HT caused a reversible depolarization and induced spike-firing in HMs

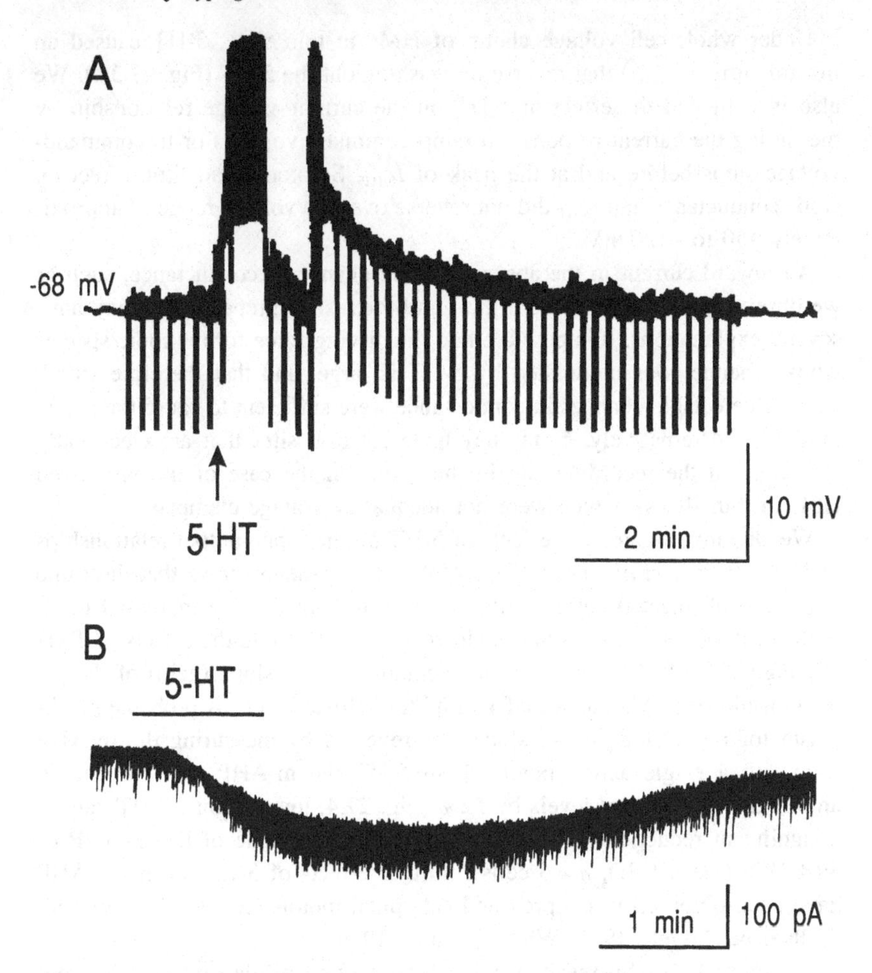

Fig. 27.3. Serotonin (5-HT) excites HMs *in vitro*. (A) 5-HT (100 μM via pressure-ejection from a broken-tip pipette) caused a reversible depolarization of HM recorded in a thick-slice preparation. Depolarization induced action potentials. Repolarization of the membrane during depolarization revealed that 5-HT caused little change in input resistance. Negative deflections of membrane potential are in response to constant amplitude hyperpolarizing current injections (0.3 nA). (B) 5-HT (10 μM) in the perfusate caused a reversible inward current during whole-cell recording of an HM in thin-slice preparation. The cell was held at −77 mV. The small, transient, inward current deflections probably reflect spontaneous synaptic currents. From Berger *et al.* (1992) with permission of Elsevier Scientific Publishers Ireland Ltd.

in the thick slice (Fig. 27.3A). At the peak of the 5-HT response, when the membrane potential was returned to its control level (−68 mV) by injecting hyperpolarizing current, the response to hyperpolarizing current pulses was similar to that observed during control conditions. These data indicate that in neonatal HMs, 5-HT had no apparent effect on input resistance.

Under whole-cell voltage clamp of HMs in thin slice, 5-HT caused an inward current ($I_{5\text{-HT}}$) that reverse on washing out the 5-HT (Fig. 27.3B). We also investigated the effect of 5-HT on the current–voltage relationship by measuring the current response to ramp command voltages or to command-voltage steps before and at the peak of $I_{5\text{-HT}}$. Serotonin had little effect on slope conductance and $I_{5\text{-HT}}$ did not reverse over the voltage range of approximately −50 to −120 mV.

An inward current in the absence of any change in conductance, such as we observed with 5-HT in both thin- and thick-slice preparations, may have several explanations. It is possible that the driving force for the ionic species whose channel was modulated by 5-HT is large, and that therefore small, yet non-detectable changes in conductance were sufficient to cause the measured $I_{5\text{-HT}}$. Alternatively, 5-HT may have acted at sites that are electrically remote from the recording site in the soma (in the case of the visualized HMs in thin slices), which were not adequately voltage clamped.

We also investigated the effects of 5-HT on the input–output relationships of HMs (Berger *et al.*, 1992). Fig. 27.4 (upper panels) shows that the same amplitude of injected current pulse resulted in considerably increased firing at the peak of the 5-HT response. However, in contrast to the effects of TRH (cf. Fig. 27.2), 5-HT caused an augmentation in the slope (gain) of the *f–I* relationship (Fig. 27.4, lower left) with little leftward shift. A probable mechanism for this change in *f–I* slope was revealed by measuring the m-AHP following a single action potential; in 5-HT the m-AHP was reduced in amplitude from control levels by 52% (Fig. 27.4, lower right). 5-HT caused a significant reduction ($P < 0.01$) in the mean amplitude of the m-AHP of $49 \pm 19\%$ (mean ± SD, $n = 6$ cells). Similar effects of 5-HT on the m-AHP have been observed in lamprey and cat spinal motoneurones (Van Dongen, Grillner & Hökfelt, 1986; White & Fung, 1989).

The m-AHP in HMs results from activation of a Ca^{2+}-dependent K^+ current that is blocked by application of the bee venom apamin (Viana *et al.*, 1993b). Blockade of the N-type Ca^{2+} channel with ω-conotoxin (Viana, Bayliss & Berger, 1993a) also results in a marked reduction in the m-AHP (Viana *et al.*, 1993b). Future experiments in neonatal HMs are required to determine whether 5-HT blocks the m-AHP indirectly by acting on Ca^{2+} channels responsible for the Ca^{2+}-activated K^+ current, or directly by inhibiting the Ca^{2+}-activated K^+ channel itself.

It appears that the response of HMs to 5-HT is also developmentally regulated. We have observed in preliminary experiments, using the thick-slice preparation, that in contrast to neonatal rat HMs, the resultant depolarization of adult rat HMs to 5-HT causes an increase in input resistance. Further, adult rat HMs do not exhibit a reduction in the m-AHP amplitude in response to 5-HT, but the *f–I* relationship does show a leftward shift. These two responses to 5-HT in adult HMs are similar to those described above to TRH.

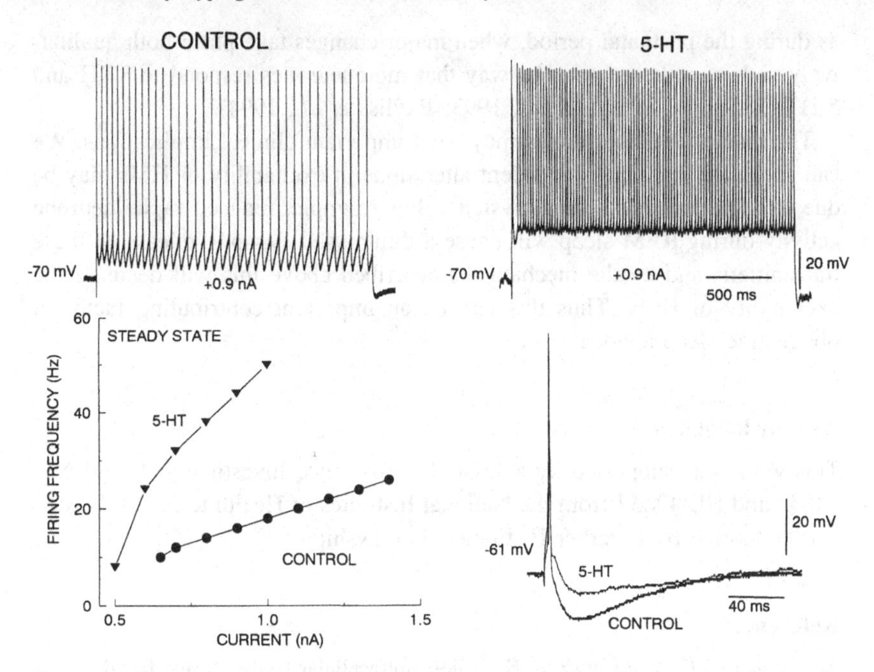

Fig. 27.4. Serotonin increases the firing response to injected current pulses by, in part, reducing the medium-duration after-hyperpolarization (m-AHP). Upper traces: Responses to depolarizing injected current pulses during control and during application of 5-HT (100 μM via bath), with the starting membrane potential returned to the same value as during control conditions by injection of constant hyperpolarizing current. Lower left: *f–I* plot in 5-HT and during control conditions revealed an increase in slope in the presence of 5-HT. Lower right: 5-HT caused a decrease in m-AHP amplitude. A single action potential was generated from same membrane potential with a brief pulse (2 ms duration) of injected current. Data were derived from HM (thick-slice preparation) recording in bridge mode. Same cell as in Fig. 27.3A, but during a different application of 5-HT. From Berger *et al.* (1992) with permission of Elsevier Scientific Publishers Ireland Ltd.

In addition to the postsynaptic actions of 5-HT on motoneurones that have been reviewed here, recent evidence has indicated that 5-HT may also have important presynaptic effects (Lindsay & Feldman, 1993). Such presynaptic actions can alter the efficacy of synaptic transmission to the target motoneurone.

Summary

This review has provided evidence that the neurotransmitters TRH and 5-HT have diverse actions on motoneurones, which alter both the sub- and suprathreshold behaviour of these cells. This diversity results from effects on different ion channels, and for 5-HT may involve differential activation of receptor subtypes. Furthermore, ontogenetic factors may contribute to the diversity, as during the postnatal period, when major changes take place both qualitatively and quantitatively in the way that motoneurones respond to TRH and 5-HT (Ziskind-Conhaim *et al.*, 1993; Bayliss *et al.*, 1994).

The results of these studies may have important clinical implications. We can speculate that state-dependent alterations in excitability of HMs may be due to TRH and 5-HT input systems. For example, reduced raphe neurone activity during REM sleep will cause a diminished synaptic release of these transmitters, and by the mechanisms described above this will decrease the excitability of HMs. Thus this can be an important contributing factor in obstructive sleep apnoea.

Acknowledgements

This work was supported by a Javits Neuroscience Investigator Award NS-14857 and HL-49657 from the National Institutes of Health to A.J.B. D.A.B. was supported by a Parker B. Francis Fellowship.

References

Aghajanian, G.K. & Rasmussen, K. (1989). Intracellular studies in the facial nucleus illustrating a simple new method for obtaining viable motoneurons in adult rat brain slices. *Synapse*, **3**, 331–338.

Barrett, E.F. & Barrett, J.N. (1976). Separation of two voltage-sensitive potassium currents, and demonstration of a tetrodotoxin-resistant calcium current in frog motoneurones. *Journal of Physiology (London)*, **255**, 737–774.

Bayliss, D.A., Viana, F. & Berger, A.J. (1992). Mechanisms underlying excitatory effects of thyrotropin-releasing hormone on rat hypoglossal motoneurons *in vitro*. *Journal of Neurophysiology*, **68**, 1733–1745.

Bayliss, D.A., Viana, F., Kanter, R.K., Szymeczek-Seay, C.L., Berger, A.J. & Millhorn, D.E. (1994). Early postnatal development of thyrotropin-releasing hormone (TRH) expression, TRH receptor binding and TRH responses in neurons of rat brainstem. *Journal of Neuroscience*, **14**, 821–833.

Berger, A.J., Bayliss, D.A. & Viana, F. (1992). Modulation of neonatal rat hypoglossal motoneuron excitability by serotonin. *Neuroscience Letters*, **143**, 164–168.

Berger, A.J. & Takahashi, T. (1990). Serotonin enhances a low-voltage-activated calcium current in rat spinal motoneurons. *Journal of Neuroscience*, **10**, 1922–1928.

Fisher, N.D. & Nistri, A. (1993). Substance P and TRH share a common effector pathway in rat spinal motoneurons: an *in vitro* electrophysiological investigation. *Neuroscience Letters*, **153**, 115–119.

Hille, B. (1992). *Ionic Channels of Excitable Membranes*. Sunderland, MA: Sinauer Associates.

Holstege, J.C. & Kuypers, H.G.J.M. (1987). Brainstem projections to spinal motoneurons: an update. *Neuroscience*, **23**, 809–821.

Jacobs, B.L. & Azmitia, E.C. (1992). Structure and function of the brain serotonin system. *Physiological Reviews*, **72**, 165–229.

Johansson, O., Hökfelt, T. , Pernow, B., Jeffcoate, S.L., White, N., Steinbusch, H.W.M., Verhofstad, A.A.J., Emson, P.C. & Spindel, E. (1981).

Immunohistochemical support for three putative transmitters in one neuron: coexistence of 5-hydroxytryptamine, substance P and thyrotropin releasing hormone-like immunoreactivity in medullary neurons projecting to the spinal cord. *Neuroscience*, **6**, 1857–1881.

Larkman, P.M. & Kelly, J.S. (1992). Ionic mechanisms mediating 5-hydroxytryptamine- and noradrenaline-evoked depolarization of adult rat facial motoneurones. *Journal of Physiology (London)*, **456**, 473–490.

Lindsay, A.D. & Feldman, J.L. (1993). Modulation of respiratory activity of neonatal rat phrenic motoneurons by serotonin. *Journal of Physiology (London)*, **461**, 213–223.

Nistri, A., Fisher, N.D. & Gurnell, M. (1990). Block by the neuropeptide TRH of an apparently novel K^+ conductance of rat motoneurones. *Neuroscience Letters*, **120**, 25–30.

Rekling, J.C. (1990). Excitatory effects of thyrotropin-releasing hormone (TRH) in hypoglossal motoneurons. *Brain Research*, **510**, 175–179.

Takahashi, T. (1985). Thyrotropin-releasing hormone mimics descending slow synaptic potentials in rat spinal motoneurons. *Proceedings of the Royal Society of London, Series B*, **225**, 391–398.

Takahashi, T. & Berger, A.J. (1990). Direct excitation of rat spinal motoneurones by serotonin. *Journal of Physiology (London)*, **423**, 63–76.

Van Dongen, P.A.M., Grillner, S. & Hökfelt, T. (1986). 5-Hydroxytryptamine (serotonin) causes a reduction in the after-hyperpolarization following the action potential in lamprey motoneurons and premotor interneurons. *Brain Research*, **366**, 320–325.

Viana, F., Bayliss, D.A. & Berger, A.J. (1993a). Calcium conductances and their role in the firing behavior of neonatal rat hypoglossal motoneurons. *Journal of Neurophysiology*, **69**, 2137–2149.

Viana, F., Bayliss, D.A. & Berger, A.J. (1993b). Multiple potassium conductances and their role in action potential repolarization and repetitive firing behavior of neonatal rat hypoglossal motoneurons. *Journal of Neurophysiology*, **69**, 2150–2163.

Wang, M.Y. & Dun, N.J. (1990a). 5-Hydroxytryptamine responses in neonate rat motoneurones *in vitro*. *Journal of Physiology (London)*, **430**, 87–103.

Wang, M.Y. & Dun, N.J. (1990b). Direct and indirect actions of thyrotropin-releasing hormone on neonatal rat motoneurons *in vitro*. *Neuroscience Letters*, **113**, 349–354.

White, S.R. & Fung, S.J. (1989). Serotonin depolarizes cat spinal motoneurons *in situ* and decreases motoneuron after-hyperpolarizing potentials. *Brain Research*, **502**, 205–213.

Ziskand-Conhaim, L., Seebach, B.S. & Gao, B.-X. (1993). Changes in serotonin-induced potentials during spinal cord development. *Journal of Neurophysiology*, **69**, 1338–1349.

28

Serotonin and central respiratory disorders in the newborn

GÉRARD HILAIRE, DIDIER MORIN, ERIC DI PASQUALE AND ROGER MONTEAU

'Biologie des Rythmes et du Développement', Département de Physiologie et Neurophysiologie, URA 1862, Faculté Saint Jérôme, Marseille, France

Development of the serotonergic system

The serotonergic system is one of the first neurochemical systems to differentiate during ontogeny: neuroblasts which contain serotonin (5-HT) are described in the fetal rat as early as day 12–13 (full term 21 days) and 5-HT may be synthesized and released within the central nervous system by 5-HT neurones well before it can be implicated in synaptic transmission. 5-HT is synthesized from its specific precursor L-tryptophan via 5-hydroxylation in 5-hydroxytryptophan (5-HTP) and subsequent decarboxylation and is removed from the synaptic cleft by re-uptake and metabolized to 5-hydroxyindoleacetic acid (5-HIAA) by monoamine oxidase. In rats, brain levels of 5-HT and 5-HIAA increased dramatically from embryonic day 15 to day 19 (Arevalo *et al.*, 1991) but brainstem 5-HT levels at birth are only 32% of the adult levels and increase progressively to reach adult levels at the end of the third postnatal week (Hamon & Bourgoin, 1982). Nutritional factors during pregnancy may affect 5-HT biosynthesis mechanisms, however, and enhanced biosynthesis at birth was reported in both newborn humans and rats malnourished *in utero* (Hernandez, Manjarrez & Chagoya, 1989).

Serotonergic system and central respiratory activity

In the adult, 5-HT has a nearly ubiquitous distribution in the brain and most of the brain functions investigated appear to be influenced by 5-HT mechanisms (thermoregulation, hormone secretion, and other autonomic functions, nociception, motor activities, mood, etc.). Even if the different authors agree that 5-HT affects a given function, conflicting results are often reported regarding the 5-HT effects on the function investigated. This is particularly obvious in the case of central respiratory activity where both excitatory and inhibitory 5-HT effects have been published (for references see Monteau & Hilaire, 1991).

In cats, 5-HT is widely distributed within the phrenic nucleus with descending 5-HT axons mainly originating from the nucleus raphe pallidus or obscurus; stimulations of the raphe obscurus have been reported to activate or to depress phrenic motoneurones but 5-HT electrophoretically applied to phrenic motoneurones has remarkably weak effects. These discrepancies may be due to 5-HT acting on several targets, since 5-HT may modify phrenic motoneurone discharge via direct action on the motoneurones, via relayed action involving the medullary respiratory centres, and via indirect actions through central interactions and peripheral loops. Another example of conflicting results about 5-HT effects on central respiratory activity in cats deals with application of the 5-HT precursor, 5-HTP, which is said to depress respiration when applied intraperitoneally but to activate respiration when applied centrally. Here again, differences in effects may reflect activation of different central and peripheral 5-HT targets. In rat, however, both peripheral application of 5-HTP and central application of 5-HT depressed respiration.

In man, some pathological observations revealed abnormal levels of 5-HT metabolite which might be linked to respiratory disorders. First, in adults, the level of 5-HIAA in the cerebrospinal fluid of patients with sleep apnoea syndrome was shown to be abnormally elevated compared with that of control patients (Cramer *et al.*, 1981). Secondly, in infants, post-mortem high-performance liquid chromatography (HPLC) analysis of the cerebrospinal fluid content of victims of Sudden Infant Death Syndrome (SIDS) revealed statistically higher levels of 5-HIAA than in control infants (Caroff *et al.*, 1992). SIDS is the main cause of infant death in industrialized countries and is likely to be due to numerous and various causes, but respiratory disorders are commonly suspected since frequent obstructive apnoeas occurred in infants who subsequently died of SIDS (Khan *et al.*, 1988). In both cases, however, it is impossible to know whether abnormal levels of 5-HIAA are the cause or the consequence of the syndrome.

Overall, neither experimental studies in animals nor observations in humans have led to precise conclusions about 5-HT effects on central respiratory activity, mainly because of the complexity of the *in vivo* analysis.

In vitro study of the 5-HT modulation of the central respiratory activity in the neonatal rat

To know whether 5-HT affects central respiratory activity, experiments were performed *in vitro* on isolated brainstem–spinal cord preparations from newborn (Suzue, 1984) and fetal (Di Pasquale, Monteau & Hilaire, 1992a) rats, which have been shown to retain the ability to generate a central respiratory activity *in vitro* for periods of several hours (see Fig. 28.1A). These reduced preparations facilitate interpretations of pharmacological experiments because of the absence of the periphery and the ability to administer drugs locally.

Replacing the normal bathing medium by medium containing 5-HT or

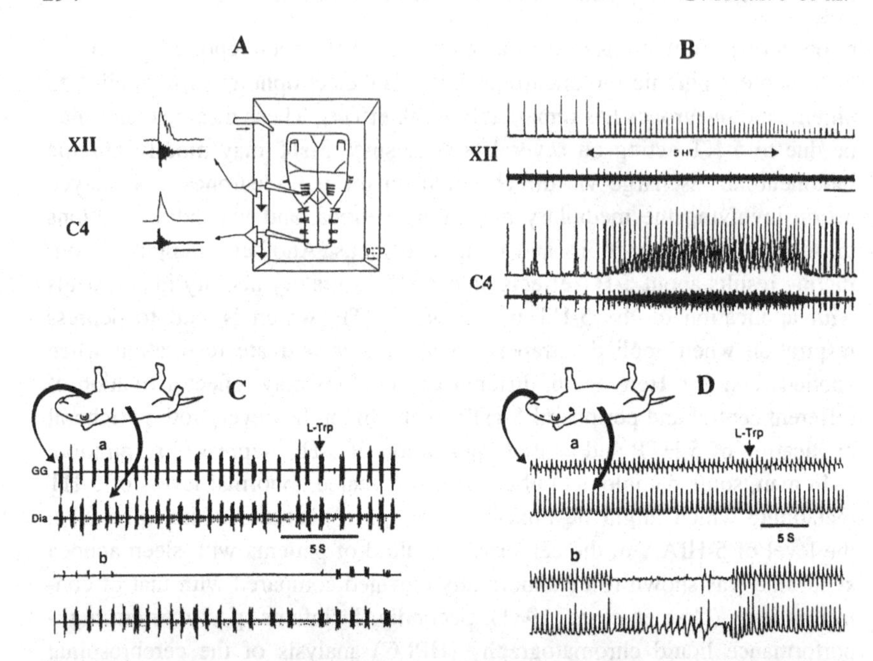

Fig. 28.1. *In vitro* and *in vivo* demonstration of 5-HT depression of hypoglossal inspiratory activity in the newborn rat. (A) and (B) are from *in vitro* experiments. (A) Schematic representation of the experimental model allowing the recording *in vitro* of inspiratory discharges (integrated and raw discharges, upper and lower beams respectively) in hypoglossal (XII) and cervical (C4) roots from isolated brainstem–spinal cord from newborn rats (ventral surface upward; time scale 1 s). (B) Same representation as in (A) (but time scale 1 minute); the normal medium was replaced by a medium containing 30 μM 5-HT as indicated by the horizontal arrow. Note the increase in respiratory frequency, the tonic C4 discharge and the depression of the integrated XII discharge. (C) and (D) are from *in vivo* experiments on anaesthetized newborn rats. (C) The EMG inspiratory bursts from the genioglossal muscle (GG, upper beam) and the diaphragm (Dia, lower beam) before (Ca) and 20 min after (Cb) L-tryptophan (L-Trp) intraperitoneal injection (50 mg/kg at the arrow). Note the silencing of GG discharge in Cb. (D) Recording from air flow (upper beam) and thoracic respiratory movements (lower beam) before (Da) and 30 minutes after (Db) L-tryptophan load. Note the obstructive apnoea in Db.

specific agonists or antagonists revealed that activation of medullary 5-HT_{1A} receptors increased the respiratory frequency, RF (see Fig. 28.1B and Morin, Monteau & Hilaire, 1991a). This effect is dose-dependent and particularly dramatic at embryonic day 18. Increasing the endogenous level of 5-HT increased the RF (stimulation of the raphe nuclei, activation of 5-HT biosynthesis by 5-HT precursors, blockade of 5-HT re-uptake by fluoxetine; Morin *et al.*, 1990, 1991a), while blocking the effect of endogenous 5-HT with the 5-HT_1 antagonist NAN-190 decreased RF. Effects are particularly dramatic in young fetuses with respiratory arrest under NAN-190. These results revealed that 5-HT biosynthesis mechanisms continue functioning *in vitro* and exert a permanent excitatory modulation on the respiratory rhythm generator, particularly potent at the fetal stage. This is consistent with some *in utero* data obtained in sheep where fetal breathing was stimulated by administration of 5-HT precursors but depressed by 5-HT antagonists (Fletcher *et al.*, 1988). In newborn rats, mimetic effects of general application were obtained by local application of 5-HT within the rostral ventrolateral medulla (Di Pasquale *et al.*, 1992b), in sites assumed by some (Smith *et al.*, 1991) to contain the primary respiratory rhythm generator.

Central respiratory activity was recorded not only on cervical roots but also on glossopharyngeal (IX), vagal (X) and hypoglossal nerves (XII). At the same time as the respiratory frequency was increased by 5-HT, a tonic discharge was elicited in cervical recordings (see Fig. 28.1B), but never in cranial nerves. The tonic discharge originated from activation of spinal postsynaptic 5-HT_2 receptors which bring previously silent non-respiratory motoneurones to firing (Morin, Monteau & Hilaire, 1991b). The amplitude of the inspiratory bursts decreased slightly in cervical, IX and X recordings (20–30%) but the XII inspiratory discharge was dramatically diminished (80%: see Fig. 28.1B and Morin, Monteau & Hilaire, 1992; Morin, 1993). Depression of the XII inspiratory discharge, which was prevented by 5-HT_2 antagonist pretreatment (ketanserin and SR46349B, Sanofi Research), could be elicited by (1) bath application of 5-HT_2 agonist (DOI) and L-tryptophan, (2) stimulation of raphe nuclei and (3) local application of 5-HT within the hypoglossal nucleus.

Intracellular recordings revealed that the inspiratory drive received by the XII motoneurones was decreased under 5-HT and a presynaptic inhibitory process was postulated. This agrees with *in vitro* experiments performed on brainstem slices where 5-HT did not depress the excitability of XII motoneurones disconnected from their central drivers (Berger, Bayliss & Viana, 1992 and this volume). In preparations from fetal rats, the XII discharge was not depressed by 5-HT and related agents (8-OH-DPAT, DOI, L-tryptophan).

These *in vitro* experiments revealed that 5-HT may therefore interact with central respiratory activity during perinatal life via different types of receptors to elicit RF changes (medullary 5-HT_{1A} receptors) and facilitation of cervical motoneurones (spinal 5-HT_2 postsynaptic receptors). After birth, 5-HT may also depress the XII inspiratory discharge via medullary 5-HT_2 presynaptic receptors.

In vivo study of respiratory disturbances elicited by a sudden increase in 5-HT biosynthesis in anaesthetized newborn rats

It could be postulated that 5-HT may, on the one hand favour ventilation via the acceleration of the RF but, on the other, disturb ventilation via obstructive apnoea due to the depression of the XII discharge, since silencing of the genioglossus (muscle of the tongue controlled by the XII nerve) has been reported to be responsible for obstructive apnoea (Sauerland & Harper, 1976). *In vivo* experiments were conducted on anaesthetized newborn rats and L-tryptophan (50 mg/kg) was injected intraperitoneally to increase the endogenous level of 5-HT (Hilaire *et al.*, 1993). HPLC measurements of cerebrospinal fluid samples confirmed significant increases in 5-HT levels 30 min after the injection.

The electromyographic activity of the diaphragm and the genioglossal muscle was recorded in freely breathing newborn rats (2–8 days old). L-Tryptophan load did not elicit reproducible changes in the RF and diaphragmatic discharges but decreased the amplitude of the inspiratory bursts of the genioglossal muscle within 20–30 min. In some rats, the genioglossal inspiratory activity was totally suppressed for several respiratory cycles (see Fig. 28.1C).

In other experiments, the respiratory displacement of the lower ribs and the resulting airflow changes were recorded. In half the experiments, L-tryptophan load elicited characteristic obstructive apnoeas (latency 20–30 min) with total cessation of airflow while respiratory rib displacements persisted (see Fig. 28.1D). These obstructive apnoeas, which lasted several respiratory cycles, recurred frequently in some animals, leading to respiratory distress and death in 10% of the experiments. In the young adult rats and half the newborn rats, however, L-tryptophan load never elicited respiratory disturbances.

Serotonin and central respiratory disorders

The preceding results suggest that a sudden activation of 5-HT biosynthesis mechanisms during the postnatal period may elicit dramatic respiratory disturbances. Taking into account that 5-HT may depress the XII inspiratory discharge in the newborn (*in vivo* and *in vitro* experiments) but in neither fetal (*in vitro* experiments) nor young adult (*in vivo* experiments) rats, it appears likely that the depression of the XII inspiratory discharge induced by 5-HT is age-related and may reflect some 5-HT postnatal developmental features. It is to be noted that obstructive apnoeas were observed only in some litters while newborn rats from half the litters were insensitive to L-tryptophan; no satisfactory explanation was found for this difference (number of pups per litter, weight, age, etc.). Since malnutrition during pregnancy disturbs postnatal 5-HT biosynthesis (Hernandez *et al.*, 1989), differences in respiratory response to L-tryptophan load might reflect nutritional differences during gestation at the breeding centre.

The experiments reported above demonstrate that 5-HT may have a key role during the perinatal period in regulating the central respiratory activity in term of RF and pattern of discharge. The temptation to draw a parallel between the respiratory disturbances elicited in newborn rats by a sudden activation of 5-HT biosynthesis and the high levels of 5-HIAA in victims of Sudden Infant Death Syndrome (Caroff *et al.*, 1992) and in patients suffering from sleep apnoea (Cramer *et al.*, 1981) must be resisted, however, since extrapolation from rat to human is hazardous. Nevertheless, it is proposed as a working hypothesis that abnormal 5-HT biosynthesis at birth may have dramatic respiratory consequences.

Acknowledgements

The experiments reported above were supported by the CNRS, the Elf–Sanofi group, French Health Ministry, INSERM, the 'Naître et Vivre' foundation, the 'Provence–Alpes–Côte d'Azur' council and the 'UIII' University of Aix-Marseille.

References

Arevalo, R., Afonso, D., Castro, R. & Rodriguez, M. (1991). Fetal brain serotonin synthesis and catabolism is under control by mother intake of tryptophan. *Life Sciences*, **49**, 53–66.

Berger, A., Bayliss, D. & Viana, F. (1992). Modulation of neonatal rat hypoglossal motoneuron excitability by serotonin. *Neuroscience Letters*, **143**, 164–168.

Caroff, J., Girin, E., Alix, D., Cann-Moisan, C., Sizun, J. & Barthelemy, L. (1992). Neurotransmission et mort subite du nourrisson: étude du liquide céphalo-rachidien. *Comptes Rendus de l'Académie des Sciences, Paris*, **314**, 451–454.

Cramer, H., Warter, J.-M., Renaud, B., Krieger, J., Marescaux, C. & Hammers, R. (1981). Cerebrospinal fluid adenosine 3′-5′-monophosphate, 5-hydroxyindolacetic acid and homovanillic acid in patients with sleep apnoea syndrome. *Journal of Neurology, Neurosurgery and Psychiatry*, **44**, 1165–1167.

Di Pasquale, E., Monteau, R. & Hilaire, G. (1992a). *In vitro* study of central respiratory-like activity of the fetal rat. *Experimental Brain Research*, **89**, 459–453.

Di Pasquale, E., Morin, E., Monteau, R. & Hilaire, G. (1992b). Serotonergic modulation of the respiratory rhythm generator at birth: an *in vitro* study in the rat. *Neuroscience Letters*, **143**, 91–95.

Fletcher, D., Hanson, M., Moore, P., Nijhuis, J. & Parkes, M. (1988). Stimulation of breathing movements by L-5-hydroxytryptophan in fetal sheep during normoxia and hypoxia. *Journal of Physiology (London)*, **404**, 575–589.

Hamon, M. & Bourgoin, S. (1982). Characteristics of 5-HT metabolism and function in the developing brain. In *Biology of Serotonergic Transmission*, ed. N. Osborne, pp. 197–220. Chichester: Wiley.

Hernandez, J., Manjarrez, G. & Chagoya, G. (1989). Newborn humans and rats malnourished in utero: free plasma L-tryptophan, neutral amino acids and brain serotonin synthesis. *Brain Research*, **488**, 1–13.

Hilaire, G., Morin, D., Lajard, A.M. & Monteau, R. (1993). Changes in serotonin metabolism may elicit obstructive apnoea in the newborn rat. *Journal of Physiology (London)*, **466**, 367–382.

Khan, A., Blum, D., Rebuffat, E., Sottiaux, M., Grosswasser, J. & Muller, M.F.

(1988). Polysomnographic studies in infants who subsequently died of sudden infant death syndrome. *Pediatrics*, **82**, 721–727.
Monteau, R. & Hilaire, G. (1991). Spinal respiratory motoneurons. *Progress in Neurobiology*, **37**, 83–144.
Morin, D. (1993). Compared effects of serotonin on the inspiratory activity of glossopharyngeal, vagal, hypoglossal and cervical motoneurons in neonatal rat brain stem–spinal cord preparations. *Neuroscience Letters*, **140**, 61–64.
Morin, D., Hennequin, S., Monteau, R. & Hilaire, G. (1990). Depressant effect of raphe stimulation on inspiratory activity of the hypoglossal nerve: *in vitro* study in the newborn rat. *Neuroscience Letters*, **116**, 299–303.
Morin, D., Monteau, R. & Hilaire, G. (1991a). 5-Hydroxytryptamine modulates central respiratory activity in the newborn rat: an *in vitro* study. *European Journal of Pharmacology*, **192**, 89–95.
Morin, D., Monteau, R. & Hilaire, G. (1991b). Serotonin and cervical respiratory motoneurones: intracellular study in the newborn rat brainstem–spinal cord preparation. *Experimental Brain Research*, **84**, 229–232.
Morin, D., Monteau, R. & Hilaire, G. (1992). Compared effects of serotonin on cervical and hypoglossal inspiratory activities: an *in vitro* study in the newborn rat. *Journal of Physiology (London)*, **451**, 605–629.
Sauerland, E. & Harper, R. (1976). The human tongue during sleep: electromyographic activity of the genioglossus muscle. *Experimental Neurology*, **51**, 160–170.
Smith, J., Ellenberger, H., Ballanyi, D., Richter, D. & Feldman, J. (1991). Pre-Bötzinger complex: a brain stem region that may generate respiratory rhythm in mammals. *Science*, **254**, 726–729.
Suzue, T. (1984). Respiratory rhythm generation in the *in vitro* brain stem–spinal cord preparation of the neonatal rat. *Journal of Physiology (London)*, **354**, 173–183.

29

Are medullary respiratory neurones multipurpose neurones?

LAURENT GRÉLOT, STÉPHANE MILANO, CHRISTIAN GESTREAU AND ARMAND L. BIANCHI

Département de Physiologie et Neurophysiologie, Faculté des Sciences et Techniques de Saint Jérôme, Marseille, France

Summary

Recent studies performed in invertebrates have challenged the classical views on the role of a neurone within a network. Indeed, it has been clearly demonstrated that a neurone can belong to different networks involved in the generation of different motor activities (Meyrand, Simmers & Moulins, 1991; Weimann, Meyrand & Marder, 1991). The activation of respiratory neurones during non-respiratory behaviours such as vomiting, coughing or swallowing suggest that such 'multipurpose or multifunctional' neurones might also exist in the central nervous system (CNS) of mammals.

Introduction

The neural origin of motor activities has been the interest of physiologists since the last century. During the 1960s, rapid progress took place with the development of invertebrate preparations performing complex motor tasks with few neurones organized in 'simple' circuitry. In parallel, vertebrate preparations permitting the study of continuing motor activities such as respiration or locomotion were developed by Cohen (1969, 1971, 1979, 1981), von Euler (1973, 1977, 1983, 1986), Sears (1964, 1971, 1990), Lund (Lund & Dellow, 1971; Lund, 1976; Lund & Enomoto, 1988), Grillner (1974, 1981, 1985) and others. As time has gone by, it has become clear that different rhythmic motor patterns result from the interaction between a neuronal network, often referred to as a central rhythm generator or central pattern generator (CPG) residing within the CNS, and afferent inputs (i.e. peripheral feedback loops) arising as the consequence of the movements themselves. This organization appears rather general, although the specific solutions differ somewhat according to the species and the motor acts (Cohen, Rossignol & Grillner, 1988).

Over the past decades, the organization and functioning of CPGs responsible for the generation and maintenance of motor activities involved in homeostatic mechanisms have been intensively investigated in vertebrate preparations. Among these homeostatic mechanisms, breathing is a vital function which is generally viewed as resulting from the activation of a medullary set of neurones forming the respiratory CPG. This CPG, which functions automatically throughout life, elaborates a motor programme distributed to a complex muscular apparatus. This respiratory motor apparatus is controlled by a large number of motoneuronal pools which consist schematically of pump muscles (the thoraco-abdominal muscles, i.e. the diaphragm, the intercostal and abdominal muscles) and of valve muscles (the upper airway, i.e. the laryngeal and pharyngeal muscles). Clearly these respiratory muscles appear to be essential also for the production of numerous non-respiratory motor activities such as the common expulsive behaviours (i.e. micturition, defecation, parturition, coughing, vomiting or sneezing) or integrated behaviours (i.e. vocalization, phonation or swallowing). The present chapter will describe the patterns of activity of the medullary respiratory neurones during several non-respiratory behaviours including those clearly antagonistic with breathing (e.g. vomiting and swallowing). The possibility that these neurones serve different functions will be discussed.

Organization of the respiratory CPG

A respiratory neurone is defined as a neurone exhibiting a rhythmic activity with a firing discharge pattern or membrane potential changes which have a steady relationship with the central respiratory command as monitored by the movements of the thorax, the electromyogram of the diaphragm or the neural activity of the phrenic nerve. There are two basic types of respiratory neurones: the inspiratory neurones which fire during the burst of discharge of the phrenic nerve, and the expiratory neurones which fire between these phrenic bursts. In both inspiratory and expiratory neuronal populations, three subcategories (augmenting, decrementing or constant) may be defined on the basis of their firing patterns or membrane potential changes. These various subcategories have been recognized by examination of their discharge patterns (augmenting, decrementing or constant/plateau) and the time in the respiratory cycle at which maximum discharge frequency occured (early or late), in addition to the phase (inspiration (I) or expiration (E)) in which the neurones fire. Hence, I augmenting (I-AUG) and I decrementing neurones (I-DEC or early-I neurones) with a peak discharge frequency in late or early inspiration, respectively, and E augmenting (E-AUG) and E decrementing (E-DEC or post-I neurones) neurones with a peak discharge frequency in late or early expiration, respectively, have been identified. Finally, I (I-CON) and E (E-CON) neurones exhibit a constant discharge frequency in either phase of the respiratory cycle (Fig. 29.1) (Cohen, 1979; Ezure, 1990).

The respiratory neurones have been classified in three functional groups on the basis of their axonal projections. The first group consists of respiratory

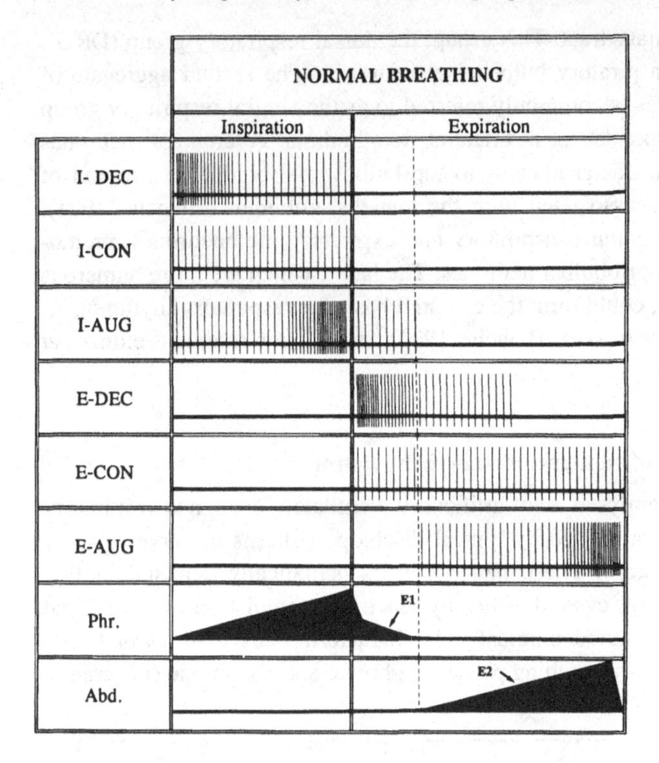

Fig. 29.1. Sketch showing the six basic firing patterns of medullary respiratory neurones. Inspiration is defined as the period of augmenting (ramp) phrenic (Phr.) nerve activity while expiration consists of the period of declining (E1, or stage 1 of expiration) and silent (E2, or stage 2 of expiration) phrenic discharge.

motoneurones providing innervation to the respiratory musculature. These motoneurones exhibit inspiratory- or expiratory-related firing activity, and are widely distributed in the medulla oblongata and the spinal cord. The second group comprises inspiratory and expiratory bulbospinal neurones sending their axons to the spinal cord. These neurones are considered as premotoneurones transmitting central information, i.e. the central respiratory drive potentials (CRDPs) (Sears, 1964), to the spinal respiratory motoneurones. Neurones of the third group possess axons which project to other neurones strictly located within the medulla. These interneurones, termed respiratory propriobulbar neurones, are viewed as essential for respiratory rhythmogenesis (Richter, 1982).

Several lines of evidence suggest that the respiratory rhythm in mammals is generated by a neural network located within the medulla oblongata (Richter, Ballantyne & Remmers, 1986). Respiratory neurones are concentrated into two well-defined distinct regions of the medulla oblongata (Feldman, 1986). The first one overlaps, in the dorsal medulla, the ventrolateral part of the nucleus of the solitary tract. This group, the dorsal respiratory group (DRG), contains mainly inspiratory bulbospinal neurones. The second aggregate of respiratory neurones is commonly referred to as the ventral respiratory group (VRG). It corresponds to a bilateral longitudinal column of neurones extending from the facial nucleus to the bulbospinal border in a region of the lateral medulla associated with the nucleus ambiguus (Bianchi, 1971). This large group contains inspiratory and expiratory motoneurones, premotoneurones and propriobulbar neurones. The latter, which are more numerous in the rostral VRG, could form the essential basis for respiratory rhythmogenesis as indicated by *in vivo* (Bianchi, 1974) and *in vitro* studies (Feldman *et al.*, 1991).

Characterization of non-respiratory behaviours

In decerebrate, paralysed and artificially ventilated cats, non-respiratory behaviours are characterized by typical discharge patterns in nerves usually regarded as respiratory. These patterns are schematically depicted in Fig. 29.2. Thus, vomiting, evoked either by electrical stimulation of abdominal vagal afferents or pharmacologically, is characterized by a series of bursts of co-activation (i.e. the retching phase) of phrenic and abdominal (i.e. cranial

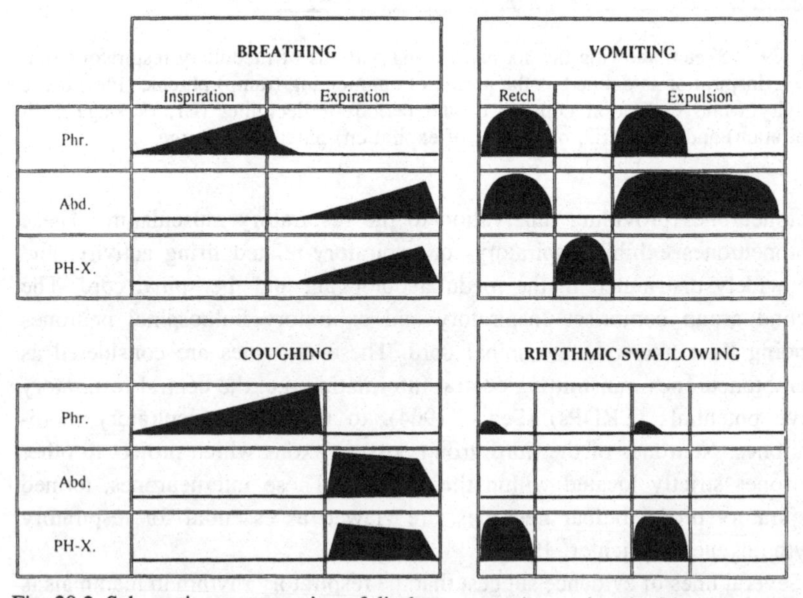

Fig. 29.2. Schematic representation of discharge patterns on phrenic (Phr.), abdominal (Abd.) and pharyngeal vagus (PH-X) nerves during breathing and non-respiratory motor activities such as vomiting, coughing and rhythmic swallowing. Vomiting is represented with a single retch and an expulsion, swallowing with two buccopharyngeal stages, and coughing with diaphragmatic and abdominal stages.

ilio-hypogastric) nerves. The retching phase usually culminates in an expulsion phase in which abdominal discharge is prolonged compared with both phrenic and abdominal discharges occurring during the preceding retching phase (Miller, Tan & Suzuki, 1987).

Both coughing and swallowing are elicited by repetitive electrical stimulation of the superior laryngeal nerves. Coughing consists of a large ramp-like activation of the phrenic nerve activity (i.e. the diaphragmatic phase) followed immediately by a large abdominal nerve discharge (i.e. the abdominal phase). Swallowing was characterized by brief bursts of activity in the pharyngeal vagus and hypoglossal nerves innervating the pharyngeal constrictors and the genioglossus, respectively.

Activity of the medullary respiratory neurones during non-respiratory motor acts

The reshaping of discharge patterns of phrenic and abdominal motoneurones involved in non-respiratory reflexes, as opposed to their discharge patterns during ventilation, sets the problem of how the activity of medullary neurones of the respiratory CPG, i.e. premotoneurones and propriobulbar neurones, may behave during these non-respiratory motor acts.

The respiratory premotoneurones

Since phrenic (i.e. inspiratory-related) and abdominal (i.e. expiratory-related) motoneurones fire simultaneously during vomiting, one might expect a common behaviour for both inspiratory and expiratory bulbospinal neurones. However, there is a striking discrepancy between the patterns of activity exhibited by inspiratory and expiratory bulbospinal neurones during vomiting. While one-third of expiratory premotoneurones fire appropriately to be involved in the control of the abdominal muscles during vomiting (Miller *et al.*, 1987), almost all (90%) inspiratory bulbospinal cells (at least in the cat) are actively inhibited, and thus exhibit patterns of activity incompatible with the control of the diaphragm (Bianchi & Grélot, 1989; Miller *et al.*, 1990). These results demonstrate that the output level of the respiratory CPG clearly differs from that of the vomiting CPG. In contrast, recent results of our group indicate that the inspiratory premotoneurones of the DRG are implicated in the production of laryngeal-induced reflex activities. Indeed, almost all (96%) of them are strongly activated during the diaphragmatic phase of coughing while the expiratory premotoneurones are also intensely recruited and activated during the abdominal phase (Jakuš, Tomori & Stránsky, 1985).

During rhythmic swallowing, 80% of inspiratory premotoneurones exhibit brief (150–200 ms) depolarizations giving an intense firing concomitantly with the buccopharyngeal stages. This is somewhat surprising since the

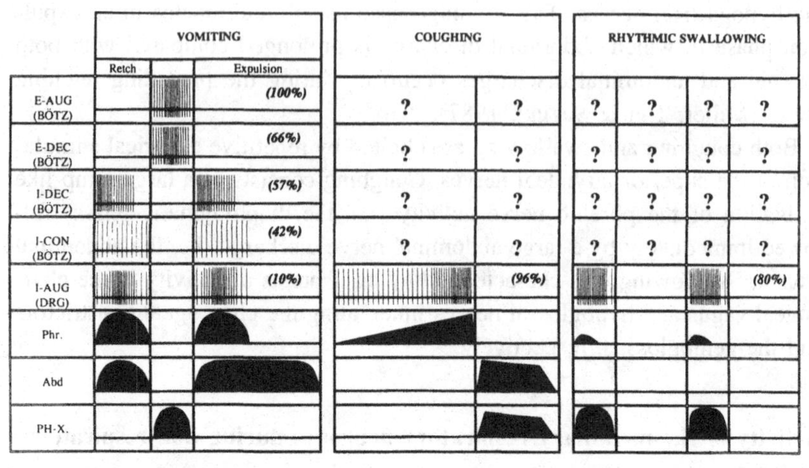

Fig. 29.3. Simplified schematic representation of firing activities of some medullary respiratory propriobulbar neurones during vomiting, coughing and swallowing. Neurones were recorded in the rostral part of the ventral respiratory group (i.e. the Bötzinger (BÖTZ) complex) and the dorsal respiratory group (DRG). Percentages indicate the ratio of the population exhibiting the depicted pattern of discharge. Question marks indicate absence of data. Abbreviations are as in previous figures and in the text.

buccopharyngeal stages of swallowing occur during an apnoeic (i.e. suspension of inspiration) period. These swallowing-related activities are responsible for only very weak activations of the phrenic nerve (i.e. phrenic breakthroughs) which interrupt rhythmically the apnoeic period (Grélot, Barillot & Bianchi, 1989). Hence, these results suggest that, unlike in vomiting, the inspiratory output level of the respiratory network also serves functions such as coughing and swallowing.

The respiratory interneurones

Recently, the patterns of activity during vomiting of the respiratory propriobulbar neurones of the rostral part of the VRG (i.e. the 'Bötzinger complex') have been investigated (Miller & Nonaka, 1990; Miller & Ezure, 1992). It appears that most of the interneurones are activated during vomiting, discharging either in phase or out of phase with the phrenic and abdominal nerve co-activations (Fig. 29.3). A striking observation is that many of the inhibitory connections revealed during normal breathing between different subpopulations of respiratory neurones seem to be used to generate the motor output pattern during vomiting. In addition, those medullary respiratory interneurones that establish synaptic connections which could impair the vomiting-related activation of cranial or spinal respiratory motoneurones are silent (i.e. disfacilitated or actively inhibited) during emesis. On the other hand, investigations in progress in our laboratory show that most (80%) of the inspiratory interneurones of the DRG are activated during the bucco-pharyngeal stages of swallowing. In addition, all of them are activated during the diaphragmatic phase of coughing, during which they exhibit a brief bell-shaped depolarization responsible for a quite intense discharge firing (Fig. 29.3).

Conclusions

These results involving central respiratory-related neurones have a more general implication extending beyond the research field of homeostatic motor mechanisms, since they provide new information on the organization and functioning of CPG of mammals. Specifically, they provide evidence for the existence of 'multipurpose or multifunctional' neurones in the CNS of

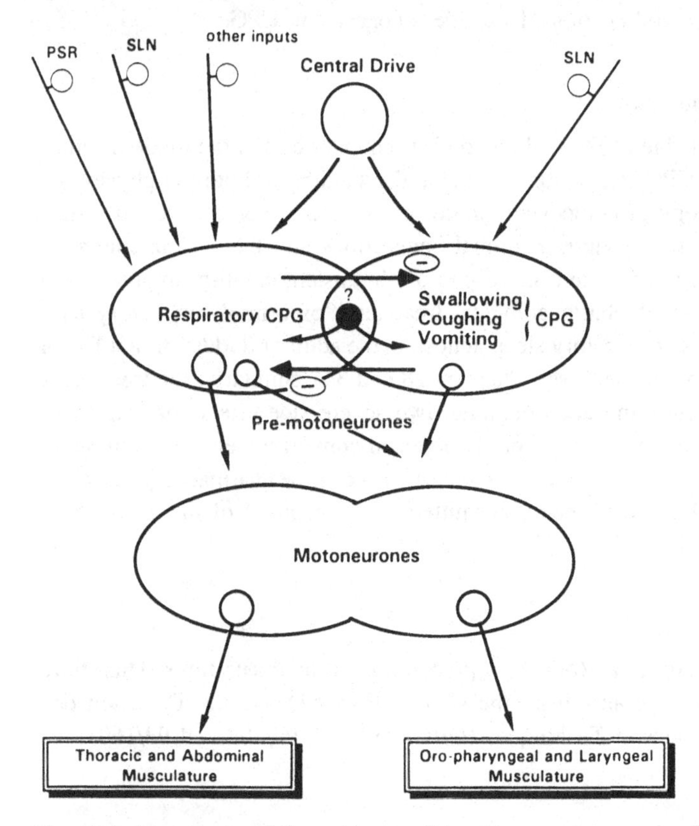

Fig. 29.4. Hypothesis of the existence of 'multipurpose' neurones (black circle) in the medulla oblongata of mammals, assumed to belong to the different central pattern generators (CPGs). Minus signs indicate synaptic inhibition. SLN, superior laryngeal nerve; PSR, pulmonary stretch receptor.

mammals, and suggest that a given neurone can be involved in more than one physiological function, even if these functions are clearly antagonistic – such as breathing and vomiting or swallowing (Fig. 29.4). Although we still do not understand how the same neurone can participate in the antagonistic functions, recent results drawn from *in vitro* studies provide a cellular basis for the existence of such 'multifunctional' neurones. In particular, some solitary tract nucleus neurones exhibit pacemaker properties conditioned by the application of *N*-methyl-D-aspartate (NMDA), an excitatory amino acid agonist (Tell & Jean, 1991, 1993). Depending on the level of the membrane potentials, NMDA-induced endogenous oscillations underlie different patterns of discharge, in particular a bursting or a tonic mode. If these membrane properties were preserved *in vivo*, the same cell could serve as a rhythm-generating neurone or as a 'follower' neurone, depending only on the balance of excitatory and inhibitory synaptic events it receives. This would imply the abandonment of the concept of respiratory, vomiting, coughing and swallowing 'centres' and a modification of our ideas concerning CPGs.

Clinical perspectives

A better knowledge of the real anatomical boundaries and the mode of operation within a CPG is essential for the understanding of both the physiology and the pathology of homeostatic motor mechanisms. For instance, determination of both the synaptic neuronal connections and the various chemical messengers involved within the respiratory brainstem circuitry might lead to the development of pharmacological treatments of central respiratory dysfunctions such as respiratory sleep apnoea in the adult or Sudden Infant Death Syndrome in the neonate (see Chapters 28 and 34). Similarly, the treatments of intractable vomiting and coughing, two severe side effects of anticancer therapies and administration of angiotensin-converting enzyme inhibitors respectively, will be better controlled when the neuropharmacological basis (i.e. the neural circuitry, neurotansmitters and receptors) of these effects is well understood.

Acknowledgments

We thank Mrs Jocelyne Roman for preparation of the illustrations. This study was supported by grants from the CNRS (ERS 61) and the Direction des Recherches, Etudes et Techniques (DRET, 87/110, 90/108 and 93/160).

References

Bianchi, A.L. (1971). Localisation et étude des neurones respiratoires bulbaires: mise en jeu antidromique par stimulation spinale ou vagale. *Journal de Physiologie (Paris)*, **63**, 5–40.

Bianchi, A.L. (1974). Modalités de décharge et propriétés anatamo-fonctionelles des neurones respiratoires bulbaires. *Journal de Physiologie (Paris)*, **68**, 555–587.

Bianchi, A.L. & Grélot, L. (1989). Converse motor output of inspiratory bulbospinal premotoneurones during vomiting. *Neuroscience Letters*, **104**, 298–302.

Cohen, A.H., Rossignol, S. & Grillner, S. (1988). *Neural Control of Rhythmic Movements in Vertebrates*. New York: Wiley.

Cohen, M.I. (1969). Discharge patterns of brain-stem respiratory neurons during Hering–Breuer reflex evoked by lung inflation. *Journal of Neurophysiology*, **32**, 356–374.

Cohen, M.I. (1971). Switching of the respiratory phases and evoked phrenic responses produced by rostral pontine electrical stimulation. *Journal of Physiology (London)*, **217**, 133–158.

Cohen, M.I. (1979). Neurogenesis of respiratory rhythm in the mammal. *Physiological Reviews*, **59**, 1105–1173.

Cohen, M.I. (1981). How is respiratory rhythm generated? *Federation Proceedings*, **40**, 2372–2377.

Euler, C. von (1973). The role of proprioceptive afferents in the control of respiratory muscles. *Acta Neurobiologiae Experimentalis*, **33**, 329–341.

Euler, C. von (1977). The functional organization of the respiratory phase-switching mechanisms. *Federation Proceedings*, **36**, 2375–2380.

Euler, C. von (1983). On the central pattern generator for the basic breathing rhythmicity. *Journal of Applied Physiology*, **55**, 1947–1959.

Euler, C. von (1986). Brain stem mechanisms for generation and control of breathing pattern. In *Handbook of Physiology*, sect. 3, *The Respiratory System*, vol. 2, *Control of Breathing*, ed. N.S. Cherniack & J.G. Widdicombe, pp. 1–67. Washington, DC: American Physiological Society.

Ezure, K. (1990). Synaptic connections between medullary respiratory neurons and considerations on the genesis of respiratory rhythm. *Progress in Neurobiology*, **35**, 429–450.

Feldman, J.L. (1986). Neurophysiology of breathing in mammals. In *Handbook of Physiology*, sect. 1, *The Nervous System IV*, ed. F.E. Bloom, pp. 463–524. Washington, DC: American Physiological Society.

Feldman, J.L. & Grillner, S. (1983). Control of vertebrate respiration and locomotion: a brief account. *The Physiologist*, **26**, 310–316.

Feldman, J.L., Smith, J.C. & Liu, G. (1991). Respiratory pattern generation in mammals: *in vitro* en bloc analyses. *Current Opinion in Neurobiology*, **1**, 590–594.

Grélot, L., Barillot, J.C. & Bianchi, A.L. (1989). Pharyngeal motoneurones: respiratory-related activity and responses to laryngeal afferents in the decerebrate cat. *Experimental Brain Research*, **78**, 336–344.

Grillner, S. (1974). On the generation of locomotion in the spinal dogfish. *Experimental Brain Research*, **20**, 459–470.

Grillner, S. (1981). Control of locomotion in bipeds, tetrapods and fish. In *Handbook of Physiology*, sect. 1, *The Nervous System*, vol. II, *Motor Control*, ed. V.B. Brooks, pp. 1179–1236. Washington, DC: American Physiological Society.

Grillner, S. (1985). Neurological bases of rhythmic motor acts in vertebrates. *Science*, **228**, 143–149.

Jakuš, J., Tomori, Z. & Stránsky, A. (1985). Activity of bulbar respiratory neurones during cough and other respiratory tract reflexes in cats. *Physiologia Bohemoslovenica*, **34**, 127–136.

Lund, J.P. (1976). Evidence for a central neural pattern generator regulating the chewing cycle. In *Mastication*, ed. D.J. Anderson & B. Matthews, pp. 204–212. Bristol: Wright.

Lund, J.P. & Dellow, P.J. (1971). The influence of interactive stimuli on rhythmical masticatory movements in rabbits. *Archives of Oral Biology*, **16**, 215–223.

Lund, J.P. & Enomoto, S. (1988). The generation of mastication by the mammalian central nervous system. In *Neural Control of Rhythmic Movements in Vertebrates*, ed. A.H. Cohen, S. Rossignol & S. Grillner, pp. 41–72. New York: Wiley.

Meyrand, P., Simmers, J. & Moulins, M. (1991). Construction of a pattern-generating circuit with neurones of different networks. *Nature*, **351**, 60–63.

Miller, A.D. & Ezure, K. (1992). Behavior of inhibitory and excitatory propriobulbar respiratory neurons during fictive vomiting. *Brain Research*, **578**, 168–176.

Miller, A.D. & Nonaka, S. (1990). Bötzinger expiratory neurones may inhibit phrenic motoneurons and medullary inspiratory neurons during vomiting. *Brain Research*, **521**, 352–354.

Miller, A.D., Nonaka, S., Lakos, S.F. & Tan, L.K. (1990). Diaphragmatic and external intercostal muscle control during vomiting: behavior of inspiratory bulbospinal neurons. *Journal of Neurophysiology*, **63**, 31–36.

Miller, A.D., Tan, L.K. & Suzuki, I. (1987). Control of abdominal and expiratory intercostal muscle activity during vomiting: role of ventral respiratory group expiratory neurons. *Journal of Neurophysiology*, **57**, 1854–1866.

Richter, D.W. (1982). Generation and maintenance of the respiratory rhythm. *Journal of Experimental Biology*, **100**, 93–107.

Richter, D.W., Ballantyne, D. & Remmers, J.E. (1986). How is the respiratory rhythm generated? *News in Physiological Sciences*, **1**, 109–112.

Sears, T.A. (1964). The slow potentials of thoracic respiratory motoneurones and their relation to breathing. *Journal of Physiology (London)*, **175**, 404–424.

Sears, T.A. (1971). Breathing: a sensorimotor act. *The Scientific Basis of Medicine Annual Reviews*, [1971], 129–147.

Sears, T.A. (1990). Central rhythm generation and spinal integration. *Chest*, **97**, 45S-51S.

Tell, F. & Jean, A. (1991). Activation of *N*-methyl-D-aspartate receptors induces endogenous rhythmic bursting activities in nucleus tractus solitarii neurones: an intracellular study on adult rat brainstem slices. *European Journal of Neuroscience*, **3**, 1353–1365.

Tell, F. & Jean, A. (1993). Ionic basis for endogenous rhythmic patterns induced by activation of *N*-methyl-D-aspartate receptors in neurons of the rat nucleus tractus solitarii. *Journal of Neurophysiology*, **70**, 2379–2389.

Weimann, J., Meyrand, P. & Marder, E. (1991). Neurons that form multiple pattern generators: identification and multiple activity patterns of gastric/pyloric neurones in the crab stomatogastric system. *Journal of Neurophysiology*, **65**, 111–122.

30

Reflex control of expiratory motor output in dogs

J. RICHARD ROMANIUK*, THOMAS E. DICK, GERALD S. SUPINSKI AND ANTHONY F. DIMARCO

Department of Medicine, Case Western Reserve University, and Division of Pulmonary and Critical Care, MetroHealth Medical Center, Cleveland, Ohio, USA

Studies concerning neural control of respiratory timing and tidal volume have focused predominantly on the control the inspiratory phase (von Euler, 1986). With application of lung inflation during the inspiratory phase, three distinct reflexes have been described: (1) inspiratory off-switch (Clark & von Euler, 1972), (2) reversible graded inhibition (Younes, Remmers & Baker, 1978) and (3) low-threshold facilitation (Bartoli *et al.*, 1975; DiMarco *et al.*, 1981). Each of these reflexes is mediated through vagal mechanisms presumably via pulmonary stretch receptors. After vagotomy, changes in lung volume result in only minor effects on phrenic nerve activity.

Previous studies related to the control of the amplitude of expiratory motor activity and expiratory time were based predominantly on the response to expiratory loads (Bishop, 1967; Bishop, Hirsch & Thursby, 1978). They demonstrated that increases in lung volume during the expiratory phase facilitate expiratory motor activity and prolong expiratory time (T_E) (Bishop *et al.*, 1978; Barrett *et al.*, 1994). More recent studies, however, showed that expiratory muscle activities may be inhibited by pulsed lung inflation (Arita & Bishop, 1983; Bajic *et al.*, 1992; Cohen, Feldman & Sommer, 1985; Younes, Vaillancourt & Milic-Emili, 1974), increases in lung volume (Chung *et al.*, 1987; Fregosi, Bartlett & St John, 1990; Polacheck, Remmers & Younes, 1978) or by stimulation of vagal afferents (Haxhiu *et al.*, 1988; Smith *et al.*, 1990). On the basis of these new data, we further examined vagal influences on expiratory motor activity and timing. In experiments performed on anaesthetized (pentobarbitone) and tracheotomized dogs, we studied the effects of positive end-expiratory pressure (PEEP), pulsed lung inflation and vagal electrical stimulation on four different expiratory motor outputs: triangularis sterni (TS) and internal intercostal (II) of the third intercostal segment, transversus abdominis (TA) and thyrohyoid (TH) electromyographic (EMG) activities. The first three muscles are thoracic or abdominal expiratory muscles (TS, II and TA), and the fourth (TH) an upper airway muscle. PEEP

* On leave of absence from the Medical Research Center, Polish Academy of Sciences, Warsaw, Poland.

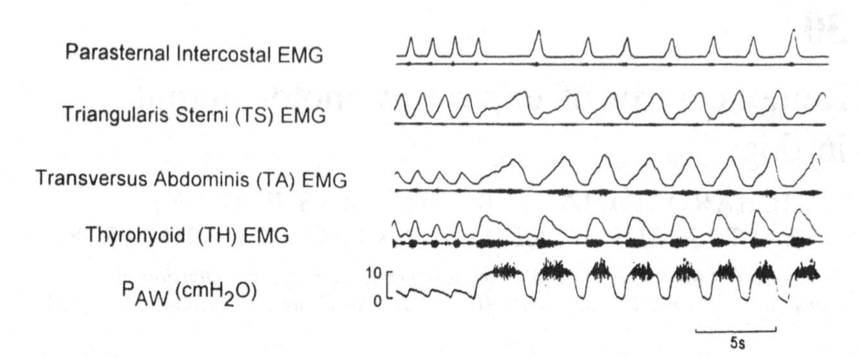

Fig. 30.1. Representative tracings of raw and integrated motor activities depicting the effect of PEEP on inspiratory and expiratory (decrementing and incrementing: Richter, 1982) activities in dogs. The bottom tracing is airway pressure (P_{AW}). A moderate PEEP evoked a biphasic response in TS and TA EMG activities. An initial inhibition was followed by an increase in EMG activities. The TH EMG was facilitated during PEEP. Time mark represents 5 s. (High-frequency noise on the P_{AW} record was produced by air bubbling through the column of water.)

and electrical vagal stimulation were maintained over several breaths (long-term stimulation), whereas lung inflation was applied within one expiratory phase (short-term stimulation).

Long-term effects of change in vagal input

The classic respiratory response to PEEP is shown in Fig. 30.1. During the expiratory load, T_E is prolonged and the amplitudes of TS and TA activities are augmented. On the basis of these responses, it was suggested that increases in lung volume reflexly 'promote' expiration (see von Euler, 1986). The first response to PEEP, however, is a decrease in the rate of rise of expiratory activity which suggests inhibitory rather than facilitatory effects of lung volume on expiratory motor activity. Since T_E is prolonged (see Fig. 30.1), expiratory activity may reach a higher peak amplitude despite a smaller rate of rise of activity. To explain the discrepancies between excitatory and inhibitory vagal influences on expiratory motor output it has been suggested (Fregosi *et al.*, 1990) that *tonic* vagal input is facilitatory for expiratory action, whereas phasic vagal input is inhibitory. Supporting this theory we have shown that threshold (Karczewski, Naslonska & Romaniuk, 1980), continuous electrical stimulation of vagus nerve may facilitate expiratory motor output (Romaniuk *et al.*, 1992). This facilitation increases with time of stimulation (see Fig. 30.2), depends on the pattern of stimulation and is not secondary to changes in chemical drive.

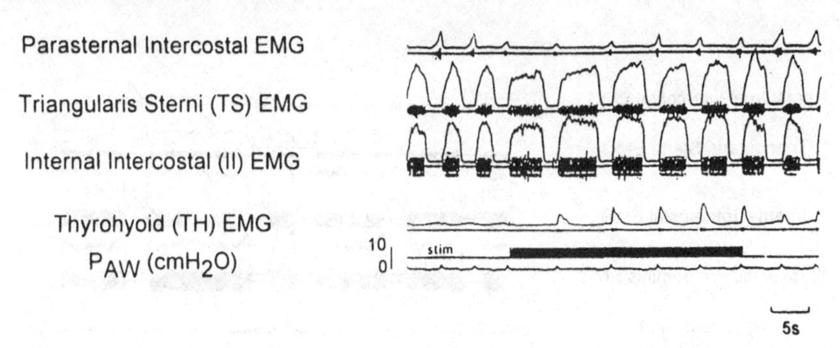

Fig. 30.2. The effect of continuous electrical stimulation on expiratory activities. Tracings are the same as in Fig. 30.1, but TA is replaced by internal intercostal (II) EMG activity. During stimulation, there was breath-by-breath increase in the amplitude of expiratory activities. The first response to stimulation was, however, an inhibition of TS EMG activity. Stim, stimulus marker.

Short-term effects of change in vagal input

Cohen *et al.* (1985) and Fregosi *et al.* (1990) showed that small lung inflations may facilitate whereas larger inflations may reversibly inhibit expiratory motor activity. We also found a similar effect of lung inflation on expiratory motor output which was dependent upon the size of inflation. The effects of pulsed lung inflation on chest wall expiratory muscle activity are shown in Fig. 30.3. In Fig. 30.3A a small lung inflation (~5 cm H_2O) enhances II and TA EMG activities and shortens T_E. In Fig. 30.3B greater inflation transiently inhibits the EMG activities of all three muscles and prolongs T_E. The facilitatory effects were observed only with small lung inflations. Following intravenous administration of 2 mg/kg pentobarbitone, expiratory facilitation could be reversed to inhibitory effects. The inhibitory effects depend on the degree of inflation, larger degrees of inflation (up to 15 cm H_2O) producing a greater decrease in expiratory motor output. We have also shown that expiratory reversible inhibition was time independent (Romaniuk *et al.*, 1991a).

In contrast to expiratory inhibition, T_E prolongation is time dependent (Knox, 1973). During expiration, the later an inflation was applied, the greater was the T_E prolongation. According to Knox (1973), inflation applied in the latter portion of expiration did not evoke any response. However, Marek, Prabhakar & Mikulski (1983), Trippenbach & Kelly (1985) and ourselves (Romaniuk *et al.*, 1991b) found that lung inflation applied in the last part of expiration may evoke a variety of responses, including T_E shortening. It was found previously, that superior laryngeal (Lewis *et al.*, 1990) or vagal (Sammon, Romaniuk & Bruce, 1993) afferent stimulation in mid-inspiration may randomly produce reversible inhibition or irreversible inspiratory termin-

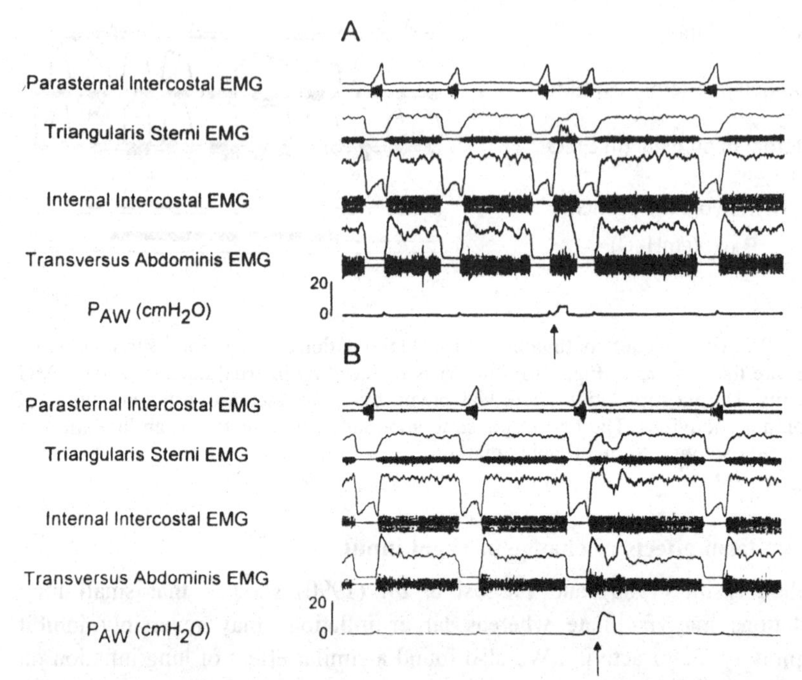

Fig. 30.3. The effects of different magnitudes of lung inflation applied at the beginning of expiratory activity on TS, II and TA EMG activities. The bottom tracing is airway pressure (P_{AW}). (A) A small lung inflation facilitated II and TA EMG activities and slightly inhibited TS EMG activity. (B) A larger lung inflation inhibited all expiratory motor activities.

ation (off-switch). We have shown that application of lung inflation in the last 40% of T_E may produce similar effects on expiratory motor activity, i.e. the expiration may be prolonged or switched into inspiratory phase (Romaniuk *et al.*, 1991b). This latter effect may represent a vagally mediated expiratory off-switch (Klages *et al.*, 1992).

Site of action of vagal input

Chest wall and upper airway expiratory activities may respond differently to peripheral input (St John & Zhou, 1989). We have shown that motor activity of upper airway expiratory muscle active during phase I of expiration is excited by lung inflation during expiration (Dick *et al.*, 1991), whereas activity of chest wall muscles is inhibited (Romaniuk *et al.*, 1991a). Fig. 30.4 shows the effects of prolonged lung inflation and its release on TS, TA and TH EMG activities. During inflation TS EMG is inhibited, TA EMG is transiently inhibited with following excitation, whereas TH EMG is facilitated. After release of inflation there is recovery of TS EMG activity, gradual

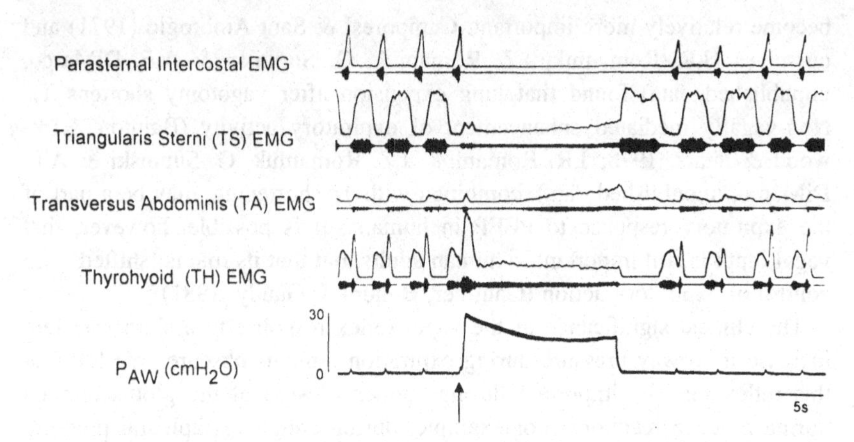

Fig. 30.4. The effect of prolonged lung inflation and its release on TS, TA and TH EMG activities. During inflation, TS EMG activity was inhibited, TA EMG activity was inhibited and subsequently facilitated, whereas TH EMG activity was enhanced. With the release of lung inflation, there was recovery of TS activity towards the control values, gradual decrease of TA EMG activity and complete cessation of TH EMG activity.

decrease of TA EMG and complete cessation of TH EMG activity. These results confirm previous suggestions (St. John & Zhou, 1989; Smith *et al.*, 1990) that peripheral feedback may change centrally generated expiratory activity differentially at upper airway, rib cage and abdominal motoneurone levels.

Functional and clinical significance

Vagal control of the expiratory phase (timing and amplitude of motor activity) is especially important in newborns. It is well known (Bodegard *et al.*, 1969; Mortola, 1987; Colin *et al.*, 1989) that newborns develop patterns of breathing in which end-expiratory volume is elevated by three different mechanisms: constriction of upper airways, action of expiratory abdominal muscles or increase in breathing frequency. All these mechanisms could be mediated vagally (Mortola, 1987; Fedorko, Kelly & England, 1988). It is possible that reflexes evoked by small lung inflations stimulate abdominal muscle activities and shorten T_E.

It was also suggested by Bishop *et al.* (1978) that there must be a reflex mechanism which, in humans, produces hyperventilation in response to PEEP – or, more specifically, reflex shortening of T_E in response to elevated lung volume. This reflex is opposite to vagally mediated prolongation of T_E during PEEP. It was shown previously (Bodegard *et al.*, 1969; Bodegard & Schwieler, 1971) that in humans, in the process of maturation, the Hering–Breuer inflation reflex declines in strength and the reflexes from thoracic wall become relatively more important. Camporesi & Sant'Ambrogio (1971) and ourselves (J.R. Romaniuk, J.Z. Romaniuk, G. Supinski & A.F. DiMarco, unpublished data) found that lung expansion after vagotomy shortens T_E. Non-vagally mediated enhancement of expiratory activity (Bainton, Kirkwood & Sears, 1978; J.R. Romaniuk, J.Z. Romaniuk, G. Supinski & A.F. DiMarco, unpublished data), combined with T_E shortening, may be a part of the expiratory response to PEEP in humans. It is possible, however, that vagal input is still important in human adults, but that its role is 'shifted' into control of expiratory action (Gauthier, Bonora & Gaudy, 1981).

The clinical significance of the vagal reflex response to a short transient increase in airway pressure during expiration remains obscure. We felt that this reflex may be important during sudden closure of the glottis elicited during defence reactions. For example, during cough, intrapleural pressure increases rapidly and transiently during expiration. However, preliminary evidence indicates that vagal afferent input is gated during cough (Romaniuk *et al.*, 1993). On the other hand, during swallowing in unconscious patients, a vagal reflex may take part in the integration of respiration and swallowing (Nishino & Hiraga, 1991). On-line control of expiratory motor output is important in the mechanism of expiratory braking in infants (Kosch & Stark, 1984). The full description of vagal reflexes during expiration may explain some of the breathing problems of premature infants during sleep (Stark et al., 1987). Therefore, even though the clinical significance of this reflex has yet to be elucidated, it has proved to be a useful tool in elucidating a control mechanism determining the phase duration of expiration.

Summary

On the basis of our experiments and recently published studies on the vagal control of expiratory phase, we suggest that the various expiratory muscle groups are controlled differentially by a set of reflexes similar to the ones earlier described for the inspiratory phase. The effects of vagal input on expiratory motor outputs can be either inhibitory or excitatory, dependent on the amount and timing of lung expansion and also on which expiratory muscles are observed.

Acknowledgement

This work was supported by the National Institutes of Health grant HL-34143

References

Arita, H. & Bishop, B. (1983). Responses of cat's internal intercostal motor units to hypercapnia and lung inflation. *Journal of Applied Physiology*, **54**, 375–386.

Bainton, C.R., Kirkwood, P.A. & Sears, T.A. (1978). On the transmission of the stimulating effects of carbon dioxide to the muscles of respiration. *Journal of Physiology (London)*, **280**, 249–272.

Bajic, J., Zuperku, E.J., Tonkovic-Capin, M. & Hopp, F.A. (1992). Expiratory bulbospinal neurones of dogs. I. Control of discharge patterns by pulmonary stretch receptors. *American Journal of Physiology*, **262**, R1075-R1086.

Barrett, J., Cerny, F., Hirsch, J.A. & Bishop, B. (1994). Control of breathing patterns and abdominal muscles during graded loads and tilts. *Journal of Applied Physiology*, **76**, 2473–2480.

Bartoli, A., Cross, B.A., Guz, A., Huszczuk, A. & Jefferies, R. (1975). The effects of varying tidal volume on the associated phrenic motoneurone output: studies of vagal and chemical feedback. *Respiration Physiology*, **25**, 135–155.

Bishop, B. (1967). Diaphragm and abdominal muscle responses to elevated airway pressures in the cat. *Journal of Applied Physiology*, **22**, 959–965.

Bishop, B., Hirsch, J. & Thursby, M. (1978). Volume, flow and timing of each breath during positive-pressure breathing in man. *Journal of Applied Physiology*, **45**, 495–501.

Bodegard, G. & Schwieler, G.H. (1971). Control of respiration in newborn babies. II. The development of the thoracic reflex response to an added respiratory load. *Acta Paediatrica Scandinavica*, **60**, 181–186.

Bodegard, G., Schwieler, G.H., Skoglund, S. & Zetterstrom, R. (1969). Control of respiration in newborn babies. I. The development of the Hering–Breuer inflation reflex. *Acta Paediatrica Scandinavica*, **58**, 567–571.

Camporesi, E. & Sant'Ambrogio, G. (1971). Influences on the respiratory rhythm originating from the lungs and the chest wall. *Pflügers Archiv*, **324**, 311–318.

Chung, K.F., Jones, P., Keyes, S.J., Morgan, B.M. & Snashall, P.D. (1987). Vagal control of end-expiratory lung volume in anaesthetized dogs. *Bulletin Européen de Physiologie Respiratoire*, **23**, 353–358.

Clark, R.F. & Euler, C. von (1972). On the regulation of depth and rate of breathing. *Journal of Physiology (London)*, **222**, 267–295.

Cohen, M.I., Feldman, J.L. & Sommer, D. (1985). Caudal medullary expiratory neurone and internal intercostal nerve discharges in the cat: effects of lung inflation. *Journal of Physiology (London)*, **368**, 147–178.

Colin, A.A., Wohl, M.E.B., Mead, J., Ratjen, F.A., Glass, G. & Stark, A.R. (1989). Transition from dynamically maintained to relaxed end-expiratory volume in human infants. *Journal of Applied Physiology*, **67**, 2107–2111.

Dick, T.E., Romaniuk, J.R., Supinski, G.S. & DiMarco, A.F. (1991). Excitation of post-inspiratory laryngeal activity by vagal input in dogs. *FASEB Journal*, **5**, A1479.

DiMarco, A.F., Euler, C. von, Romaniuk, J.R. & Yamamoto, Y. (1981). Positive feedback facilitation of external intercostal and phrenic inspiratory activity by pulmonary stretch receptors. *Acta Physiologica Scandinavica*, **113**, 375–386.

Euler, C. von (1986). Brainstem mechanisms for generation and control of breathing pattern. In *Handbook of Physiology*, sect. 3, *The Respiratory System*, vol. II, *Control of Breathing*, part I, ed. N.S. Cherniack & J.G. Widdicombe, pp. 1–67. Bethesda, MD: American Physiological Society.

Fedorko, L., Kelly, E.N. & England, S.J. (1988). Importance of vagal afferents in determining ventilation in newborn rats. *Journal of Applied Physiology*, **65**, 1033–1039.

Fregosi, R.F., Bartlett, D. Jr & St John, W.M. (1990). Influence of phasic volume feedback on abdominal expiratory nerve activity. *Respiration Physiology*, **82**, 189–200.

Gauthier, H., Bonora M. & Gaudy, J.H. (1981). Breuer–Hering reflex and breathing pattern in anesthetized humans and cats. *Journal of Applied Physiology*, **51**, 1162–1168.

Haxhiu, M.A., van Lunteren, E., Deal, E.C. Jr & Cherniack, N.S. (1988). Effect of stimulation of pulmonary C-fibre receptors on canine respiratory muscles. *Journal of Applied Physiology*, **65**, 1087–1092.

Karczewski, W.A., Nasłońska, E. & Romaniuk, J.R. (1980). Respiratory responses to stimulation of vagal afferent fibres. *Acta Neurobiologiae Experimentalis (Warsaw)*, **40**, 543–562.

Klages, S., Bellingham, M.C., Schwarzacher, S.W. & Richter, D.W. (1992). The termination of expiration in cats. In *Rhythmogenesis in Neurons and Networks*, ed. N. Elsner & D.W. Richter, p. 68. Stuttgart: Thieme.

Knox, C.K. (1973). Characteristics of inflation and deflation reflexes during expiration in the cat. *Journal of Neurophysiology*, **36**, 284–295.

Kosch, P.C. & Stark, A.R. (1984). Dynamic maintenance of end-expiratory lung volume in full-term infants. *Journal of Applied Physiology*, **57**, 1126–1133.

Lewis, J.L., Bachoo, M., Polosa, C. & Glass, L. (1990). The effects of superior laryngeal nerve stimulation on respiratory rhythm: phase-resetting and aftereffects. *Brain Research*, **517**, 44–50.

Marek, W., Prabhakar, N.R. & Mikulski, A. (1983). The influence of chemosensory, laryngeal and vagal afferents on respiratory phase-switching mechanisms and the generation of in- and expiratory efferent activities. In *Central Neurone Environment*, ed. M.E. Schläfke, H.P. Koepchen & W.R. See, pp. 197–203. Berlin: Springer.

Mortola, J.P. (1987). Dynamics of breathing in newborn mammals. *Physiological Reviews*, **67**, 187–243.

Nishino, T. & Hiraga, K. (1991). Coordination of swallowing and respiration in unconscious subjects. *Journal of Applied Physiology*, **70**, 988–993.

Polacheck, J., Remmers, J. & Younes, M. (1978). Effect of volume on expiratory neural output. *Federation Proceedings*, **37**, 806.

Richter, D.W. (1982). Generation and maintenance of the respiratory rhythm. *Journal of Experimental Biology*, **100**, 93–107.

Romaniuk, J.R., Dick, T.E., Supinski, G. & DiMarco, A.F. (1991a). Inhibition of expiratory muscle activity by vagal input in dogs. *FASEB Journal*, **5**, A1479.

Romaniuk, J.R., Dick, T.E., Bruce, E., Supinski, G. & DiMarco, A.F. (1991b). Shortening of the expiratory phase by lung inflation (LI) in dogs. *Society for Neuroscience Abstracts*, **17**, 621 (246.10).

Romaniuk, J.R., Dick, T.E., Supinski, G., Kotas, P. & DiMarco, A.F. (1992). Vagally mediated long-term facilitation of expiratory muscle activity in dogs. *FASEB Journal*, **6**, A1754.

Romaniuk, J.R., Dick, T.E., Sundaresan, V., Supinski G., Hudgel, D. & DiMarco, A.F. (1993). Vagal modulation of expiratory motor output during cough in anesthetized dogs. *FASEB Journal*, **7**, A403.

St John, W.M. & Zhou, D. (1989). Differing control of neural activities during various portions of expiration in the cat. *Journal of Physiology (London)*, **418**, 189–204.

Sammon, M., Romaniuk, J.R. & Bruce, E.N. (1993). Bifurcation of respiratory pattern produced with phasic vagal stimulation in the rat. *Journal of Applied Physiology*, **75**, 912–926.

Smith, C.A., Ainsworth, D.M., Henderson, K.S. & Dempsey, J.A. (1990). The influence of carotid body chemoreceptors on expiratory muscle activity. *Respiration Physiology*, **82**, 123–136.

Stark, A.R., Cohlan, B.A., Waggener, T.B., Frantz, I.D. III & Kosch, P.C. (1987). Regulation of end-expiratory lung volume during sleep in premature infants. *Journal of Applied Physiology*, **62**, 1117–1123.

Trippenbach, T. & Kelly, G. (1985). Expiratory effects of vagal stimulation in newborn kittens. *Journal of Applied Physiology*, **59**, 218–222.

Younes, M., Remmers, J. & Baker, J. (1978). Characteristics of inspiratory inhibition by phasic volume feedback in rats. *Journal of Applied Physiology*, **45**, 80–86.

Younes, M., Vaillancourt, P. & Milic-Emili, J. (1974). Interaction between chemical factors and duration of apnea following lung inflation. *Journal of Applied Physiology*, **36**, 190–201.

31

Abnormal thoraco-abdominal movements in patients with chronic lung disease

M. GOLDMAN

Department of Medicine, School of Medicine, University of California, Los Angeles, California, USA

Abnormal thoraco-abdominal movements have been used to infer disturbances of respiratory muscle actions for more than a half century. The present chapter reviews some significant advances in analysis of respiratory muscle action, focusing largely on human studies. We review a particular contribution of neuroscience: the integration of neurophysiological principles into conventional mechanical analyses. This integration has led to our present understanding of abnormal thoraco-abdominal movements in humans. While earlier qualitative interpretations emphasized descriptions of respiratory muscle 'disco-ordination' or 'asynchrony', the present discussion develops a new perspective of co-ordinated thoraco-abdominal movements in patients with chronic lung disease (chronic airflow obstruction, CAO). We conclude that apparently abnormal thoraco-abdominal movements are accounted for by passive motions and/or local non-uniformities of thoracic or abdominal movements which occur in the context of well co-ordinated neural commands to the various respiratory muscles.

Early descriptions of respiratory muscle actions were based on visual observation of patients. While mechanical measurements of respiratory pressure and volume changes significantly improved clinicians' understanding of respiratory muscle action, it was uniquely the inclusion of neurophysiological measurements done simultaneously with mechanical studies which yielded more comprehensive advances in understanding. Important mechanical studies were contributed by several workers (Rahn *et al.*, 1946; Campbell & Green, 1955; Campbell, 1958; Agostoni & Rahn, 1960; Agostoni & Mognoni, 1966; Grimby, Bunn & Mead, 1966). Significant studies which provided the neurophysiological infrastructure for analysis of respiratory muscle actions appeared during the 1950s and 1960s (Campbell & Green, 1953a, b; Petit, Milic-Emili & Delhez, 1960; Taylor, 1960; Delhez & Petit, 1966; Sears & Newsom-Davis, 1968).

Detailed analysis of respiratory muscle action developed from improvements in manometric and volumetric respiratory measurements and innovative presentation of mechanical data in the form of the pressure–volume diagram. When neural control of respiratory muscles was assessed simultaneously by addition of electromyographic (EMG) measurements, an integrated neuromechanical analysis of respiration was developed. Notable contributions to a comprehensive description of the act of breathing performed by co-ordinated actions of the various respiratory muscles resulted from the collaboration of neurophysiologists and respiratory 'mechanics'. Recent measurements of respiratory muscle EMG activity and mechanical changes in patients with CAO permit a new view of respiratory muscle 'co-ordination' or 'synchronization' in such patients. The viewpoint emphasized in the present discusson is that expiratory muscle activation during spontaneous breathing at rest is a prominent feature of the normal response to chronic obstructive lung disease. As a result, the normal phasic relationship of thoracic and abdominal movements is altered by the emergence of local non-uniformities of thoracic and/or abdominal wall movements, most notable in the supine posture.

The mechanical actions of the respiratory muscles may be analysed by the relationship between respiratory pressures and lung volume at different lung volumes: the PV diagram (Rahn *et al.*, 1946). The interchange of elastic potential energy between the lung and the thorax was clarified by analysing pressures measured inside the thorax (oesophagus) at different lung volumes during relaxation of the respiratory muscles relative to those measured during breathing (Campbell, 1958). This PV diagram represents the three-dimensional equivalent of the length–tension diagram of muscle. The departures during breathing from the relaxed PV diagram are thus equivalent to the 'active' tension developed by the muscle. In this way, the fundamental mechanical tool used in respiratory mechanical analyses may be viewed as a logical extension of the conventional neurophysiological description of the mechanics of muscle contraction.

A more comprehensive basis for understanding the actions of respiratory muscles is offered by integrating measurements of respiratory muscle EMG activity (in particular, of the abdominal muscles) with mechanical measurements, including gastric pressure and lung volume changes (Campbell & Green, 1953 a, b). This analysis was extended by using both oesophageal and gastric pressures and lung volume measurements (Agostoni & Rahn, 1960). Abdominal muscle actions were graphically represented by adding quantitative measurements of the separate movements of the rib cage and abdomen (Grimby *et al.*, 1966; Konno & Mead, 1967, 1968). This was the first use of an 'abdominal' PV diagram to characterize the mechanics of abdominal muscle contraction.

Mechanical analyses of respiratory muscle actions were further refined by quantifying the mechanical interaction between the diaphragm and rib cage (Goldman & Mead, 1973; Goldman, Grimby & Mead, 1976; Grimby, Goldman & Mead, 1976). This permitted separate graphical representations for the PV diagram of the diaphragm, intercostal and accessory muscles of the thorax, and the abdominal muscles. In particular, it provided for a mechanical analysis of thoracic muscle contraction independent of the diaphragm. Thus, improved understanding of respiratory muscle actions over a number of decades largely followed improvements in manometric and volumetric measurements of the respiratory system, and innovative presentation of mechanical data in the form of the pressure–volume diagram.

A unique collaboration between Sears and the Mead group then provided the basis for new contributions of neuroscience to an improved understanding of respiratory muscle action (Goldman *et al.*, 1978; Grassino *et al.* 1978). Quantitative measurements of EMG activity permitted analysis of respiratory muscle function without voluntary or pharmacological relaxation of the respiratory muscles. Human subjects could be assessed and compared quantitatively under different conditions during breathing at rest, during exercise or sleep, and in the presence of respiratory or neuromuscular disease.

One of the first descriptions of abnormal 'paradoxical' chest wall movements was that of inward inspiratory movements of the lower rib cage in patients with advanced cardiac and/or chronic lung disease (Hoover, 1920). This was ascribed to contraction of the diaphragm which was abnormally low and flat, secondary to cardiac enlargement and/or hyperinflation of the lung. Another early description of 'disturbance' of respiratory muscle action emphasized 'upper costal' movements without lateral expansion of the lower ribs and 'hypertonic' contraction of the sternocleidomastoid and scalene muscles during expiration in patients with chronic lung disease (Cournand *et al.*, 1936). These authors noted that what remained of 'costodiaphragmatic' respiration 'failed to synchronize with' the upper costal respiration.

A remarkable biphasic pattern of abdominal movements was reported during spontaneous breathing in patients with advanced chronic lung disease (Ashutosh *et al.*, 1975). Thoracic movements appeared to be normal in these individuals, and the abnormal abdominal movements were attributed to elevation of the rib cage by increased activation of thoracic inspiratory muscles, and straightening of the spine near end-inspiration. Abnormal 'paradoxical' abdominal movements were clearly demonstrated in patients with paralysed or paretic diaphragms (Newsom-Davis *et al.*, 1976). These movements were interpreted in the context of a normal pattern of neural commands to the respiratory muscles.

It was the work of Sears and his colleagues which first clearly elucidated abnormal thoraco-abdominal movements with particular reference to respiratory muscle EMG activity (Da Silva *et al.*, 1977). This group showed in anaesthetized cats that 'abnormal' or 'paradoxical' movements of the rib cage could occur in the face of physiological, normally synchronized neural respiratory commands. Lateral rib cage movements showed a clearly non-uniform distribution along the cephalocaudal axis which corresponded with the distribution of 'passive' (mechanical, structural-dependent) resistance to applied muscle forces. Interestingly, the sternum moved in the opposite direction of the overall thoracic cage movements. This passive response to respiratory muscle forces applied to the anterior ends of the ribs occurred as a result of the 'floating' nature of the sternum, suspended by the costal cartilage from the costochondral articulations. This model of non-uniform thoracic movements which occur in response to normally synchronized respiratory muscle efforts and neural respiratory commands provides the rationale for the present analyses of measurements in humans. This author's own experience recapitulates the thematic evolution of our present understanding of respiratory muscle actions in the production of 'abnormal' or 'asynchronous' thoraco-abdominal movements.

During voluntary attempts in humans to expire below resting lung volume, a clear difference had been observed between abdominal wall movements in the erect and supine postures: when sitting or standing, the abdominal wall moved inwards (normally) during active expiration below resting lung volume, whereas in the supine posture, it moved paradoxically outwards (Goldman & Mead, 1973). The clinical relevance of this observation became clear only after collaborative studies at the Institute of Neurology in London (Newsom-Davis *et al.*, 1976), when a patient with Shy–Drager syndrome was observed to show inward movements of the anterior walls of *both* the abdomen and the rib cage during inspiration while asleep in the supine posture. Marked inspiratory stridor occurred during sleep because of failure of vocal cord abduction, and close scrutiny of the patient's thorax and abdomen revealed a marked change in shape during his forced inspiration through the high glottic resistance: the lateral dimensions of thorax and upper abdomen increased sharply in response to strong inspiratory muscle contraction, while the anteroposterior dimension decreased.

Subsequently, recordings in supine anaesthetized dogs in the author's laboratory (Nochomovitz *et al.*, 1981) were observed to show paradoxical outward movement of the anterior abdominal wall during expiration, associated with EMG activity of internal intercostal muscles. During subsequent inspiration the anterior abdominal wall moved paradoxically inwards. Lateral movements of the abdominal wall were synchronous with those of the thorax: both structures moved normally outwards during inspiration and inwards during expiration. It was concluded that the apparent abdominal 'paradox' reflected a local change of abdominal shape: during active expiration (driven by activity of thoracic expiratory muscles), the abdomen became more circular in cross-section, with a decrease in the lateral and increase in the anteroposterior diameters, while during inspiration, it returned to a more elliptical shape. In this circumstance, local paradoxical movements were attributable simply to changes of shape of the abdominal wall, a conclusion supported by the pattern of pressures, volumes and respiratory muscle EMG activities.

These measures provided a comprehensive description of normally coordinated respiratory muscle commands leading to local 'asynchrony.'

Subsequent unpublished observations in critically ill human patients with CAO showed the same pattern of thoraco-abdominal movements in the supine posture. It was inferred that active expiration with movements of the rib cage predominating, caused the obvious change in shape of the abdomen, leading to local paradox of the anterior abdominal wall during both phases of respiration.

The present observations were made in humans by integrating respiratory muscle EMGs with volume and pressure measurements. It has recently been shown in children with bronchopulmonary dysplasia that a biphasic abdominal respiratory pattern occurs during non-REM (but not REM) sleep (Goldman et al., 1993). During each inspiratory phase of the rib cage cycle, abdominal motion is paradoxical (in the expiratory direction) at the onset of inspiration, and then is in phase (inspiratory) during the second half of inspiration. The abdomen remains in phase with airflow during the first portion of expiration, but moves paradoxically outwards during the latter half to two-thirds of expiration. Thus the apparent respiratory frequency of abdominal movements is twice that of rib cage movements. The paradoxical abdominal expiratory movements occur despite brisk abdominal EMG activity in some patients, suggesting that they are passive, driven by simultaneous contraction of thoracic expiratory muscles causing an increase in abdominal pressure. In contrast, during REM sleep, with a marked decrease in expiratory and accessory inspiratory muscle activity, motion of the abdomen is clearly in phase with airflow at all times, and is driven by active contraction (and relaxation) of the diaphragm. The rib cage manifests paradoxical movements at the onset of both inspiration and expiration secondary to its increased compliance relative to intrathoracic pressure changes.

These interpretations are examined further in the data in Fig. 31.1A and B, from a child with bronchopulmonary dsyplasia during REM and non-REM sleep, respectively, showing (from the top down) rib cage and abdominal movements, gastric pressure, and EMG from the lower lateral thoracic surface. The thoracic surface electrodes are placed anterior to the anterior margin of the external intercostal muscles in order to record diaphragmatic activity with minimal surface interference from inspiratory intercostal muscles. In Fig. 31.1A, the onset of inspiration is shown by the brisk increase in diaphragmatic EMG activity, and sharply defined increases in gastric pressure and outward movement of the abdomen. Rib cage movements are paradoxical at the onset of inspiration before moving normally outwards during the latter portion of inspiration. At the onset of expiration, there is an abrupt cessation of diaphragmatic EMG activity, and sharp falls in gastric pressure and abdominal tracings. The rib cage continues to move outwards briefly before moving normally inwards again for most of the latter portion of expiration.

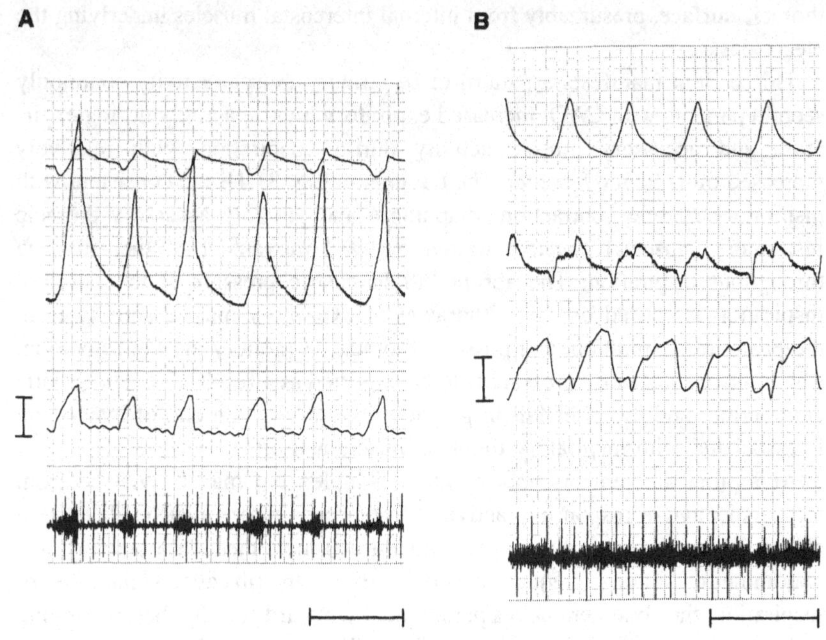

Fig. 31.1. From top to bottom, tracings show rib cage movements, abdominal movements, intragastric pressure changes, and electromyographic (EMG) activity recorded from the lower lateral thoracic surface in a child with bronchopulmonary dysplasia. (A) REM sleep; (B) non-REM sleep. Rib cage and abdominal movements were measured by respiratory inductance plethysmography, with increasing circumference upward on the tracings. Calibrations: pressure, 20 cm H_2O; time, 5 s. See text for discussion.

These tracings are consistent with passive movements of both the abdomen and rib cage throughout both phases of respiration, driven by activity of the diaphragm alone. In REM sleep, diaphragmatic activity is preserved, while activity of expiratory and accessory inspiratory muscles is markedly decreased or absent. The paradoxical movement of the rib cage occurs passively, in response to changes in intrathoracic pressure.

In constrast, during non-REM sleep expiratory and accessory inspiratory muscles may show brisk activity in patients with CAO. In Fig. 31.1B, the lateral thoracic surface EMG signal shows activity in both phases of respiration. The onset of inspiration is marked by clear outward movement of the rib cage, in phase with airflow (not shown) and a large fall in gastric pressure. Immediately thereafter, there is a brief burst of diaphragmatic EMG activity, associated with a small rise in gastric pressure. In late inspiration there is a brief fall in gastric pressure associated with a decrease in diaphragmatic EMG activity. End-inspiration is marked by the peak of the rib cage signal. Throughout expiration, the rib cage moves inwards and gastric pressure increases, associated with a second increase in EMG activity from the lateral thoracic surface, presumably from internal intercostal muscles underlying the electrodes.

This patient manifests a pattern of respiratory muscle activity commonly seen in patients with CAO: increased expiratory muscle activity during expiration and accessory muscle activity during inspiration, with relatively increased thoracic movements. The tracings in Fig. 31.1B are consistent with inspiratory muscle contraction (diaphragm and possibly accessory thoracic inspiratory muscles) causing normal outward movement of the rib cage throughout inspiration. The abrupt fall in gastric pressure at the onset of inspiration is caused by the withdrawal of expiratory muscle activity from the preceding expiration, and possibly thoracic inspiratory muscle activation. The brief rise in gastric pressure is to be expected as a result of diaphragmatic contraction, and the brief fall in gastric pressure near the end of inspiration is associated with inspiratory thoracic muscle activity.

Abdominal movements appear rather complex but may be viewed from the perspective of response to activity of other respiratory msucles developed above. With respect to the rib cage, abdominal movements are biphasic, i.e. abdominal respiratory 'frequency' is twice that of the rib cage. At the onset of inspiration, the abdomen moves paradoxically inwards briefly, before moving outwards in response to contraction of the diaphragm. At the onset of expiration, there is another brief period of paradoxical (outward) movement before moving inwards for the majority of the duration of expiration. The movements of the abdomen throughout inspiration are passive, since intragastric pressure and abdominal dimensions decrease, and then increase and decrease together. In the early part of expiration, the brief outward movement of the abdomen occurs passively in response to the abrupt rise in gastric pressure. Thereafter, the abdomen moves actively inwards, with active contraction of expiratory muscles. During the latter three-quarters of expiration, the active movement of the abdomen is signalled by the increase in gastric pressure with inward movement of the abdomen. Thus although the thoracic surface EMG is not precise in defining which muscles are active, it appears in Fig. 31.1B that activity of abdominal and thoracic expiratory muscles is occurring simultaneously.

In summary, integration of neurophysiological measurements with conventional mechanical data offers increased understanding of abnormal thoraco-abdominal movements during respiration. In the examples shown, the apparently complex and 'unco-ordinated' movements of the abdomen may be readily explained by passive and active movements in response to normally co-ordinated respiratory muscle activation. The unusual appearance occurs because of increased accessory inspiratory and expiratory muscle activity in response to chronic lung disease. The development of a comprehensive neuromechanical analysis of respiratory muscle actions is still evolving. Patients with chronic lung disease have only recently been analyzed by incorporating the mechanical insights provided by Sears and his co-workers into conventional mechanical and EMG data analysis. The present state of our knowledge suggests that paradoxical abdominal movements may appear relatively early in the course of neuromuscular disease, with paresis of the diaphragm. In contrast, abdominal paradox during quiet breathing corresponds to a relatively more severe degree of disease progression in patients with chronic lung disease, with either (or both) substantial compromise of the diaphragm or an abnormal drive to active expiration, due to increased carbon dioxide levels. More extensive measurements in patients with chronic lung disease are needed. It is this author's view that the neuromechanical insights first elucidated by Sears and his co-workers will lead to improved understanding of respiratory motor control and degree of disease progression in patients with chronic lung disease in a way analogous to previously defined contributions to clinical neurological patients.

References

Agostoni, E. & Mognoni, P. (1966). Deformation of the chest wall during breathing efforts. *Journal of Applied Physiology*, **21**, 1827–1832.

Agostoni, E. & Rahn, H. (1960). Abdominal and thoracic pressures at different lung volumes. *Journal of Applied Physiology*, **15** 1087–1092.

Ashutosh, K., Gilbert, R., Auchincloss, J.H. & Peppi, D. (1975). Asynchronous breathing movements in patients with chronic obstructive pulmonary disease. *Chest*, **67**, 553–557.

Campbell, E.J.M. (1958). *The Respiratory Muscles and the Mechanics of Breathing*. London: Lloyd-Luke (Medical Books).

Campbell, E.J.M. & Green, J.H. (1953a). The expiratory function of the abdominal muscles in man: an electromyographic study. *Journal of Physiology (London)*, **120**: 409–418.

Campbell, E.J.M & Green, J.H. (1953b). The variations in intra-abdominal pressure and activity of the abdominal muscles during breathing: a study in man. *Journal of Physiology (London)*, **122**, 282–290.

Campbell, E.J.M & Green, J.H. (1955). The behaviour of the abdominal muscles and the intra-abdominal pressure during quiet breathing and increased pulmonary ventilation: a study in man. *Journal of Physiology (London)*, **127**, 423–426.

Cournand, A., Brock, H.J., Rappaport, J. & Richards, D.W, Jr (1936). Disturbance of action of respiratory muscles as a contributing cause of dyspnea. *Archives of Internal Medicine*, **57**, 1008–1011.

Da Silva, K.M.C., Sayers, B.M.A., Sears, T.A. & Stagg, D.T. (1977). The changes in configuration of the rib cage and abdomen during breathing in the anaesthetized cat. *Journal of Physiology (London)*, **266**, 499–521.

Delhez, L. & Petit, J.M. (1966). Données actuelles de l'electromyographie respiratoire chez l'homme normal. *Electromyography*, **6**, 101–146.

Goldman, M.D., Grassino, A., Mead, J. & Sears, T.A. (1978). Mechanics of the human diaphragm during voluntary contraction: dynamics. *Journal of Applied Physiology: Respiratory, Environmental, Exercise Physiology*, **44**: 840–848.

Goldman, M.D., Grimby, G. & Mead J. (1976). Mechanical work of breathing derives from rib cage and abdominal V–P partitioning. *Journal of Applied Physiology*, **41**, 752–763.

Goldman, M.D. & Mead, J. (1973). Mechanical interaction between the diaphragm and rib cage. *Journal of Applied Physiology*, **35**, 197–204.

Goldman, M.D., Pagani, M., Trang, H., Praud, J., Sartene, R. & Gaultier, C. (1993). Asynchronous chest wall movements during non-rapid eye movement and rapid eye movement sleep in children with bronchopulmonary dysplasia. *American Review of Respiratory Disease*, **147**, 1175–1184.

Grassino, A., Goldman, M.D., Mead, J. & Sears, T.A. (1978). Mechanics of the human diaphragm during voluntary contraction: statics. *Journal of Applied Physiology: Respiratory, Environmental, Exercise Physiology*, **44**, 829–839.

Grimby, G., Bunn, J. & Mead, J. (1966). Relative contribution of rib cage and abdomen to ventilation during exercise. *Journal of Applied Physiology*, **24**, 159–166.

Grimby, G., Goldman, M. & Mead, J. (1976). Respiratory muscle action inferred from rib cage and abdominal V–P partitioning. *Journal of Applied Physiology*, **41**, 739–751.

Hoover, C.F. (1920). Definitive percussion and inspection in estimating size and contour of heart. *Journal of the American Medical Association*, **75**, 1626–1630.

Konno, K. & Mead, J. (1967). Measurement of the separate volume changes of rib cage and abdomen during breathing. *Journal of Applied Physiology*, **22**, 407–422.

Konno, K. & Mead, J. (1968). Static volume–pressure characteristics of the rib cage and abdomen. *Journal of Applied Physiology*, **24**, 544–548.

Newsom-Davis, J., Goldman, M., Loh, L. & Casson, M. (1976). Diaphragm function and alveolar hypoventilation. *Quarterly Journal of Medicine*, **45**, 87–100.

Nochomovitz, M.L, Goldman, M.D., Mitra, J. & Cherniack, N.S. (1981). Respiratory responses in reversible diaphragm paralysis. *Journal of Applied Physiology: Respiratory, Environmental, Exercise Physiology*, **51**, 1150–1156.

Petit, J.M., Milic-Emili, G. & Delhez, L. (1960). Role of the diaphragm in breathing in conscious normal man: an electromyographic study. *Journal of Applied Physiology*, **15**, 1101–1106.

Rahn, H.A., Otis, B., Chadwick, L.E. & Fenn, W.O. (1946). The pressure-volume diagram of the thorax and lung. *American Journal of Physiology*, **146**, 161–178.

Sears, T.A. & Newsom-Davis, J. (1968). The control of respiratory muscles during voluntary breathing. *Annals of the New York Acadamy of Sciences*, **155** 183–190.

Taylor, A. (1960). The contribution of the intercostal muscles to the effort of respiration in man. *Journal of Physiology (London)*, **151**, 390–402.

32

Respiratory rhythms and apnoeas in the newborn

BERNARD DURON

Laboratoire de Neurophysiologie, URA 1331 CNRS 10, Amiens, France

In newborns of many species, breathing patterns are characterized by a very irregular rhythm interrupted by high-frequency respiratory periods (Mortola, 1984) and by the development of spontaneous apnoeas. Inspiratory activities of the phrenic nerve and diaphragm in newborns and particularly in preterm babies (Duron, Khater-Boidin & Wallois, 1991) consist of bursts of action potentials of short duration (50–60 ms) and low frequency (4–6 Hz). Moreover, in kittens, during eupnoea there exists relatively weak neuronal inspiratory activity (Bystrzycka, Nail & Purves, 1975; Marlot & Duron, 1976; Goldberg & Milic-Emili, 1977). In the newborn kitten, the inspiratory time is of short duration and the average duration of phrenic motor unit discharge does not exceed 500 ms (Duron & Marlot, 1979). The inspiratory time progressively lengthens during postnatal development, at the same time as the discharge pattern of phrenic motor units changes.

At birth, with regard to early inspiratory motor units (Hilaire, Monteau & Dussardier, 1972), we observed a very rapid increase in discharge frequency, which reached values of around 60 Hz, very clearly higher than those found in the adult animal. The end of the discharge is sudden, suggesting the intervention of powerful inhibitory mechanisms (Duron & Marlot, 1979). Moreover, in various experimental procedures (anaesthetized or decerebrate preparations), bilateral vagotomy, which in adults reinforces central inspiratory activity, induces prolongation and reinforcement of expiration in newborn animals (Marlot & Duron, 1979a). As shown in Fig. 32.1, bilateral vagotomy not only increases the expiratory time but also provokes the appearance of electrical activity in the expiratory muscles. During the postnatal period, a progressive prolongation of the inspiratory time appears, with a concomitant shortening of the expiratory time after bilateral vagotomy. Thus, at birth, the respiratory cycle is characterized by a weakness of the central inspiratory activity which contrasts with the powerful activity of the expiratory muscles.

The predominance of expiratory activity can be seen as a part of the general motor behaviour, which in the newborn is essentially characterized by

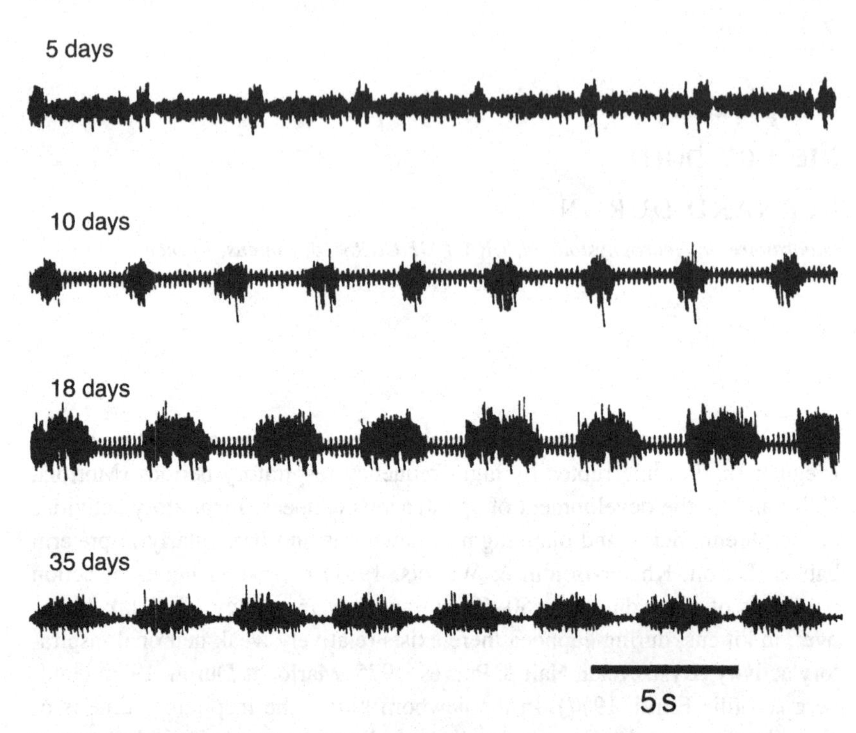

Fig. 32.1. Respiratory effects of bilateral vagotomy during postnatal development. Electromyograms of the second intercartilaginous muscle in decerebrated bilaterally vagotomized kittens are shown between the ages of 5 days and 35 days. In the 5-day-old kitten one can observe the important activity of the triangularis sterni picked up by the electrodes implanted in the second intercartilaginous muscle. Note the progressive increase in the inspiratory time and the progressive decrease in the expiratory time.

activation of the flexor muscle group to which the expiratory muscles could belong. As Weed (1917) and Skoglund (1960) showed, decerebrate newborn kittens exhibit a flexor activity in all muscles. During the postnatal period, extensor rigidity follows a cephalo-caudal direction, appearing in the upper limbs during the fourth week of postnatal life. The thoracic level is reached during the fifth and sixth weeks, at the same time as respiratory function begins to attain adult features (Duron & Marlot, 1980). Thus, one can postulate that in newborns the nervous control of breathing is largely dominated by the non-specific control of the reticular formation which controls predominantly the flexor muscles. The weakness of inspiratory activity could result either from an immaturity of the central inspiratory network or from a powerful inhibitory action of various peripheral afferent systems.

Indeed, in spite of marked vagus nerve immaturity observed histologically (Marlot & Duron, 1979b), discharge patterns of afferent units are similar in kittens and in adult cats (Marlot & Duron, 1979a). Nevertheless, the

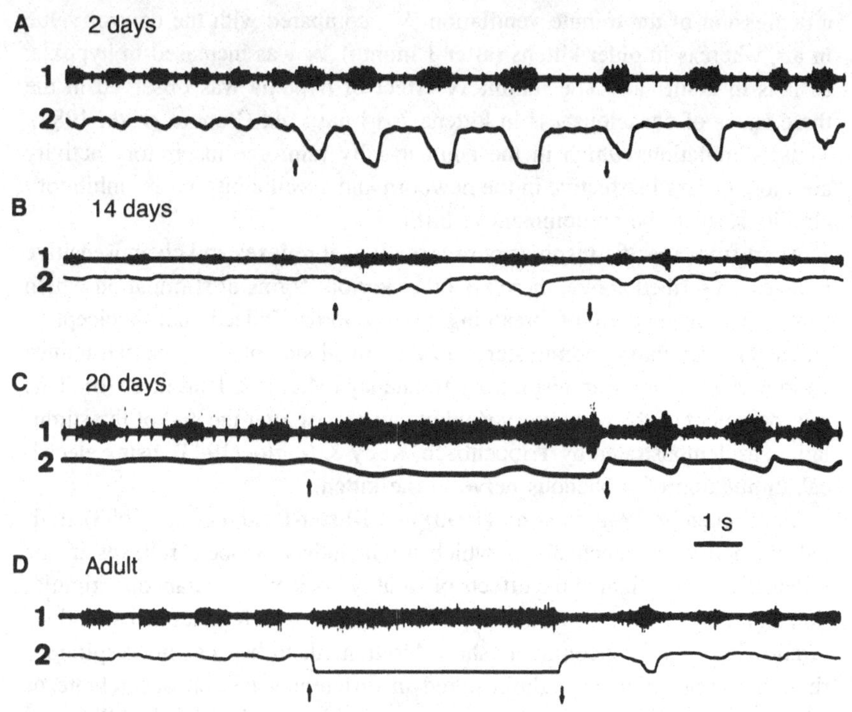

Fig. 32.2. Postnatal evolution of the Hering–Breuer deflation reflex. In each sequence the arrows indicate the onset and the end of the deflation. 1, diaphragm electromyogram; 2, oesophageal pressure. Note the progressive transition from a 'phasic' response in the 2-day-old kitten (A) to the 'tonic' response of the adult cat (D).

inhibitory action of slowly adapting pulmonary stretch receptors, which exists in all species studied (Trippenbach, Kelly & Marlot, 1985), is stronger at birth than later during maturation (Bodegard *et al.*, 1969; Marlot & Duron, 1979a; Javorka, 1984). On the other hand, published results concerning the Hering–Breuer deflation reflex are somewhat conflicting. Indeed in newborn rabbits (Dawes & Mott, 1959; Trippenbach *et al.*, 1985) and in kittens (Trippenbach & Kelly, 1985), lung deflation induces an increase in respiratory frequency and an augmentation of phrenic activity. In our experiments on anaesthetized or decerebrate kittens, tracheal occlusion performed at the end of expiration or strong pulmonary deflation (−10 to −20 mmHg pressure) provoked a reinforcement of inspiratory muscle activity accompanied by a lengthening of the expiratory time and often by a reinforcement of expiratory muscle activity. This 'phasic' response differs from the 'tonic' inspiratory reinforcement generally observed in adult cats which is present throughout lung deflation. During the postnatal period, the adult response appears progressively around the third week (Fig. 32.2).

Similarly, the ventilatory response to hypoxia is characterized at birth by a depression of the minute ventilation, V_E, compared with the control value in air, whereas in older kittens (after 1 month) V_E was increased in hypoxia, as it is in adult cats. The inhibitory effect of hypoxia was observed in the three states of consciousness in kittens 7–14 days old (Bonora *et al.*, 1984). Thus, stimulations which in the adult usually reinforce inspiratory activity are more or less ineffective in the newborn and specific inspiratory inhibitory circuits seem to be predominant at birth.

Apart from specific respiratory reflexes (vagal reflexes and chemosensitive reflexes) described above, there are other various forms of stimulation which can affect the pattern of breathing in newborns. Indeed, non-nociceptive cutaneous stimulation administered to the dorsal side of the neck can induce an important increase in respiratory frequency (Marlot & Duron, 1976). This reinforcement of the respiratory rhythm can persist after the end of the stimulation, as demonstrated by Trippenbach, Kelly & Marlot (1983), using electrical stimulation of saphenous nerve in the kitten.

More recently, Wallois *et al.* (1993) and Khater-Boidin *et al.* (1993) studied the non-vagal mechanisms which might induce apnoeic reflexes in the kitten. They investigated the effects of weakly nociceptive cutaneous stimulation and those of various forms of oral stimulation which in adults can induce swallowing and concomitant short-duration disturbances in respiratory rhythm. Stimulation was administered in different ways, on decerebrate or anaesthetized animals. Cutaneous stimulation by pinching and pulling the skin was applied on the ventral side of the neck, the thorax and the abdomen. Oral stimulation was made either by a small cotton-wool tip or by water instilled in the mouth (0.05–0.2 ml). In 10 animals, one of the two lingual nerves was stimulated electrically using bipolar silver electrodes. Electrical stimulation consisted of a series of pulses at 20 Hz, 50 μs at 3 V for 1 s.

Electrical activity of respiratory muscles was recorded by means of bipolar electrodes as previously described (Duron & Caillol, 1971). Inspiratory activity was detected by electrodes implanted in the second juxtasternal intercartilaginous muscle. This technique avoids laparotomy and the risk of pneumothorax following implantation of electrodes in the diaphragm. In addition, oesophageal pressure was recorded by means of a small balloon connected to a manometric transducer. This procedure enabled the indirect detection of the possible persistence of diaphragmatic movements, especially during inhibition of intercartilaginous muscle inspiratory activity. Electrodes inserted in the lateral internal intercostal muscles of the ninth or tenth intercostal spaces picked up expiratory activity. Electromyograms of two muscles innervated by the hypoglossal nerve (sternohyoid and geniohyoid muscles) were simultaneously recorded by means of electrodes similar to those used for the chest wall muscles.

At birth and during the first 10 days of life, regardless of the preparation (anaesthetized or decerebrate), such cutaneous stimulation systematically

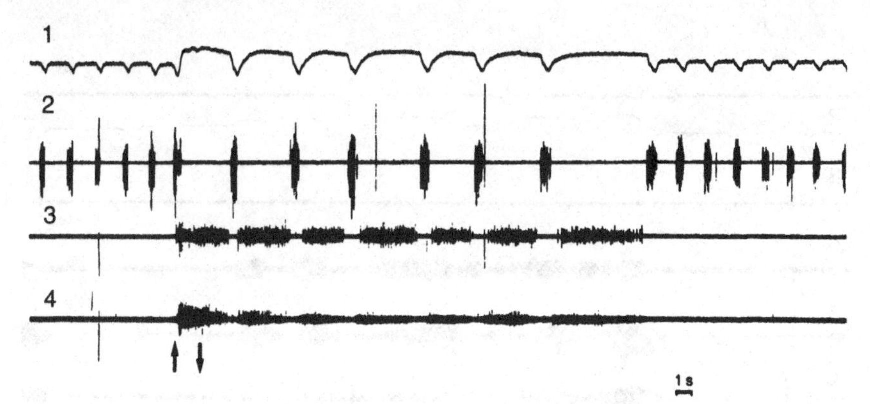

Fig. 32.3. Effects observed on both chest wall and hypoglossal innervated muscle activities in a 1-day-old kitten after weakly nociceptive stimulation of the ventral side of the abdomen. Arrows indicate onset and termination of stimulation. 1, oesophageal pressure; 2, second intercartilaginous muscle inspiratory activity; 3, ninth internal intercostal muscle expiratory activity; 4, sternohyoid muscle activity.

induced expiratory apnoeas of from 3 to 15 s duration, i.e. as much as 9 times the control expiratory time. Apnoeas were characterized by an inhibition of inspiratory activities which were observed both in intercartilaginous muscle activity and in the oesophageal barogram (Fig. 32.3). At the same time a reinforcement both of expiratory muscle activity and of hypoglossal-innervated muscle activity developed. After 2 weeks of age the effect of cutaneous stimulation, when present, was progressively less marked. Apnoeas gave way to bradypnoeas. The effect disappeared at 3–4 weeks of age.

Since no marked differences have been observed between anaesthetized or decerebrate animals, the general predominant expiratory pattern observed after cutaneous stimulation cannot be regarded as motor behaviour implying cerebral control but results from the activation of reflexes in which the brain-stem and spinal cord are involved. Thus, the reinforcement of expiratory activity observed with nociceptive cutaneous stimulation could be considered as a part of the more general flexor afferent reflex which is one of the main characteristics of motor behaviour in newborn mammals. Inhibition of inspiratory activity observed with nociceptive cutaneous stimulation contrasts with previous results showing that non-nociceptive cutaneous stimulation accelerate respiratory rhythm and reinforce inspiratory activity (Trippenbach *et al.*, 1983). Moreover, the effect of cutaneous stimulation could depend not only on the type of afferent fibres activated but also on the body site of stimulation, namely the dorsal or the ventral side.

As showed by Wallois *et al.* (1993), small amounts of water (0.05–0.15 ml) instilled in the mouth of young (0–14 days) anaesthetized or decerebrate kittens, in the prone or supine position respectively, are sufficient to provoke prolonged apnoeas (Fig. 32.4A). During the first week of postnatal life the

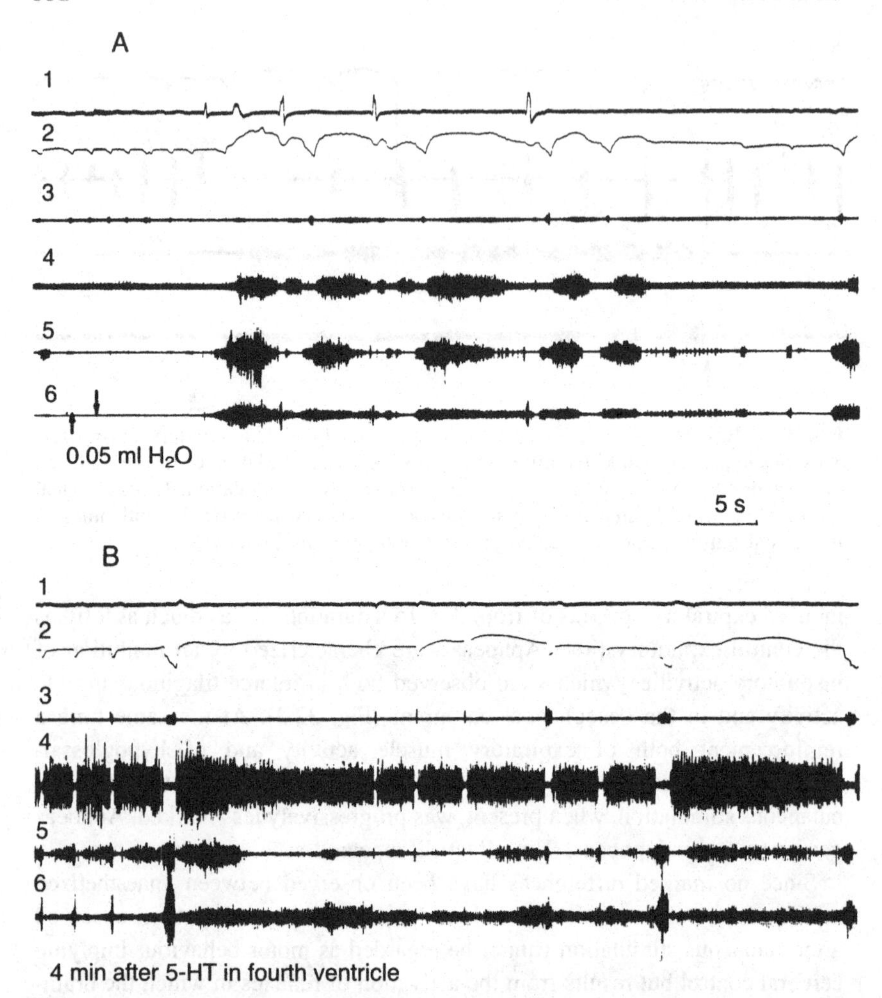

Fig. 32.4. Comparative effects in a 5-day-old kitten of oral stimulation by water (A) and an application of 0.1 ml 5-HT 0.1 M to the floor of the fourth ventricle (B). In each sequence: 1, retrolaryngeal pressure; 2, oesophageal pressure; 3, second intercartilaginous muscle inspiratory activity; 4, ninth internal intercostal muscle expiratory activity; 5, sternohyoid muscle activity; 6, geniohyoid muscle activity. Note the similar effects observed for the two types of stimulation.

duration of apnoeas can reach 60 s, i.e. approximately 60 times the control expiratory time. Depending on the level of anaesthesia, apnoeas occur between the first and the tenth respiratory cycles after the onset of the stimulation. Moreover, the duration of apnoeas induced by such oral stimulations seems to be linked not only to age but also to the amount of water instilled in the mouth. Nevertheless, in some experiments, prolonged apnoeas developed after weak stimulation (0.05 ml water). In kittens older than 3 weeks, regardless of the amount of water instilled in the mouth, instead of prolonged apnoeas (which became rare), bradypnoea occurred and eventually repetitive fits of coughing with no inhibition of inspiratory activity.

The marked and prolonged apnoeas observed in the youngest animals must be considered as a characteristic response in the postnatal period and could be similar to apnoeic reflexes resulting from nasal stimulation (Kratschmer, 1870), including the particular diving response (Korpas & Tomori, 1979) which characterizes the respiratory response in birds and aquatic mammals following peri- and intranasal stimulation. Long-duration apnoeas observed in kittens during the first days of the postnatal period following weak stimulation (0.05 ml water) could reflect the transient persistence of a lower phylogenetic stage.

It is difficult to confirm, particularly when animals are in the supine position, that water stimulates the buccal cavity only, without any action on the oropharynx. Indeed, apnoeas have been observed following laryngeal and epiglottic stimulation (Storey, 1975; Javorka, Tomori & Zavarska, 1980). Nevertheless, the similar effects observed following cotton-wool stimulation or lingual nerve electrical stimulation indicate that the buccal cavity itself constitutes a trigger zone for the apnoeic reflex and implies a role for trigeminal afferents. Moreover, the results of Wallois *et al.* (1993) suggest that the mechanisms which co-ordinate swallowing and breathing are immature at birth.

Thus, the irregularity of the respiratory rhythm and particularly the apnoeas which characterize the breathing pattern in the newborn, could result from the conjunction of an immaturity of the central inspiratory network and a potent inhibitory action of various afferent peripheral systems which simultaneously activate the central expiratory drive.

Data obtained in various experimental conditions in cat or in *in vitro* preparations such as the brainstem–spinal cord preparation (Morin *et al.*, 1990; Lindsay & Feldman, 1993) or slices (Berger, Bayliss & Viana, 1992) of the newborn rat, underline the role played by serotonin (5-HT) in both the central control of respiration and the control of the hypoglossal complex (see also Chapter 28). We have recently attempted to investigate *in vivo* the central action of 5-HT on the nervous control of breathing in the kitten during the postnatal period (Khater-Boidin *et al.*, 1993). A few micromoles of 5-HT (solution of 5-HT 0.1 M in deionized water with dimethylsulphoxide) was superfused onto the floor of the fourth ventricle 2 hours after dissipation of fluothane anaesthesia in decerebrate kittens.

At birth and during the first week of postnatal life, 5-HT applied to the floor of the fourth ventricle induced an important reinforcement of both the expiratory muscle activity and that of the hypoglossal innervated muscles, which appears with a short delay (30–60 s) after the injection time. At the same time prolonged apnoeas appear (Fig. 32.4B). The breathing pattern in

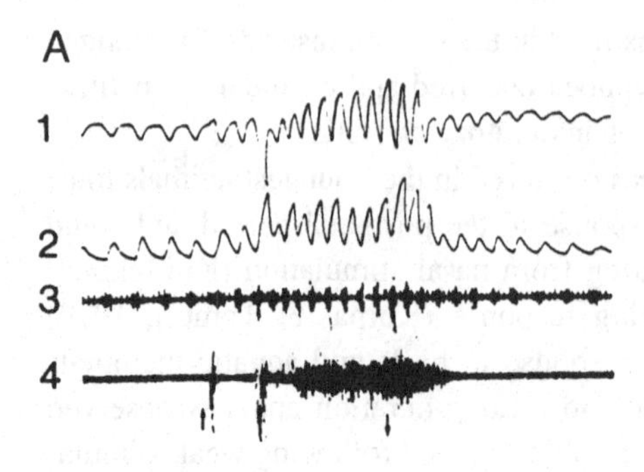

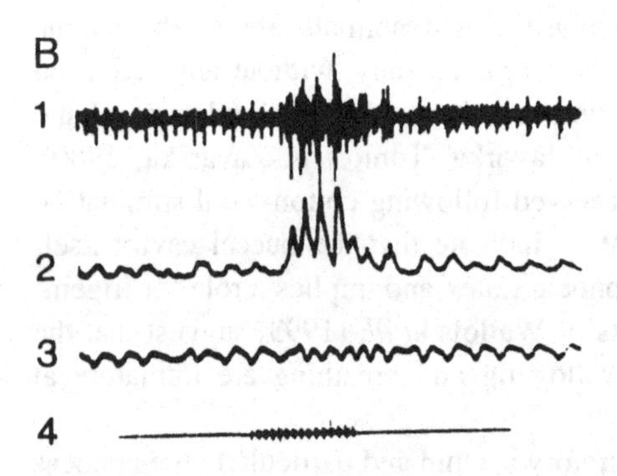

Fig. 32.5. Comparative effects of non-nociceptive cutaneous stimulation of the back of a 1-day-old kitten (A) and of a premature baby (B). (A) 1, oesophogeal pressure; 2, integrated diaphragm electromyogram; 3, diaphragm electromyogram; 4, hindlimb flexor muscle activity. (B) 1, diaphragm electromyogram; 2, integrated diaphragm electromyogram; 3, nasal thermogram; 4, period of stimulation. Note the similar increases in respiratory rhythms and reinforcements of inspiratory activities.

such experimental conditions was similar to that observed after both weakly nociceptive cutaneous or oral stimulation. Control tests without 5-HT did not produce any modification of the breathing pattern, leading one to think that a serotoninergic relay is involved in the mechanisms of newborn kitten apnoeas. Similar results obtained in recent experiments in our laboratory with 1-(2,5-dimethoxy-4-iodophenyl)-2-aminopropane hydrochloride (DOI) suggest that 5-HT_2 receptors are particularly involved. However, our results are opposite to those observed in the *in vitro* preparation of newborn rat brainstem–spinal cord (Morin *et al.*, 1990). This discrepancy may reflect either differences in experimental conditions or species differences. Nevertheless, it seems important to note that in newborn kittens the probable central effect of 5-HT mimics the effects obtained by various forms of stimulation and particularly those of oral stimulation.

Our results obtained both in full-term babies and in kittens indicate that the newborn kitten appears to be a valid model for analysis of the mechanisms of the nervous control of breathing in the human newborn, especially in the premature baby. It is interesting to note, as shown in Fig. 32.5, that non-nociceptive cutaneous stimulation of the back can induce a strong increase in the respiratory rhythm of both premature babies and newborn kittens. We could therefore suppose that, in a similar way, apnoeas could be induced by various forms of cutaneous or oral stimulation in preterm or full-term newborns. Ethical considerations limit direct attempts to answer this question. It appears difficult to extrapolate the data obtained in newborn animals so as to explain the respiratory disorders which are probably playing an important role in the tragic and undoubtedly multifactorial accident at present named Sudden Infant Death Syndrome (SIDS). This very particular pathology appears in healthy babies in infancy and is strongly correlated to sleep. During the postnatal development of the kitten, we observe a progressive disturbance of spontaneous and induced apnoeas and a concomitant decrease in the respiratory effect of 5-HT. It thus seems necessary to determine the developmental stage at which kittens or other young animals exhibit a respiratory pattern similar to that of the infant. In such a model we could study the various forms of stimulation able to induce sudden death and analyse the respiratory pattern during sleep in the postnatal period. Indeed, few data are available to explain the mechanisms of central and obstructive apnoeas linked to changes of states of vigilance.

References

Berger, A.J., Bayliss, D.A. & Viana, F. (1992). Modulation of neonatal rat hypoglossal motoneuron excitability by serotonin. *Neuroscience Letters*, **143**, 164–168.

Bodegard, G., Schwieler, G.H., Skoglund, S. & Zetterstrom, R. (1969). Control of respiration in newborn babies. I. The development of Hering–Breuer inflation reflex. *Acta Paediatrica Scandinavica*, **58**, 567–571.

Bonora, M., Marlot, D., Gautier, H. & Duron, B. (1984). Effects of hypoxia on ventilation during postnatal development in conscious kittens. *Journal of Applied Physiology*, **56**, 1464–1471.

Bystrzycka, E., Nail, B.S. & Purves, M.J. (1975). Central and peripheral neural respiratory activity in the mature sheep and foetus and newborn lamb. *Respiration Physiology*, **25**, 199–215.

Dawes, G.S. & Mott, J.C. (1959). Reflex respiratory activity in the newborn rabbit. *Journal of Physiology (London)*, **145**, 85–97.

Duron, B. & Caillol, M.C. (1971). Activité électrique des muscles intercostaux au cours du frisson thermique chez le chat anesthesié. *Journal de Physiologie (Paris)*, **63**, 523–538.

Duron, B., Khater-Boidin, J. & Wallois, F. (1991). Diaphragmatic activity in preterm babies during wakefulness and sleep. *Sleep Research*, **20A**, 84.

Duron, B. & Marlot, D. (1979). Postnatal evolution of inspiratory activity in the

kitten. In *Central Nervous Control Mechanisms in Breathing*, ed. C. von Euler & H. Lagercrantz, pp. 327–336. Oxford: Pergamon Press.
Duron, B. & Marlot, D. (1980). Nervous control of breathing during postnatal development in the kitten. *Sleep*, **3**, 323–330.
Goldberg, M.S., & Milic-Emili, J. (1977). Effect of pentobarbital sodium on respiratory control in newborn rabbits. *Journal of Applied Physiology*, **4**, 845–851.
Hilaire, G., Monteau, R. & Dussardier, M. (1972). Modalités de recrutement des motoneurones phréniques. *Journal de Physiologie (Paris)*, **64**, 457–478.
Javorka, K. (1984). Respiratory reflexes in newborns. *Folia Medica Martiniana*, **10**, 127–145.
Javorka, K., Tomori, Z. & Zavarska, L. (1980). Protective and defensive airway reflexes in premature infants. *Physiologia Bohemoslovenica*, **29**, 29–35.
Khater-Boidin, J., Wallois, F., Toussaint, P. & Duron, B. (1994). Non-vagal reflex apnea in the newborn kitten and during the early postnatal period. *Biology of the Neonate*, **65**, 41–50.
Korpas, J. & Tomori, Z. (1979). Cough and other respiratory reflexes. *Progress in Brain Research*, **12**, pp. 251–263.
Kratschmer, F. (1870). Über reflex von der Nassenschleimhaut auf Atmung und Kreislauf. *Sitzungsberichte der Akademie der Wissenschaften zu Wien, Mathematisch-Naturwissenschaffliche Klasse, Abteilung 3*, **62**, 147–170.
Lindsay, A.D. & Feldman, J.L. (1993). Modulation of respiratory activity of neonatal rat phrenic motoneurones by serotonin. *Journal of Physiology (London)*, **461**, 223–233.
Marlot, D. & Duron, B. (1976). Cutaneous stimulations and spontaneous respiratory activity in the newborn kitten. In *Respiratory Centres and Afferent Systems*, ed. B. Duron, pp. 273–279. Paris: INSERM Publications.
Marlot, D. & Duron, B. (1979a). Postnatal development of vagal control of breathing in the kitten. *Journal de Physiologie (Paris)*, **75**, 891–900.
Marlot, D. & Duron, B. (1979b). Postnatal maturation of phrenic, vagus, and intercostal nerves in the kitten. *Biology of the Neonate*, **36**, 264–272.
Morin, D., Hennequin, S., Monteau, R. & Hilaire, G. (1990). Serotoninergic influences on central respiratory activity: an *in vitro* study in the newborn rat. *Brain Research*, **535**, 281–287.
Mortola, J.P. (1984). Breathing pattern in newborns. *Journal of Applied Physiology*, **56**, 1533–1540.
Skoglund, S. (1960). On the postnatal development of postural mechanisms as revealed by electromyography and myography in decerebrate kittens. *Acta Physiologica Scandinavica*, **49**, 299–317.
Storey, A.T. (1975). Laryngeal water receptors initiating apnea in the lamb. *Experimental Neurology*, **47**, 42–55.
Trippenbach, T. & Kelly, G. (1985). Expiratory effects of vagal stimulation in newborn kitten. *Journal of Applied Physiology*, **59**, 218–222.
Trippenbach, T., Kelly, G. & Marlot, D. (1983). Respiratory effects of stimulation of intercostal muscles and saphenous nerve in kitten. *Journal of Applied Physiology*, **54**, 1736–1744.
Trippenbach, T., Kelly, G. & Marlot, D. (1985). Effects of tonic vagal input on breathing pattern in newborn rabbits. *Journal of Applied Physiology*, **59**, 223–228.
Wallois, F., Khater-Boidin, J., Dusaussoy, F. & Duron, B. (1993). Oral stimulations induce apnoea in newborn kitten. *Neuroreport*, **4**, 903–906.
Weed, L.H. (1917). The reactions of kittens after decerebration. *American Journal of Physiology*, **43**, 131–157.

33

Cardiorespiratory interactions during apnoea

C.P. SEERS

Sobell Department of Neurophysiology, Institute of Neurology, National Hospital for Neurology and Neurosurgery, Queen Square, London, UK

The act of breathing fulfils the essential need for acquiring oxygen and removing carbon dioxide. But it is only with the close co-operation of the cardiovascular system that oxygen can reach every tissue in the body. For many years it has been known that cardiac and respiratory control systems interact and influence one another to maintain homeostasis in the face of changing demands for increased blood flow and oxygenation. Changes in environment (e.g. diving) and the demands of exercise require that the response of these control systems is quick and efficient.

As well as these major cardiovascular adjustments more subtle interactions also occur. These include changes in the cardiorespiratory interactions by temperature and long-term blood pressure fluctuations (see Kitney & Rompelman, 1980, for review). The best known interaction, and one which has been used extensively in clinical investigations, is the breath by breath modulation of the heart rate, called respiratory sinus arrhythmia (RSA). This can be defined as the rapid increase in heart rate which occurs during the inspiratory phase of respiration, and the subsequent slowing of the heart during expiration.

Fig. 33.1 shows two examples of RSA recorded in the barbiturate anaesthetized cat. The top trace in each panel shows a low-pass filtered version of the instantaneous heart rate (Hyndman, 1980). Below are integrated ('leaky integrator' low-pass filter) electromyographic (EMG) recordings from the external (inspiratory, I_{EMG}) and internal (expiratory, E_{EMG}) intercostal muscles recorded from intercostal spaces T4 and T8 respectively.

During inspiration the heart rate rises rapidly. As inspiration ends, as indicated by the I_{EMG} the heart rate falls until its slowing is curtailed by the next inspiration. As the ratio of the inspiratory time (T_I) to the expiratory time (T_E) approaches 1 : 1 RSA can have a distinctive sinusoidal appearance. In Fig. 33.1 the RSA is also modulated at a lower frequency than the respiratory rate, which is seen as a slow change in the mean heart rate.

In contrast, the lower panel of Fig. 33.1 shows a very different respiratory pattern. In this recording, morphine sulphate (1 mg/kg intravenously) was

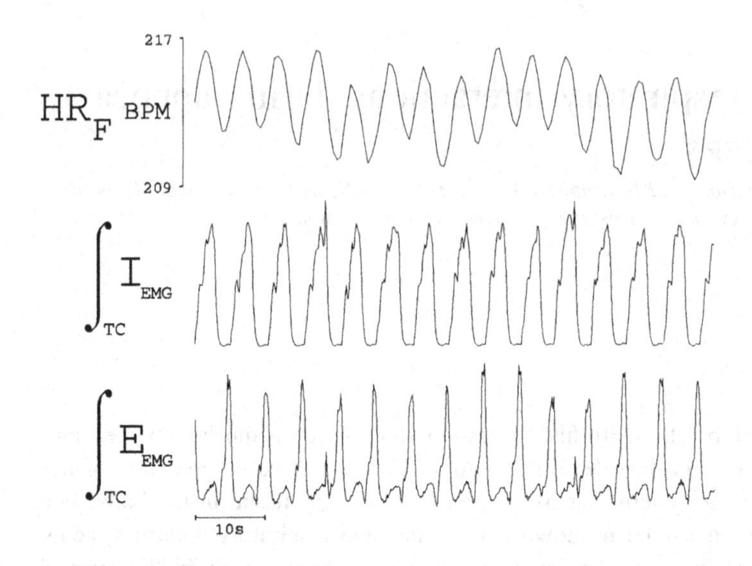

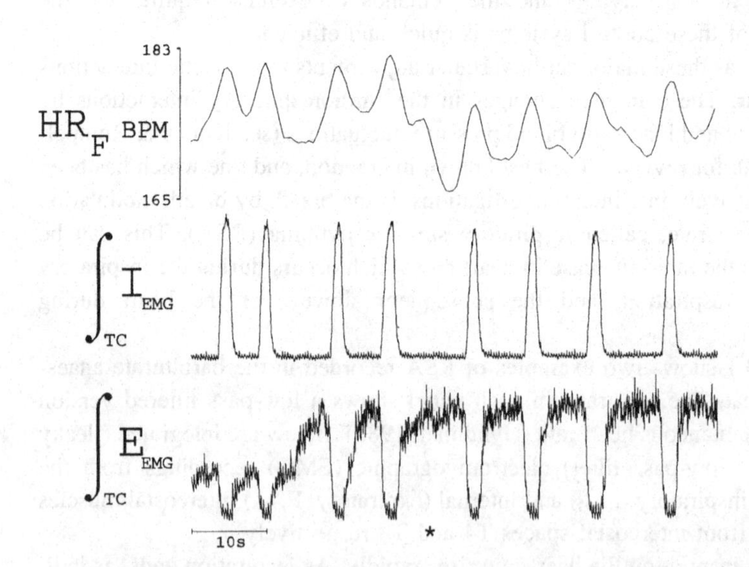

Fig. 33.1. Respiratory sinus arrythmia in the spontaneously breathing anaesthetized cat. Top panel: Filtered version of the instantaneous heart rate, HR_F; integrated (low-pass filtered) electromyograms from the external (inspiratory, I_{EMG}) and internal (expiratory, E_{EMG}) intercostal muscles. Bottom panel: After 1 mg/kg intravenous morphine sulphate, showing the expiratory modulation of the heart rate during the prolonged expirations, especially during the longest one (asterisk).

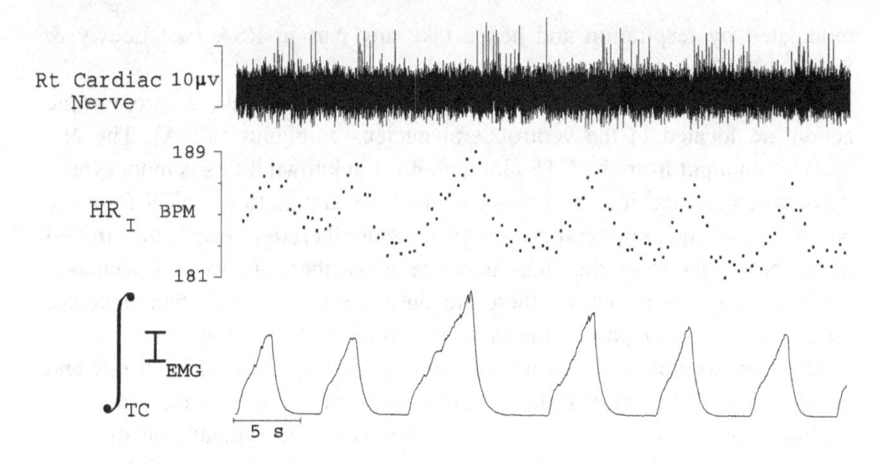

Fig. 33.2. Efferent recording from the right cardiac vagal branch of the cat. Top trace, neurogram; middle trace, instantaneous heart rate, HR_I; bottom trace, integrated inspiratory EMG. Efferent spikes fire maximally in the post-inspiratory period, with some firing throughout the rest of expiration. Cardiac vagal activity is completely inhibited during inspiration.

given to slow respiration, and to elevate the CO_2 threshold for respiratory rhythm generation (see Howard & Sears, 1991). The respiratory cycle is markedly changed, T_I is now shorter, but the inspiratory tachycardia still retains the shape seen in the top panel.

The modulation of the heart rate in both amplitude and morphology is often very closely linked with pattern of breathing, and the degree of slowing which occurs in the expiratory period is to some extent dependent upon the expiratory duration, T_E. For example, when T_E is exceptionally long (Fig. 33.1, marked with an asterisk) the heart rate clearly shows a strong modulation, with a similar shape to the E_{EMG}.

RSA is almost completely blocked by atropine or vagotomy, whereas β-blockade has little effect on RSA. It is therefore believed that RSA is due entirely to the vagal drive to the heart sino-atrial node. Blockade by atropine also increases the mean heart rate to a value which is equal to, or greater than, the tachycardia seen during the inspiratory phase, which suggests that RSA is due to tonic vagal efferent activity which is inhibited during inspiration. Direct recordings of cardiac efferent activity in the dog (Iriuchijima & Kumada, 1963; Jewett, 1964) and cat (Kunze, 1972) and recordings in our own laboratory (Fig. 33.2) indicate that this is probably correct.

Work by numerous authors has shown the cardiomotor neurones of the vagus are located in two regions of the brainstem. One group is in the dorsal motor nucleus of the vagus (DMV), close to the nucleus tractus solitarius (NTS). However, there is still considerable debate as to whether these efferents have a chronotropic effect on the heart and are capable of being modulated by respiration and hence take any part in RSA (see Loewy & Spyer, 1990, for review).

The majority of cardiovagal motoneurones known to have a chronotropic action are located in the ventrolateral nucleus ambiguus (vlNA). The NA receives an input from the NTS although it is not known if this is monosynaptic or whether local interneurones are involved. Inputs to the NTS from the baroreceptors and peripheral chemoreceptors are therefore readily transmitted to the NA. Section of the sinus nerve removes these inputs and decreases the amount of vagal output; therefore the baroreceptor and chemoreceptor inputs form a major part of the drive which is the 'vagal tone'.

The baroreceptor input has a major role in the regulation of heart rate and blood pressure. Any rise in the systemic blood pressure increases the vagal cardiomotor output and simultaneously decreases the sympathetic drive to the heart. This is the classic 'baroreflex' autonomic response, with one arm of the autonomic output to the heart (and other organs) working reciprocally. Fig. 33.3A shows an example of this response with direct simultaneous recordings from the vagal and sympathetic cardiac nerves (Koizumi & Kollai, 1992).

Anrep, Pascual & Rössler (1936a, b) proposed two mechanisms to account for the occurrence of RSA. First, it could occur as a result of a reflex inhibition of vagal tone by lung inflation and the stimulation of lung stretch afferents. Subsequent work by Daly, Potter, McCloskey and others (see Daly, 1986, for review) has confirmed that lung inflation does have a powerful inhibitory effect on the two main drives to the cardiovagal motoneurones. This can be so effective that baroreceptor or chemoreceptor stimulation given during lung inflation, which would normally evoke a marked bradycardia in expiration, fails to have any effect on heart rate (Gandevia, McCloskey & Potter, 1978). The site in the brainstem of the inhibition from the lung is unknown, with no evidence for inhibition by lung stretch receptors neurones in either the NTS or vlNA which might account for the inhibition (see Spyer, 1990, for review). The magnitude of this mechanism's contribution to RSA during eupnoea is unknown. However, in circumstances where there is a greater tidal volume and chest movement, for example in exercise or in disease, the lung afferent inputs might contribute significantly to the generation of RSA.

Anrep *et al.*'s (1936a, b) second mechanism for RSA is a central inhibition of the vagal tone by the 'respiratory centre'. Work by Gilbey *et al.* (1984) has shown this occurs at the cardiovagal motoneurones in the vlNA. Intracellular recordings from identified cardiovagal (chronotropic) neurones show a postsynaptic inhibition of the neurones by chloride mediated inhibitory postsynaptic potentials during inspiration. The cardiovagal motoneurones were active during stage I expiration (post-inspiratory) and showed a weak and variable wave of inhibition through stage II expiration (cf. Fig. 33.2).

The cardiovagal motoneurones in the vlNA are close to, and intermingled

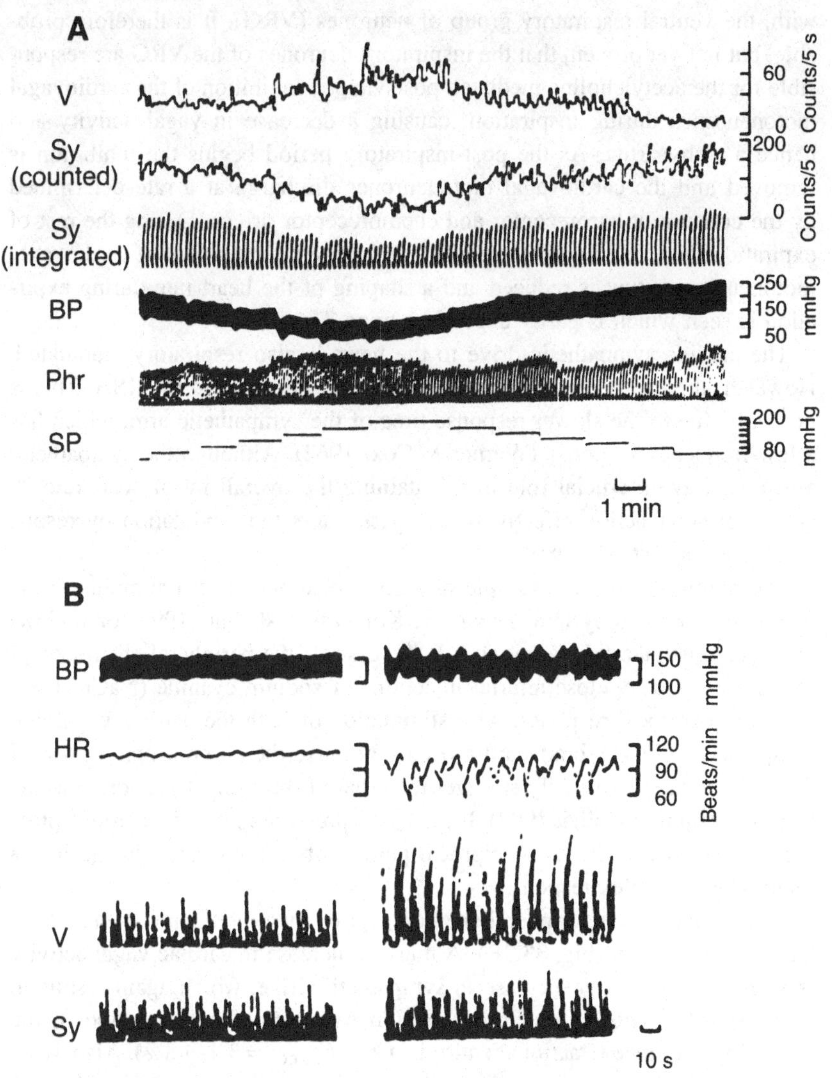

Fig. 33.3. (A) Example of the baroreflex. Simultaneous neurograms from the cardiac vagus (V) and cardiac sympathetic (Sy) nerves, blood pressure (BP), sinus pressure (SP) and phrenic nerve (Phr). (B) Non-reciprocal activity in the autonomic nervous system. Left panel: Normoxic control (P_{a,O_2} =90 mmHg). Right panel: Hypoxic (P_{a,O_2} = 34 mmHg) stimulation of peripheral chemoreceptors showing increased cardiac vagal and sympathetic neurograms and RSA. Reproduced with permission from Koizumi & Kollai (1992).

with, the ventral respiratory group of neurones (VRG). It is therefore probable, but not yet proven, that the inspiratory neurones of the VRG are responsible for the acetylcholine mediated postsynaptic inhibition of the cardiovagal motoneurones during inspiration, causing a decrease in vagal activity and hence a tachycardia. As the post-inspiratory period begins the inhibition is removed and the cardiovagal motoneurones discharge at a rate determined by the concurrent baroreceptor and chemoreceptor drives. During the rest of expiration there is a variable amount of inhibition, the amount of cardiovagal motoneurone output is reduced and a shaping of the heart rate during expiration is seen which is partly dependent upon T_E.

The cardiac sympathetic drive to the heart is also respiratory-modulated. However, there is no evidence that it contributes directly to the RSA. This is probably due to the slower response time of the sympathetic arm, which has a low-frequency response (Warner & Cox, 1962). Although the sympathetic arm does have a crucial role in maintaining the overall mean heart rate its slower effector action effectively low-pass filters the modulation by respiration (see Saul *et al.*, 1989).

The baroreflex is one example of a reciprocal action to a stimulus in the autonomic nervous system. However, Koizumi & Kollai (1992 for review) have demonstrated this is not always the case. If the peripheral chemoreceptors are excited by close arterial injections of sodium cyanide (NaCn) there is a non-reciprocal response, with stimulation of both the cardiac vagal and sympathetic arms. A bradycardia occurs because the chronotropic action of the cardiac vagus is usually stronger than that of the sympathetic drive to the heart (Koizumi & Kollai, 1992). If the blood pressure increases, baroreceptors are stimulated and the non-reciprocal pattern of response can change into a reciprocal baroreflex response.

Stimulation of the peripheral chemoreceptors by hypoxia at constant P_{a,CO_2} has a similar action (Fig. 33.3B). A marked increase in cardiac vagal activity is seen, with a smaller increase in sympathetic drive, which again results in a bradycardia. This non-reciprocal action can occur for stimulation in the physiological range (fractional end-tidal CO_2, $F_{ET,CO_2} = 3.4–4.8\%$). Also, RSA is very markedly enhanced with the increase in vagal tone and the enhanced respiratory drive due to the hypoxia (Fig. 33.3B).

Non-reciprocal stimulation by the chemoreceptors has a particular relevance when one considers the action of CO_2 on the respiratory system. In hyperventilation-induced hyperoxic–hypocapnic apnoea there is cessation of the central respiratory rhythm generation. However, the respiratory system is not 'quiet' but shows a tonic activation of the expiratory side of the system. As P_{a,CO_2} is increased below the threshold for rhythm generation there is a graded increase in the tonic expiratory activity (Bainton, Kirkwood & Sears, 1978) which is analogous to the cardiomotor response to CO_2. When the CO_2 threshold for rhythm generation is reached and inspiratory activity begins,

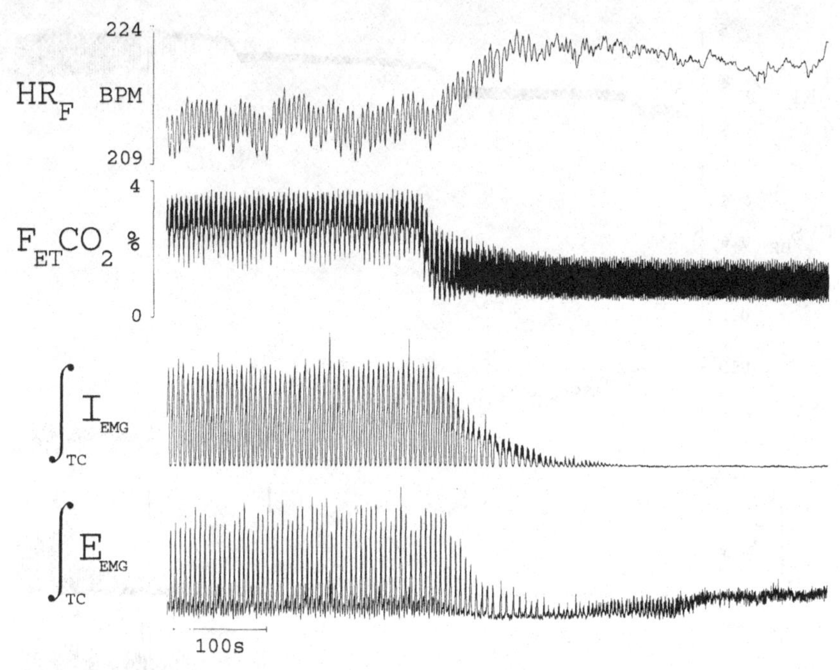

Fig. 33.4. Induced hypocapnic apnoea in the ventilated anaesthetized cat. Apnoea was induced by lowering the inspired CO_2 so that the F_{ET,CO_2} is reduced to 2%. Heart rate (HR_F) shows a tachycardia and both the inspiratory EMG (I_{EMG}) and expiratory EMG (E_{EMG}) decline.

the tonic expiratory activity is phasically inhibited. Thus, the central pattern of respiration is 'sculpted' from the underlying tonic drives (Sears, 1990).

In the barbiturate-anaesthetized, non-paralysed cat hyperventilation at a high frequency (58/min) and low tidal volume causes an apnoea and a tachycardia. Fig. 33.4 shows recordings made in an animal ventilated with air and CO_2 ($F_{ET,CO_2} = 3.8\%$). There are strong I_{EMG} and E_{EMG} activities with some contribution from the animal to the ventilation seen as dips in the F_{ET,CO_2}. A marked RSA can also be seen in the heart rate. Hypocapnic apnoea is induced by reducing the inspired CO_2 concentration so that the F_{ET,CO_2} is 2%. This is reflected in the decreasing EMG output and finally apnoea. Phasic E_{EMG} is replaced by tonic activity (bottom trace). The heart rate shows a tachycardia with the marked RSA being replaced by a noisy low-frequency modulation of the heart rate. The tachycardia in apnoea reflects a decrease in both the vagal and sympathetic drives to the heart with lowering in P_{a,CO_2}, but in this example the decrease in vagal tone has a significantly greater effect than the loss of sympathetic drive.

Apnoea can be regarded as an infinitely long expiratory period, during which the 'gate' at the cardiovagal neurones remains open. This means the

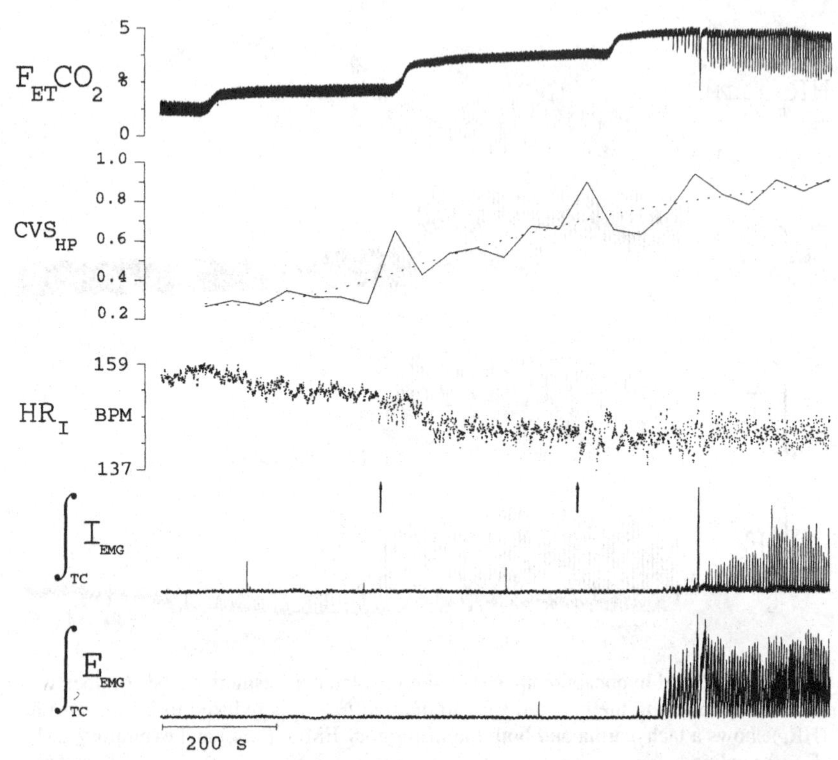

Fig. 33.5. CO_2 titration from hypocapnic apnoea over three levels of F_{ET,CO_2}. Heart rate (HR_I) shows a bradycardia and an increase in the heart rate variability (HRV) measured by the coefficient of variability, CVS_{HP}. Bursts of HRV (marked with arrows) can be seen which are not related to respiration.

prevailing central and peripheral chemosensitive drives are reflected in the vagal output, with no curtailment of the drive by inspiration. Simultaneously the CO_2 dependent drive to the cardiac sympathetic arm is free from respiratory modulation.

If the chemical drive is now increased by adding CO_2 in a series of steps (CO_2 titration) to the inspired gases a graded bradycardia develops. Fig. 33.5 illustrates a typical response as the $F_{ET}CO_2$ is increased over three levels. In this example the sympathetic arm has been blocked (propranolol 1 mg/kg intravenously) so the resulting drive to the heart is only vagal. The mechanical ventilation has not changed, and therefore the heart rate and the variability in the heart rate (HRV) are dependent upon the drive through the 'open gate' to the cardiomotor neurones in the NA. Not only does this drive force the heart rate down, but it also increases the beat-to-beat HRV. A gradual increase in HRV is seen throughout the CO_2 titration. There are periods of more marked variability (marked with arrows in Fig. 33.5) which, although

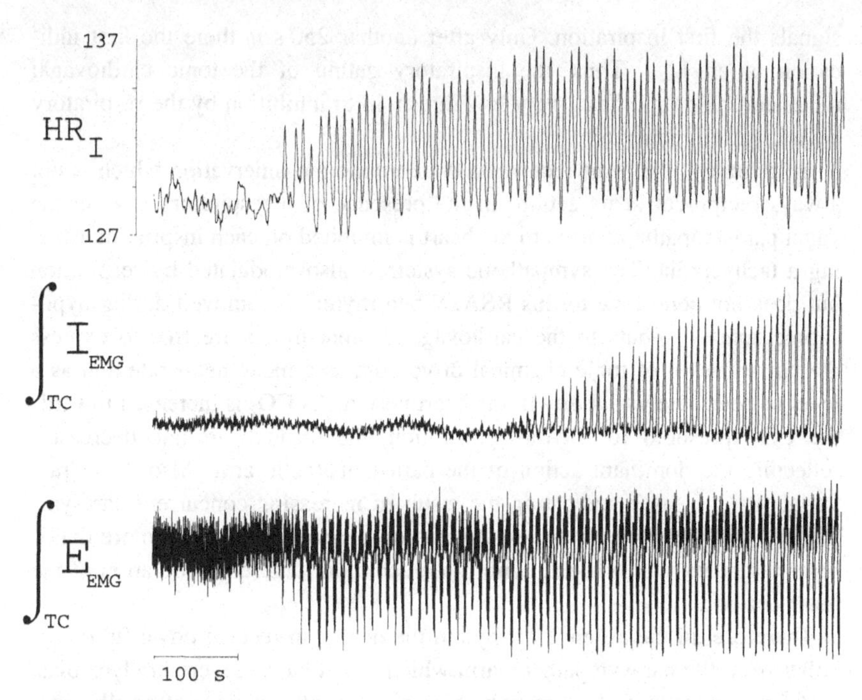

Fig. 33.6. The onset of the respiratory rhythm at the end of a series of increments in F_{ET,CO_2} (CO_2 titration). RSA appears before the first signs of respiratory rhythm generation in the I_{EMG} record. Tonic E_{EMG} is phasically interrupted concurrent with the appearance of RSA.

resembling RSA, are of a lower frequency, and at a time when the F_{ET,CO_2} is well below the threshold for rhythm generation.

Fig. 33.5 also shows the HRV quantified by calculating the coefficient of variation (CVS_{HP}) (Välimäki *et al.*, 1988) of the beat-to-beat intervals in 50 s epochs. As the heart rate decreases due to the increasing CO_2 drive then the amount of variability in the heart period increases. When the CO_2 threshold for rhythm generation is reached the random variability in the heart period again becomes highly modulated as RSA resumes.

Recent work in this laboratory shows that β-blockade of the heart can eliminate a portion of this noise or HRV, which means that a part of the HRV is attributable to the sympathetic drive. However, most of the HRV and the bradycardia are eliminated when atropine is added to block the vagal drive, and the heart rate becomes metronome-like with little or no variability.

Inhibition of the tonic expiratory activity during apnoea is a clear indication of the onset of rhythm (Bainton & Kirkwood, 1979). However, RSA can also clearly mark the onset of rhythm. Fig. 33.6 shows the end of a CO_2 titration, an incomplete inhibition of the E_{EMG} being followed immediately by a significant tachycardia from the baseline of HRV as the onset of RSA signals the first inspiration. Only after another 250 s is there the first indication of an I_{EMG}. Thus, the inspiratory gating of the tonic cardiovagal motoneurone output has a very low threshold to inhibition by the respiratory central pattern generator.

In summary, the heart receives a dual autonomic innervation which is not always reciprocal in its action. In the presence of a respiratory rhythm the vagal parasympathetic drive to the heart is inhibited on each inspiration causing a tachycardia. The sympathetic system is also modulated by respiration but does not contribute to this RSA. When rhythm is removed during hypocapnic apnoea, inputs to the cardiovagal motoneurones are free to express the current level of tonic chemical drive both as a mean heart rate and as a level of heart rate variability in the heart period. As CO_2 is increased towards the CO_2 threshold for rhythm generation, the mean heart rate decreases, reflecting the dominant action of the parasympathetic arm. Also, heart rate variability increases, indicating the result of increasing concurrent parasympathetic and sympathetic drives. As rhythm resumes, RSA once more dominates the heart rate, bringing with it a much more ordered modulation of the heart period.

Therefore, in the absence of rhythm the heart can receive powerful drives either from the parasympathetic arm which might cause severe bradycardias, or from the sympathetic arm which is capable of causing potentially fatal ventricular arrhythmias (Eckberg *et al.*, 1993). However, the non-reciprocal action of the neural drives to the heart during apnoea allows each one to protect against excess activity in the other.

References

Anrep, G.V., Pascual, W. & Rössler, R. (1936a). Respiratory variations of the heart rate. I. The reflex mechanism of the respiratory arrhythmia. *Proceedings of the Royal Society of London, Series B*, **119**, 191–217.

Anrep, G.V., Pascual, W. & Rössler, R. (1936b). Respiratory variations of the heart rate. II. The central mechanism of the respiratory arrhythmia and the inter-relations between the central and the reflex mechanisms. *Proceedings of the Royal Society of London, Series B*, **119**, 218–232.

Bainton, C.R. & Kirkwood, P.A. (1979). The effect of carbon dioxide on the tonic and rhythmic discharges of expiratory bulbospinal neurones. *Journal of Physiology (London)*, **296**, 291–314.

Bainton, C.R., Kirkwood, P.A. & Sears, T.A. (1978). On the transmission of the stimulating effects of carbon dioxide to the muscles of respiration. *Journal of Physiology (London)*, **280**, 249–272.

Daly, M. de B. (1986). Interactions between respiration and circulation. In *Handbook of Physiology*, sect. 3, *The Respiratory System*, vol. II, *Control of Breathing*, part 2, ed. N.S. Cherniack & J.G. Widdicombe, pp. 529–594. Bethesda, MD: American Physiological Society.

Eckberg, D.L., Halliwill, J.R., Smith, M.L. & Minisi, A.L. (1993). Autonomic complicity in catastrophic cardiac rhythms. In *Cardiovascular Reflex Control in*

Health and Disease, ed. R. Hainsworth & A.L. Mark. London: W.B. Saunders.
Gandevia, S.C., McCloskey, D.I. & Potter, E.K. (1978). Inhibition of baroreceptor and chemoreceptor reflexes on heart rate by afferents from the lungs. *Journal of Physiology (London)*, **276**, 369–381.
Gilbey, M.P., Jordan, D., Richter, D.W. & Spyer, K.M. (1984). Synaptic mechanisms involved in the inspiratory modulation of vagal cardio-inhibitory neurones in the cat. *Journal of Physiology (London)*, **356**, 65–78.
Howard, R.S. & Sears, T.A. (1991). The effects of opiates on the respiratory activity of thoracic motoneurones in the anaesthetized and decerebrate rabbit. *Journal of Physiology (London)*, **437**, 181–199.
Hyndman, B.W. (1980). Cardiovascular recovery to psychological stress: a means to diagnose in man and task? In *The Study of Heart-rate Variability*, ed. R.I. Kitney & O. Rompelman, pp. 191–224. Oxford: Clarendon Press.
Hyndman, B.W., Kitney, R.I. & Sayers, McA. B. (1971). Spontaneous rhythms in physiological control systems. *Nature*, **223**, 339–341.
Iriuchijima, J. & Kumada, M. (1963). Efferent cardiac vagal discharge of the dog in response to electrical stimulation of sensory nerves. *Japanese Journal of Physiology*, **13**, 599–605.
Jewett, D.L. (1964). Activity of single efferent fibres in the cervical vagus nerve of the dog, with special reference to possible cardio-inhibitory fibres. *Journal of Physiology (London)*, **175**, 321–357.
Kitney, R.I. & Rompelman, O. (eds.) (1980). *The Study of Heart-rate Variability*. Oxford: Clarendon Press.
Koizumi, K. & Kollai, M. (1992). Multiple modes of operation of cardiac autonomic control: development of ideas from Cannon and Brooks to the present. *Journal of the Autonomic Nervous System*, **41**, 19–30.
Kunze, D.L. (1972). Reflex discharge patterns of cardiac vagal efferent fibres. *Journal of Physiology (London)*, **222**, 1–15.
Loewy, A.D. & Spyer, K.M. (1990). Vagal preganglionic neurones. In *Central Regulation of Autonomic Functions*, ed. A.D. Loewy & K.M. Spyer, pp. 68–87. Oxford: Oxford University Press.
Saul, J.P., Berger, R.D., Chen, M.H. & Cohen, R.J. (1989). Transfer function analysis of autonomic regulation. II. Respiratory sinus arrhythmia. *American Journal of Physiology*, **256**, H153-H161.
Sayers, McA. B. (1980). Signal analysis of heart-rate variability. In *The Study of Heart-Rate Variability*, ed. R.I. Kitney & O. Rompelman, pp. 27–58. Oxford: Clarendon Press.
Sears, T.A. (1990). Central rhythm generation and spinal integration. *Chest*, **97**, 45S-51S.
Spyer, K.M. (1990). The central nervous organization of reflex circulatory control. In *Central Regulation of Autonomic Functions*, ed. A.D. Loewy & K.M. Spyer, pp. 168–188. Oxford: Oxford University Press.
Välimäki, I.A., Nieminen, T., Antila, K.J. & Southall, D.P. (1988). Heart-rate variability and SIDS. Examination of heart-rate patterns using an expert system generator. In *The Sudden Infant Death Syndrome. Cardiac and Respiratory Mechanisms and Interventions. Annals of the New York Academy of Sciences*, vol. 533, ed. P. Schwartz, D.P. Southall & M. Valdes-Dapena, pp. 228–237. New York: New York Academy of Sciences.
Warner, H.R. & Cox, A. (1962). A mathematical model of heart rate control by sympathetic and vagus efferent information. *Journal of Applied Physiology*, **17**, 349–355.

34

Impairment of respiratory control in neurological disease

ROBIN S. HOWARD

The Harris Unit, National Hospital for Neurology and Neurosurgery, Queen Square, and The Lane Fox Respiratory Unit, St Thomas' Hospital, London, UK

It has proved difficult to attribute precise respiratory function to localized anatomical substrates in man because lesions are rarely isolated and, even with newer imaging techniques, ante-mortem localization is imprecise. Furthermore accurate diagnosis of respiratory insufficiency has led to earlier therapeutic intervention with controlled ventilation. Finally there is probably considerable redundancy and plasticity of neural function where progressive and destructive mass lesions can have little or no functional consequence whilst acute discrete lesions in a similar distribution may lead to profound respiratory impairment.

It has been conventional to consider neural control of respiration to depend on two anatomically and functionally independent pathways (Plum, 1970). Metabolic (automatic) respiration is the homeostatic pathway by which ventilation may be mediated to maintain acid–base status and oxygenation to the metabolic requirements. The behavioural (voluntary) system operates during wakefulness and allows voluntary modulation of respiration in response, for example, to speaking, singing, breath-holding and straining. This has proved a valuable and durable model to explain and predict patterns of respiratory insufficiency; however, there is increasing evidence for interactions between the two systems (Orem & Netick, 1986; Murphy *et al.*, 1990).

Techniques of central motor stimulation and functional imaging have confirmed that the respiratory muscles are activated behaviourally with phrenic motoneurones being controlled by rapidly conducting, oligosynaptic pathways from the contralateral motor cortex. Inspiratory muscles have a direct representation in the primary motor cortex, premotor cortex, supplementary motor area and thalamus, whilst for active expiration the areas are more extensive and involve limbic cortex (Gandevia & Rothwell, 1987; Macefield & Gandevia, 1991; Colebach *et al.*, 1991; Maskill *et al.*, 1991; Ramsay *et al.*, 1993). Stimulation of large areas of the limbic cortex inhibits respiration by slowing the respiratory rate (Kaarda, 1960), whilst both rate and tidal volume are increased by simultaneous bilateral stimulation of the anterior amygdala (Pool & Ranshoff, 1949). Apnoea and prolonged expiratory apneusis are common during complex partial and generalized seizures. This may be associated with upper airway obstruction, laryngospasm and masseter spasm leading to hypoxaemia and cyanosis. Isolated apnoea may be an ictal phenomenon requiring prolonged ventilation (Nelson & Ray, 1968; Coulter, 1984; Shorvon, 1993).

The effects of brainstem disorders on respiration are determined by the pathology, localization and speed of onset of the lesion. Baker, Matzke & Brown (1950) showed that acute respiratory failure in infants with acute bulbar poliomyelitis was associated with consistent and selective focal necrosis and neuronal damage in the region of the nucleus ambiguus. Further clinical and experimental studies have confirmed the importance in respiratory rhythm generation of localized areas in the dorsolateral tegmentum of the pons and medulla in the region of the nucleus tractus solitarius and nucleus retroambigualis (for review see Howard & Newsom-Davis, 1992). As a consequence of lesions in this area automatic respiratory control is disrupted; the patient is voluntarily able to maintain the respiratory pattern and breathes normally whilst awake and alert, but during sleep there is a sudden or progressive decline in tidal volume and respiratory rate culminating in sleep apnoea. The commonest cause of brainstem lesions which disrupt respiration is cerebrovascular disease. Unilateral or bilateral infarcts involving the lateral medullary tegmentum lead to disruption of the CO_2 responsiveness and subsequently acute failure of the automatic pathway (Devereaux, Keane & Davis, 1973; Levin & Margolis, 1977; Bogousslavsky *et al.*, 1990). Other clinical causes of automatic respiratory failure include acute encephalitides, other infections of the central nervous system such as *Borrelia* and *Listeria*, postinfectious encephalomyelitis, and malignant disease, either primary or secondary or as a paraneoplastic brainstem encephalitis.

Abnormal patterns of rate and rhythm are also often a reflection of impaired automatic ventilatory control. Many of these patterns were first described by Plum as manifestations of progressive central brainstem herniation and were considered to have fairly precise localizing value (Plum & Posner, 1983). However, these phenomena are now seen less commonly because controlled ventilation is instituted earlier in patients with neurogenic respiratory insufficiency. Primary central neurogenic hyperventilation is considered to be present if rapid, regular hyperventilation persists in the face of alkalosis, elevated P_{O_2}, low P_{CO_2} and in the absence of any pulmonary or airway disorder. It is extremely rare and the few cases in the literature have been associated with either lymphoma or infiltrating glioma (Rodriguez *et al.*, 1982; Pauzner *et al.*, 1989). However, hyperventilation in the seriously ill patient is common but is due to intrinsic pulmonary involvement leading to ventilation : perfusion mismatch, pulmonary shunting and increased vagally mediated reflexes (Mazzara, Ayres & Grace, 1974; North & Jennett, 1974; Leigh & Shaw, 1976). In apneustic breathing there are sustained inspiratory cramps with a prolonged pause at full inspiration or alternating brief end-inspiratory and expiratory pauses. The pattern has been associated with bilateral tegmental infarcts or demyelinating lesions in the pons. In cluster breathing respiration occurs in irregular bursts separated by variable periods of apnoea, the regularity and decrescendo–crescendo pattern of Cheyne–Stokes respiration is absent and the cycle time is much shorter. Ataxic respiration is characterized by a completely irregular respiratory cycle of variable frequency and tidal volume alternating with periods of apnoea, it is particularly associated with medullary compression due to rapidly expanding lesions and may be an important sign of impending respiratory arrest. Complex abnormalities of automatic control also occur in mitochondrial cytopathy, in particular the phenotype of Leigh's disease. In our experience these patients show complex patterns of respiratory insufficiency associated with failure of automatic control, bulbar weakness and cardiac and respiratory muscle weakness in addition to multiple metabolic abnormalities including recurrent lactic acidosis.

Selective interruption of the voluntary pathways in man is much less common. It leads to a strikingly regular and unvarying respiratory pattern during wakefulness as well as sleep, and loss of the ability to take a deep breath, hold the breath, cough voluntarily or initiate any kind of volitional respiratory movement. However, tidal volume remains responsive to CO_2 and a reflex cough is preserved. Voluntary control is impaired by lesions affecting the descending corticospinal tract in the cervical cord (Howard & Newsom-Davis, 1992); this pattern may be seen also in association with destructive lesions of the basal pons or of the medullary pyramids and adjacent ventromedial portion. These lesions may result in the 'locked in' syndrome in which there is preserved consciousness and vertical eye movements but quadriplegia and loss of remaining cranial nerve function due to interruption of descending corticospinal and corticobulbar pathways, the lower cranial nerve nuclei remaining intact. Munschauer *et al.* (1991) described a patient with locked in syndrome due to infarction of the basal pons. This led to loss of voluntary control with normal CO_2 responses but preservation of respiratory modulation to emotional stimuli including laughing, coughing and anxiety. These findings imply that descending limbic influences on automatic respiration are anatomically and functionally independent of the voluntary respiratory system. It was suggested that such a descending pathway, mediating limbic control of respiration, lay either in the pontine tegmentum or the lateral basis pontis.

In patients with bulbar lesions, particularly vascular or demyelinating, the combination of impaired swallow, abnormalities of the respiratory rhythm, reduced vital capacity and reduced or absent triggering of a cough reflex all increase the risk of aspiration pneumonia (Howard *et al.*, 1992). A further respiratory abnormality seen in patients with acute demyelinating brainstem lesions is the tonic seizures described by Matthews (1958). These attacks may take the form of paroxysmal hyperventilation precipitated by changes in posture (Howard *et al.*, 1992).

Hiccups consist of brief bursts of intense inspiratory activity involving the diaphragm and inspiratory intercostal muscles with reciprocal inhibition of the expiratory intercostals (Newsom Davis, 1970; Howard, 1992). Glottic closure occurs almost immediately after the onset of diaphragmatic contraction, thus minimizing the ventilatory effect. Intractable hiccups may be the result of structural or functional disturbances of the medulla or afferent or efferent connections with the respiratory muscles. They may be associated with structural lesions of the medulla including infarction in the territory of the posterior inferior cerebellar artery, tumour, tuberculoma, abscess, syrinx, haematoma and demyelination. The development of hiccups in this context may anticipate the development of irregularities of the respiratory rhythm culminating in respiratory arrest.

Cheyne–Stokes respiration (CSR) is a pattern of periodic breathing in which phases of hyperpnoea regularly alternate with apnoea. The breathing waxes from breath to breath in a smooth crescendo and then, once a peak is reached, wanes in an equally smooth decrescendo. CSR occurs almost exclusively with non-REM sleep. It is associated with neurological dysfunction at the bilateral subcortical level, raised intracranial pressure and cardiac failure (Tobin & Snyder, 1984).

It is conventional to divide sleep apnoea into obstructive and central types, the former occurring when there is upper airway obstruction despite normal movement of the intercostals and diaphragm and the latter with the absence of all respiratory phased movements. There is considerable evidence that similar mechanisms may underlie different forms of apnoea (Douglas, 1986). It is important to emphasize that the upper airway respiratory muscles (genioglossus, tensor palatini and stylopharyngeus) behave as more conventional respiratory muscles with activities phasically modulated during inspiration and expiration and by classical respiratory stimuli. Thus obstructive sleep apnoea (OSA) is thought to occur when the motor drive to the upper airway muscles is inadequate despite the presence of effective diaphragm contractions, while central sleep apnoea (CSA) is a closely related but more severe phenomenon, with inadequate drive to both the upper airway muscles and the diaphragm. OSA is exacerbated by coexisting factors including obesity, structural abnormalities of the upper airway, depressant drugs and alcohol, supine position, endocrine, metabolic, musculoskeletal and autonomic disturbances. In addition to the disturbance of sleep architecture caused by microarousals OSA leads to progressive hypoxia and hypercapnia, eventually untreated hypercapnic respiratory failure develops associated with pulmonary and systemic hypertension, congestive cardiac failure and polycythemia (Stradling & Phillipson, 1986).

A number of patterns of involuntary movements of the respiratory musculature may interfere with ventilation. These include laryngeal dyskinesiae/dystonia which are characterized by paradoxical adduction of the vocal cords, often precipitated by neuroleptics; patients may present with wheezing, stridor and exertional dyspnoea. In diaphragm flutter there are frequent spontaneous contractions of the diaphragm unrelated to inspiration, otherwise the diaphragm moves normally. It is associated with dyspnoea, pain in the thoracic and abdominal wall and epigastric pulsation. In respiratory myoclonus, which is often associated with palatal myoclonus, there are frequent but irregular bilateral diaphragmatic contractions superimposed on normal movements. Paroxysmal nocturnal dystonia presents with stereotypic body movements during sleep with tachycardia, respiratory irregularities and frequent, repetitive transient arousals; patients complain of excessive daytime somnolence (Maccario & Lustman, 1990). Dystonia of the abdominal wall ('belly dancer's dystonia') may occasionally present with breathlessness; it is abolished by breath-holding and deep inspiration (Iliceto *et al.*, 1990).

Multisystem atrophy is a global term which includes many neurodegenerative disorders (Quinn, 1989). A characteristic feature is paresis of the vocal cord abductors (posterior cricoarytenoids) – the cords lie closely opposed, leading to severe upper airway limitation during sleep and giving rise to the characteristic presenting feature of severe nocturnal stridor (Bannister *et al.*, 1981). Other factors also contribute to the development of respiratory insufficiency. These include abnormalities of rate, rhythm and amplitude during sleep, a reduction in central respiratory drive leading to OSA due to upper airway occlusion and CSA due to loss of automatic control. A further important factor is the accompanying autonomic failure which contributes to impaired cardiorespiratory control mechanisms (Chokroverty, 1992).

Foramen magnum lesions are an important cause of acute or subacute respiratory insufficiency. Cerebellar ectopia and syringomyelia may present with either progressive nocturnal hypoventilation or sudden respiratory arrest, usually precipitated by some intercurrent event (Fish *et al.*, 1988). In a series of patients with rheumatoid atlantoaxial dislocation we have shown that clinically unsuspected hypoventilation and sleep apnoea are common in patients with severe medullary compression and may contribute to the high mortality of the condition (Howard *et al.*, 1994). The respiratory complications may be due to abnormalities in automatic ventilatory control, involvement of descending ventrolateral automatic or corticospinal voluntary pathways, upper airway obstruction or aspiration associated with bulbar weakness.

Disorders of the lower motor neurone, i.e. affecting anterior horn cell, peripheral nerve, neuromuscular junction and muscle, may all cause respiratory insufficiency. Although many different factors contribute in individual conditions the most important mechanism is respiratory muscle weakness. These conditions form the vast majority of patients with neurogenic respiratory failure and accounted for 92% of our recent review of 319 consecutive patients at the Lane Fox respiratory unit at St Thomas' Hospital with a primarily neurological cause for chronic respiratory insufficiency. This series reflects a specialized interest in long-term and domiciliary respiratory support (Howard & Spencer, 1993).

During acute poliomyelitis respiratory insufficiency occurs as a result of respiratory muscle weakness or involvement of the central respiratory control mechanisms. Respiratory insufficiency may develop many years after poliomyelitis, even in the absence of any respiratory involvement during the acute illness or convalescent phase. The commonest pattern of post-polio respiratory insufficiency is progressive nocturnal hypoventilation due to respiratory muscle weakness and fatigue, particularly with decreased compliance due to kyphoscoliosis. Other contributory factors include pregnancy, alcohol and sedative drugs (Howard, Wiles & Spencer, 1988).

An important proportion of patients with motor neurone disease may develop respiratory insufficiency early in the course of the disease and may present with respiratory failure or even respiratory arrest. The commonest cause of respiratory insufficiency is generalized respiratory muscle weakness, often with predominant diaphragm paresis. Nocturnal hypoventilation and both central and obstructive sleep apnoea have been demonstrated in the pseudobulbar form of motor neurone disease (Howard, Wiles & Loh, 1989).

The incidence of acute respiratory failure in Guillain–Barré syndrome is between 20% and 30%. Any of the respiratory muscles may be involved but prolonged phrenic nerve conduction time may be the most sensitive predictor of impending ventilatory failure, morbidity and mortality. The associated bulbar weakness and autonomic instability contribute to the necessity for controlled ventilation (Ropper, 1985).

Extensive generalized respiratory muscle involvement leads to the respiratory failure which may occur in acute porphyric neuropathy. Predominant phrenic nerve involvement may occur in neuropathies associated with underlying carcinoma, diphtheria, herpes zoster-varicella, post-immunisation and neuralgic amyotrophy. Acute respiratory failure is also a feature of vasculitic and toxic neuropathies (Hughes & Bihari, 1993). Phrenic nerve involvement occurs most commonly as a result of trauma during thoracic surgery, hypothermia (Efthimiou *et al.*, 1992) or direct involvement by neoplasm.

Respiratory failure in myasthenia gravis often results from a myasthenic crisis (usually precipitated by infection) but is also associated with cholinergic crisis, thymectomy or steroid myopathy. Associated bulbar weakness predisposes to aspiration and acute respiratory failure necessitating urgent intubation and ventilation. Respiratory muscle weakness is a common cause of morbidity and mortality in muscular dystrophies (Smith, Edwards & Calverley, 1991). In Duchenne muscular dystrophy diaphragmatic weakness is not prominent but chronic respiratory insufficiency is due to intercostal weakness, scoliosis, reduced lung compliance, aspiration and repeated infections (Smith *et al.*, 1987). In limb girdle dystrophies the major pattern of respiratory involvement is gradual, progressive global weakness of the respiratory muscles compounded by scoliosis. Respiratory tract infections are common despite the absence of clinically overt bulbar weakness (Howard *et al.*, 1993). In contrast, acid maltase deficiency is characterized by early and selective diaphragm weakness often with minimal involvement of other respiratory and bulbar musculature (Trend *et al.*, 1985). In dystrophia myotonica respiratory involvement is multifactorial. Respiratory muscle weakness may affect both the diaphragm and expiratory muscles leading to a poor cough, restrictive lung defect and alveolar hypoventilation; there is little evidence that myotonia of the respiratory muscles is a significant factor. A central abnormality may contribute to alveolar hypoventilation, as may reduced ventilatory response to CO_2 in the absence of CO_2 retention, hypersomnolence and an undue sensitivity to anaesthetics and sedatives. Respiratory symptoms develop late in the course of the disease in association with severe respiratory muscle and bulbar weakness, myotonia and systemic features. Hypersomnolence is the commonest presenting symptom and this frequently occurs in the absence of hypoventilation. Cardiomyopathy may also be important in sudden unexpected death in these patients (Howard *et al.*, 1993). Respiratory muscle weakness also occurs in the inflammatory myopathies such as polymyositis, dermatomyositis and inclusion body myopathy, and in endocrine myopathies including Addison's disease, acromegaly, thyrotoxicosis and myxoedema (Laroche, Moxham & Green, 1992).

Respiratory involvement due to primary neurological disease in man is probably common but often remains unsuspected because of its insidious course. The clinical manifestations may be variable and unpredictable due to the multifocal nature of the underlying disease and the presence of coexisting pulmonary, cardiovascular or autonomic factors.

References

Baker, A.B., Matzke, H.A. & Brown J.R. (1950). Poliomyelitis. III. Bulbar poliomyelitis: a study of medullary function. *Archives of Neurology and Psychiatry*, **63**, 257–281.

Bannister, R., Gibson, W., Michaels, L. & Oppenheimer, D.R. (1981). Laryngeal abductor paralysis in multiple system atrophy. *Brain*, **104**, 351–368.

Bogousslavsky, J., Khurana, R., Deruaz, J.P., Hornung, J.P., Regli, F., Janzer, R. & Perret, C. (1990). Respiratory failure and unilateral caudal brainstem infarction. *Annals of Neurology*, **28**, 668–673.

Chokroverty, S. (1992). The assessment of sleep disturbance in autonomic failure. In *Autonomic Failure*, 3rd edn., ed. R. Bannister & C.J. Mathias, pp. 442–461. Oxford: Oxford Medical Publications.

Colebach, J.G., Adams, L., Murphy, K., Martin, A.J., Lammertsma, A.A., Tochon-Danguy, H.J., Clark, J.C., Friston, K.J. & Guz, A.(1991). Regional cerebral blood flow during volitional breathing in man. *Journal of Physiology (London)*, **443**, 91–103.

Coulter, D.L. (1984). Partial seizures with apnoea and bradycardia. *Archives of Neurology*, **41**, 173–174.

Devereaux, M.W., Keane, J.R. & Davis, R.L. (1973). Automatic respiratory failure associated with infarction of the medulla. *Archives of Neurology*, **29**, 46–52.

Douglas, N.J. (1986). Breathing during sleep in adults. In *Recent Advances in Respiratory Medicine 4*, ed. D.C. Flenley & T.L. Petty, pp. 231–248. Edinburgh: Churchill Livingstone.

Efthimiou, J., Butler, J., Woodham, C., Westaby, S. & Benson, M.K. (1992). Phrenic nerve and diaphragm function following open heart surgery: a prospective study with and without topical hypothermia. *Quarterly Journal of Medicine*, **85**, 845–853.

Fish, D.R., Howard, R.S., Wiles, C.M. & Symon, L. (1988). Respiratory arrest: a complication of cerebellar ectopia in adults. *Journal of Neurology, Neurosurgery and Psychiatry*, **51**, 714–717.

Gandevia, S.C. & Rothwell, J.C. (1987). Activation of the human diaphragm from the motor cortex. *Journal of Physiology (London)*, **384**, 109–118.

Howard, R.S. (1992). The causes and treatment of intractable hiccups. *British Medical Journal*, **305**, 1237–1238.

Howard, R.S. & Newsom-Davis, J. (1992). The neural control of respiratory function In *Neurosurgery: The Scientific Basis of Clinical Practice*, 2nd edn, ed. A. Crockard, R. Hayward & J.T. Hoff, pp. 318–333. Oxford: Blackwell Scientific.

Howard, R.S., Henderson, F., Hirsch, N.P., Stevens, J.M., Kendall, B.E. & Crockard, H.A. (1994). Respiratory abnormalities due to craniovertebral junction compression in rheumatoid disease. *Annals of the Rheumatic Diseases*, **53**, 134–136.

Howard, R.S. & Spencer, G.T. (1993). Neurogenic respiratory failure. In *Neurological Rehabilitation*, ed. R. Greenwood, M.P. Barnes, T.M. McMillan & C.D. Ward. pp. 299–310. Edinburgh: Churchill Livingstone.

Howard, R.S., Wiles, C.M., Hirsch, N.P., Loh, L., Spencer, G.T. & Newsom Davis, J. (1992). Respiratory involvement in multiple sclerosis. *Brain*, **115**, 479–494.

Howard, R.S., Wiles, C.M., Hirsch, N.P. & Spencer, G.T. (1993). Respiratory involvement in primary muscle disorders: assessment and management. *Quarterly Journal of Medicine*, **86**, 175–189.

Howard, R.S., Wiles, C.M. & Loh, L. (1989). Respiratory complications and their management in motor neurone disease. *Brain*, **112**, 1155–1170.

Howard, R.S., Wiles, C.M. & Spencer, G.T. (1988). The late sequelae of poliomyelitis. *Quarterly Journal of Medicine*, **66**, 219–232.

Hughes, R.A.C. & Bihari, D. (1993). Acute neuromuscular respiratory paralysis. *Journal of Neurology, Neurosurgery and Psychiatry*, **56**, 334–343.

Iliceto, G., Thompson, P.D., Day, B.L., Rothwell, J.C., Lees, A.J. & Marsden, C.D. (1990). Diaphragmatic flutter, the moving umbilicus syndrome, and 'belly dancer's ' dyskinesia. *Movement Disorders*, **5**, 15–22.

Kaarda, B.R. (1960). Cingulate, posterior orbital, anterior insular and temporal pole cortex. In *Handbook of Physiology*, sect. 1, *Neurophysiology*, vol. 2, ed. H.W. Magoun, pp. 1345–1372. Washington, DC: American Physiological Society.

Laroche, C.M., Moxham, J. & Green, M. (1989). Respiratory muscle weakness and fatigue. *Quarterly Journal of Medicine*, **71**, 373–397.

Leigh, R.J. & Shaw, D.A. (1976). Rapid, regular respiration in unconscious patients. *Archives of Neurology*, **33**, 356–361.

Levin, B.E. & Margolis, G. (1977). Acute failure of automatic respiration secondary to a unilateral brainstem infarct. *Annals of Neurology*, **1**, 583–586.
Maccario, M. & Lustman, L.I. (1990). Paroxysmal nocturnal dystonia presenting as excessive daytime somnolence. *Archives of Neurology*, **47**, 291–294.
Macefield, G. & Gandevia, S.C. (1991). The cortical drive to human respiratory muscles in the awake state assessed by premotor cerebral potentials. *Journal of Physiology (London)*, **439**, 545–558.
Maskill, D., Murphy, K., Mier, A., Owen, M. & Guz, A. (1991). Motor cortical representation of the diaphragm in man. *Journal of Physiology (London)*, **443**, 105–121.
Matthews, W.B. (1958). Tonic seizures in disseminated sclerosis. *Brain*, **81**, 193–206.
Mazzara, J.T., Ayres, S.M. & Grace, W.J. (1974). Extreme hypocapnia in the critically ill patient. *American Journal of Medicine*, **56**, 450–456.
Munschauer, F.E., Mador, M.J., Ahuja, A. & Jacobs, L. (1991). Selective paralysis of voluntary but not limbically influenced automatic respiration. *Archives of Neurology*, **48**, 1190–1192.
Murphy, K., Mier, A., Adams, L. & Guz, A. (1990). Putative cerebral cortical involvement in the ventilatory response to inhaled CO_2 in conscious man. *Journal of Physiology (London)*, **420**, 1–18.
Nelson, D.A. & Ray, C.D. (1968). Respiratory arrest from seizure discharges in limbic system. *Archives of Neurology*, **19**, 199–207.
Newsom Davis, J. (1970). An experimental study of hiccup. *Brain*, **93**, 851–872.
North, J.B. & Jennett, S. (1974). Abnormal breathing patterns associated with acute brain damage. *Archives of Neurology*, **31**, 338–344.
Orem, J. & Netick, A. (1986). Behavioral control of breathing in the cat. *Brain Research*, **366**, 238–253.
Pauzner, R., Mouallem, M., Sadeh, M., Tadmor, R. & Farfel, Z. (1989). High incidence of primary cerebral lymphoma in tumour-induced central neurogenic hyperventilation. *Archives of Neurology*, **46**, 510–512.
Plum, F. (1970). Neurological integration of behavioural and metabolic control of breathing. In *Breathing*. Hering–Breuer Centenary Symposium, ed. R. Parker. London: Churchill.
Plum, F. & Posner, J.R. (1983). Diagnosis of stupor and coma. Philadelphia: F.A. Davis.
Pool, J.L. & Ranshoff, J. (1949). Autonomic effects on stimulating rostral portion of cingulate gyri in man. *Journal of Neurophysiology*, **12**, 385–392.
Quinn, N.P. (1989). Multiple system atrophy: the nature of the beast. *Journal of Neurology, Neurosurgery and Psychiatry*, **52**, (Special Supplement), 78–89.
Ramsay, S.C., Adams, L., Murphy, K., Corfield, D.R., Grootoonk, S., Bailey, D.L., Frackowiack, R.S.J. & Guz A. (1993). Regional cerebral blood flow during volitional expiration in man: a comparison with volitional inspiration. *Journal of Physiology (London)*, **461**, 85–101.
Rodriguez, M., Beale, P.L., Marsh, H.M. & Okazaki, H. (1982). Central neurogenic hyperventilation in an awake patient with brainstem astrocytoma. *Annals of Neurology*, **11**, 625–628.
Ropper, A.H. (1985). Guillain–Barré syndrome: managenment of respiratory failure. *Neurology*, **35**, 1662–1665.
Shorvon, S.J. (1993). Tonic clonic status epilepticus. *Journal of Neurology, Neurosurgery and Psychiatry*, **56**, 125–134.
Smith, P.E.M., Calverley, P.M.A., Edwards, R.H.T., Evans, G.A. & Campbell, E.J.M. (1987). Practical problems in the respiratory care of patients with muscular dystrophy. *New England Journal of Medicine*, **316**, 1197–1205.
Smith, P.E.M., Edwards, R.H.T. & Calverley, P.M.A. (1991). Mechanisms of sleep

disordered breathing in chronic neuromuscular disease: implications for management. *Quarterly Journal of Medicine*, **81**, 961–973.
Stradling, J.R. & Phillipson, E.A. (1986). Breathing disorders during sleep. *Quarterly Journal of Medicine*, **58**, 3–18.
Tobin, M.J. & Snyder J.V. (1984). Cheyne–Stokes respiration revisited: controversies and implications. *Critical Care Medicine*, **12**, 882–887.
Trend, P.StJ., Wiles, C.M., Spencer, G.T., Morgan Hughes, J., Lake, B.D. & Patrick, A.D. (1985). Acid maltase deficiency in adults. *Brain*, **108**, 845–860.

35

The respiratory muscles in neurological disease

GARY H. MILLS AND MALCOLM GREEN
Respiratory Muscle Laboratory, Royal Brompton National Heart and Lung Hospital, London, UK

Respiratory muscle involvement in neurological disease may be underestimated because of the reduced mobility that may occur in these conditions. As the respiratory muscles are vital for life, formal tests of respiratory muscle function have an important role in neurological disorders. Moderate degrees of impairment may be associated with nocturnal hypoventilation, whereas more severe involvement leads to respiratory failure. If acute this can lead to respiratory arrest, and if of gradual onset may be associated with signs of cor pulmonale. Weakness of the expiratory muscles may impair coughing and bulbar weakness may lead to aspiration.

Respiratory muscle weakness may be produced by lesions affecting the central nervous system, the peripheral nerves, the neuromuscular junction or the muscles themselves.

Central nervous system

Respiratory dysfunction is seen in patients after hemiplegia secondary to cerebrovascular disease (De Troyer, Zegers de Beyl & Thirion, 1981). Patients with a unilateral corticospinal lesion demonstrate reduced activity of both the intercostal muscles and the diaphragm on the side of the paresis during inspiration.

Parkinson's disease may lead to low vital capacity and a reduction in maximum inspiratory effort (Bogaard *et al.*, 1989). Bruin and colleagues recently concluded that anti-parkinsonian medication improved the co-ordination of upper airways and chest wall musculature, particularly during rapid events such as coughing and swallowing, and therefore might reduce the incidence of aspiration (De Bruin *et al.*, 1993). Chorea may impose involuntary efforts on the normal respiratory pattern (Newsom-Davis, 1970). Cerebellar atrophy has also been reported in a small number of cases as reducing voluntary respiratory muscle strength, although electrical stimulation of the phrenic nerves produced normal twitch transdiaphragm pressures (Mier & Green,

1988). Multiple sclerosis is a rare cause of respiratory failure. Cooper descibed a case requiring positive pressure ventilation following an acute exacerbation. Ventilation via a tracheostomy was required for 8 weeks. Subsequently the tracheostomy was closed and the patient was able to return to work (Cooper, Trend & Wiles, 1985).

Spinal cord lesions above the C3–5 phrenic outflow lead to paralysis of the diaphragm and intercostal muscles. These patients may be ventilated via a tracheostomy with positive pressure ventilation or with the aid of diaphragm pacing (Moxham & Schneerson, 1993), although this is technically difficult and not without risk. Diaphragm pacing has also been employed in the central apnoea syndromes, which may be either idiopathic (Ondine's curse) or secondary to organic lesions in the brainstem, such as strokes or encephalitis (Glenn *et al.*, 1978). De Troyer & Heilporn (1980) investigated respiratory mechanics in quadraplegia in patients with intact diaphragm function. Intercostal paralysis leads to loss of stability of the rib cage with paradoxical inward movement of the upper chest during inspiration. This leads to loss of pulmonary compliance, increase in the work of breathing, and an increase in the tendency towards alveolar collapse. They also identified a group with persistent intercostal electromyographic (EMG) activity who did not demonstrate these paradoxical movements, suggesting that intercostal EMG recordings should be made in patients with quadraplegia to determine those most at risk from alveolar collapse and resultant chest infections.

Lower motor neurone lesions

Herpes zoster can rarely cause phrenic paralysis (Brostoff, 1966). More commonly anterior horn cell damage due to polio causes respiratory failure in the initial stages of the disease, which can later be complicated by chest wall deformity leading to a chronic deterioration in respiratory function, sometimes after decades.

Amyotrophic lateral sclerosis (ALS, or motor neurone disease) has presented purely as respiratory failure (Fromm, Wisdom & Block, 1977; Al-Shaikh *et al.*, 1986). This, or similar diseases, may affect both upper and lower motor neurones and also lead to lung infections secondary to aspiration as bulbar function becomes impaired (Small *et al.*, 1993). In a study of the natural history of ALS, Ringel *et al.* (1993) found that the rate of decline in respiratory muscle strength was the best predictor of prognosis and life-span.

Guillain–Barré syndrome may lead rapidly to respiratory failure. Deterioration may not be clear from clinical examination or peak expiratory flow rate measurements. Frequent vital capacity measurements can monitor the speed of deterioration in respiratory muscle strength and provide a good indication

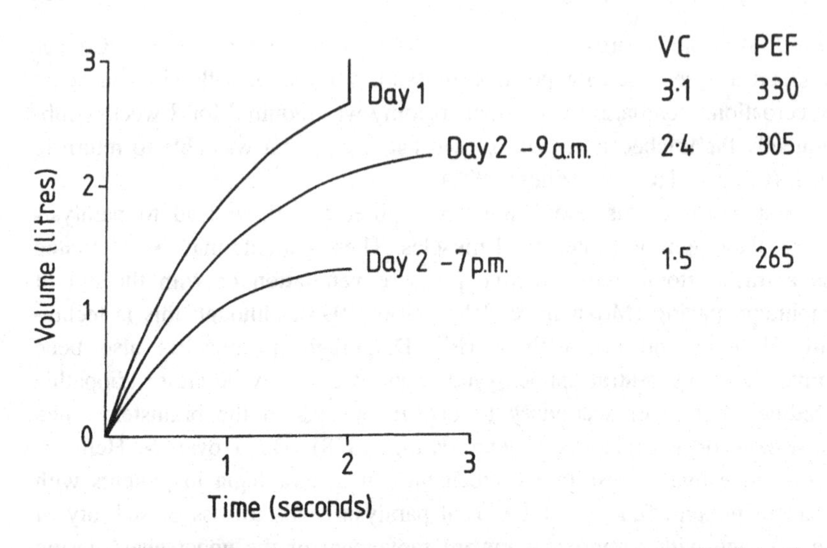

Fig. 35.1. Vital capacity (VC) measured over 2 days in a woman with Guillain–Barré syndrome, resting in bed. Her clinical status changed little over this period, but there was a substantial fall in VC due to increasing weakness of her respiratory muscles. She was eventually ventilated on the evening of day 2, but her deterioration would have been missed by relying on clinical observation or on peak flow recordings (PEF).

of incipient respiratory failure (Fig. 35.1). This allows early transfer to an intensive care unit, with intubation and mechanical ventilation as appropriate (Chevrolet & Deleamont, 1991).

A number of infections are associated with a polyneuropathy. These include mycoplasma, infectious mononucleosis, cytomegalovirus, mumps, measles and rubella (Schonberg *et al.*, 1981). Diphtheria may lead to a demyelinating polyneuropathy.

Bilateral diaphragm paralysis has occurred as part of a paraneoplastic syndrome (Thomas *et al.*, 1984).

Hereditary neuropathies

Severe bilateral diaphragm weakness has been descibed in Charcot Marie Tooth disease, but abnormalities in oxygenation during sleep were minor (Laroche *et al.*, 1988c). In infants the respiratory muscles may be severely impaired in Werdnig Hoffmann disease (progressive spinal muscular atrophy). The acute infantile form produces severe generalized muscle weakness beginning before the age of 4 months. In the absence of respiratory support, this leads to a fatal outcome secondary to respiratory failure in 95% of cases by 18 months. A chronic infantile form allows longer survival, but affected children usually develop scoliosis, which further compromises respiratory function (Campbell & Liversedge, 1981).

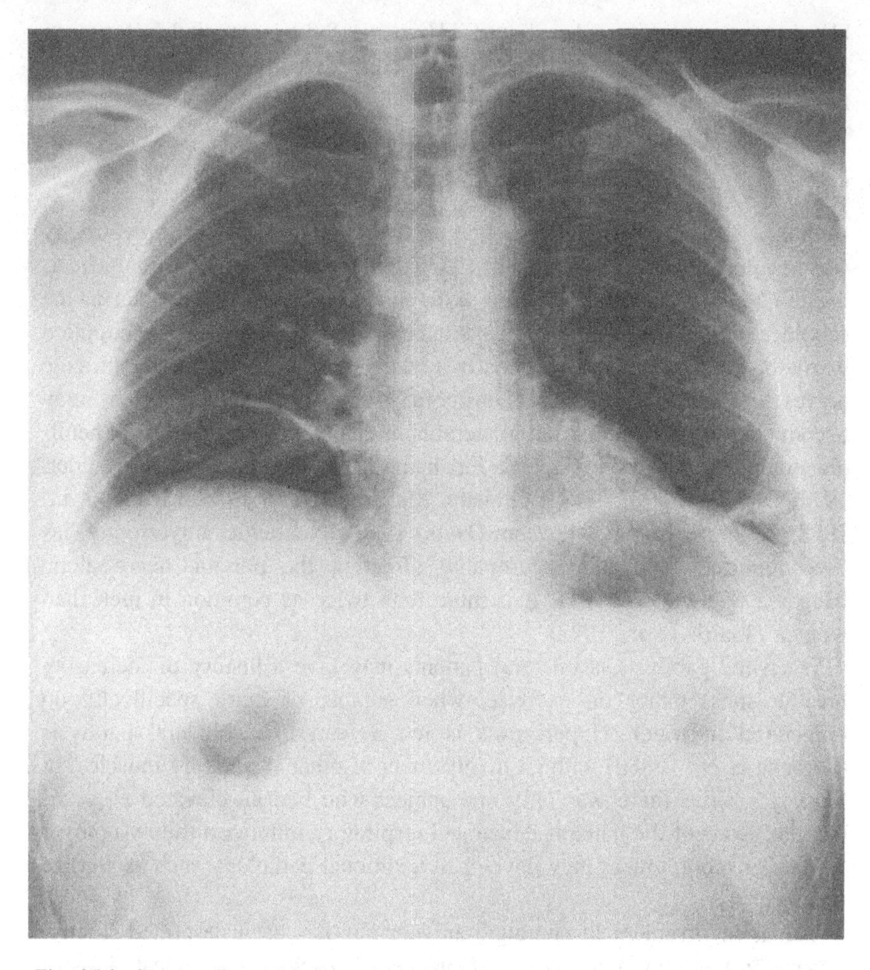

Fig. 35.2. Chest radiograph of a patient with bilateral diaphragm paralysis illustrating a reduction in lung volume and some areas of basal atelectasis.

Isolated phrenic nerve palsy

Isolated hemidiaphragm paralysis is usually asymptomatic, although breathlessness may be noted at the onset of the paralysis, or if there are other complicating factors such as lung disease or obesity. Bilateral phrenic palsy causes breathlessness on exercise and on lying flat.

Radiological studies may be helpful in unilateral hemidiaphragm paralysis as a conspicuously raised hemidiaphragm may be seen on the chest radiograph with some atelectasis in the lung field immediately above the paralysed hemidiaphragm. However, in bilateral paralysis there is no such contrast between the two sides and the only feature may be a matching reduction in size of both lung fields, making radiological diagnosis more difficult (Fig. 35.2). Arterial blood gases may be normal or show mild hypoxia. Subjects studied during sleep show periods of hypoxia overnight, which coincide with episodes of obstructive apnoea or hypopnoea.

Neuralgic amyotrophy

Neuralgic amyotrophy may cause phrenic nerve palsy. Neuralgic amyotrophy was first described in 46 patients in 1943 (Spillane, 1943) as a disorder affecting the brachial plexus, presenting with sudden onset of severe pain in the shoulder region, followed 1–28 days later by weakness of muscles supplied by the brachial plexus. More recently it has been recognized that the phrenic nerves and rarely the vagus (Dinsmore, Irvine & Callender, 1985) may become involved, resulting in unilateral hemidiaphragmatic or bilateral hemidiaphragmatic paralysis (Cape & Fincham, 1965; Tsairis, Dyck & Mulder, 1972; Dinsmore *et al.*, 1985; Graham, Martin & Haas, 1985; Pieters *et al.*, 1988; Gregory, Loh & Newsom-Davis, 1990). Neuralgic amyotrophy has been reported as a mononeuropathy affecting the phrenic nerve alone (England & Sumner, 1987). It is more than twice as common in men than women (Tsairis *et al.*, 1972).

When the paralysis is bilateral patients may give a history of increasing breathlessness either on exercise, when supine, or more specifically on immersion in water. Hypercapnia is not a feature of bilateral paralysis (Laroche *et al.*, 1988b) without involvement of other respiratory muscles. In Mulvey's series there was only one subject who had an elevated P_{a,CO_2} of 6.6 kPa. None of the patients developed respiratory failure on follow-up over 5 years or more, unless they developed additional pathology such as cardiac failure.

Bulbar involvement in neuralgic amyotrophy has been described (Pierre, Laterre & Van den Bergh, 1990). However, there were no clinical signs of this in a series of 16 patients reported by Mulvey *et al.* (1993).

Prognosis for neuralgic amyotrophy where diaphragm involvement is not a feature has been reported as 75–90% recovery at 3 years (Tsairis *et al.*, 1972; Subramony, 1988). However, recovery of diaphragm function seems less likely and recurrent episodes have been reported (Gregory *et al.*, 1990). Mulvey followed up nine patients for 2–4 years. Eight reported some symptomatic improvement, but only three of these showed improvement in transdiaphragmatic pressures during a maximal sniff or following transcutaneous electrical stimulation of the phrenic nerve (Mulvey *et al.*, 1993).

Nerve trauma

Cardiothoracic surgery may be complicated by phrenic nerve injury secondary to trauma or cold. This may require prolonged postoperative ventilation, especially in paediatric cardiac surgery (Chandler *et al*., 1982; Large *et al*., 1985). Electrical transcutaneous stimulation of the phrenic nerves may be useful to allow bedside assessment of phrenic nerve conduction (Fox *et al*., 1989; Launois *et al*., 1993).

The phrenic nerve may also be damaged by malignancies in the neck or anterior mediastinum.

Diaphragm flutter is rare and may be secondary to unilateral or bilateral infiltration or trauma of the phrenic nerves. Contraction occurs at a rate of 1–10 per second, producing cogwheel respiration (Philipps & Eldridge, 1973).

Hiccups are much more common, and usually resolve spontaneously. Chronic hiccup can be debilitating and may be associated with cerebral tumours, myocardial infarction, renal failure or abdominal surgery. Treatment should be aimed at the underlying cause. Symptomatic relief may be possible with metaclopramide and chlorpromazine. Baclofen has been reported as effective in the suppression of chronic hiccup (Launois *et al*., 1993).

Autonomic neuropathy

Platypnoea has been descibed in patients with pulmonary shunts, intracardiac shunts and severe lung disease, and recently in autonomic failure. Platypnoea is breathlessness brought on by assuming an upright posture and relieved by assuming a recumbent one. Autonomic failure may produce inadequate perfusion of the upper part of the lung, resulting in a large increase in dead space, which is improved by fluid loading (Fox *et al*., 1989).

The Shy–Drager syndrome may result in periodic respiratory gasps and irregular respirations (Bannister & Oppenheimer, 1972; Lockwood, 1976), multiple apnoeic spells and dysrhythmic breathing when erect (Chokroverty, Sharp & Barron, 1978), together with upper airway obstruction (Israel & Marino, 1977; Gilmartin *et al*., 1984).

Autonomic dysfunction due to diabetes mellitus is associated with respiratory muscle impairment, and with a fall in vital capacity and inspiratory muscle strength, although this may also be due to impaired glucose utilization or microangiopathy (Wanke *et al*., 1990).

The neuromuscular junction

Mouth pressures may be reduced in patients with myasthenia gravis, and improve with neostigmine (Rinqvist & Rinqvist, 1971). Mier-Jedrzejowicz *et al*. (1988) found that limb strength was not closely related to respiratory muscle strength. They also noted that phrenic nerve stimulation at 1 Hz resulted in normal twitch transdiaphragmatic pressures in all but the most severely affected subjects, although higher frequencies did lead to failure of diaphragm contraction. Phrenic stimulation at 3–5 Hz is probably optimal for detection of diaphragm involvement in myasthenia (Mier-Jedrzejowicz *et al.*, 1988; Mier *et al.*, 1992) (Figs. 35.3, 35.4).

Lambert–Eaton syndrome is often associated with an underlying small cell carcinoma of the lung (O'Neill, Murray & Newsom-Davis, 1988). IgG antibodies are formed against presynaptic Ca^{2+} channels, thus interfering with the release of acetylcholine (Lang *et al.*, 1981). Weakness sufficient to cause respiratory failure has been described (O'Neill *et al.*, 1988).

Drugs and toxins

Respiratory muscle impairment may occur on exposure to drugs and toxins that affect transmission at the synapse. Toxins that have a presynaptic action include tetanus and botulinum toxin (Wilcox *et al.*, 1989) and α-latrotoxin from the black widow spider (Sperelakis, Suszkiw & Hackett, 1985). Neuromuscular blockade can be precipitated by tic bites including the wood tic and dog tic (*Dermasentor*) and the Australian tic (*Icsodes*), and rarely by wasp stings. Snake venom, particularly from the South American rattlesnake, affects pre- and postsynaptic transmission and may prove fatal (Swift, 1981).

Certain drugs such as aminoglycosides (especially neomycin), penicillamine, chloroquine, procainamide, phenytoin and lithium may impair neuromuscular transmission by a postsynaptic action and may uncover latent myasthenia or produce prolonged postoperative neuromuscular blockade.

Organophosphate compounds are used in farming and in chemical weapons. These substances may irreversibly phosphorylate cholinesterase, so rendering it permanently inactive. As a result the concentration of acetylcholine rises in the synapse producing long-lasting depolarization of the postsynaptic membrane which can present as acute respiratory failure.

Muscle disorders

Muscular dystrophies

Duchenne muscular dystrophy eventually leads to severe respiratory muscle weakness and death. Assisted ventilation, especially using nasal intermittent positive pressure, may be appropriate if the respiratory muscles are affected earlier than the limb muscles. Ventilation via tracheostomy has been recommended (Raphael *et al.*, 1991), but has many complications and may not improve quality of life. Respiratory muscle strength is frequently impaired in limb girdle muscular dystrophy, and the degree of limb involvement may not correlate with the level of respiratory impairment (Stubgen *et al.*, 1994). Respiratory muscle weakness has also been described in facio-scapulo-humeral dystrophy although the exact incidence is unknown (Wolf *et al.*, 1981).

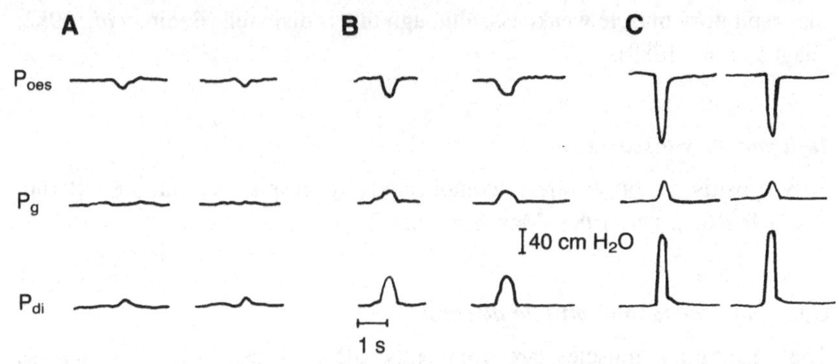

Fig. 35.3. Sniff pressures recorded in a 52-year-old man who presented with orthopnoea and progressive respiratory failure. (A) Investigation revealed very severe weakness of his diaphragm and other respiratory muscles, which was later attributed to myasthenia gravis. He was treated appropriately with substantial improvement in his respiratory muscle strength, shown here at 3 months (B) and 7 months (C). Indicated pressures: P_{oes}, oesophogeal; P_g, gastric; P_{di}, transdiaphragmatic.

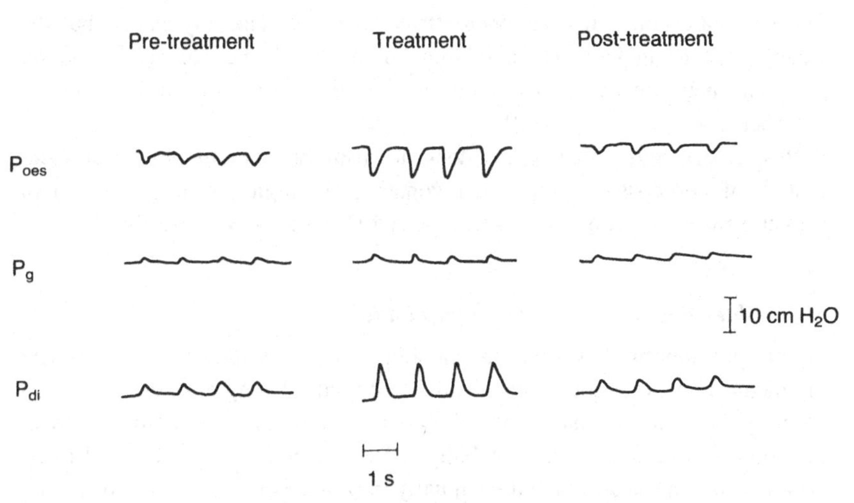

Fig. 35.4. Transient improvement in oesophogeal, gastric and transdiaphragmatic pressures (P_{oes}, P_g, P_{di}) during treatment with an anticholinesterase, tested with bilateral electrical phrenic stimulation in a patient with myasthenia.

Dystrophia myotonica

Respiratory muscle dysfunction is common in dystrophia myotonica and may lead to respiratory failure or recurrent pulmonary infection. Patients frequently show hypoxia during sleep, especially if overweight (Finnimore *et al.*,1994). Alveolar hypoventilation has been said to be out of proportion to the respiratory muscle weakness, although this is disputed (Begin *et al.*, 1982; Clague *et al.*, 1989).

Inflammatory diseases

Polymyositis is only rarely complicated by respiratory failure (Braun, Arora & Rochester, 1983; Martin *et al.*, 1985).

Congenital metabolic muscle diseases

The respiratory muscles are frequently affected early in acid maltase deficiency, and diaphragm paralysis may be the presenting feature (Rosinow & Engel, 1978; Sivak *et al.*, 1981). Respiratory muscle weakness may also occasionally occur with mitochondrial myopathy, nemoline myopathy and carnitine deficiency (Laroche & Green, 1990).

Drug-induced myopathies

Prolonged weakness after neuromuscular blockade has been described on the intensive care unit. This has been attributed to an acute myopathy, but the exact cause is unclear. There is some indirect evidence of an interaction between neuromuscular blocking drugs and corticosteroids (Hansen-Flaschen, Cowen & Raps, 1993).

Respiratory muscle weakness does not usually seem to be a prominent feature of corticosteroid-induced myopathy, although low maximal mouth pressure measurements have been reported (Decramer & Stas, 1992).

Myopathies due to electrolyte abnormalities

Electrolyte abnormalities such as calcium (Aubier *et al.*, 1985), phosphate or magnesium depletion reduce muscle strength (Dhingra *et al.*, 1984). In the periodic paralyses intermittent diffuse muscle weakness is associated with abnormalities of potassium metabolism. Respiratory muscles are said to be spared in the initial stages, but eventually may also become paralysed (Engel, 1981).

Endocrine myopathies

Respiratory muscle function may be impaired in endocrine disorders such as hyperthyroidism (Mier *et al.*, 1989), hypothyroidism (Weiner, Chausow & Szidon, 1986; Laroche *et al.* 1988a) and Addison's disease (Mier *et al.*, 1988). This is thought to be part of the generalized myopathy and in a few documented cases it has improved with treatment of the underlying condition.

Hypothyroidism can also be associated with phrenic neuropathy (Hamley *et al.*, 1975).

Conclusion

The respiratory muscles are frequently involved in neurological disease. The resulting impairment of respiratory function contributes to morbidity and mortality, and may go unrecognised in the early stages of the disease.

References

Al-Shaikh, B., Kinnear, W., Higenbottam, T.W., Smith, H.S., Schneerson, J.M. & Wilkinson, I. (1986). Motor neurone disease presenting with respiratory failure. *British Medical Journal*, **292**, 1325–1326.

Aubier, M., Viires, N., Piquet, J., Murciano, D., Blanchet, T., Marity, C., Gherardi, P. & Pariente, R. (1985). Effects of hypocalcaemia on diaphragm strength generation. *Journal of Applied Physiology*, **58**, 2054–2061.

Bannister, R. & Oppenheimer, D. (1972). Degenerative diseases of the nervous system associated with autonomic failure. *Brain*, **95** 457–474.

Begin, R., Bureau, M.A., Lupien, L., Bernier, J.P. & Lemieux, B. (1982). Pathogenesis of respiratory insufficiency in myotonic dystrophy: the mechanical factors. *American Review of Respiratory Disease*, **125**, 312–318.

Bogaard, J.M., Hovestadt, A., Meerwaldt, J., van der Meche, V.D.A. & Stigt, J. (1989). Maximal expiratory and inspiratory flow volume curves in Parkinson's disease. *American Review of Respiratory Disease*, **199**, 610–614.

Braun, N.M., Arora, N.S. & Rochester, D.F. (1983). Respiratory muscle and pulmonary function in polymyositis and other proximal myopathies. *Thorax*, **38**, 616–623.

Brostoff, J. (1966). Diaphragmatic paralysis after herpes zoster. *British Medical Journal*, **ii**, 1571–1572.

Campbell, M.J. & Liversedge, L.A. (1981). The motor neurone diseases (including the spinal muscular atrophies). In *Disorders of Voluntary Muscle*, ed. J. Walton, pp. 736–752. Edinburgh: Churchill Livingstone.

Cape, C.A. & Fincham, R.W. (1965). Paralytic brachial neuritis with diaphragmatic paralysis: contralateral recurrence. *Neurology*, **15**, 191–193.

Chandler, K.W., Rozas, C.J., Kory, R.C. & Goldman, A.L. (1982). Bilateral diaphragmatic paralysis complicating local cardiac hypothermia during open heart surgery. *American Review of Respiratory Disease*, **125**, 98.

Chevrolet, J.C. & Deleamont, P. (1991). Repeated vital capacity measurements as predictive parameters for mechanical ventilation need and weaning success in Guillain–Barré syndrome. *American Review of Respiratory Disease*, **144**, 814–818.

Chokroverty, S., Sharp, J.T. & Barron, K.D. (1978). Periodic respiration in the erect posture in Shy Drager syndrome. *Journal of Neurology, Neurosurgery and Psychiatry*, **41**, 980–986.

Clague, J.E., Carter, J., Coakley, J., Edwards, R.H.T. & Calverley, P.M.A. (1989). Ventilatory control and inspiratory effort sensation in dystrophia myotonica. *American Review of Respiratory Disease*, **139**, 349.

Cooper, C.B., Trend, P.S.J. & Wiles, C.M. (1985). Severe diaphragm weakness in multiple sclerosis. *Thorax*, **40**, 633–634.

De Bruin, P.F.C., De Bruin, V.M.S., Lees, A.J. & Pride, N.B. (1993). Effect of treatment on airway dynamics and respiratory muscle strength in Parkinson's disease. *American Review of Respiratory Disease*, **148**, 1576–1580.

De Troyer, A. & Heilporn, A. (1980). Respiratory mechanics in quadraplegia: the respiratory function of the intercostal muscles. *American Review of Respiratory Disease*, **122**, 591–600.

De Troyer, A., Zegers de Beyl, D. & Thirion, M. (1981). Function of the respiratory muscles in acute hemiplegia. *American Review of Respiratory Disease*, **123**, 631–632.

Decramer, M. & Stas, K.J. (1992). Corticosteroid induced myopathy involving respiratory muscles in patients with chronic obstructive pulmonary disease or asthma. *American Review of Respiratory Disease*, **146**, 800–802.

Dhingra, S., Solven, F., Wilson, A. & McCarthy, D. (1984). Hypomagnesemia and respiratory muscle power. *American Review of Respiratory Disease*, **129**, 497–498.

Dinsmore, W.W., Irvine, A.K. & Callender, M.E. (1985). Recurrent neuralgic amyotrophy with vagus and phrenic nerve involvement. *Clinical Neurology and Neurosurgery*, **87**, 39–40.

Engel, A.G. (1981). Metabolic and endocrine myopathies. In *Disorders of Voluntary Muscle*, ed. S.J. Walton, pp. 664–711. Edinburgh: Churchill Livingstone.

England, J.D. & Sumner, A.J. (1987). Neuralgic amyotrophy: an increasingly diverse entity. *Muscle and Nerve*, **10**, 60–68.

Finnimore, A.J., Jackson, R.V., Morton, A. & Lynch, E. (1994). Sleep hypoxia in myotonic dystrophy and its correlation with awake respiratory function. *Thorax*, **49**, 66–70.

Fox, J.L., Brown, E., Harrison, J.K., Williams, J. & Terry, P.B. (1989). Platypnoea-orthodeoxia and progressive autonomic failure. *American Review of Respiratory Disease*, **140**, 1802–1804.

Fromm, G.B., Wisdom, P.J. & Block, A.J. (1977). Amyotrophic lateral sclerosis presenting with respiratory failure. *Chest*, **71**, 612–614.

Gilmartin, J.J., Wright, A.J., Cartilidge, N.E. F. & Gibson, G.J. (1984). Upper airway obstruction complicating the Shy–Drager syndrome. *Thorax*, **39**, 313–314.

Glenn, W.W., Gee, J.B., Cole, D.R., Farmer, W.C., Shaw, R.K. & Meckman, C.B. (1978). Combined central alveolar hypoventilation and upper airway obstruction: treatment by tracheostomy and diaphragm pacing. *American Journal of Medicine*, **64**, 50–60.

Graham, A.N., Martin, P.D. & Haas, L.F. (1985). Neuralgic amyotrophy with bilateral diaphragmatic palsy. *Thorax*, **40**, 635–636.

Gregory, R.P., Loh, L. & Newsom-Davis, J. (1990). Recurrent isolated alternating phrenic nerve palsies: a variant of brachial neuritis? *Thorax*, **45**, 420–421.

Hamley, F.H., Timms, R.T., Mihn, V.D. & Moser, K.M. (1975). Bilateral phrenic nerve paralysis in myxoedema. *American Review of Respiratory Disease*, **111**, 911.

Hansen-Flaschen, J., Cowen, J. & Raps, E.C. (1993). Neuromuscular blockade in the ICU. *American Review of Respiratory Disease*, **147**, 234–236.

Israel, R.H. & Marino, J.M. (1977). Upper airway obstruction in the Shy–Drager syndrome. *Annals of Neurology*, **2**, 83.

Lang, B., Newsom-Davis, J., Wray, D., Vincent, A. & Murray, N. (1981). Autoimmune aetiology for myasthenic (Eaton–Lambert) syndrome. *Lancet*, **ii**, 224–226.

Large, S.R., Heywood, L.J., Flower, C.D., Cory-Pierce, R., Wallwork, J. & English, T.A.H. (1985). Incidence and aetiology of a raised hemidiaphragm after cardiopulmonary bypass. *Thorax*, **40**, 444–447.

Laroche, C.M., Cairns, T., Moxham, J. & Green, M. (1988a). Hypothyroidism presenting with respiratory muscle weakness. *American Review of Respiratory Disease*, **138**, 472–474.

Laroche, C.M., Carroll, N., Moxham, J. & Green, M. (1988b). Clinical significance of severe isolated diaphragm weakness. *American Review of Respiratory Disease*, **138**, 862–886.

Laroche, C.M., Carroll, N., Moxham, J., Stanley, N.N., Courtney Evans, R.J. & Green, M. (1988c). Diaphragm weakness in Charcot Marie Tooth Disease. *Thorax*, **43**, 478–479.

Laroche, C.M. & Green, M. (1990). Respiratory muscle involvement in systemic disease. In *Problems in Respiratory Care*, ed. M. Tobin, pp. 409–422. Phiadelphia: J.B. Lippincott.

Launois, S., Bizec, J.L., Whitelaw, W.A., Cabane, J. & Derenne, J.P. (1993). Hiccup in adults: an overview. *European Respiratory Journal*, **6**, 563–575.

Lockwood, A.H. (1976). Shy–Drager syndrome with abnormal respiration and antidiuretic hormone release. *Archives of Neurology*, **33**, 292–295.

Martin, L., Chalmers, I.M., Dhingra, S., McCarthy, D. & Hunter, T. (1985). Measurements of maximum respiratory pressures in polymyositis and dermatomyositis. *Journal of Rheumatology*, **12**, 104–107.

Mier, A., Brophy, C., Moxham, J. & Green, M. (1992). Repetitive stimulation of phrenic nerves in myasthenia gravis. *Thorax*, **47**, 640–644.

Mier, A., Brophy, C., Wass, J.A., Besser, G.M. & Green, M. (1989). Reversible respiratory muscle weakness in hyperthyroidism. *American Review of Respiratory Disease*, **139**, 529–533.

Mier, A. & Green, M. (1988). Respiratory muscle weakness associated with cerebellar atrophy. *American Review of Respiratory Disease*, **137**, 673–677.

Mier, A., Laroche, C., Wass, J. & Green, M. (1988). Respiratory muscle weakness in Addison's disease. *British Medical Journal*, **297**, 457–458.

Mier-Jedrzejowicz, A.K., Brophy, C., Moxham, J. & Green, M. (1988). Respiratory muscle function in myasthenia gravis. *American Review of Respiratory Disease*, **138**, 867–873.

Moxham, J. & Schneerson, J.M. (1993). Diaphragmatic pacing. *American Review of Respiratory Disease*, **148**, 533–536.

Mulvey, D.A., Aquilina, R.J., Elliot, M.W., Moxham, J. & Green, M. (1993). Diaphragmatic dysfunction in neuralgic amyotrophy: an electrophysiologic examination of 16 patients presenting with dyspnoea. *American Review of Respiratory Disease*, **147**, 66–71.

Newsom-Davis, J. (1970). Diseases of the nervous system. In *The Respiratory Muscles: Mechanics and Neural Control*, 2nd edn, ed. E.J.M. Campbell, E. Agostoni & J. Newsom-Davis, p. 314. London: Lloyd-Luke.

O'Neill, J.H., Murray, N.M.F. & Newsom Davis, J. (1988). The Lambert–Eaton myasthenic syndrome: a review of 50 cases. *Brain*, **111**, 577–596.

Philipps, J.R. & Eldridge, F.L. (1973). Respiratory myoclonus (Leeuwenhoek's disease). *New England Journal of Medicine*, **289**, 1390–1392.

Pierre, P.A., Laterre, C.E. & Van den Bergh, P.Y. (1990). Neuralgic amyotrophy with involvement of cranial nerves 9, 10, 11 and 12. *Muscle and Nerve*, **13**, 704–707.

Pieters, T., Lambert, M., Huaux, J.P. & Nagant de Deuxchaisnes, C. (1988). Hemidiaphragmatic paralysis, an unusual presentation of Parsonage–Turner syndrome. *Clinical Rheumatology*, **7**, 402–405.

Raphael, J.-C., Chevret, S., Chastang, C. & Bouvet, F. (1991). A prospective multicentre study of home mechanical ventilation in Duchenne de Boulogne muscular dystrophy. In *Recommendations for Home Mechanical Ventilation*, ed. C.F. Donner, P. Howard & D. Robert, pp. 312–316. Arona, Italy:

European Respiratory Society Rehabilitation and Chronic Care Scientific Group.

Ringel, S.P., Murphy, J.R., Alderson, M.D., Bryan, W., England, J.D., Miller, R.G., Petajan, J.H., Smith, S.A., Roelofs, R.I., Ziter, F., Lee, M.Y., Brinkmann, J.R., Almada, A., Gappmaier, E., Graves, J., Herbelin, L., Mendoza, M., Mylar, D., Smith, P. & Yu, P. (1993). The natural history of amyotrophic lateral sclerosis. *Neurology*, **43**, 1316–1322.

Rinqvist, I. & Rinqvist, T. (1971). Respiratory mechanics in untreated myasthenia gravis with special reference to the respiratory forces. *Acta Medica Scandinavica*, **190**, 499.

Rosinow, E.G. & Engel, A.G. (1978). Acid maltase deficiency in adults presenting as respiratory failure. *American Journal of Medicine*, **64**, 485–491.

Schonberger, L.B., Hurwitz, E.S., Katona, P., Holman, R.C. & Bregman, D.J. (1981). Guillain–Barré syndrome: its epidemiology and association with influenza vaccination. *Annals of Neurology*, **9** (Supplement), 31–38.

Sivak, E.D., Salanga, V.D., Wilbourn, A.J., Mitsumoto, H. & Golish, J. (1981). Adult onset acid maltase deficiency presenting as diaphragmatic paralysis. *Annals of Neurology*, **9**, 613–615.

Small, G.A., Pullman, S., Emerson, R. & Lange, D.J. (1993). Central motor delay in motor neuron diseases. *Electroencephelography and Clinical Neurophysiology*, **2**, s71.

Sperelakis, N., Suszkiw, J.B. & Hackett, J.T. (1985). The neuromuscular junction. In *The Thorax*, ed. C. Roussos & P.T. Macklem, pp. 175–177. New York Marcel Dekker.

Spillane, J.D. (1943). Localised neuritis of the shoulder girdle: a report of 46 cases in the MEF. *Lancet*, **ii**, 532–535.

Stubgen, J.-P., Ras, G.J., Schultz, C.M. & Crowther, G. (1994). Lung and respiratory muscle function in limb girdle muscular dystrophy. *Thorax*, **49**, 61–65.

Subramony, S.H. (1988). AAEE case report no 14: neuralgic amyotrophy (acute brachial neuropathy). *Muscle and Nerve*, **11**, 39–44.

Swift, T.R. (1981). Disorders of neuromuscular transmission other than myasthenia gravis. *Muscle and Nerve*, **4**, 334–353.

Thomas, N.E., Passamonte, P.M., Sunderrajan, E.V., Andelin, J.B. & Ansbacher, L.E. (1984). Bilateral diaphragmatic paralysis as a possible paraneoplastic syndrome from renal cell carcinoma. *American Review of Respiratory Disease*, **129**, 507–509.

Tsairis, P., Dyck, P.J. & Mulder, D.W. (1972). Natural history of brachial plexus neuropathy. *Archives of Neurology*, **27**, 109–117.

Wanke, T., Formanek, D., Auinger, M., Popp, W., Zwick, H. & Irsigler, K. (1990). Inspiratory muscle performance and pulmonary function changes in insulin dependent diabetes mellitus. *American Review of Respiratory Disease*, **143**, 97–100.

Weiner, M., Chausow, A. & Szidon, P. (1986). Reversible respiratory muscle weakness in hypothyroidism. *British Journal of Diseases of the Chest*, **80**, 391–395.

Wilcox, P., Andolfatto, G., Fairburn, M.S. & Pardy, R.L. (1989). Long-term follow-up of symptoms, pulmonary function, respiratory muscle strength, and exercise performance after botulism. *American Review of Respiratory Disease*, **139**, 157–163.

Wolf, E., Shochina, M., Ferber, I. & Gonen, B. (1981). Phrenic nerve and diaphragmatic involvement in progressive muscular dystrophy. *Electromyography and Clinical Neurophysiology*, **21**, 35–53.

Part IV

Development, survival, regeneration and death

Individual neurones may show plasticity or die during the course of the normal development of the nervous system as well as in disease or following injury. In this final section we bring some of these issues together. Thus Diamond and Gloster describe influences on the plasticity of peripheral nerves, which they argue represent, at least in part, influences at work in the normal CNS. The chapters by Johnson and by Munson and Nishimura also concern influences from the periphery, but this time largely on the properties of the central parts of the neurones, either their connections or their ultrastructure: an important issue in both cases is the restitution of function following injury. In the next chapter Nicholls describes a new preparation: the CNS of the neonatal opossum maintained in tissue culture. The opossum emerges from the womb at a very immature stage, in which it seems that central nerve fibres can regenerate following injury, like peripheral nerves but unlike those of the adult CNS. This preparation should provide a new base for direct study of the influences operating on plasticity in the CNS.

The final two chapters, by Pullen and by Krieger, extend the theme of neuronal survival from Johnson's chapter to the context of motor neurone diseases, conditions characterized by the selective degeneration of neurones in cortico-motor pathways. It is likely that only by understanding those factors that normally allow these neurones to survive and to repair following injury or insult will we understand why in these devastating diseases they do not.

36

Axonal growth and plasticity in the adult nervous system

JACK DIAMOND AND ANDREW GLOSTER*

Department of Biomedical Sciences, McMaster University, Hamilton, Ontario, Canada

Two types of peripheral nerve growth can restore a functional innervation to deprived target tissues

We are interested in the functional reinnervation of target tissues that can occur spontaneously, or be promoted surgically, after nerves are compromised by damage or disease. To learn more of the mechanisms involved in such recoveries–mechanisms which we believe are also involved in the intrinsic plasticity of the adult nervous system–we have studied the sensory and sympathetic nerves supplying the skin of the adult rat. We have focused on their *collateral sprouting*, an arborizing growth emanating from undamaged nerve axons, and axonal *regeneration*, an elongating growth induced by axotomy and originating at the region of injury. *In vivo*, collateral sprouting almost invariably occurs at or within the target tissue of the nerves concerned; experimentally it is usually evoked by partially denervating the target tissue (in our investigations, the skin), following which the remaining undamaged fibres begin to sprout into the deprived areas (Diamond *et al.*, 1976). Axons regenerating after a nerve crush, the preferred injury in experiments to study this phenomenon, almost invariably grow within and along the degenerating nerve trunk to reach their former target regions, where they undergo branching much as in collateral sprouting. Despite an apparently identical morphological and functional end result, the surprising finding which emerged from our animal studies was that collateral sprouting and axonal regeneration of the same nerves are quite differently regulated (Diamond *et al.*, 1987, 1992a, b; Gloster & Diamond, 1992).

The experiments involved the 'mapping' of the sensory and sympathetic skin territories, or fields, supplied by a selected cutaneous nerve (the medial branch of the 13th thoracic dorsal cutaneous nerve, or mDCN-T13). Electrophysiological nerve recordings were used to map the area subserved by the light touch modality, and nociceptive reflexes to map the heat and mechano-nociceptive

* Present address: Montreal Neurological Institute, McGill University, 3801 rue University, Montreal, Quebec, Canada H3A 2B4.

fields of the much thinner Aδ and C fibres. Direct electrical stimulation of the nerve was done to reveal the sympathetically innervated area; this activated the pilomotor muscles, leading to a fine ruffling of the skin within an area which defines the sympathetic pilomotor field. To induce collateral sprouting neighbouring cutaneous nerves were removed, leaving only the mDCN-T13 intact, with its innervation field surrounded by a relatively vast area of denervated skin; over the next 2–4 weeks this skin became progressively reinnervated by sprouts emanating from the spared nerve. To study regeneration, the same denervations were done initially but the spared nerve was then immediately crushed. Regenerating axons reached the now totally denervated skin of the rat some 10–14 days later; the same approaches used in the collateral sprouting studies were employed to detect the appearance and progressive expansion of the sensory and sympathetic fields established by the regenerating axons.

The role of nerve growth factor (NGF) in collateral sprouting

The collateral sprouting of intact cutaneous mechano-nociceptive (Aδ) and heat nociceptive (C) fibres into adjacent denervated skin (the Aα fibres mediating light touch do not sprout in adult mammals) (Jackson & Diamond, 1984) was completely inhibited by daily systemic (subcutaneous) injections of anti-NGF sera (Diamond *et al.*, 1987; Diamond, Coughlin & Holmes, 1992a). The results obtained using field expansions as measures of collateral sprouting were confirmed morphologically, using light and electron microscopy to identify the newly sprouted nerves in the skin. If the anti-NGF treatment was initiated after the sprouting had begun, this rapidly ceased, and further field expansion was prevented. These effects of anti-NGF were completely reversible; after cessation of the antibody treatment the fields expanded normally. In addition, when examined after 9 days of daily intradermal injections of exogenous NGF into normal untreated and unoperated animals, the nociceptive field sizes had increased in the already innervated skin, presumably by collateral sprouting of the undamaged nociceptive nerves.

Following a report that failed to find increases in the receptive fields of Aδ mechano-nociceptive fibres after 4 days of NGF injections (Lewin, Ritter & Mendell, 1993) we repeated these experiments, focusing on the time course of the response; we find that the field expansion does not become apparent until at least 5 or more days after beginning the daily injections, thus explaining the apparent discrepancy. We have also carefully examined the sizes of nociceptive fields during the early period of NGF administration when reductions in nociceptive thresholds were reported (Lewin *et al.*, 1993), a finding we have confirmed. We found no changes in the behaviourally measured field areas, and thus these early-induced threshold changes do not contribute to the later expansions caused by collateral sprouting. Analogous results to those obtained on sensory nerves were obtained with the sympathetic pilomotor nerves (Gloster &

Diamond, 1992). Daily anti-NGF injections completely, and reversibly, inhibited the collateral sprouting of sympathetic fibres, and if the treatment was initiated after sympathetic sprouting had already begun, further field expansion was prevented.

We conclude that the collateral sprouting of both nociceptive and sympathetic pilomotor nerves following partial denervation of skin is entirely dependent on endogenous NGF.

Regeneration occurs independently of NGF

In contrast to the findings with collateral sprouting, daily injections of anti-NGF did not prevent the regeneration and consequent recovery of function by any of the three classes of cutaneous sensory fibres examined (Diamond *et al.*, 1987, 1992b) (those subserving light touch regenerated perfectly normally, despite their failure to undergo collateral sprouting), or by sympathetic pilomotor fibres (Gloster & Diamond, 1992).

This result was obtained even when the anti-NGF treatment regimes were 3- to 8-fold greater than that which totally prevented the collateral sprouting of sensory fibres (Diamond *et al.*, 1987, 1992b). Moreover, when the collateral sprouting and regeneration paradigms were established concurrently on opposite sides of animals receiving anti-NGF treatment, the collateral sprouting of nociceptive nerves was completely prevented on the one side while regeneration occurred successfully on the other. The treatment not only failed to prevent regeneration, but failed even to decrease its rate. Nor was the rate of sensory nerve regeneration affected by systemic NGF injections. Finally, sensory fibres regenerated normally during anti-NGF treatment in animals which had received dorsal rhizotomies, ruling out the possibility that the regeneration was being supported by a central source of neurotrophin acquired by way of the dorsal roots. The continuity of the endoneurial tubes in the nerve was not critical to this NGF-independent regeneration, which occurred successfully whether the nerves were cut or crushed.

Regenerating sympathetic fibres had the same characteristics as regenerating sensory ones (Gloster & Diamond, 1992). The extent and rate of the sympathetic regrowth was not affected by daily anti-NGF treatment and, as with the regenerating sensory nerves, the treatment continued to be ineffective even when the sympathetic axons were regenerating within the denervated skin surrounding the original field areas; at such a time the invading fibre growth resembled collateral sprouting, a growth that was totally prevented by anti-NGF treatment. Numerous observations indicate that the failure of anti-NGF treatment to influence regenerating axons is not due to an inability of the administered antibodies to reach the elongating axons (Diamond *et al.*, 1987, 1992b). Of special relevance is the finding that in the skin, regenerating and collaterally sprouting axons use the same perineurial pathways along which to grow, although only the latter are affected by anti-NGF treatment. Another powerful argument relating to the 'access' question comes from the sympathetic nerve studies (see below) in which the 6-OHDA lesions destroyed only the nerve terminals, where the perineurial sheath is open-ended (Burkel, 1967) and the blood–nerve barrier is absent (Low, 1976).

The trophic dependency of the sympathetic recovery following 6-OHDA treatment

Six-hydroxy-DOPA (6-OHDA) administration to adult animals results in the destruction of the terminal and preterminal regions of sympathetic fibres, followed by a regrowth of these terminals (Tranzer & Thoenen, 1968; Tranzer *et al.*, 1969; de Champlain, 1970). This regrowth has one important feature that puts it into the same category as regeneration induced by nerve cut or crush, namely that it is evoked by axonal injury. However, there is also a resemblance to collateral sprouting, in that the regrowth of sympathetic fibres after 6-OHDA treatment occurs exclusively within the target tissue.

We studied this intriguing situation by challenging the regrowth after 6-OHDA treatment with anti-NGF treatment (cf. Bjerre, Bjorklund & Edwards, 1974), specifically to see whether the character of the recovery would fit a regenerative growth response or a collateral sprouting one (Gloster & Diamond, 1992). In the event, an NGF-independent component of growth was revealed, like regeneration, and an NGF-dependent one, like collateral sprouting. When the nerve field was 'isolated' by surrounding denervation prior to the single 6-OHDA injection, and in the absence of anti-NGF treatment, there was a smoothly continuous recovery of the pilomotor fields, extending far into the denervated skin beyond the original field borders. During daily anti-NGF treatment, however, the recovery of the pilomotor fields began normally but expansion ceased when their sizes reached about two-thirds of their original values. Further growth was totally NGF-dependent.

We hypothesize that the initial NGF-independent phase of recovery corresponds to a regenerative mode of sympathetic fibre growth, and the second, NGF-dependent phase to collateral sprouting. The regenerative phase is not an all-or-nothing phenomenon, but is related to the degree of axonal injury. A minimum injury is achieved by the 'chemical clipping' of the sympathetic terminals by 6-OHDA, inducing a small amount of NGF-independent regenerative growth. After nerve crush, a more traumatic injury, the regenerative response induced is so great that it continues to drive axonal growth independently of NGF throughout the entire experimental period. We anticipated that multiple 6-OHDA lesions would induce a greater regenerative response than a single injection, and a longer period of NGF-independent growth before the axons would revert to their collateral sprouting mode. This proved to be the case. In animals which received four 6-OHDA injections over a 12 day period, the entire normal field of pilomotor innervation was re-established before the regrowth was halted by the concurrent anti-NGF treatment. This kind of relation between the vigour of the regenerative response in axons and the degree of axonal injury has been described previously (Howe & Bodian, 1941; McQuarrie & Grafstein, 1973; McQuarrie, Grafstein & Gershon, 1977; but see McQuarrie *et al.*, 1978).

Nerve growth and plasticity

The 6-OHDA results together with the anti-NGF findings summarized earlier, point to a difference between the regulation of regeneration and that of collateral sprouting. Molecular studies too have contributed to this view; during collateral sprouting there is an upregulation of both the high- (trk A) and the low- (p75NGFR) affinity NGF receptors on the dorsal root ganglion neurones, but essentially no change in these expressions occurs during axonal regeneration (Mearow *et al.*, 1994). A further distinction between the two types of growth came from studies of the effects on them of impulses in the growing axons. A quite minor degree of impulse activity in undamaged axons can dramatically reduce the latency of their sprouting into adjacent denervated skin ('precocious sprouting') (Nixon *et al.*, 1984; Doucette & Diamond,1987). In contrast, driving the axons electrically had no effect on the rate or extent of their regeneration after injury (Diamond *et al.*, 1992b). The regulation of collateral sprouting by trophic factors and by impulses turn out to be interrelated. The induction of precocious sprouting in cutaneous nociceptive nerves is prevented in rats which receive an anti-NGF 'umbrella' at the time of the nerve activation (Diamond *et al.*, 1992a). The sprouting then proceeds with the normal (non-precocious) time course. Thus, the induction of precocious sprouting, as well as the normal process of collateral sprouting of the cutaneous sensory nerves, is NGF-dependent (Diamond *et al.*, 1992a). If interactions like these between nerve activity and trophic factors occur in the central nervous system, they could make an important contribution to plasticity within it. It is important to note that while regeneration is a reparative response to nerve injury, largely limited to the peripheral nervous system, collateral sprouting probably represents a continuing dynamic growth behaviour expressed throughout the entire normal nervous system (Diamond *et al.*, 1987); in this context it is not surprising that different mechanisms have evolved to regulate these two growth responses.

References

Bjerre, B., Björklund, A. & Edwards, D.C. (1974). Axonal Regeneration of peripheral adrenergic neurones: effects of anti-serum to nerve growth factor in mouse. *Cell and Tissue Research*, **148**, 441–476.

Burkel, W.E. (1967). The histological fine structure of perineurium. *Anatomical Record*, **158**, 177–190.

de Champlain, J. (1970). Degeneration and regrowth of adrenergic nerve fibres in the rat peripheral tissues after 6-hydroxydopamine. *Canadian Journal of Physiology and Pharmacology*, **49**, 345–355.

Diamond, J., Cooper, E., Turner, C. & MacIntyre, L. (1976). Trophic regulation of nerve sprouting. *Science*, **193**, 371–377.

Diamond, J., Coughlin, M. & Holmes, M. (1992a). Endogenous NGF and nerve impulses regulate the collateral sprouting of sensory axons in the skin of the adult rat. *Journal of Neuroscience*, **12**, 1454–1466.

Diamond, J., Coughlin, M., MacIntyre, L., Holmes, M. & Visheau, B. (1987). Evidence that endogenous nerve growth factor is responsible for the collateral sprouting, but not the regeneration, of nociceptive axons in adult rats. *Proceedings of the National Academy of Sciences, USA*, **84**, 6596–6600.

Diamond, J., Holmes, M., Foerster, A. & Coughlin, M. (1992b). Sensory nerves in adult rats regenerate and restore sensory function to the skin independently of endogenous NGF. *Journal of Neuroscience*, **12**, 1467–1476.

Doucette, R. & Diamond, J. (1987). The normal and precocious sprouting of heat nociceptors in the skin of adult rats. *Journal of Comparative Neurology*, **261**, 592–603.

Gloster, A. & Diamond, J. (1992). Sympathetic nerves in adult rats regenerate normally and restore pilomotor function during an anti-NGF treatment that prevents their collateral sprouting. *Journal of Comparative Neurology*, **326**, 363–374.

Howe, H.A. & Bodian, D. (1941). Refractoriness of nerve cells to poliomyelitis virus after interruption to their axons. *Johns Hopkins Hospital Bulletin*, **69**, 92–103.

Jackson, P. & Diamond, J. (1984). Temporal and spatial constraints on the collateral sprouting of low-threshold mechanosensory nerves in the skin of rats. *Journal of Comparative Neurology*, **226**, 336–345.

Lewin, G.R., Ritter, A.M. & Mendell, L.M. (1993). Nerve growth factor-induced hyperalgesia in the neonatal and adult rat. *Journal of Neuroscience*, **135**, 2136–2148.

Low, F.N. (1976). The perineurium and connective tissue of peripheral nerve. In *The Peripheral Nerve*, ed. D.N. Landon, pp. 159–187. London: Chapman and Hall.

McQuarrie, I.G. & Grafstein, B. (1973). Axon outgrowth enhanced by a previous injury. *Archives of Neurology*, **29**, 53–55.

McQuarrie, I.G., Grafstein, B., Dreyfus, C.F. & Gershon, M.D. (1978). Regeneration of adrenergic axons in rat sciatic nerve: effects of conditioning lesions. *Brain Research*, **141**, 21–34.

McQuarrie, I.G., Grafstein, B. & Gershon, M.D. (1977). Axonal regeneration in the rat sciatic nerve: effect of a conditioning lesion and dbcAMP. *Brain Research*, **132**, 444–453.

Mearow, K.M., Kril, Y., Gloster, A. & Diamond, J. (1994). Expression of NGF receptor and GAP-43 mRNA in DRG neurones during collateral sprouting and regeneration of dorsal cutaneous nerves. *Journal of Neurobiology*, **25**, 127–142.

Nixon, B.J., Doucette, R., Jackson, P. & Diamond, J. (1984). Impulse activity evokes precocious sprouting of nociceptive nerves into denervated skin. *Somatosensory Research*, **2**, 97–126.

Tranzer, J.P. & Thoenen, H. (1968). An electron microscopic study of selective, acute degeneration of sympathetic nerve terminals after administration of 6-hydroxydopamine. *Experientia*, **24**, 155–156.

Tranzer, J.P., Thoenen, H., Snipes, R.L. & Richards, J.G. (1969). Recent developments on the ultrastructural aspect of adrenergic nerve endings in various experimental conditions. *Progress in Brain Research*, **31**, 33–46.

37

Target dependence of motoneurones

I.P. JOHNSON
Department of Anatomy and Developmental Biology, Royal Free Hospital School of Medicine, London, UK

Target dependence of motoneurones

The role of the peripheral target in the maintenance of motoneurones has been highlighted recently by reports that certain neurotrophic factors can promote motoneuronal survival (Yan, Elliot & Snider, 1992; Oppenheim *et al.*, 1992; Sendtner *et al.*, 1992a; Henderson *et al.*, 1993; Lindsay, 1995). While these studies have considerable potential to advance our understanding of the mechanisms underlying the regeneration and degeneration of motoneurones, they are concerned principally with the developing neuromuscular system and with motoneurones *in vitro*. This chapter considers briefly the effect of age on the target-dependence of motoneurones *in vivo*.

Motoneurones in developing and neonatal animals

The critical role of the periphery for developing motoneurones was highlighted by Hamburger (1934, 1977), who showed that limb bud removal in chick embryos increased the number of motoneurones lost above that expected for programmed cell death, while grafting a supernumerary limb reduced this number. Studies of the molecular nature of the interaction between motoneurones and their targets were facilitated by the serendipitous discovery of nerve growth factor (NGF), a target-derived polypeptide promoting the maintenance and survival of neural-crest-derived neurones and able to ameliorate the retrograde effects of axotomy on them (Hamburger & Levi-Montalcini, 1949; Rich *et al.*, 1987; Purves, 1988). While NGF was not found to have trophic effects on motoneurones (Oppenheim, Maderut & Tytell, 1982; Yan *et al.*, 1988), characterization of its molecular structure allowed the subsequent identification of three other structurally related polypeptides (neurotrophins) which do (see Mudge, 1993, for review). These neurotrophins (brain-derived neurotrophic factor, neurotrophin-3 and neurotrophin-4/5) are produced by target tissues at appropriate times in development and can influence motoneurones through either retrograde axonal transport or the activation of specific membrane receptors associated with tyrosine kinase (Kaplan, Martin-Zana & Parada, 1991; Meakin & Shooter, 1992; Ip *et al.*, 1993). Although the identification of specific molecules has contributed greatly to our understanding of one aspect of the trophic maintenance of motoneurones, it is likely that motoneurone target-dependence in development is mediated through several mechanisms. These include the degree of muscle activity and motoneuronal activation (Okado & Oppenheim, 1984; Oppenheim, 1991; Lowrie & Vrbova, 1992; Navarrete & Vrbova, 1993), as well as soluble factors unrelated to the neurotrophin family, such as ciliary neurotrophic factor (Sendtner, Kreutzberg & Thoenen, 1990; Masu *et al.*, 1993), fibroblast growth factors (Grothe & Unsiker, 1992), glial-cell-line-derived neurotrophic factor (Yan, Matheson & Lopez, 1995; Oppenheim *et al.*, 1995) and other factors which can be extracted from skeletal muscle (McManaman & Oppenheim, 1993). In addition, the survival of certain motoneuronal types appears to be secondary to the effects of hormones on their targets (Forger *et al.*, 1992). Thus, a complicated picture emerges of multiple factors operating to maintain motoneurones in developing and neonatal animals. Whether the trophic effects of these factors *in vivo* are shared, synergistic or antagonistic, and whether they can serve to maintain motoneurones in the long term, remains to be determined.

At the cellular level, peripheral nerve injury in developing and neonatal animals often leads to a florid retrograde response of motoneurones. This affects the nucleus in particular and usually leads to motoneuronal death in a few days (Nissl, 1894; Romanes, 1946; LaVelle & LaVelle, 1958; Borke, 1983; Clarke, Jones & LaVelle, 1991). Natural and axotomy-induced degeneration of embryonic chick motoneurones can be delayed for up to 12 hours by the inhibition of RNA and protein synthesis (Oppenheim *et al.*, 1990), indicating that it is an active process that requires gene expression. The finding that some degenerating motoneurones show ultrastructural changes characteristic of programmed cell death (apoptosis) corroborates this interpretation (Chu-Wang & Oppenheim, 1978). The presence of a ladder-like pattern of oligonucleosomal-sized bands on electrophoresis of the DNA of cultured sympathetic neurones deprived of NGF (see Altman, 1993, for review), further supports the view that the degeneration of immature neurones is apoptotic, and from recent studies of apoptotic thymocytes and lymph node cells (Peitsch *et al.*, 1993) it might be inferred that the programmed cell death of motoneurones is due in part to the expression of DNase I.

In summary, interruption of motoneurone–target interaction in developing and neonatal animals results in the rapid degeneration of motoneurones. Such degeneration may in part be due to the activation of a cell death programme and this can be variously prevented, ameliorated or delayed by a variety of target-derived factors.

Motoneurones in adult animals

In contrast to motoneurones in developing and neonatal animals, motoneurones in adult animals show much less propensity to die on interruption of their contact with the periphery (Romanes, 1951; Carlsson, Lais & Dyck, 1979; Schmalbruch, 1984). Instead, most motoneurones respond by switching their metabolism away from that associated with neurotransmission and towards that associated with axonal elongation and the re-establishment of neuromuscular contact. The switch from a target-dependent state to an almost target-independent state occurs within the space of a day or two towards the end of the first postnatal week in rodents (Lowrie & Vrbová, 1992). This switch may be related to the postnatal elimination of polyneuronal innervation of muscle, increased muscle activity, or age-related changes in the responsiveness of motoneurones to trophic factors (Oppenheim & Nunez, 1982; Burls *et al.*, 1990; Grothe & Unsiker, 1992; Navarrete & Vrbová, 1993).

The metabolic changes induced by axotomy are often accompanied by morphological alterations of motoneurones, although this varies according to species, system and type of nerve injury (Lieberman, 1974; Søreide, 1981). Even among motoneurone subtypes there are differences: for example, Nissl body disintegration, 'chromatolysis', is a classical alteration associated with the retrograde response to axotomy of large (alpha) motoneurones, but not of small (gamma) motoneurones (Johnson & Sears, 1989). The importance of the peripheral target in regulating the response to axotomy is highlighted by the fact that, in general, the retrograde changes induced by axotomy subside with the reinnervation of muscle (Grafstein & McQuarrie, 1978; Kreutzberg, 1993).

For those motoneurones that do degenerate following axotomy, the process in the adult takes months rather than days (Aldskogius, Barron & Regal, 1980; Costen & Johnson, 1991; Snider, Elliot & Yan, 1992) and the response in the stages prior to frank degeneration primarily affects structures in the motoneuronal cytoplasm.

Ultrastructurally, axotomy-induced motoneuronal degeneration is associated with a mixture of changes characteristic of both apoptosis and necrosis (Fernando, 1973; Søreide, 1981), although recent studies of muscle in moths during metamorphosis indicate that programmed cell death can take on a variety of morphological appearances (Schwartz *et al.*, 1993). It has been shown that the inhibition of DNA-dependent RNA synthesis at the time of axotomy can delay the onset of the retrograde response of rat facial motoneurones (Torvik & Hedding, 1969). However, the long time-course for motoneuronal degeneration in the adult makes it impossible to determine whether such treatment, which will itself cause cell death in the long term, can prevent degeneration. Also, in view of the fact that only a proportion of injured

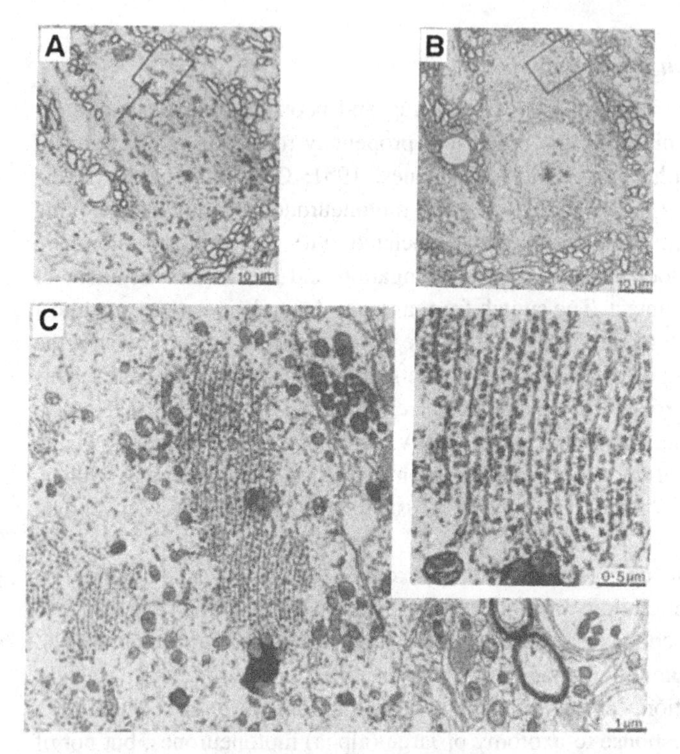

Fig. 37.1. (A) Light micrograph (LM) of a Nissl body (arrow) in a normal cat thoracic motoneurone. (B) Electron micrograph (EM) of the same motoneurone. (C) Enlargement of the box shown in (A). Note the regular arrays of rough endoplasmic reticulum (RER) which characterize the Nissl bodies of normal motoneurones.

motoneurones die, it is unfortunate that there is at present no reliable means of identifying those motoneurones destined to survive or die following axotomy.

In this respect it is interesting that cranial motoneurones are more likely than spinal motoneurones to degenerate following axotomy (Schmalbruch, 1984; Snider & Thanedar, 1989). The site of axotomy is in general closer to the perikarya of cranial motoneurones, and may be expected to provoke a more vigorous retrograde response (Lieberman, 1974). This feature alone, however, does not account for the survival of other cranial motoneurones within the same nucleus whose axons have been interrupted at the same level. Differences of neuropeptide expression (Moore, 1989; Rethelyi, Metz & Lund, 1989; Arvidsson *et al.*, 1993), extent of synaptic terminal loss (Sumner 1975; Chen, 1978; Johnson & Sears, 1989), or capacity to establish collateral synaptic contacts (Havton & Kellerth, 1987) are other possible factors which may contribute to this system's difference in motoneuronal vulnerability following axotomy. Differential vulnerability of certain neuronal populations is

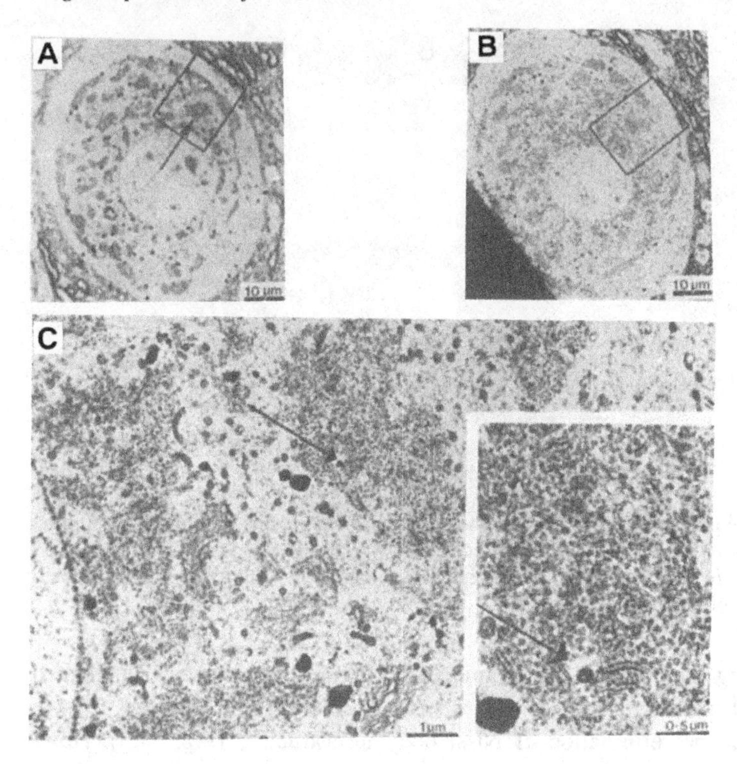

Fig. 37.2. (A) LM of a Nissl body (arrow) in a cat thoracic motoneurone 64 days following nerve transection and ligation. (B) EM of the same motoneurone. (C) Enlargement of the box shown in (A). Note that the Nissl body is composed primarily of randomly sited polyribosomes and short fragments of RER.

a common feature of human neuronal degenerative conditions and the possibility exisits that insights into the mechanisms underlying this might be gained from studies of this phenomenon in the relatively controlled conditions of the axotomy model.

In the absence of significant axotomy-induced motoneuronal degeneration in the adult, we studied the role of the peripheral target in the normal maintenance of cat thoracic motoneurones. Intercostal nerve crush, versus nerve transection and ligation was employed to create conditions suitable or unsuitable for axonal regeneration and muscle reinnervation, respectively, and particular attention was paid to the ultrastructural organization of Nissl bodies as a measure of the integrity of motoneuronal–target interaction (Johnson, Pullen & Sears, 1985; Sears, 1987). We found that the normal Nissl bodies of motoneurones whose contact with muscle was intact, were composed of highly ordered lamellae of rough endoplasmic reticulum (RER). This orderliness was lost following axotomy and only restored with the restoration of neuromuscular contact (Figs. 37.1, 37.2).

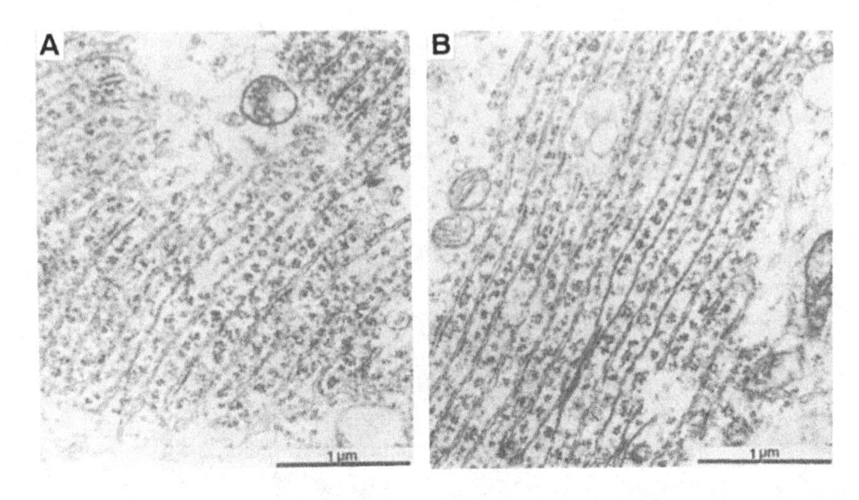

Fig. 37.3. Nissl bodies of motoneurones after allowing axons ending in a neuroma to reinnervate denervated (A) or already innervated (B) intercostal muscle for 64 days. Even in the absence of functional reinnervation (B), normal, highly ordered Nissl body ultrastructure is restored.

Allowing the cut proximal stump of a previously transected and ligated intercostal nerve to reinnervate a newly denervated intercostal muscle resulted in the reformation of Nissl body ultrastructure (Fig. 37.3). However, Nissl body ultrastructure also regained its normal, highly ordered appearance, even if the host muscle retained its original innervation and no respiratory-phased electromyographic activity could be ascribed to activation by the transplanted 'foreign' nerve (Johnson & Sears, 1992). Interestingly, Munson and Nishimura (this volume) describe the restoration of electrophysiological properties of axotomized cat spinal motoneurones which were allowed to regenerate back to skin. It would appear, therefore, that functional reinnervation of muscle is not an absolute requirement for the return to normal of all the properties of motoneurones following axotomy.

To determine whether a soluble factor from muscle was responsible for the restoration of Nissl body orderliness, the proximal stumps of intercostal nerves were placed in polysulphone dialysis tubes with a molecular weight cut-off of 100 000 and the tubes placed in the substance of innervated muscle, denervated muscle or scar tissue (Johnson & Sears, 1992). Surprisingly, in none of these instances did axonal regeneration occur over more than 5 mm, even when nerves were placed in tubes that were open distally (Figs. 37.4, 37.5). While endothelial cell and fibroblasts proliferated within the dialysis tubes, it remains possible that the exclusion of macrophages from the growing tips of the nerves may have impeded axonal elongation (Jenq, Jenq & Coggeshall, 1987; Perry & Brown, 1992). In all cases where the proximal nerves

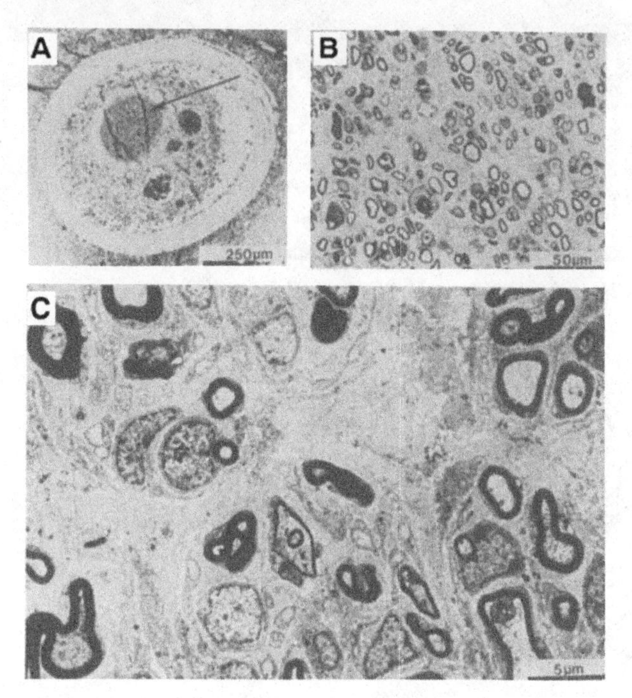

Fig. 37.4. (A) At 5 mm distal to the site of an intercostal nerve crush performed 64 days previously, with the nerve (arrow) placed in a polysulphone dialysis tube, containing 20 mm of the nerve distal to the crush. (B) and (C) Enlargements of the area arrowed in (A), showing many myelinated axons surrounded by Schwann cells and cells of the nerve sheaths.

were placed in dialysis tubes, the cell bodies of motoneurones had Nissl bodies with a disorganized ultrastructure (Fig. 37.6) and the nature of the putative signal from muscle therefore remains obscure.

To gain some insight into the functional significance of target-dependent changes in Nissl body ultrastructure, we studied the relationship of ultrastructural changes to changes in the expression of calcitonin gene-related peptide (CGRP), a 37 amino acid neuropeptide produced by alternative tissue-specific splicing of the calcitonin gene (Amara *et al.*, 1982). Axotomy, resulting in Nissl body disorganization, was associated at 1 week with increased CGRP expression (Figs. 37.7, 37.8). Axonal regeneration and muscle reinnervation at 64 days was associated with the restoration of ordered Nissl body ultrastructure and the amelioration of differences in CGRP expression by motoneurones ipsilateral and contralateral to the axotomy. In contrast, nerve transection and ligation, which at 64 days was still associated with the presence of disorganized Nissl bodies, was also associated with elevated levels of CGRP expression (Johnson, Sears & Hunter, 1993).

CGRP is normally transported to and released from the neuromuscular

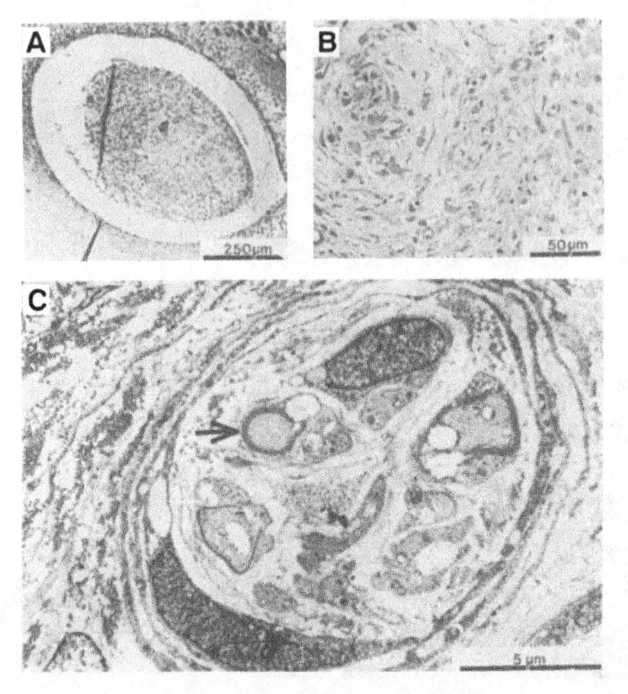

Fig. 37.5. (A) At 10 mm distal to the site of an intercostal nerve crush, performed 64 days previously, with the nerve placed in a polysulphone dialysis tube, containing 20 mm of the nerve distal to the crush. (B) and (C) enlargements of (A), showing that the tube is now occupied primarily by non-neuronal cells, although some very small fascicles can be seen (C), which have thinly myelinated axons (arrow); but these disappear within 5 mm.

junction (Kashihara, Sakaguchi & Kuno, 1989; Uchida *et al.*, 1990), and even after axotomy it continues to be transported to the axonal growth cone where its release is probable (Raivich *et al.*, 1992). CGRP may thus be regarded as a secretory peptide. However, our studies show that increased expression of CGRP correlates with a reduction in the linear length of RER in disorganized Nissl bodies (Johnson *et al.*, 1985). For CGRP, therefore, the classical relationship between RER and secretory protein synthesis (Palade, 1975; Smeekens, Chan & Steiner, 1992) does not hold. These results urge caution in the direct extrapolation to the intact cell of concepts of the roles of free and bound ribosomes in the process of protein synthesis which have been derived principally from the study of cell-free systems.

In summary, motoneurones in the adult animal rarely die after the interruption of peripheral target contact. Instead, a primary function of the peripheral target in the adult appears to be the regulation of motoneuronal metabolism so that stable nerve–muscle interaction is maintained. Axotomy generally

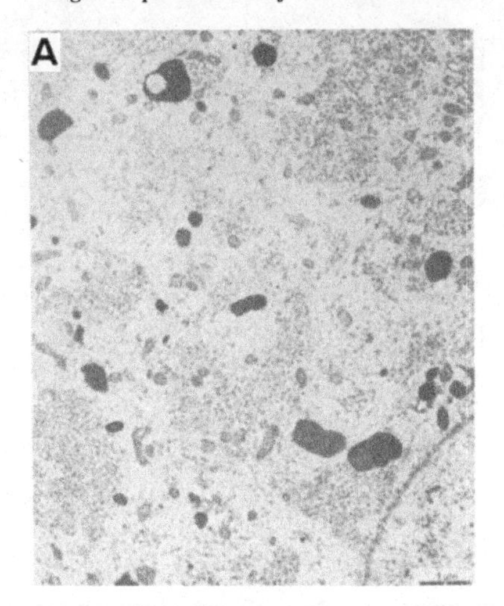

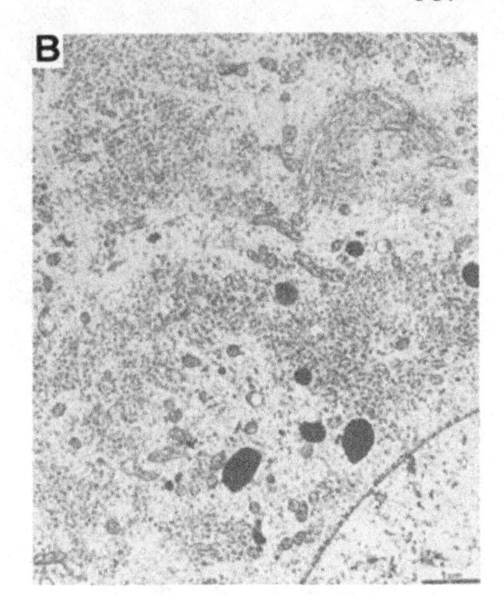

Fig. 37.6. Nissl bodies of motoneurones whose axons ended either in a tube which was sealed distally (A) or one which was open distally (B); both tubes were placed in denervated muscle. After 64 days, Nissl bodies in both cases are composed of randomly sited polyribosomes and short fragments of RER.

induces metabolic changes directed towards the re-establishment of neuromuscular contact, and when axotomy-induced death does occur it is restricted to subpopulations of motoneurones and has a longer timecourse than for motoneurones in developing and neonatal animals.

Motoneurones in ageing animals

Motoneurones in ageing animals accumulate lipofuscin, show changes in dendritic morphology, have decreased conduction velocities and some may die (Chase *et al.*, 1985; Haschizume, Kanda & Burke, 1988; Sturrock, 1990; Ramirez & Ulfhake, 1992). In addition, muscle atrophy, and reductions in the area of the neuromuscular junction, have been reported for ageing rodents (Gutmann & Hanzlikova, 1966; Fahim, & Robbins, 1982; Ansved & Edström, 1991). As a result of these age-related alterations, changes in the target-dependence of motoneurones might be expected. Certainly, differences in nuclear morphology and the activity of certain enzymes have been found between axotomized motoneurones in 3-month-old versus 15-month-old rats (Vaughan, 1989, 1990), and decreased rates of axonal regeneration and capacities for collateral sprouting have been described (Drahota & Gutmann, 1961; Pestronk, Drachman & Griffin, 1980), which provide a basis for assuming that motoneurone target-dependence in the ageing animal may differ from that in the adult.

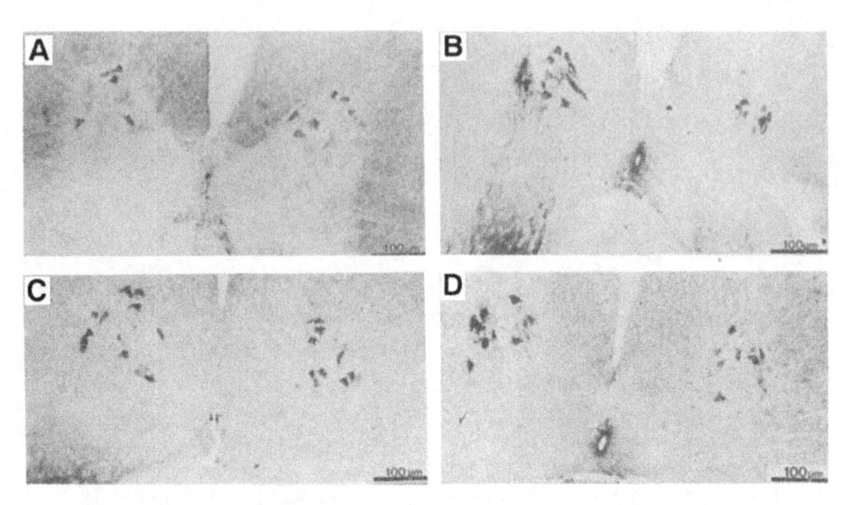

Fig. 37.7. CGRP-like immunoreactivity in motoneurones. No difference between the left and right ventral horns is seen for normal animals (A) or those 64 days following intercostal nerve crush (C), while increased immunostaining is seen ipsilaterally at 7 days (B) and 64 days (D) following nerve transection and ligation. Polyclonal antibody was visualized with diaminobenzidine, using the peroxidase–anti-peroxidase method.

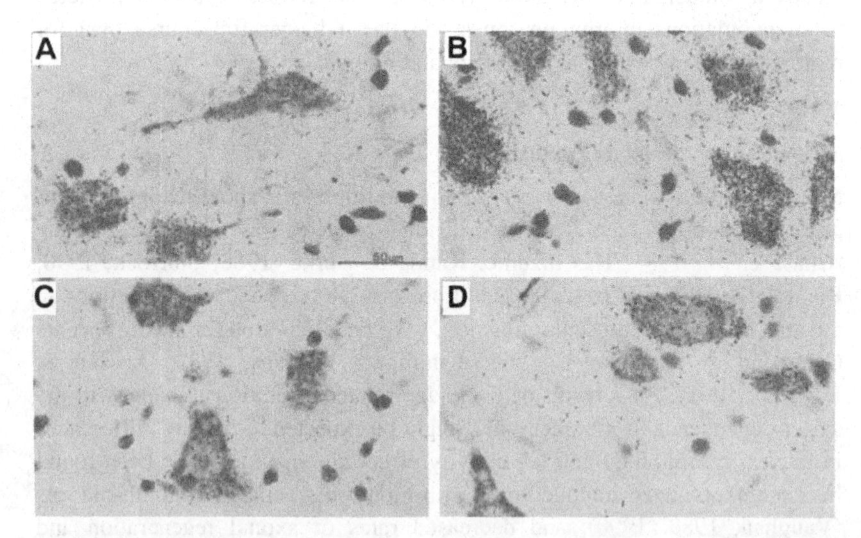

Fig. 37.8. CGRP mRNA in motoneurones. Compared with motoneurones in the non-operated ventral horn (A, C), motoneurones ipsilateral to intercostal nerve transection and ligation at 7 days (B) and 64 days (D) show increased signals for CGRP mRNA. *In situ* hybridization was visualized by autoradiography, using ^{35}S-labelled synthetic oligodeoxynucleotide probes.

Extending this idea, it has also often been assumed that ageing motoneurones are more vulnerable to injury than adult motoneurones, due either to intrinsic mechanisms or to the loss of target-derived trophic support (Appel, 1981; Tandan & Bradley, 1985). However, this assumption does not appear to have been tested directly. Therefore, to determine whether motoneurones in ageing animals showed increased vulnerability to injury, we again used the paradigm of nerve crush versus nerve transection and ligation in adult (1–2 year) and ageing (10–15 year) cats. We found, with the exception of some microglial changes (Johnson, Sears & Hunter, 1991a) and differences in the extent of synaptic terminal loss from the motoneurones (Johnson, *et al.*, 1991b), that the retrograde response of motoneurones in ageing cats was almost identical to that seen in adult cats. In particular, the target-dependent changes in Nissl body ultrastructure and CGRP immunoreactivity were identical for both adult and ageing cats and no evidence for any axotomy-induced motoneuronal degeneration was found (Johnson *et al.*, 1991a, c).

In summary, while changes in the motor unit occur as a result of ageing, for cat thoracic motoneurones this does not appear to be reflected in any major change in motoneuronal target-dependence compared with motoneurones in the adult.

Conclusions

In the developing and neonatal animal, motoneurones depend on the peripheral target for their survival. In contrast, in the adult and ageing animal, the peripheral target serves more to regulate motoneuronal metabolism in a way that is appropriate for maintaining the stability of neuromuscular interactions. In particular, our studies of the ageing cat fail to support a commonly held assumption that motoneurones in the ageing animal exhibit a greater dependency on their peripheral targets than those in the adult. Care therefore needs to be exercised when concepts of neurotrophism based on studies of the developing neuromuscular system are extrapolated to neurodegenerative conditions of the adult and aged.

References

Aldskogius, H., Barron, K.D. & Regal, R. (1980). Axon reaction in dorsal motor vagal and hypoglossal neurones of the adult rat: light microscopy and RNA cytochemistry. *Journal of Comparative Neurology*, **193**, 165–177.

Altman, J. (1993). Programmed cell death: the paths to suicide. *Trends in Neurosciences*, **15**, 278–280.

Amara, S.G., Jones, J., Rosenfeld, M.G., Ong, E.S. & Evans, R.M. (1982). Alternative RNA processing in calcitonin gene expression generates mRNA encoding different gene products. *Nature*, **298**, 240–224.

Ansved, T. & Edström, L. (1991). Effects of age on fibre structure, ultrastructure and expression of desmin and spectrin in fast- and slow-twitch rat muscles. *Journal of Anatomy*, **174**, 61–79.

Appel, S.H. (1981). A unifying hypothesis for the cause of amyotrophic lateral sclerosis, Parkinsonism and Alzheimer disease. *Annals of Neurology*, **10**, 449–505.

Arvidsson, U., Piehl, F., Johnson, H., Ulfhake, B., Cullheim, S. & Hökfelt, T. (1993). The peptidergic motoneurone. *NeuroReport*, **4**, 849–856.

Borke, R. (1983). Intrasomatic changes in the maturing hypoglossal nucleus after injury. *Journal of Neurocytology*, **12**, 873–883.

Burls, A., Subramanian, K., Lowrie, M.B. & Vrbová, G. (1990). Absence of nerve–muscle interaction influences the survival of developing motoneurones. *European Journal of Neuroscience*, **3**, 216–221.

Carlsson, J., Lais, A.C. & Dyck, P.J. (1979). Axonal atrophy from persistent peripheral axotomy in the adult cat. *Journal of Neuropathology and Experimental Neurology*, **38**, 579–585.

Chase, M.H., Morales, F.R., Boxer, P.A. & Fung, S.J. (1985). Aging of motoneurons and synaptic processes in the cat. *Experimental Neurology*, **90**, 471–478.

Chen, D.H. (1978). Qualitative and quantitative study of synaptic displacement in chromatolysed spinal motoneurons of the cat. *Journal of Comparative Neurology*, **177**, 635–664.

Chu-Wang, I.W. & Oppenheim, R.W. (1978). Cell death of motor neurons in the chick embryo spinal cord. I. A light and electron microscopic study of naturally occurring and induced cell loss during development. *Journal of Comparative Neurology*, **177**, 33–58.

Clarke, P.G.H., Jones, K.J. & LaVelle, A. (1991). Ultrastructural changes in the nucleolus of facial motor neurons following axotomy during an early critical period in development. *Journal of Comparative Neurology*, **197**, 33–58.

Costen, M. & Johnson, I.P. (1991). Neuronal loss and calcitonin gene-related peptide-like immunoreactivity (CGRP-LI) in the rat hypoglossal nucleus following axotomy. *European Journal of Neuroscience*, **4** (Supplement), 4230.

Drahota, Z. & Gutmann, E. (1961). The influence of age on the time course of reinnervation of muscle. *Gerontologie*, **5**, 88–109.

Fahim, M.A. & Robbins, N. (1982). Ultrastructural studies of young and old mouse neuromuscular junctions. *Journal of Neurocytology*, **11**, 641–656.

Fernando, D.A. (1973). Ultrastructural observations on retrograde degeneration in perikarya of the hypoglossal nucleus. *Acta Anatomica*, **86**, 191–204.

Forger, N.G., Hodges, L.L., Roberts, S.L. & Breedlove, S.M. (1992). Regulation of motoneurone death in the spinal nucleus of the bulbocavernosus. *Journal of Neurobiology*, **23**, 1192–1203.

Grafstein, B. & McQuarrie, I.G. (1978). Role of the nerve cell body in axonal regeneration. In *Neuronal Plasticity*, ed. C.W. Cotman, pp. 155–195. New York: Raven Press.

Grothe, C. & Unsiker, K. (1992). Basic fibroblast growth factor in the hypoglossal system: specific retrograde transport, trophic and tissue-related responses. *Journal of Neuroscience Research*, **32**, 317–328.

Gutmann, E. & Hanzlikova, V. (1966). Motor unit in old age. *Nature*, **209**, 921–992.

Hamburger, V. (1934). The effects of wing bud extirpation on the development of the central nervous system in chick embryos. *Journal of Experimental Zoology*, **68**, 449–494.

Hamburger, V. (1977). The developmental history of the motor neuron. *Neuroscience Program Bulletins*,**15**, 1–37.

Hamburger, V. & Levi-Montalcini, R. (1949). Proliferation, differentiation and degeneration in the spinal ganglia of the chick embryo under normal and experimental conditions. *Journal of Experimental Zoology*, **111**, 457–501.

Haschizume, K., Kanda, K. & Burke, R.E. (1988). Medial gastrocnemius motor nucleus in the rat: age-related changes in the number and size of motoneurones. *Journal of Comparative Neurology*, **269**, 425–430.

Havton, L. & Kellerth, J.-O. (1987). Regeneration by supernumerary axons with synaptic terminals in spinal motoneurons of cats. *Nature*, **325**, 711–714.

Henderson, C.E., Camu, W., Mettling, C., Gouin, A., Coulsen, K., Karihaloo, M., Rullamas, J., Evans, T., McMahon, S.B., Armanini, M.P., Berkmeier, L., Philips, H. & Rosenthal, A. (1993). Neurotrophins promote motor neuron survival and are present in embryonic limb bud. *Nature*, **363**, 266–270.

Ip, N.Y., Stitt, T.N., Tapley, P., Klein, R., Glass, D.J., Fandl, J., Greene, L.A., Barbacid, M. & Yancopoulos, G.D. (1993). Similarities and differences in the way neurotrophins interact with the *trk* receptors in neuronal and non-neuronal cells. *Neuron*, **10**, 137–149.

Jenq, C.-B., Jenq, L.L. & Coggeshall, R.E. (1987). Nerve regeneration changes with filters of different pore size. *Experimental Neurology*, **97**, 622–671.

Johnson, I.P., Pullen, A.H. & Sears, T.A. (1985). Target dependence of Nissl body ultrastructure in cat thoracic motoneurones. *Neuroscience Letters*, **61**, 201–205.

Johnson, I.P. & Sears, T.A. (1989). Ultrastructure of axotomized alpha and gamma motoneurones in the cat thoracic spinal cord. *Neuropathology and Applied Neurobiology*, **15**, 149–163.

Johnson, I.P. & Sears, T.A. (1992). Differences in the ultrastructure of Nissl bodies in cat thoracic motoneurones after exposing intercostal nerves to different environments. *Journal of Anatomy*, **180**, 377.

Johnson, I.P., Sears, T.A. & Hunter, A.S. (1991a). Retrograde response to axotomy of motoneurons in the thoracic spinal cord of the aging cat. *Neurobiology of Ageing*, **12**, 151–160.

Johnson, I.P., Sears, T.A. & Hunter, A.S. (1991b). Synaptic terminal changes following long-term axotomy of motoneurones in adult and ageing cats. *European Journal of Neuroscience*, **4** (Supplement), 3021.

Johnson, I.P., Sears, T.A. & Hunter, A.S. (1991c). Target-dependent changes in calcitonin gene-related peptide (CGRP)-like immunoreactivity in thoracic motoneurones of the ageing cat. *European Journal of Neuroscience*, **4** (Supplement), 4247.

Johnson, I.P., Sears, T.A. & Hunter, A.S. (1993). Differences in CGRP expression associated with particular patterns of Nissl body ultrastructure in axotomised motoneurones. *Journal of Anatomy*, **183**, 193.

Kaplan, D.R., Martin-Zana, D. & Parada, L.F. (1991). Tyrosine phosphorylation and tyrosine kinase activity of the *trk* proto-oncogene products induced by NGF. *Nature*, **350**, 158–160.

Kashihara, Y., Sakaguchi, M. & Kuno, M. (1989). Axonal transport and distribution of endogenous calcitonin gene-related peptide in rat peripheral nerve. *Journal of Neuroscience*, **9**, 3796–3802.

Kreutzberg, G.W. (1993). Dynamic changes in motoneurones during regeneration. *Restorative Neurology and Neuroscience*, **5**, 59–60.

LaVelle, A. & LaVelle, F.W. (1958). Neuronal swelling and chromatolysis as influenced by the state of cell development. *American Journal of Anatomy*, **102**, 219–241.

Lieberman, A.R. (1974). Some factors affecting the retrograde responses to axonal lesions. In *Essays on the Nervous System. A Festschrift for Professor J.Z. Young*, ed. R. Bellairs & E.G. Gray, pp. 71–105. Oxford: University Press.

Lindsay, R.M. (1995). Neuron saving schemes. *Nature*, **373**, 289–290.

Lowrie, M.B. & Vrbová, G. (1992). Dependence of postnatal motoneurones on their targets: a review and hypothesis. *Trends in Neurosciences*, **15**, 80–84.
Masu, Y., Wolf, E., Holtmann, B., Sendtner, M., Brem, G. & Thoenen, H. (1993). Disruption of the CNTF gene results in motor neuron degeneration. *Nature*, **365**, 27–32.
McManaman, J. & Oppenheim, R.W. (1993). Skeletal muscle-derived neurotrophic factors and motoneuron development. In *Neurotrophic Factors*, ed. S.E. Loughlin & J.H. Fallon, pp. 475–487, London: Academic Press.
Meakin, S.O. & Shooter, E.M. (1992). The nerve growth factor family of receptors. *Trends in Neurosciences*, **15**, 323–331.
Moore, R.Y. (1989). Cranial motoneurones contain either galanin- or calcitonin gene-related peptide-like immunoreactivity. *Journal of Comparative Neurology*, **282**, 512–522.
Mudge, A.W. (1993). Motor neurons find their factors. *Nature*, **363**, 213–214.
Navarrete, R. & Vrbová, G. (1993). Activity-dependent interactions between motoneurones and muscles: their role in the development of the motor unit. *Progress in Neurobiology*, **41**, 93–124.
Nissl, F. (1894). Über eine neue Untersuchungmethode des Centralorgans speciell zur Festellung der Localisation der Nervezellen. *Zeitschrift für Nervenheilkunde und Psychiátrié*, **17**, 337–354.
Okado, N. & Oppenheim, R.W. (1984). Cell death of motor neurons in the chick embryo spinal cord: the loss of motor neurons following removal of afferent input. *Journal of Neuroscience*, **4**, 1639–1652.
Oppenheim, R.W. (1991). Cell death during development of the nervous system. *Annual Review of Neurobiology*, **14**, 453–501.
Oppenheim, R.W., Lucien, J.H., Johnson, J.E., Lin, L.-F.H., Li, L., Lo, A.C., Newsome, A.L., Prevette, D.M. & Wang, S. (1995). Developing motor neurons rescued from programmed and axotomy-induced cell death by GDNF. *Nature*, **373**, 344–346.
Oppenheim, R.W., Maderut, J.L. & Tytell, M. (1982). Cell death of motoneurons in the chick embryo spinal cord. VI. Reduction of naturally occurring cell death in the thoracolumbar column of Terni by nerve growth factor. *Journal of Comparative Neurology*, **210**, 174–189.
Oppenheim, R.W. & Nunez, R. (1982). Electrical stimulation of hindlimb increases neuronal cell death in chick embryo. *Nature*, **295**, 57–59.
Oppenheim, R.W., Prevette, D., Tytell, M. & Homma, S. (1990). Naturally occurring and induced neuronal death in the chick embryo *in vivo* requires protein and RNA synthesis. *Developmental Biology*, **138**, 104–113.
Oppenheim, R.W., Qin-Wei, Y., Prevette, D. & Yan, Q. (1992). Brain-derived neurotrophic factor rescues developing motoneurons from cell death. *Nature*, **360**, 755–757.
Palade, G. (1975). Intracellular aspects of the process of protein synthesis. *Science*, **189**, 347–358.
Peitsch, M.C., Polzar, B., Stephan, H., Crompton, T., MacDonald, H.R., Mannherz, H.G. & Tschopp, J. (1993). Characterization of the endogenous deoxyribonuclease involved in nuclear DNA degradation during apoptosis (programmed cell death). *EMBO Journal*, **12**, 371–377.
Perry, V.H. & Brown, M.C. (1992). Macrophages and nerve regeneration. *Current Opinion in Neurobiology*, **2**, 679–682.
Pestronk, A., Drachman, D.B. & Griffin, J.W. (1980). Effects of aging on nerve sprouting and regeneration. *Experimental Neurology*, **70**, 65–82.
Purves, D. (1988). *Body and Brain. A Trophic Theory of Neural Connections*. Cambridge, MA: Harvard University Press.
Raivich, G., Dumoulin, F.L., Streit, W.J. & Kreutzberg, G.W. (1992). Calcitonin

gene-related peptide (CGRP) in the regenerating rat sciatic nerve. *Restorative Neurology and Neuroscience*, **4**, 107–115.

Ramirez, V. & Ulfhake, B. (1992). Anatomy of dendrites in motoneurones supplying the intrinsic muscles of the foot sole in the aged cat: evidence for dendritic growth and neo-synaptogenesis. *Journal of Comparative Neurology*, **316**, 1–16.

Rethelyi, M., Metz, C.B. & Lund, P.K. (1989). Distribution of neurons expressing calcitonin gene-related peptide mRNAs in the brain stem, spinal cord and dorsal root ganglia of rat and guinea pig. *Neuroscience*, **29**, 225–239.

Rich, K., Luszczynski, J.R., Osborne, P.A. & Johnson, E.M. (1987). Nerve growth factor protects adult sensory neurons from cell death and atrophy caused by nerve injury. *Journal of Neurocytology*, **16**, 261–268.

Romanes, G.J. (1946). Motor localisation and the effects of nerve injury on the ventral horn cells of the spinal cord. *Journal of Anatomy*, **80**, 117–131.

Romanes, G.J. (1951). The motor cell columns of the lumbo-sacral spinal cord of the cat. *Journal of Comparative Neurology*, **94**, 313–363.

Sears, T.A. (1987). Structural changes following axotomy. *Journal of Experimental Biology*, **132**, 93–109.

Schmalbruch, H. (1984). Motor neuron death after sciatic nerve section in newborn rats. *Journal of Comparative Neurology*, **224**, 252–258.

Schwartz, L.M., Smith, S., Jones, M.E.E. & Osborne, B.A. (1993). Do all programmed cell deaths occur via apoptosis? *Proceedings of the National Academy of Sciences, USA*, **90**, 980–984.

Sendtner, M., Holtmann, B., Kolbeck, R., Thoenen, H. & Barde, Y.-A. (1992a). Brain-derived neurotrophic factor prevents the death of motoneurons in newborn rats after nerve section. *Nature*, **360**, 757–759.

Sendtner, M., Kreutzberg, G.W. & Thoenen, H. (1990). Ciliary neurotrophic factor prevents the degeneration of motor neurons after axotomy. *Nature*, **345**, 440–441.

Sendtner, M., Schmalbruch, H., Stöckli, K.A., Carrol, P., Kreutzberg, G.W. & Thoenen, H. (1992b). Ciliary neurotrophic factor (CNTF) prevents degeneration of motoneurons in the mouse mutant Progressive Motor Neuronopathy (pmn). *Nature*, **358**, 502–504.

Smeekens, S.P., Chan, S.J. & Steiner, D.F. (1992). The biosynthesis and processing of neuropeptides: identification of protein convertases involved in intravesicular processing. *Progress in Brain Research*, **92**, 238–246.

Snider, W.D. & Thanedar, S. (1989). Target dependence of hypoglossal motor neurons during development and in maturity. *Journal of Comparative Neurology*, **279**, 489–498.

Snider, W.D., Elliot, J.L. & Yan, Q. (1992). Axotomy-induced neural death during development. *Journal of Neurobiology*, **23**, 1231–1246.

Soreide, A.J. (1981). Variations in the axon reaction after different types of nerve lesion. *Acta Anatomica*, **110**, 173–188.

Sturrock, R.R. (1990). A comparison of age-related changes in neuron number in the dorsal motor nucleus of the vagus and the nucleus ambiguus of the mouse. *Journal of Anatomy*, **173**, 169–176.

Sumner, B.E.H. (1975). A quantitative analysis of the response of presynaptic boutons to postsynaptic motor neuron axotomy. *Experimental Neurology*, **46**, 605–616.

Tandan, R. & Bradley, W.G. (1985). Amyotrophic lateral sclerosis. 2. Etiopathogenesis. *Annals of Neurology*, **18**, 419–431.

Torvik, A. & Hedding, A. (1969). Effect of actinomycin D on retrograde nerve cell reaction: further observations. *Acta Neuropathologica*, **14**, 62–71.

Uchida, S., Yamamoto, H., Ito, S., Matsumoto, N., Wang, X.B., Yonehara, N.,

Imai, Y., Inoki, R. & Yoshida, H. (1990). Release of calcitonin gene-related peptide-like immunoreactive substance from neuromuscular junctions by nerve excitation and its action on striated muscle. *Journal of Neurochemistry*, **54**, 1000–1003.

Vaughan, D.W. (1989). Age effects on AChE and cytochrome oxidase enzyme activities in axotomised rat facial motor neurons. *Anatomical Record*, **223**, 118.

Vaughan, D.W. (1990). Effects of advancing age on the central response of rat facial neurons to axotomy: light microscope morphometry. *Anatomical Record*, **228**, 211–219.

Yan, Q., Elliot, J. & Snider, W.D. (1992). Brain-derived neurotrophic factor rescues spinal motor neurons from axotomy-induced cell death. *Nature*, **360**, 753–755.

Yan, Q., Matheson, C. & Lopez, D. (1995). In vivo neurotrophic effects of GDSNF on neonatal and adult facial motor neurons. *Nature*, **373**, 241–244.

Yan, Q., Snider, W.D., Pinzone, J.J. & Johnson, E.M. (1988). Retrograde transport of nerve growth factor (NGF) in motor neurons of developing rat: assessment of potential neurotrophic effects. *Neuron*, **1**, 335–343.

38

Rescue of neurones cross-regenerated into foreign targets

JOHN B. MUNSON AND HIROYUKI NISHIMURA*

Department of Neuroscience, University of Florida College of Medicine, University of Florida, Gainesville, Florida, USA

Introduction

Injury to motor and sensory neurones of the peripheral nervous system alters their electrical properties. In addition, chronically axotomized group Ia muscle afferents no longer generate excitatory postsynaptic potentials in spinal motoneurones. However, peripheral neurones readily regenerate and reinnervate their native tissue, following which they recover largely or completely their normal properties and capabilities. In recent years we have been interested in the ability of neurones of the peripheral nervous system to innervate *foreign* tissue (i.e. motoneurones and muscle afferents into skin; cutaneous afferents into muscle) and in the consequences of such innervation on their properties and capabilities.

The description below reviews the results of these series of experiments. The methods used were generally similar, starting with a sterile surgical procedure in which the distal connections of nerves of adult cats' hindlimbs were altered. The muscle nerve to the medial gastrocnemius (MG) may have been axotomized and capped, or partly or fully cross-united distally to the caudal cutaneous sural (CCS) nerve (Fig. 38.1). The cutaneous sensory CCS nerve may have been axotomized and capped, or cross-united distally to part of the MG nerve. Properties of the cross-united afferents and motoneurones were studied in acute electrophysiological experiments (e.g. Zengel *et al.*, 1985) 3–36 months later.

Response by motoneurones

Motoneurones of the cat's normal MG muscle have predictable values for conduction velocity (CV), input resistance (R_N), rheobase (I_{rh}) and afterhyperpolarization (AHP; Zengel *et al.*, 1985). Following chronic axotomy, CV, I_{rh} and the range of AHP values decline, and R_N increases (Fig. 38.2; reviewed in Titmus & Faber, 1990). Following regeneration into the MG muscle (or lateral gastrocnemius (LG), but *not* soleus) the MG motoneurones

* Present address: Department of Neurosurgery, Yamaguchi Red Cross Hospital Yohatababa 53–1, Yamaguchi City, Yamaguchi ken, 753 Japan.

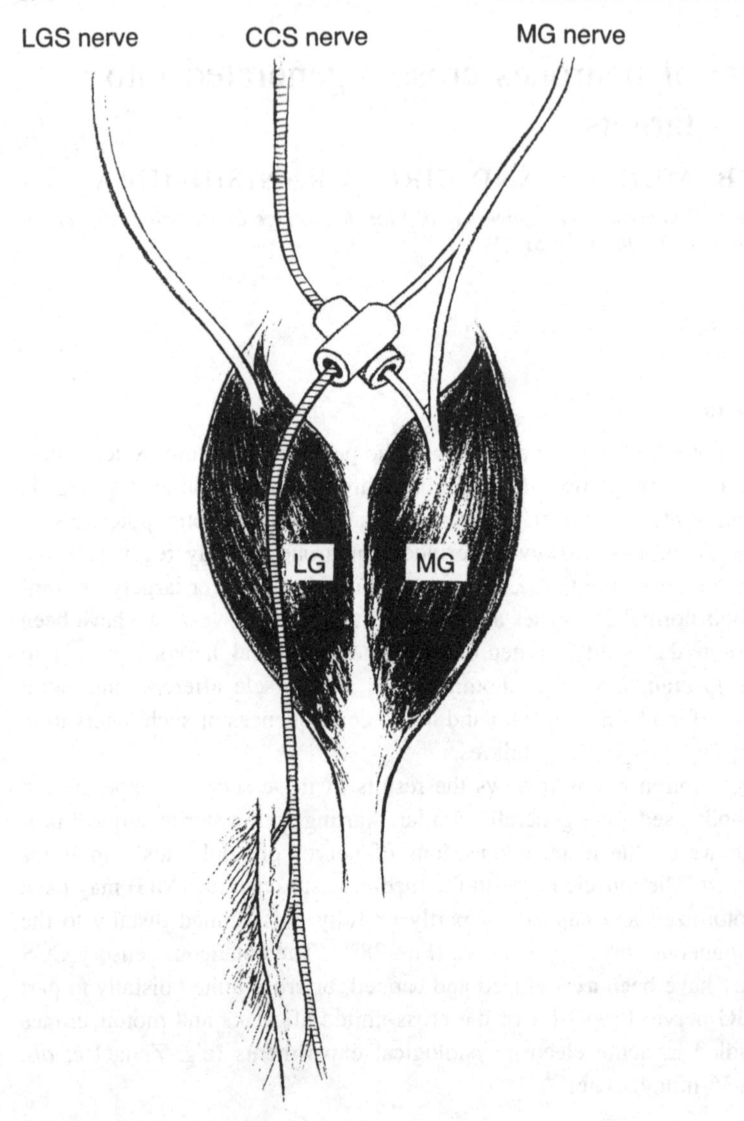

Fig. 38.1. Schematic of the surgical preparation. See text for abbreviations and details.

recover normal or near-normal values (Fig. 38.2; Foehring, Sypert & Munson, 1986; Foehring & Munson, 1990).

We find the axotomized MG motoneurones will regenerate through CCS (MG⟶CCS) and apparently terminate in hairy skin, although they do not then acquire low-threshold sensitivity to sensory stimuli (Nishimura, Johnson & Munson, 1991). Electrical properties of these MG⟶CCS motoneurones are nearly normal following regeneration of their axons through CCS, even though there is no evident functional contact with skeletal muscle. These

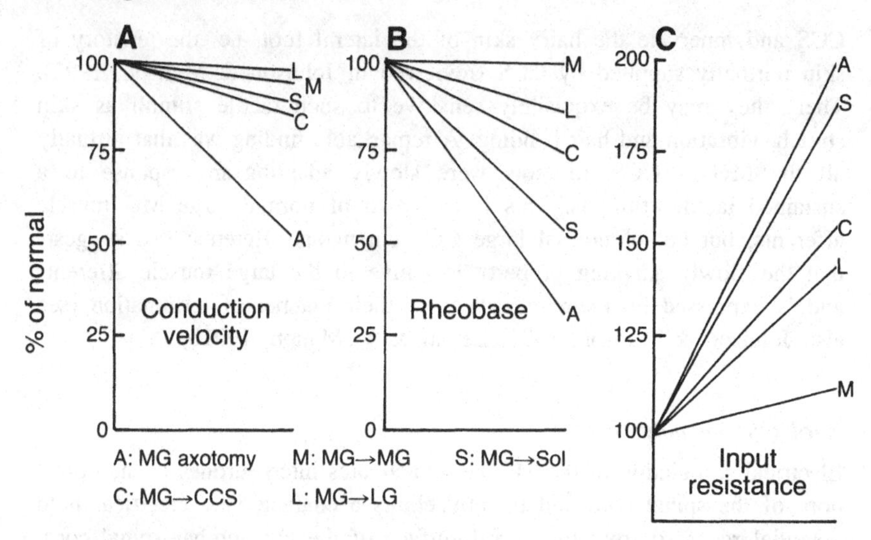

Fig. 38.2. Effects of axotomy and of cross-regeneration of MG nerve into foreign targets on electrical properties of MG motoneurones. Values for normal MG motoneurones are presented as 100% at the left of each plot; the value at the right of each plot is the value for long-term operated MG motoneurones. Note that recovery of normal values occurs best with regeneration into MG or LG, and that regeneration into the cutaneous nerve CCS may permit better recovery than does regeneration into the muscle–nerve soleus. MG⟶CCS. MG nerve cross-regenerated into CCS nerve, etc.

data are summarized in Fig. 38.2. Note that although regeneration of MG into CCS results in less than total recovery, so also does regeneration of the largely fast MG motoneurones into the slow soleus, which in turn is less restorative than is regeneration into MG or the similar muscle LG. The precise manner in which the MG⟶CCS motor axons terminate remains to be determined; however, we know that regeneration into the CCS nerve alone is not sufficient to permit rescue, because there was no recovery in an experiment in which the CCS nerve was ligated distally, thus preventing the cross-connected MG motor axons from reaching a distal target. We conclude that skin as well as skeletal muscle is a source of target-derived trophic support for adult spinal motoneurones.

Response of muscle afferents

Sensory properties

Following injury, large MG muscle afferents will regenerate and reinnervate MG muscle (e.g. Collins, Mendell & Munson, 1986) where they may or may not reinnervate their native receptor structure (i.e. muscle spindle, tendon organ). We find that muscle afferents will also regenerate through

CCS and innervate the hairy skin of the lateral foot, i.e. the territory of skin normally supplied by CCS (reviewed in Johnson & Munson, 1992). There they may be exquisitely sensitive to such tactile stimuli as skin stretch, vibration and hair bending. A remarkable finding was that virtually all the MG⟶CCS afferents were slowly adapting in response to a sustained tactile stimulus. This is true also of normal large MG muscle afferents, but not of normal large CCS cutaneous afferents; this suggests that the slowly adapting property is native to the large muscle afferents and is expressed by them regardless of their manner of termination (see also Johnson & Munson, 1991; Lewin & McMahon, 1991b).

Cord dorsum potentials

Electrical stimulation of the MG nerve activates interneurones in the dorsal horn of the spinal cord and thereby elicits a characteristic electrical field potential recorded from the dorsal surface of the low lumbar spinal cord: the cord dorsum potential (CDP). At group I strength the CDP consists of an afferent volley spike and a short-latency negative wave; at group III strength (~10 times group I threshold) a late negative wave appears (Fig. 38.3A; Bernhard, 1953). Chronic axotomy of MG delays and attenuates the early spike and attenuates or abolishes the negative slow waves (Fig. 38.3C). Cross-regeneration of MG into CCS (as into MG) restores normal latency (but not amplitude) of the spike and normal configuration of the negative waves (Fig. 38.3E). MG afferents cross-innervating CCS generate a CDP typical of normal MG (cf. Fig. 38.3A) and unlike the CDP elicited by normal CCS (Fig. 38.4A); i.e. we see no evidence of respecification of central connections of MG afferents.

Postsynaptic potentials

Electrical stimulation of the MG nerve at group I strength elicits monosynaptic excitatory postynaptic potentials (EPSPs) in LG and soleus motoneurones, examples of which are shown in Fig. 38.3B. Following chronic axotomy (~1 year) of the MG nerve (and thus of both its motor axons and afferent fibres) the MG afferents no longer elicit such EPSPs (Fig. 38.3D; Goldring *et al.*, 1980). Cut MG afferents which regenerate into MG muscle following a post-degeneration delay of up to 6 months elicit EPSPs of normal amplitude (Goldring *et al.*, 1980).

We find that the ability of injured MG muscle afferents to elicit EPSPs is rescued by their regeneration through CCS (Fig. 38.3F). Whereas the intracellular potentials generated by stimulation of the long-term axotomized MG are characterized by an absence of depolarization and by a profound hyperpolarization (Fig. 38.3D), PSPs of normal configuration (i.e. unlike those elicited by CCS stimulation: Fig. 38.4B) are produced by

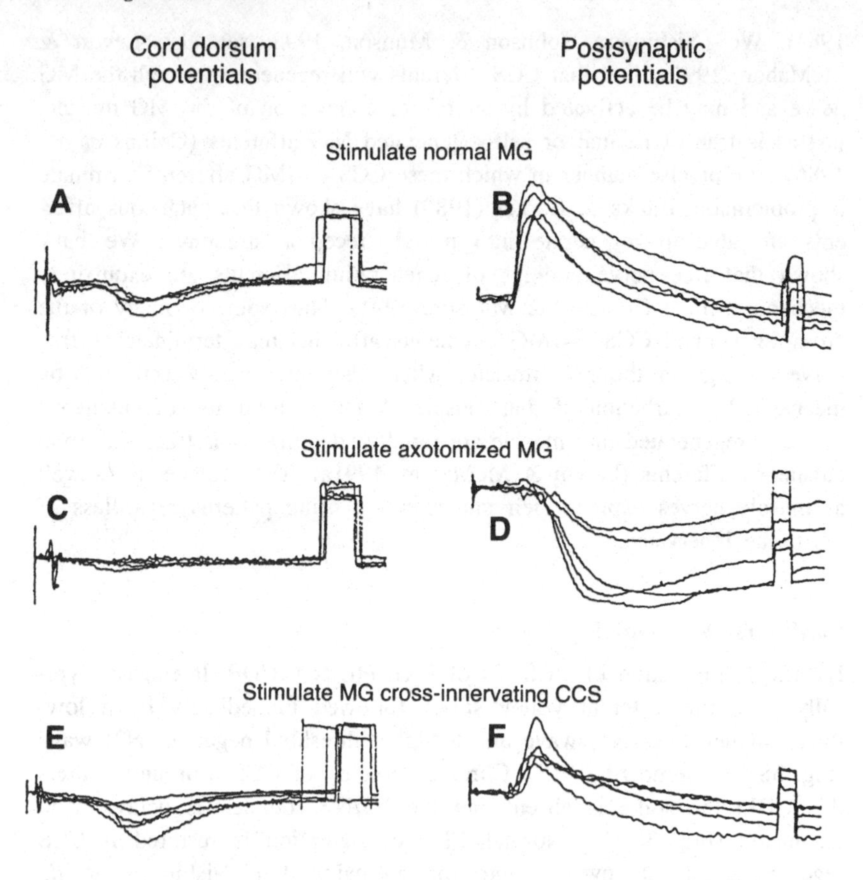

Fig. 38.3. Effects of chronic axotomy and of cross-regeneration of MG nerve into CCS nerve on cord dorsum and postsynaptic potentials elicited by stimulation of MG nerve. Each set of five records in the left-hand column (A, C, E) is from five different experiments. Each set of five records in the right-hand column (B, D, F) is from one experiment. Note the profound effects of axotomy on the configurations of potentials, and the recovery of normal configuration following cross-regeneration. Calibrations: (A), (C) and (E) 100 μV, 5 ms; (B), (D) and (F) 1 mV, 1 ms.

stimulation of the cross-regenerated MG⟶CCS nerve (Fig. 38.3F). Thus it appears that group Ia muscle afferents may require target-derived trophic support for their normal synaptic function and that the support may be derived readily from muscle or from skin.

Response of cutaneous afferents

Sensory properties

Cutaneous afferents of the cat's sural (CCS) nerve will regenerate and reinnervate their native hairy skin of the lateral foot (Horch & Lisney,

1981). We (Nishimura, Johnson & Munson, 1993; see also Lewin & McMahon, 1991a) find that CCS afferents will regenerate through the MG nerve and may be activated by stretch or contraction of the MG muscle, just as is true of normal or self-regenerated MG afferents (Collins *et al.*, 1986). The precise manner in which these CCS⟶MG afferents terminate is problematic. Banks & Barker (1989) have shown that cutaneous afferents are able to regenerate into muscle receptor structures. We have shown that free nerve endings of regenerating afferents are exquisitely mechanosensitive (Johnson & Munson, 1991). Thus some or many of the cross-regenerated CCS⟶MG cutaneous afferents may terminate as free nerve endings in the MG muscle, where they are readily activated by mechanical perturbation of that muscle. A large proportion of cutaneous afferents regenerated into muscle are rapidly adapting, as is true of normal cutaneous afferents (Lewin & McMahon, 1991a). Thus cutaneous as well as muscle nerves express their characteristic firing patterns regardless of the tissue innervated.

Cord dorsum potentials

Electrical stimulation of CCS, as of MG, elicits a CDP. It consists typically of a tiny afferent volley spike, followed immediately by a low-threshold negative (N1) wave and a higher threshold negative (N2) wave (Fig. 38.4A; Bernhard, 1953). Chronic axotomy of CCS profoundly alters this CDP: the spike is absent and the N1/N2 complex is delayed and attenuated (Fig. 38.4C). Normal CDP configuration is restored if CCS regenerates into its own or into foreign hairy skin (Nishimura *et al.*, 1993). We find that normal CDP configuration is also restored if CCS regenerates into MG (Fig. 38.4E); note especially that the configuration is that of the normal CCS and is clearly different from that of the normal MG (Fig. 38.3A); i.e. we see no evidence for respecification of dorsal horn neuronal circuitry.

Postsynaptic potentials

Electrical stimulation of CCS elicits complex plurisynaptic PSPs in motoneurones of the triceps surae. These PSPs differ in accordance with the type of target muscle (e.g. Nishimura *et al.*, 1993). In contrast to the above-described MG-afferent elicited potentials, those elicited by CCS stimulation are quite insensitive to chronic axotomy of the CCS nerve (Fig. 38.4D). The only unequivocal alteration resulting from such axotomy of CCS is a 2–3 ms increase in latency, which is explainable on the basis of slowed conduction of the cut afferents. It thus appears that in contrast to the monosynaptic pathway from MG group Ia afferents to triceps surae

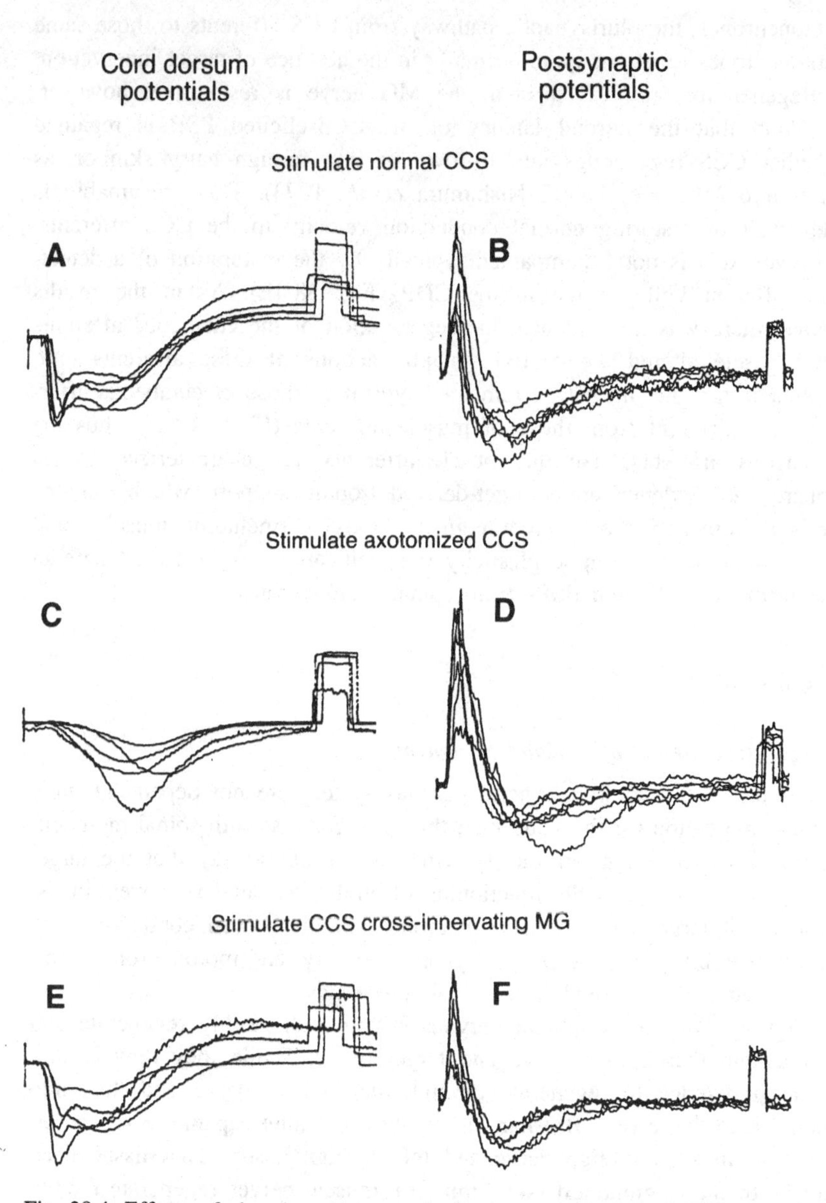

Fig. 38.4. Effects of chronic axotomy and of cross-regeneration of CCS nerve into MG nerve on cord dorsum and postsynaptic potentials elicited by stimulation of CCS nerve. Each set of five records in the left-hand column (A, C, E) is from five different experiments. Each set of five records in right-hand column (B, D, F) is from one experiment. Axotomy of CCS profoundly alters cord dorsum potentials, but alters only the latency of postsynaptic potentials. Both are restored by regeneration of CCS into MG. Calibrations: (A), (C) and (E) 100 (small) or 200 (large) μV, 5 ms; (B), (D) and (F) 1 mV, 10 ms.

motoneurones, the plurisynaptic pathway from CCS afferents to those same motoneurones functions quite normally in the absence of target innervation.

Regeneration of CCS through the MG nerve is restorative, however. We find that the normal latency of the CCS-elicited PSP is regained whether CCS regenerates into its own or into foreign hairy skin or, as here, into MG (Fig. 38.4F; Nishimura *et al.*, 1993). This presumably is the result of restoring normal conduction velocity to the CCS afferents; however, this is not accompanied typically by the restoration of a detectable afferent volley spike in the CDP (Fig. 38.4E). As in the results above, there was no evidence that regeneration of the cutaneous afferents into muscle altered the central synaptic actions of these afferents onto motoneurones, i.e. the CDPs remained typical of those of cutaneous afferents and different from those of muscle afferents (Fig. 38.3B). Thus the cutaneous afferents, like the muscle afferents, are characterized by an apparent dependence upon target-derived trophic support, which may be derived from native or from foreign target tissue (including muscle), and by an absence of synaptic plasticity (respecification) as measured here in the form of CDPs and PSPs from spinal motoneurones.

Discussion

Target-dependence of peripheral neurones

Neurones of the adult peripheral nervous system are not dependent upon target innervation for their survival; this is in contrast with spinal motoneurones of the neonatal nervous system. This is not to say that the target tissue plays no role in the functioning of adult peripheral neurones: in the absence of target tissue sensory neurones exhibit slowed conduction and an altered ability to generate postsynaptic activity, and motoneurones exhibit altered electrical (and histological) properties.

Neurones of the peripheral nervous system will readily regenerate and reinnervate their native nerve and target tissue. While they may exhibit some preference to regenerate through their native nerve and thus into their native tissue (e.g. Brushart, 1993) they are quite capable of regenerating also through foreign nerve and into foreign tissue. That tissue need not be of the original sort (skin, muscle); muscle nerves regenerate readily into skin, and cutaneous nerves regenerate readily into muscle.

Peripheral nerves appear to require some form of target-derived trophic support, based upon the fact that their properties and capabilities are altered in the absence of target contact but then recover following restoration of contact with their target tissue. We find that the native target tissue is apparently not an exclusive source of trophic support: following cross-regeneration into foreign targets, cutaneous and muscle sensory nerves recovered virtually completely with regard to the measures studied here. Recovery of normal motoneurone properties was somewhat less complete following regeneration into skin than following regeneration into a similarly constituted (i.e. 'fast') muscle, but recovery is incomplete also following regeneration into the 'slow' soleus muscle. Thus skin as well as muscle provides trophic support for MG motoneurones, although it may not substitute completely.

Current research has identified several neurotrophins as potential candidates for the putative trophic factors(s) (reviewed in Raivich & Kreutzberg, 1993, and in Mudge, 1993). These include brain-derived neurotrophic factor (BDNF), neurotrophin-3 (NT-3), neurotrophin-4/5 (NT-4/5), and ciliary neurotrophic factor (CNTF). The facts that the neurotrophins act on both sensory and motor neurones and that BDNF and NT-3 are both expressed in skin and muscle are suggestive of likely candidate trophins in these tissues, as is the fact that mature motoneurones express trkB and trkC, i.e. the preferred receptors for BDNF and NT-3.

Clinical importance

Damage to peripheral nerves is a common occurrence. Given proper conditions robust regeneration and reinnervation and some return of sensory and motor function can be expected. However, the exuberance of such reinnervation may constitute a problem as well as an attribute. Muscles may become innervated and activated by motoneurones which normally function antagonistically to the newly innervated muscle. Sensory neurones may innervate and be activated by target tissue having no relation to the tissue normally innervated, and at least some of the normal central synaptic actions of those neurones are retained. We saw no evidence for plasticity of central connections from these low-threshold afferents. A challenge for the nervous system is to utilize the anomalous and perhaps maladaptive actions of these misdirected components.

Acknowledgements

We thank our many colleagues for their contributions to these experiments: Drs J.W. Fleshman, R.C. Foehring, R.D. Johnson, L.M. Mendell, G.W. Sypert, J.S. Taylor and J.E. Zengel. Research support has been provided by NIH grants NS15913 and NS27511.

References

Banks, R.W. & Baker, D. (1989). Specificities of afferents reinnervating cat muscle spindles after nerve section. *Journal of Physiology (London)*, **408**, 345–372.

Bernhard, C.G. (1953). The spinal cord potentials in leads from the cord dorsum in relation to peripheral source of afferent stimulation. *Acta Physiologica Scandinavica*, **29**, Supplement 106, 1–29.

Brushart, T.M.E. (1993). Motor axons preferentially innervate motor pathways. *Journal of Neuroscience*, **13**, 2730–2738.

Collins, W.F. III, Mendell, L.M. & Munson, J.B. (1986). On the specificity of sensory reinnervation of cat skeletal muscle. *Journal of Physiology (London)*, **375**, 587–609.

Foehring, R.C. & Munson, J.B. (1990). Motoneuron and muscle-unit properties following long-term direct innervation of soleus muscle by medial gastrocnemius nerve in the cat. *Journal of Neurophysiology*, **64**, 847–861.

Foehring, R.C., Sypert, G.W. & Munson, J.B. (1986). Properties of self-reinnervated motor units of medial gastrocnemius of the cat. 1. Long-term reinnervation. *Journal of Neurophysiology*, **55**, 931–946.

Goldring, J.M., Kuno, M., Nunez, R. & Snider, W.D. (1980). Reaction of synapses on motoneurones to section and restoration of peripheral sensory connexions in the cat. *Journal of Physiology (London)*, **309**, 185–198.

Horch, K.W. & Lisney, S.J.W. (1981). On the number and nature of regenerating myelinated axons after lesions of cutaneous nerves in the cat. *Journal of Physiology (London)*, **313**, 275–286.

Johnson, R.D. & Munson, J.B. (1991). Regenerating sprouts of axotomized cat muscle afferents express characteristic firing patterns to mechanical stimulation. *Journal of Neurophysiology*, **66**, 2155–2158.

Johnson, R.D. & Munson, J.B. (1992). Specificity of regenerating sensory neurones in adult mammals. In *Sensory Neurons: Diversity, Development, and Plasticity*, ed. S.S. Scott, pp. 384–403. New York: Oxford Press.

Lewin, G.R. & McMahon, S.B. (1991a). Physiological properties of primary sensory neurons appropriately and inappropriately innervating skin in the adult rat. *Journal of Neurophysiology*, **66**, 1205–1217.

Lewin, G.R. & McMahon, S.B. (1991b). Physiological properties of primary sensory neurons appropriately and inappropriately innervating skeletal muscle in adult rats. *Journal of Neurophysiology*, **66**, 1218–1231.

Mudge, A.W. (1993). Motor neurons find their factors. *Nature*, **363**, 213–214.

Nishimura, H., Johnson, R.D. & Munson, J.B. (1991). Rescue of motoneurons from the axotomized state by regeneration into a sensory nerve in cats. *Journal of Neurophysiology*, **66**, 1462–1470.

Nishimura, H., Johnson, R.D. & Munson, J.B. (1993). Rescue of neuronal function by cross-regeneration of cutaneous afferents into muscle in cats. *Journal of Neurophysiology*, **70**, 213–222.

Raivich, G. & Kreutzberg, G.W. (1993). Peripheral nerve regeneration: role of growth factors and their receptors. *International Journal of Developmental Neuroscience*, **11**, 311–324.

Titmus, M.J. & Faber, D.S. (1990). Axotomy-induced alterations in the electrophysiological characteristics of neurons. *Progress in Neurobiology*, **35**, 1–51.

Zengel, J.E., Reid, S.A., Sypert, G.W. & Munson, J.B. (1985). Membrane electrical properties and prediction of motor-unit type of medial gastrocnemius motoneurons in the cat. *Journal of Neurophysiology*, **53**, 1323–1344.

39

Development and repair of neonatal mammalian spinal cord in culture

JOHN G. NICHOLLS AND ZOLTAN VARGA

Biocenter, Department of Pharmacology, Basel, Switzerland

Introduction

There are several lines of evidence to suggest that the mammalian central nervous system (CNS) might regenerate better after a lesion in an embryo than in an adult. For one thing the environment is favourable for neurite outgrowth since the CNS is still forming and intrinsic growth programmes are still in effect. For another, inhibitory factors in extracellular fluid or associated with membranes would probably not have developed at early stages (Björklund, 1991; Schwab, 1991).

It was Norman Saunders, then at Southampton, now in Tasmania, who introduced the South American Opossum, *Monodelphis domestica* (Fig. 39.1), as a valuable preparation for physiological studies of development to our laboratory at the Biocenter in 1989. He with Kjeld Møllgard (at Copenhagen) had used this animal to great advantage for studies of cortical development and blood–brain barrier (Saunders *et al.*, 1989). What makes the animal so useful? Firstly, it is about the size of a small rat so it can be bred in the laboratory more easily than a kangaroo or a North American Opossum. Secondly, the pups, 2–12 in number, are born at an immature stage corresponding to a 15-day-old rat embryo with no cortex or higher functions. The newborn animal is virtually decerebrate. For example, the cerebellum only starts to cover the fourth ventricle at about 4 days. Yet, the pups can breathe and suck. Moreover fictive respiration continues *in vitro*.

In our studies we have isolated the entire CNS (Fig. 39.2) and studied reflexes (by recording electrically), cell division (by measuring incorporation of bromodeoxyuridine, BrdU) and lesions to the spinal cord in younger and older animals. A suitable medium for long-term culture has been basal medium, Eagle's (BME) at room temperature (Nicholls *et al.*, 1990; Stewart *et al.*, 1991; Zou *et al.*, 1991; Treherne *et al.*, 1992; Woodward *et al.*, 1993). A brief summary of our major findings is presented in the following paragraphs.

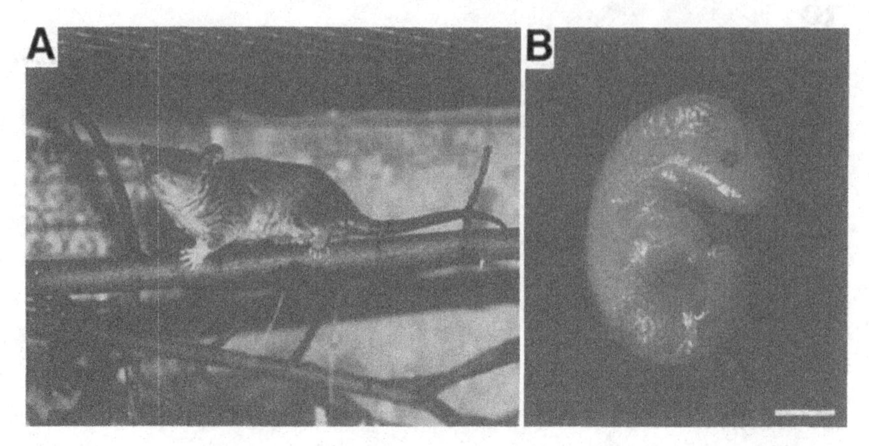

Fig. 39.1. (A) *Opossum monodelphis domestica* in the animal house at the Biocenter. The adult is about 9 cm long, excluding the tail. (B) Newborn pup aged 2 days. Note the poorly developed eyes and limbs, as well as the absence of ears or hair. The animal cannot walk or right itself. Scale bar represents 2 mm.

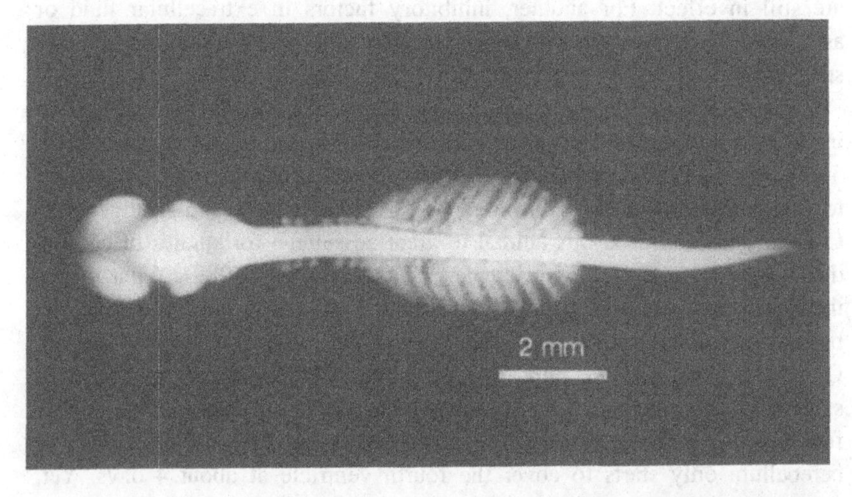

Fig. 39.2. Entire CNS of a 4-day-old opossum dissected together with the ribs. The forebrain consists of a vesicle surrounded by a thin sheet of cells. The cerebellum is still rudimentary and does not cover the fourth ventricle. Preparations such as this continue to survive, develop and produce electrical signals in culture over periods of a week or more when maintained in suitable medium.

Survival and development in culture

The isolated CNS continues to display fictive respiration in culture; if ribs are left attached they show regular inspiratory and expiratory movements for up to 2 days (Fig. 39.3). Dorsal root–ventral root reflexes persist, as do direct and synaptically mediated conduction through the spinal cord. Thus, after 10

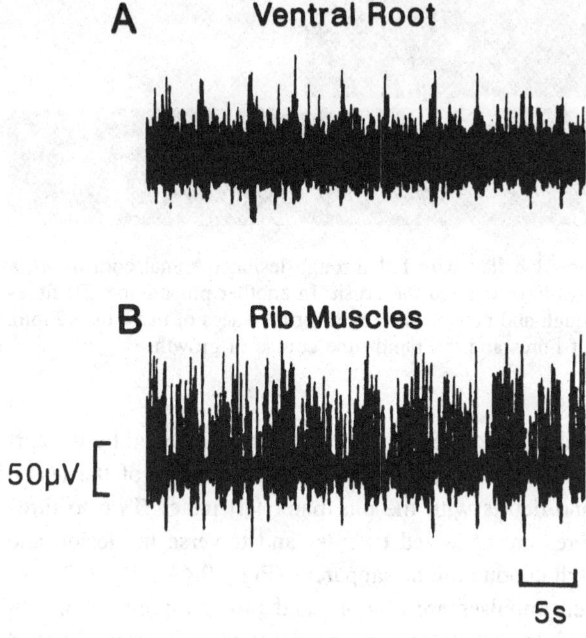

Fig. 39.3. Fictive respiration in the isolated CNS of a 2-day-old opossum. The preparation resembles that in Fig. 39.2. (A) Recording from ventral root with suction electrode shows periodic bursts of impulses that occur in phase with contractions of rib muscles and the electromyogram (B).

days in culture, electrical stimulation of one end of the cord results in short-latency and delayed volleys at the other. Maturation proceeds, albeit more slowly than in the animal. Two hours of exposure to radiolabelled BrdU after 3 days in culture produces clear labelling of cells around the ventricles and spinal canal that had didvided and susequently undergone migration. In addition radial glial cells become more prominent over 5 days in culture (Möllgard *et al.*, 1994). A valuable feature of the isolated CNS is that ions and small molecules penetrate and are washed out rapidly. Tetrodotoxin, for example, blocks all conduction within 1 min; 5 min after washing it out action potentials and reflexes are once again apparent. Similarly, gamma-aminobutyric acid (GABA), as well as glycine and *N*-methyl-D-aspartic acid (NMDA), reversibly block or modulate synaptic transmission when applied to the bathing fluid in appropriate concentrations (approx. 1–10 μM) (Zou *et al.*, 1991)

Repair after injury

When one side of the isolated spinal cord in culture is crushed by forceps at the cervical level all axons are broken. This has been shown by light and

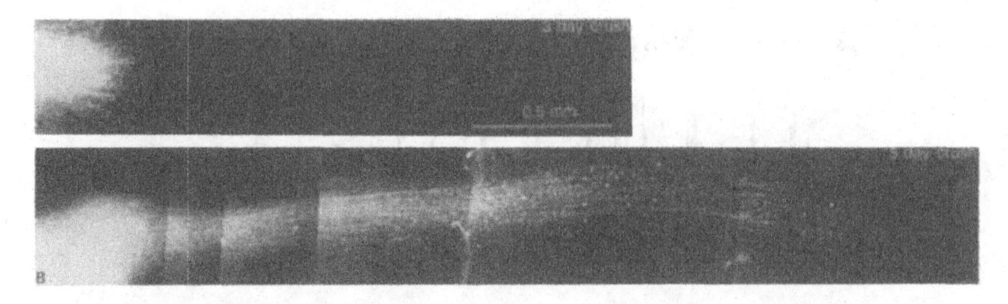

Fig. 39.4. Growth of fibres labelled with DiI through lesioned spinal cord *in vitro*. At 3 days (A) a few fibres have entered the crush. In another preparation (B) fibres at 5 days had grown through and beyond the crush for distances of more than 2 mm. Note the large number of fibres and the rapid time course of growth.

electron microscopy and by recording electrically. Fibres labelled by the carbocyanine dye DiI are seen to be broken and thorough scans of the lesion reveal only vesicles and debris with the pia mater left intact. Two to three days later, labelled fibres are observed to enter and traverse the lesion and the first signs of through conduction are apparent (Fig. 39.4A). By 4–5 days (Fig. 39.4B), conducted impulses are obvious and profuse numbers of labelled fibres are seen in light and electron micrographs to have grown beyond the crush, often as far as 2–3 mm. In a series of 66 young animals aged 3–6 days, 40% showed recovery as assessed by recording electrically, or by labelling fibres by DiI, or both (Treherne *et al.*, 1992; Woodward *et al.*, 1993). To remove any lingering doubt that these results represent growth and repair rather than the sparing of undamaged fibres at the time the crush was made, our colleague Dr H. Vischer has recently cut the spinal cord into two separate pieces, reattached them and observed clear outgrowth of labelled fibres from one piece of tissue to the other.

Age dependence of repair

Adult mammalian CNS does not exhibit comparable repair after a lesion. To determine at what stage this property becomes lost, crushes have been made in the spinal cords of older animals aged 11–14 days. The isolated CNS at this stage still survives when in culture for 5 days. In 53 of these animals only 5 showed any sign of repair and those were the youngest (10–11 days of age), with only marginal growth of one or two axons into the crush.

Embryonic rat CNS in culture

As a test of the generality of these findings, embryonic rat CNS has been isolated, lesioned and maintained in culture. In 15-day-old rat embryos the CNS is very similar to that of the new-born opossum although somewhat larger and more developed. The CNS survives well in culture and shows very similar repair as assessed by morphological and physiological criteria after an injury. In our experience older and younger rat embryos are less suitable for providing isolated CNS preparations (Saunders *et al.*, 1992).

Where do we go from here?

Numerous open questions remain. For example, does the outgrowth across the lesion in a young animal represent true regeneration of axons that sprout after having been transected or does it occur through fibres growing towards the crush site and continuing through it as a part of normal development? Double labelling experiments are now in progress to distinguish between these two mechanisms. A second question is whether newly grown fibres form synapses and if so whether they find their correct targets. This we are approaching by physiological recordings and electron microscopy. Of particular interest at the moment are the changes that occur between 9 and 12 days when growth across the lesion abruptly stops. We are searching for changes in the fine structure of the nervous system, the identity of cells that invade the crush site and the molecules that promote or inhibit growth and are expressed there during this brief critical period. Although the appearance of oligodendrocytes and the first wrappings of myelin are observed at about 9–11 days, their possible role in preventing regeneration in the neonatal animal has not been established. What seems encouraging is the range of experiments that become possible in this isolated mammalian CNS in culture.

Acknowledgments

We wish to thank our colleague Dr W. Adams for his essential help in every phase of these experiments. One of the authors (J.G.N) owes a particular debt to Tom Sears not only for his friendship and wisdom throughout the years but for planting the idea of using isolated mammalian CNS preparations during a sabbatical year in 1979. This work has been generously supported by the Swiss Nationalfond 31–362.62.92 and a grant from the International Research Institute for Paraplegia.

References

Björklund, A. (1991). Neural transplantation: an experimental tool with clinical possibilities. *Trends in Neurosciences*, **14**, 319–322.

Møllgard, M., Balslef, Y., Janas, I.M.S., Treherne, J.M., Saunders, N.R. & Nicholls, J.G. (1994). Development of the isolated CNS of a neonatal mammal (opossum *Monodelphis domestica*) maintained in long-term culture. *Journal of Neurocytology*, **23**, 151–165.

Nicholls, J.G., Stewart, R.R., Erulkar, S.D. & Saunders, N.R. (1990). Reflexes, fictive respiration and cell division in the brain and spinal cord of the newborn opossum, *Monodelphis domestica*, isolated and maintained *in vitro*. *Journal of Experimental Biology*, **152**, 1–15.

Saunders, N.R., Adam, E., Reader, M. & Møllgard (1989). *Monodelphis domestica* (grey short-tailed opossum): an accessible model for studies of early neocortical development. *Anatomica et Embryologica*, **173**, 81–94.

Saunders, N.R., Balkwill, P., Knott, G., Habsgood, M.D., Møllgard, K., Treherne, J.M. & Nicholls, J.G. (1992). Growth of axons through a lesion in the intact CNS of fetal rat maintained in long-term culture. *Proceedings of the Royal Society of London, Series B*, **250**, 171–180.

Schwab, M.E. (1991). Nerve fibre regeneration after traumatic lesions of the CNS: progress and problems. *Philosophical Transactions of the Royal Society of London, Series B*, **331**, 303–306.

Stewart, R.R., Zou, D.-J., Treherne, J.M., Møllgard, K., Saunders, N.R. & Nicholls, J.G. (1991). The intact central nervous system of the newborn opossum in long-term culture: fine structure and GABA-mediated inhibition of electrical activity. *Journal of Experimental Biology*, **161**, 25–41.

Treherne, J.M., Woodward, S.K.A., Varga, Z.M., Ritchie, J.M. & Nicholls, J.G. (1992). Restoration of conduction and growth of axons through injured spinal cord of neonatal opossum in culture. *Proceedings of the National Academy of Sciences, USA*, **89**, 431–434.

Woodward, S.K.A., Treherne, J.M., Knott, G.W., Fernandez, J., Varga, Z.M. & Nicholls, J.G. (1993). Development of connections by axons growing through injured spinal cord of neonatal opossum in culture. *Journal of Experimental Biology*, **176**, 77–88.

Zou, D.-J., Treherne, J.M., Stewart, R.R., Saunders, N.R. & Nicholls, J.G. (1991). Regulation of $GABA_B$ receptors by histamine and neuronal activity in the isolated spinal cord of neonatal opossum in culture. *Proceedings of the Royal Society of London, Series B*, **246**, 77–82.

40

Selective neuronal vulnerability in motor neurone diseases with reference to sparing of Onuf's nucleus

A.H. PULLEN
Sobell Department of Neurophysiology, Institute of Neurology, London, UK

Clinical background

Sporadic motor neurone disease (MND) is progressive, unremitting, and affects about 1/100 000 of the population with a prevalence of between 2 and 7/100 000 and a 1.2 to 1.5/1 male to female ratio. Outside the United Kingdom it is known as amyotrophic lateral sclerosis (ALS) since its major characteristics are focal muscle wasting and degeneration of the lateral corticospinal tracts (i.e. both upper and lower motoneurone involvement). The relative degree of upper and lower motor neurone involvement varies between patients.

Focal wasting of limb muscles results from degeneration of specific groups of spinal motor neurones (seen alone this 'lower motor neurone' disorder is progressive muscular atrophy). Progressive bulbar palsy results from degeneration of lower cranial motor nuclei, and hyperreflexia is consequent upon death of cortical Betz cells (seen alone this 'upper motor neurone' disorder is primary lateral sclerosis). While sensory systems are considered generally unaffected in disease of shorter duration, sensory involvement may occur in some cases (Swash & Schwartz, 1992). MND therefore is primarily selective for motor neurones controlling striated muscle (i.e. 'somatic' motor neurones) and synaptically coupled to cortico-spinal systems.

However, in MND of shorter duration, preservation of eye movement, and bladder and anal sphincter function, coupled with post-mortem evidence of sparing of oculomotor neurones in the 3rd, 4th and 5th cranial nuclei, and sacral sphincteric motor neurones in Onuf's nucleus (Mannen *et al.*, 1977; Tokoyura, 1977), suggests that not all motor neurones die. The motor system therefore demonstrates a selective neuronal vulnerability to MND. Reasons for sparing of sphincteric motor neurones are unclear and the subject of the debate reviewed below, which focuses on the anatomical and functional identity of Onuf's nucleus.

Onuf's nucleus

First described by Onufrowicz (1890), 'nucleus X' in the human is a discrete group of neurones located in the most ventral region of the ventral horn in segments S1–S3, and is now eponymously called Onuf's nucleus (Schroder, 1981; Mannen *et al.*, 1977, 1982; Konno *et al.*, 1986). Mammalian homologues of Onuf's nucleus occur in cat, dog, rabbit and subhuman primates (Sato, Mizuno & Konishi, 1978; Nagashima *et al.*, 1979; Kuzuhara, Kanazawa & Nakanishi, 1980; Roppolo, Nadelhaft & DeGroat, 1985). In rat the nucleus is split between a ventral region identical in position to the single nucleus found in the human, and a more dorso-medial region bordering lamina X (Schroder, 1980; McKenna & Nadelhaft, 1986). Retrograde horseradish peroxidase (HRP) or fluorescent labelling experiments in each of these species show motor neurones in Onuf's nucleus projecting to the external urethral and external anal sphincters (Sato *et al.*, 1978; Nagashima *et al.*, 1979; Kuzuhara *et al.*, 1980; Roppolo *et al.*, 1985; Pullen, 1988b). Onuf's nucleus was originally classified as 'somatic' because it innervated striated muscle, but others have viewed it as autonomic or even a hybrid. Since identity has direct implication for the observed difference in response of this nucleus in neurodegenerative diseases, the evidence for these differing viewpoints is worth summarizing.

Evidence suggesting a 'somatic' identity

Firstly, across the species, retrograde cytochemical labelling experiments such as those cited above show that motor neurones located in Onuf's nucleus innervate striated muscles of the external urethral and external anal sphincters. Secondly, the 'double' somatic–autonomic innervation of the urinary bladder and associated external urethral sphincter is paralleled by the 'double' innervation of the colon and its associated external anal sphincter (Bishop *et al.*, 1956). Urinary bladder and colon receive autonomic innervation via the pelvic nerve and inferior mesenteric ganglion. Preganglionic parasympathetic motoneurones reside in the intermedio-lateral grey matter (DeGroat *et al.*, 1981). The external urethral and external anal sphincters are innervated by the pudendal nerve, which contains both sensory and motor components, and parent motor neurones are located within the motor pools of the sacral ventral horn (reviewed by DeGroat & Steers, 1990). Thirdly, both sphincters are under partial 'voluntary' control.

Evidence suggesting an autonomic identity

Evidence for an autonomic identity derives from anatomical, pharmacological and clinical observations. Firstly, Rexed (1954) identified a narrow band of neurones between Onuf's nucleus in the ventral horn and the intermedio-lateral horn, and interpreted this as evidence for Onuf's nucleus being an extension of the intermedio-lateral preganglionic parasympathetic nucleus. This view has since been echoed by Gibson *et al.* (1988).

Secondly, immunocytochemistry in a number of species has revealed a rich peptidergic innervation in the immediate neuropil around Onuf's nucleus, and peptidergic-containing axon terminals synapsing directly with sphincteric motor neurones (Schroder, 1984; Katagiri *et al.*, 1986; Kawatani, Nagel & DeGroat, 1986; Gibson *et al.*, 1988; Kawatani *et al.*, 1989; Tashiro *et al.*, 1989), analogous to the peptidergic nerves and terminals associated with the preganglionic innervation of the urinary bladder and colon (Glazer & Basbaum, 1980; Schroder, 1984; Kawatani *et al.*, 1986, 1989).

Thirdly, anterograde and retrograde axonal tract labelling experiments reveal a direct mid-brain synaptic projection to Onuf's nucleus analogous to the spinal input from autonomic hypothalamic 'micturition centres' (Holstege *et al.*, 1986; Holstege & Tan, 1987). Fourthly, there is sparing of Onuf's nucleus in MND, which kills somatic motor neurones in other segments (Mannen *et al.*, 1977; Tokoyura, 1977; Pullen, Martin & Swash, 1992), but co-degeneration of Onuf's neurones and autonomic nuclei in Fabry's disease and the Shy–Drager form of multiple system atrophy (Shy & Drager, 1960; Sung, Mastri & Segal, 1979; Konno *et al.*, 1986), which has been cited as major evidence for an autonomic classification of Onuf's nucleus.

Resolution of these conflicting viewpoints may lie in a third viewpoint: that Onuf's motor neurones are hybrids (specialized forms of somatic motor neurone with some autonomic-like characteristics: Gibson *et al.*, 1988).

Observations made during experimental studies originally designed to examine the post-injury responses of axon terminals presynaptic to cat intercostal motor neurones provided both an alternative entrée into the debate and a method of examining sphincteric motor neurones independent of any previously used.

Investigations of the synaptology of the α-motor neurone

Ultrastructural investigations of altered presynaptic input to thoracic motor neurones located caudal to a spinal hemisection focused on their response during the recovery phase following the partial central deafferentation (Pullen & Sears, 1978, 1983). Morphological criteria introduced by Conradi (1969a) distinguished five different morphological classes of axon terminal synapsing with the motor neurones, based on differences in terminal size, synaptic vesicle diameter and shape, and synaptic site ultrastructure. Terminals were named in accordance with Conradi's terminology: S, F, T, M and C. Of particular significance, the C-type terminal (Fig. 40.1b) demonstrated a selective increase in presynaptic territory coupled with growth of the

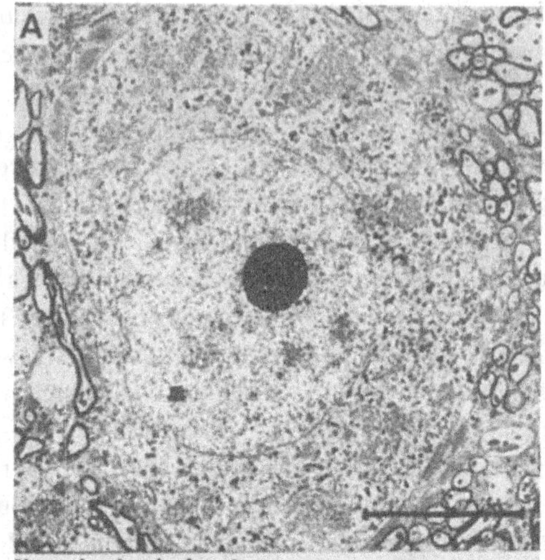

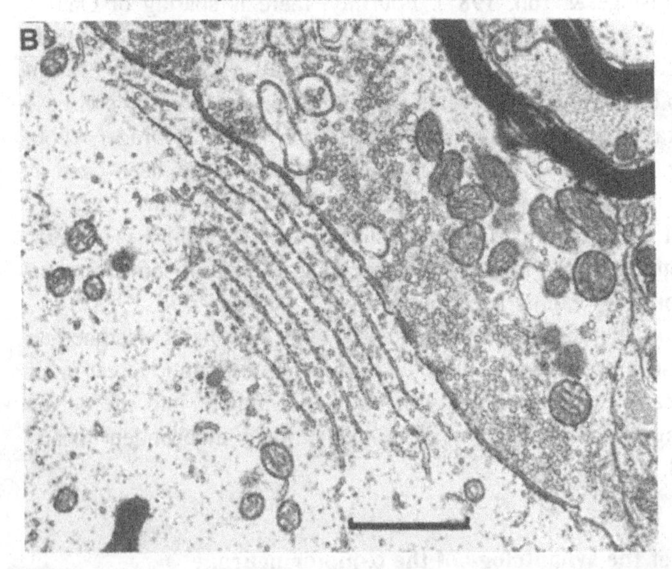

Fig. 40.1. (A) A section through the mid-nuclear plane of a spinal α-motoneurone in a cat which was perfusion-fixed with phosphate-buffered 2.5% glutaraldehyde. (B) A C-type terminal synapsing with a cat α-motoneurone. Note the high packing density of electron-lucent synaptic vesicles, and the subsynaptic cistern and Nissl body which characterize this synapse. Scale bars represent 10 μm in (A) and 1 μm in (B).

postsynaptic cistern and Nissl body (Pullen & Sears, 1978, 1983). This terminal has since become a focus of attention (Pullen, 1988a, b, 1990).

The C-type axon terminal

The C-type synapse has a unique morphology characterized presynaptically by large size (4–8 μm appositional length) and high packing density of electron-lucent 50 nm diameter synaptic vesicles. Postsynaptically, the C-type synapse is characterized by a 15 nm wide cistern located 100 nm beneath the postsynaptic membrane (the subsynaptic cistern) and extending the entire length of the presynaptic terminal. The subsynaptic cistern is contiguous with the rough endoplasmic reticulum (rER) of a subjacent Nissl body. The Nissl body comprises a highly organized multilamellated structure of alternate long rER lamellae and linear arrays of membrane-associated polyribosomes. The pre- and postsynaptic membrane 'densities' are not visible in aldehyde-fixed, osmicated cat spinal cord, but are selectively revealed by ethanolic phosphotungstic acid (Pullen, 1988a). While C-type terminals number only 1–4% of total terminal numbers in cat thoracic and lumbar motor neurones, they have the largest presynaptic occupancy of the motor neurone surface due to their size (Conradi, 1969a; Pullen & Sears, 1983). Failure to invoke degeneration of 'C' synapses following cortical ablation, dorsal root section or near or remote spinal hemisection, together with other cytochemical labelling evidence, suggests C-type terminals derive intrasegmentally, from short-axon propriospinal pathways (Conradi, 1969b; McLaughlin, 1972a, b; Matsushita & Ikeda, 1973).

C-type synapses characterize somatic α-motor neurones, and synapse with hypoglossal (Boone & Aldes, 1984), trigeminal (Hamos & King, 1980), phrenic (Goshgarian & Rafols, 1984), cervical (Matsushita & Ikeda, 1973), thoracic (Pullen & Sears, 1978) and lumbar (Conradi, 1969a) motor nuclei. C-type synapses have not been found associated with γ-motor neurones, Renshaw cells or segmental interneurones (Lagerback & Ronnevi, 1982; Lagerback, 1985; Johnson, 1986). Of special significance to the debate on the identity of sphincteric motor neurones is the fact that C-type axon terminals were not identified among the ultrastructural classes found presynaptic to cytochemically labelled preganglionic parasympathetic motor neurones in cat (Mawe, Bresnahan & Beattie, 1986; Leedy *et al.*, 1988).

Synaptology of sacral autonomic motor neurones

Mawe *et al.* (1986) and Leedy *et al.* (1988) demonstrated three main classes of terminal synapsing with autonomic motor neurones. None was identical to those characteristic of somatic motor neurones; in particular one class was characterized by dense-core synaptic vesicles, and another by small-diameter granular vesicles and synaptic dense bodies. No terminal displayed subsynaptic cisterns associated with Nissl bodies (i.e. 'C' synapses).

Such unequivocal differences in the ultrastructure of synaptic terminals of somatic and autonomic motor neurones, in particular the presence of C-type synapses on somatic motor neurones, provided a new approach to analysing sphincteric motor neurones in Onuf's nucleus, first in cat and then in the human.

Ultrastructural analysis of axon terminals synapsing with feline sphincteric motor neurones retrogradely labelled with HRP

Sphincteric motor neurones were identified in 70 μm thick transverse sections of the spinal cord using retrograde axonal transport of cytochemical label (HRP) following its injection into the left-hand hemicircle of the external anal sphincter of deeply anaesthetized cats (Pullen, 1988b). Cytochemically labelled sections of spinal cord were further processed for electron microscopy. Salient results are illustrated in Figs. 40.2 and 40.3.

Three major groups of large neurones within the ventral horn of S1–S2 were visible in 0.5 μm 'plastic' sections (Fig. 40.2): an extreme ventro-lateral group (VL), and extreme ventro-medial group (VM), and between them a discrete group of neurones conforming in position to previous descriptions of Onuf's nucleus (ON) (e.g. Sato *et al.*, 1978). HRP-labelled motor neurones were restricted to the superior (dorso-lateral) portion of ON. All other neurones were unlabelled (Fig. 40.2).

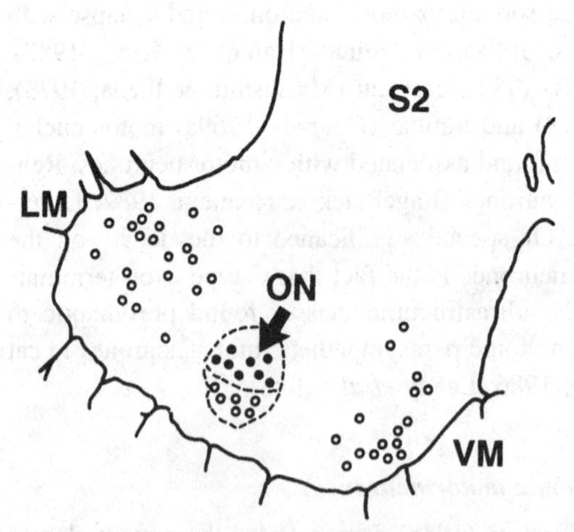

Fig. 40.2. The positions of neurones in the major nuclei of S2 in the cat are shown: VM, ventro-medial nucleus; VL, ventro-lateral nucleus; ON, Onuf's nucleus. Filled circles denote labelled neurones following injection of horseradish peroxidase (HRP) into the external anal sphincter.

Axon terminals synapsing with the perikaryon of HRP-labelled sphincteric motor neurones in ON were classified in accord with ultrastructural criteria given by Conradi (1969a), Mawe *et al.* (1986) and Leedy *et al.* (1988).

As shown in Fig. 40.3, terminals formed five morphological classes (or 'types') identical to the S, F, T, M and C types described by Conradi (1969a) and previously identified synapsing with intercostal motor neurones (Pullen & Sears, 1978). Of particular significance was the presence of the morphologically distinctive C-type synapse (Fig. 40.3). No classes similar to those reported by Mawe *et al.* (1986) or Leedy *et al.* (1988) for preganglionic parasympathetic motor neurones synapsed with HRP-labelled sphincteric motor neurones.

The presence of the C-type synapse indicated that feline sphincteric motor neurones in Onuf's nucleus are somatic α-motor neurones, rather than preganglionic parasympathetic motor neurones. However, the absence of any published data regarding the ultrastructure of axon terminals presynaptic to adult human spinal motor neurones (at any segmental level) prevented any analogy being made between Onuf's nucleus in the human and its feline homologue.

Synaptology of motor neurones in Onuf's nucleus in the human: controls and MND

Subsequent investigations of Onuf's nucleus in the human therefore examined the ultrastructure of axon terminals presynaptic to sphincteric motor neurones (Pullen *et al.*, 1992). Spinal cords were obtained during routine autopsy from non-neurological 'controls' and from patients who had died from MND. Since use of existing methods for preparing post-mortem spinal cord for electron microscopical examination proved inadequate for reliable ultrastructural identification of synaptic 'classes', new methods were devised based on those developed for the animal studies described above. Representative motoneurones are shown in Fig. 40.4.

Three factors were crucial to the success of the study. The first was reduction of post-mortem delay from the more usual periods of >20 hours to 3–6 hours. Secondly, immersion-fixation of thin (5 mm) slices of spinal cord in appropriate fixatives was necessary immediately upon removal of tissue from the body. Lastly, only the first millimetre depth of tissue from either face of the 5 mm cord-slice was used. These slices were further sectioned at 70 μm thickness prior to processing for electron microscopy.

Table 40.1 summarizes the patient details. Onuf's nucleus in 'controls' was identified with reference to published criteria. The results may be summarized as follows.

1. The well-documented loss of motor neurones in VL and VM nuclei in spinal segments S1 and S2, coupled with preserved numbers of motor neurones in ON, was confirmed (Fig. 40.5).

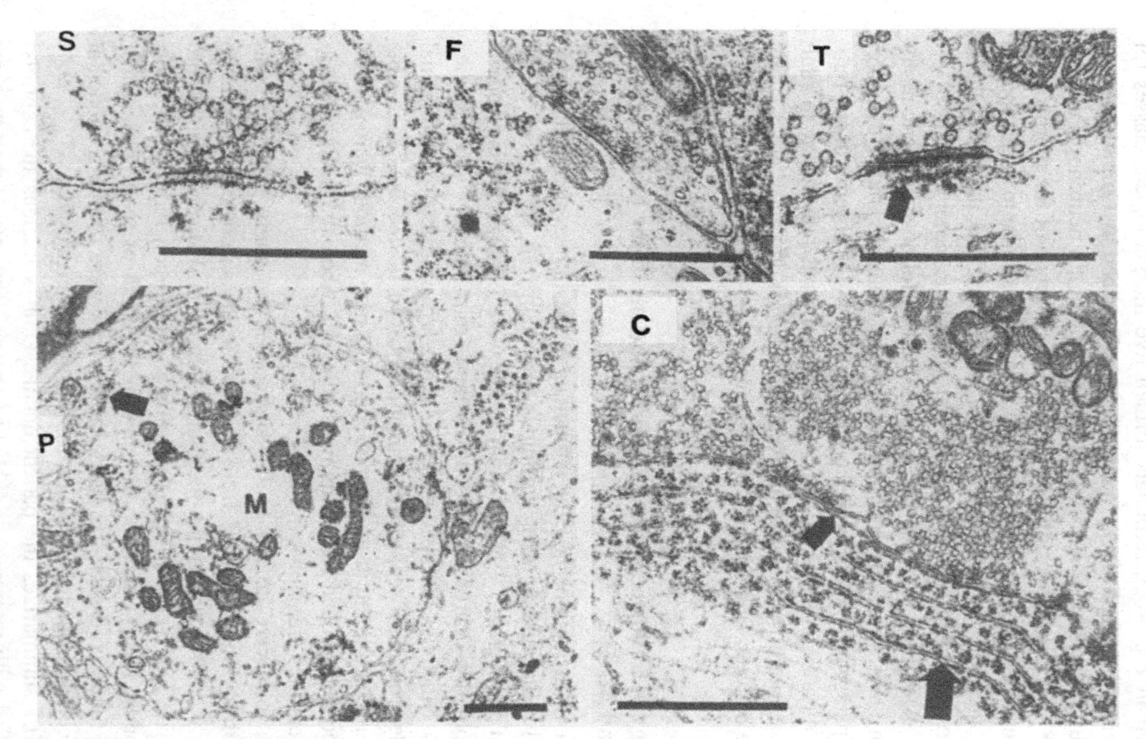

Fig. 40.3. Five ultrastructural 'types' of presynaptic terminal synapsed with HRP-labelled sphincteric motoneurones in the cat. S terminals are characterized by spherical vesicles, F terminals by flattened or polymorphic vesicles. T terminals display a subsynaptic row of electron-dense Taxi bodies. The M terminal exhibits a small presynaptic (P-) terminal on its surface opposing the neuropil. The C-synapse possesses a subsynaptic cistern (small arrow) and associated Nissl body (large arrow). Scale bars in each illustration represent 1 μm.

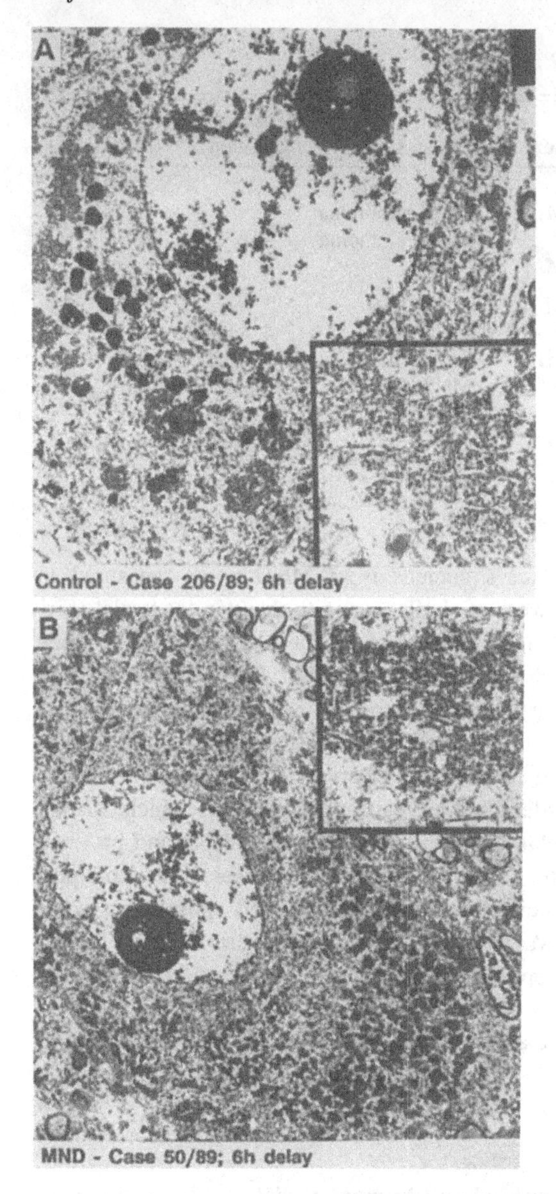

Fig. 40.4. Sacral motoneurones in Onuf's nucleus at S2 in spinal cords obtained 6 hours post-mortem from a control (A) and MND subject (B). Segments were immersion-fixed in buffered glutaraldehyde and show reasonable preservation of cell and nuclear membranes, and rER organisation of the Nissl bodies (inset photographs). Age-related electron-dense lipofuscin is a prominent feature of motoneurones in both subjects.

Table 40.1. *Details of MND patients*

	Age (years)	Sex	Diagnosis	Disease duration	PM delay (hours)
PM100/89	65	F	Renal failure	—	5
PM182/89	61	M	Lung carcimona	—	3
PM206/69	45	F	Lung carcinoma	—	6.5
PM45/91	70	M	—	—	6
PM50/89	79	M	MND/ALS	5 years	6.5
PM122/89	59	F	MND/ALS	4 years	17
PM200/89	75	M	MND/ALS	19 years	3.5
PM220/89	70	F	MND/ALS	8 months	14.5
PM223/89	72	F	MND/ALS	2 years	20
PM12/90	38	M	MND/ALS	3 years	3

PM, post-mortem.

2. Neurone diameters in Onuf's nucleus from MND cords measured in 0.5 μm 'plastic' sections were of similar range to those in 'controls' (Fig. 40.5; controls: range 17.8–71 μm; mean 38.6 + 12.9 μm. MND: range 16.1–77.8 μm; mean 45 ± 15.8 μm, $p > 0.1$), but larger than predicted from published qualitative descriptions based on paraffin-embedded tissues, which show significant tissue shrinkage (e.g. Mannen *et al.*, 1977).
3. Neurones in control spinal cords in the superior region of Onuf's nucleus at S2 exhibited clearly identifiable presynaptic terminals which, on the basis of differences in relative size, synaptic vesicle shape, and pre- and postsynaptic ultrastructure, conformed to the five classes of terminal previously identified as associated with cat thoracic and sphincteric motor neurones (i.e. S, F, T, M, and C; Fig. 40.6). No terminals on human control motor neurones appeared similar to those associated with cat preganglionic parasympathetic neurones.
4. All five morphological classes of presynaptic terminal synapsed with motor neurones in Onuf's nucleus in spinal cords from MND patients, suggesting there was no selective absence of any class.

General conclusions from animal and human studies

The identification of recognizable motor neurones bearing classifiable synaptic terminals in the MND patients indicates that at least some (if not all) sphincteric motor neurones survive this disease. The possibility that a proportion do die cannot be totally excluded, since judgements on preservation are mostly qualitative without verification by differential neuronal counts. Recent counts of neuronal numbers in serial sections through entire Onuf's nuclei obtained post-mortem from nine non-neurological controls provide a

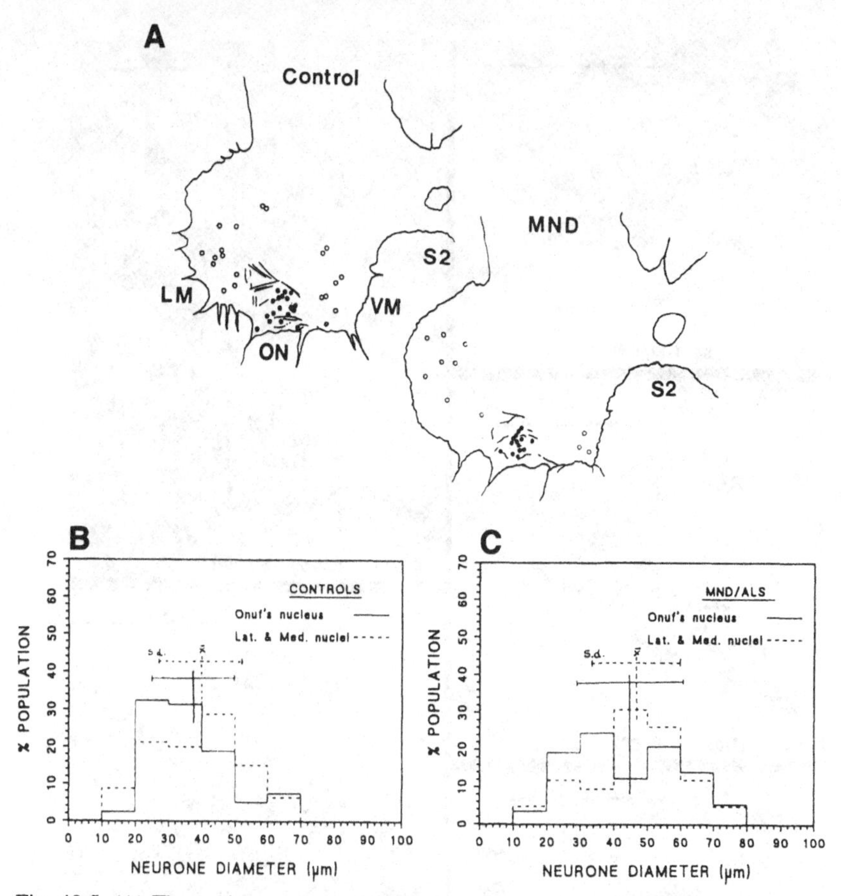

Fig. 40.5. (A) The positions of neurones in the major nuclei are compared in controls and MND patients. In each group of subjects, neuronal positions found in five random sections through S2 have been superimposed onto a common 'map'. Relative to controls, fewer neurones were recorded in LM and VM nuclei of MND subjects, but those in ON were as numerous as in control subjects. (B) and (C) The relative diameters of neurones in ON, VM and LM nuclei are compared in control and MND subjects respectively.

baseline for further investigations of Onuf's nucleus in MND (Tucker, Pullen & Martin, 1994). The common identification of S, F, T and, in particular, C-terminals on sphincteric motor neurones in cat and human suggests all mammalian sphincteric motor neurones are probably somatic. For MND this implies, firstly, that hypotheses relating the survival of sphincteric motor neurones in MND to an intrinsic autonomic property are unfounded and, secondly, that factors other than identity are responsible for their sparing. In MND, the oculomotor nucleus is also spared. Whether 'C'-type synapses characterize human oculomotor neurones is unknown, although in view of these studies of Onuf's nucleus, and the identification of 'C'-synapses in cat

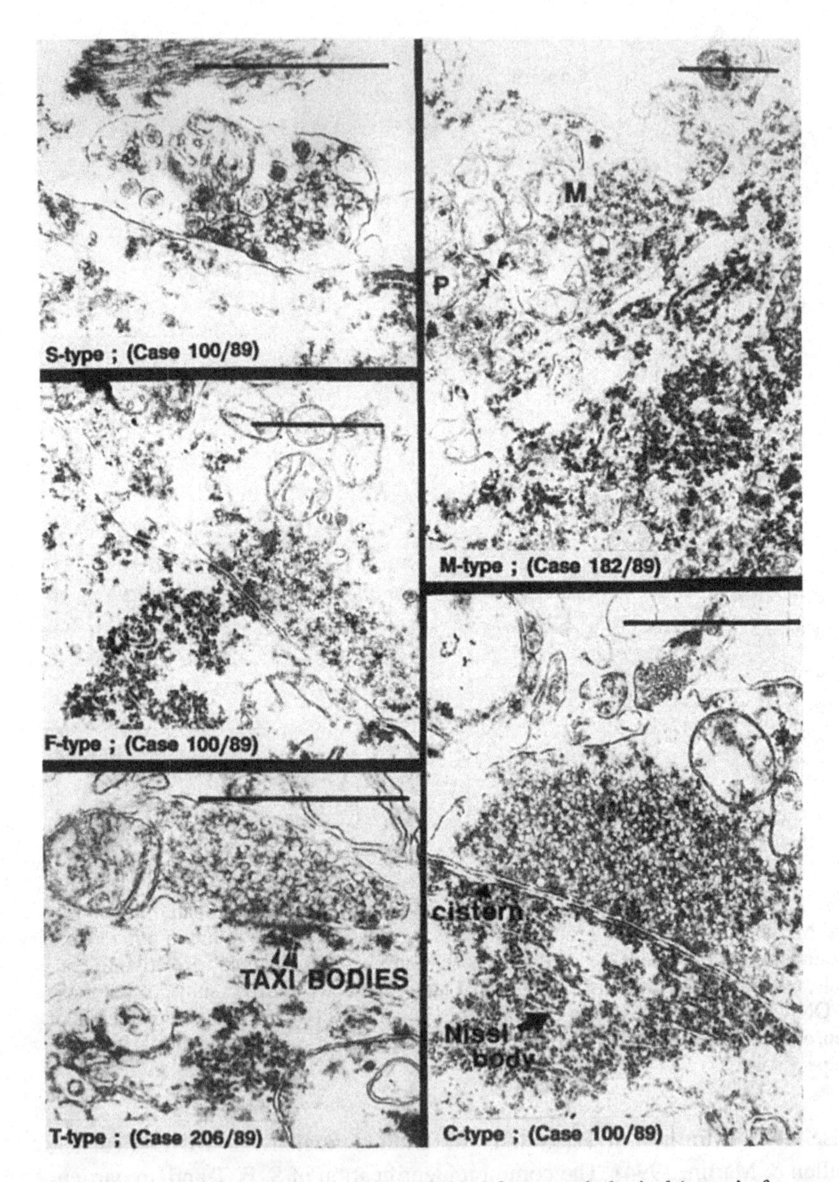

Fig. 40.6. Ultrastructural examinations revealed five morphological 'types' of presynaptic terminal synapsing with human sphincteric motoneurones which were identical to those found in cat (see Fig. 40.3), including the C-type synapse. Scale bars in each micrograph represent 1 μm.

oculomotor nuclei by Tredici, Pizzini & Milanesi (1976), it would be reasonable to predict their occurrence in the human.

Current thinking on pathogenic mechanisms of sporadic MND include excitotoxicity (see Krieger, this volume), disordered calcium and free radical metabolism, and a descending cortico-spinal degenerative process (Eisen, Kim & Pant, 1992). Within this context it is possible that the mechanism of sparing of sphincteric and oculomotor nuclei resides in the neurone. For example a unique combination of excitatory and inhibitory membrane receptors could limit the degree of neuronal excitation and thereby the possibility of excitotoxicity. Immunocytochemical evidence from the cat and the human, of differences between neighbouring spincteric motoneurones in expression of the enzyme nitric oxide synthase (Pullen & Humphreys, 1995), indicates possible differences between motoneurones within the nucleus to regulate free radicals. It is therefore also possible that the relative 'vunerability' of a motor nucleus to disease is determined by the particular balance between those motoneurones maintaining good control of free-radical metabolism and those motoneurones in which control is poor. Validation of this idea, however, must await future investigation. The unique synaptic relationships between Onuf's nucleus and the associated sacral autonomic motor nuclei and the mid-brain nuclei involved in somato-vesical reflexes (Mackel, 1979; McMahon, Morrison & Spillane, 1982; Holstege & Tan, 1987) also need to be considered.

The original tenet that Onuf's nucleus is spared in MND is founded on the frequent finding of preservation of both sphincter function and neuronal number. Despite the presence of both these features, Pullen & Martin (1995) demonstrated a loss of normal Nissl body and Golgi complex ultrastructure in most sphincteric motoneurones in 6 cases of MND (disease duration 8 months to 19 years) and they identified Bunina and filamentous skein-like cytoskeletal inclusions, both 'hall-marks' of MND, in some sphincteric motoneurones as early as 8 months after presentation. These findings indicate firstly that lack of histopathology does not preclude development of ultrastructural cytopathology, and secondly, that some motoneurones in Onuf's nucleus do not entirely resist MND. How many motoneurones need to degenerate before a deficit of sphincter tone can be detected remains to be determined.

Acknowledgement

The support of the Motor Neurone Disease Association (UK) and the Wellcome Trust is gratefully acknowledged.

References

Bishop, B., Garry, R.C., Roberts, T.D.M. & Todd, J.K. (1956). Control of the external sphincter of the anus in the cat. *Journal of Physiology (London)*, **134**, 229–240.

Boone, T.B. & Aldes, I.D. (1984). Synaptology of the hypoglossal nucleus of the rat. *Experimental Brain Research*, **57**, 22–32.

Conradi, S. (1969a). Ultrastructure and distribution of neuronal and glial elements on the motoneurone surface in the lumbosacral spinal cord of the adult cat. *Acta Physiologica Scandanavica, Supplement*, **332**, 5–48.

Conradi, S. (1969b). Ultrastructure of dorsal root boutons on lumbosacral motoneurones of adult cat revealed by dorsal root section. *Acta Physiologica Scandanavica, Supplement*, **332**, 85–111.

DeGroat, W.C., Nadelhaft, I., Milne, R.J., Booth, C.M. & Thor, K. (1981). Organisation of the sacral parasympathetic reflex pathways to the urinary bladder and large intestine. *Journal of the Autonomic Nervous System*, **3**, 133–142.

DeGroat, W.C. & Steers, W.D. (1990). Autonomic regulation of the urinary bladder and sexual organs. In *Central Regulation of Autonomic Function*, ed. A.D. Loewy & K.M. Spyer, pp. 310–333. Oxford: Oxford University Press.

Eisen, A., Kim, S. & Pant, B (1992). Amyotrophic lateral sclerosis: a phylogenetic disease of the corticomotoneuron. *Muscle and Nerve*, **15**, 219–228.

Gibson, S.L., Polak, J.M., Katagiri, T., Su, H., Weller, R.O., Brownell, D.B., Holland, S., Hughes, J.T., Kikuyama, S., Ball, J., Bloom, S.R. & Clifford-Rose, F. (1988). A comparison of the distribution of eight peptides in spinal cord from normal controls and cases of motor neurone disease with special reference to Onuf's nucleus. *Brain Research*, **474**, 255–278.

Glazer, E.J. & Basbaum, A.I. (1980). Leucine-enkephalin: localisation in and axoplasmic transport by sacral parasympathetic preganglionic neurons. *Science*, **208**, 1479–1480

Goshgarian, H.G. & Rafols, J.A. (1984). The ultrastructure and synaptic architecture of phrenic motor neurones in the spinal cord of the adult rat. *Journal of Neurocytology*, **13**, 85–109.

Hamos, J.E. & King, J.S. (1980). The synaptic organisation of the trigeminal nerve in the opossum. *Journal of Comparative Neurology*, **194**, 441–463.

Holstege, G., Griffiths, D., Wall, H.D. & Dalm, E. (1986). Anatomical and physiological observations on supraspinal control of bladder and urethral sphincter muscles in the cat. *Journal of Comparative Neurology*, **250**, 449–461.

Holstege, G. & Tan, J. (1987). Supraspinal control of motoneurons innervating the striated muscles of the pelvic floor including urethral and anal sphincters in the cat. *Brain*, **110**, 1323–1344.

Johnson, I.P. (1986). A quantitative ultrastructural comparison of alpha and gamma motoneurones in the thoracic region of the spinal cord of the adult cat. *Journal of Anatomy*, **147**, 55–72.

Katagiri, T., Gibson, S.J., Su, H.C. & Polak, J.M. (1986). Composition and central projections of the pudendal nerve in the rat investigated by combined peptide immunocytochemistry and retrograde fluorescent labelling. *Brain Research*, **372**, 313–322.

Kawatani, M., Nagel, J. & DeGroat, W.C. (1986). Identification of neuropeptides in pelvic and pudendal afferent pathways to the sacral spinal cord of the cat. *Journal of Comparative Neurology*, **249**, 117–132.

Kawatani, M., Shioda, S., Nakai, Y., Takeshige, C. & deGroat, W.C. (1989). Ultrastructural analysis of enkephalinergic terminals in parasympathetic ganglia innervating the urinary bladder of the cat. *Journal of Comparative Neurology*, **288**, 81–91.

Konno, H., Yamamoto, T., Iwasaki, Y. & Iizuka, H. (1986). Shy–Drager syndrome and amyotrophic lateral sclerosis: cytoarchitectonic and morphometric studies of sacral autonomic neurones. *Journal of the Neurological Sciences*, **73**, 193–204.

Kuzuhara, S., Kanazawa, I. & Nakanishi, T. (1980). Topographical localisation of the Onuf 's nuclear neurones innervating the rectal and vesicle striated sphincter muscles: a retrograde fluorescent double labelling study in cat and dog. *Neuroscience Letters*, **16**, 125–130.

Lagerback, P.A. (1985). An ultrastructural study of cat lumbosacral gamma motoneurones after retrograde labelling with horseradish peroxidase. *Journal of Comparative Neurology*, **240**, 256–264.

Lagerback, P.A. & Ronnevi, L.O. (1982). An ultrastructural study of serially sectioned Renshaw cells. II. Synaptic types. *Brain Research*, **246**, 181–192.

Leedy, M.G., Bresnahan, J.C., Mawe, G.M. & Beattie, M.S. (1988). Differences in synaptic inputs to preganglionic neurons in the dorsal and lateral band subdivisions of the cat sacral parasympathetic nucleus. *Journal of Comparative Neurology*, **268**, 84–90.

Mackel, R. (1979). Segmental and descending control of the external urethral and anal sphincters in the cat. *Journal of Physiology (London)*, **294**, 105–122.

Mannen, T., Iwata, M., Toyokura, Y. & Nagashima, K. (1977). Preservation of a certain motoneurone group of the sacral cord in amyotrophic lateral sclerosis: its clinical significance. *Journal of Neurology, Neurosurgery and Psychiatry*, **40**, 464–469.

Mannen, T., Iwata, M., Toyokura, Y. & Nagashima, K. (1982). The Onuf 's nucleus and the external anal sphincter muscles in amyotrophic lateral sclerosis and Shy–Drager syndrome. *Acta Neuropathologica (Berlin)*, **58**, 255–260.

Matsushita, M. & Ikeda, M. (1973). Propriospinal fiber connections of the cervical motor nuclei in the cat: a light and electron microscope study. *Journal of Comparative Neurology*, **150**, 1–32.

Mawe, G.M., Bresnahan, J. & Beattie, M.S. (1986). A light and electron microscope analysis of the sacral parasympathetic nucleus after labelling primary afferent and efferent elements with HRP. *Journal of Comparative Neurology*, **250**, 33–57.

McKenna, K.E. & Nadelhaft, I. (1986). The organisation of the pudendal nerve in the male and female rat. *Journal of Comparative Neurology*, **248**, 532–549.

McLaughlin, B.J. (1972a). Dorsal root projections to the motor nuclei in the cat spinal cord. *Journal of Comparative Neurology*, **144**, 461–474.

McLaughlin, B.J. (1972b). Propriospinal and supraspinal projections to the motor nuclei in the cat spinal cord. *Journal of Comparative Neurology*, **144**, 474–500.

McMahon, S.B., Morrison, J.F.B. & Spillane, K. (1982). An electrophysiological study of somatic and visceral convergence in the reflex control of the external sphincters. *Journal of Physiology (London)*, **328**, 379–387.

Nagashima, T., Beppu, M., Uono, M. & Yamada, H. (1979). Demonstration of neuronal localisation of Onufrowicz's cell group X in rabbit by double labelling method. *Acta Histochemica Cytochemica*, **12**, 369–391.

Onufrowicz, B. (1889). Notes on the arrangement and function of the cell groups of the sacral region of the spinal cord. *Journal of Nervous and Mental Disorders*, **26**, 498–504.

Onuf [Onufrowicz], B. (1890). On the arrangement and function of the cell groups of the sacral region of the spinal cord in man. *Archives of Neurology and Psychopathology*, **3**, 387–411.

Pullen, A.H. (1988a). Feline 'C'-type terminals possess synaptic sites associated with a hypolemmal cistern and Nissl body. *Neuroscience Letters*, **84**, 143–148.

Pullen, A.H. (1988b). Quantitative synaptology of feline motoneurones to external anal sphincter muscle. *Journal of Comparative Neurology*, 269, 414–424.

Pullen, A.H. (1990). Morphometric evidence from 'C'-synapses for a phased Nissl body response in alpha-motoneurones retrogradely intoxicated with diphtheria toxin. *Brain Research*, **509**, 8–16.

Pullen, A.M. & Humphreys, P. (1995). Diversity in localisation of nitric oxide synthase antigen and NADPH-diaphorase histochemical staining in sacral somatic motor nuclei of the cat. *Neuroscience Letters*, **196**, 33–36.

Pullen, A.H. & Martin, J.E. (1995). Ultrastructural abnormalities with inclusions in Onuf's nucleus in motoneurone disease (amyotrophic lateral sclerosis). *Neuropathology and Applied Neurobiology*, **21**, 327–340.

Pullen, A.H., Martin, J.E. & Swash, M. (1992). Ultrastructure of presynaptic input to motor neurones in Onuf's nucleus: controls and motor neurone disease. *Neuropathology and Applied Neurobiology*, **18**, 213–231.

Pullen, A.H. & Sears, T.A. (1978). Modification of 'C' synapses following partial central deafferentation of thoracic motoneurones. *Brain Research*, **145**, 141–146.

Pullen, A.H. & Sears, T.A. (1983). Trophism between 'C'-type axon terminals and thoracic motoneurones in the cat. *Journal of Physiology (London)*, **337**, 373–388.

Rexed, B.A. (1954). A cytoarchitectonic atlas of the spinal cord in the cat. *Journal of Comparative Neurology*, **100**, 297–379.

Roppolo, J.R., Nadelhaft, I. & DeGroat, W.C. (1985). The organisation of pudendal motoneurones and primary afferent projections in the spinal cord of the rhesus monkey revealed by horseradish peroxidase. *Journal of Comparative Neurology*, **234**, 475–488.

Sato, M., Mizuno, N. & Konishi, A. (1978). Localisation of motoneurones innervating peroneal muscles: a HRP study in cat. *Brain Research*, **140**, 149–154.

Schroder, H.D. (1980). Organisation of the motoneurons innervating pelvic muscles of the male rat. *Journal of Comparative Neurology*, **192**, 567–587.

Schroder, H.D. (1981). Onuf's nucleus X: a morphological study of a human spinal nucleus. *Anatomica et Embryologica (Berlin)*, **162**, 443–453.

Schroder, H.D. (1984). Somatostatin in the caudal spinal cord: an immunohistochemical study of the spinal centers involved in the innervation of the pelvic organs. *Journal of Comparative Neurology*, **223**, 400–414.

Shy, G.M. & Drager, G.A. (1960). A neurological syndrome associated with orthostatic hypotension: a clinical-pathologic study. *Archives of Neurology*, **2**, 511–527.

Sung, J.H., Mastri, A.E. & Segal, E. (1979). Pathology of Shy–Drager syndrome. *Journal of Neuropathology and Experimental Neurology*, **38**, 353–368.

Swash, M. & Schwartz, M.S. (1992). What do we really know about amyotrophic lateral sclerosis? *Journal of the Neurological Sciences*, **113**, 4–16.

Tashiro, T., Sadota, T., Matsushima, R. & Mizuno, N. (1989). Convergence of serotonin-, enkephalin- and substance P-like immunoreactive afferent fibres on single pudendal motoneurones in Onuf's nucleus of the cat: a light microscope study combining the triple immunocytochemical staining technique with the retrograde HRP-tracing method. *Brain Research*, **481**, 392–398.

Tokoyura, Y. (1977). Amyotrophic lateral sclerosis: a clinical and pathological study of the 'negative features' of the disease. *Japanese Journal of Medicine*, **16**, 269–273.

Tredici, G., Pizzini, G. & Milanesi, S. (1976). The ultrastructure of the nucleus of the oculomotor nerve (somatic efferent portion) of the cat. *Anatomica et Embryologica*, **149**, 323–346.

Tucker, D., Pullen, A.H. & Martin, J.E. (1994). A morphological and quantitative study of Onuf's nucleus in the human spinal cord. *Journal of Pathology*, **172** (Supplement), p145A.

41

Excitotoxicity in motor neurone diseases

CHARLES KRIEGER

Division of Neurology, Department of Medicine, University of British Columbia, Vancouver, BC, Canada

Introduction

There is considerable evidence that neuronal death will result following the exposure of neurones or central nervous system (CNS) tissue to excitatory amino acids (EAA) either *in vivo* or *in vitro* (Choi, 1988). This toxicity appears to be mediated by overstimulation of neurones through activation of EAA receptors ('excitotoxicity') (Choi, 1988; Rothman, 1992). Several EAA such as glutamate and aspartate are candidate neurotransmitters at synapses of the corticospinal tracts, as well as at other synapses in the CNS (Young *et al.*, 1983). These observations have been central to hypotheses suggesting that the neurodegenerative disorder amyotrophic lateral sclerosis (ALS), also known as motor neurone disease, could be associated with impaired glutamate function or excitotoxicity. ALS is a progressive disorder of unknown cause affecting motoneurones and descending spinal cord pathways.

At least five independent lines of evidence have been presented to suggest that 'glutamatergic dysfunction' or excitotoxicity may in part be responsible for the development of ALS. These lines of evidence include:

1. The association between exposure to known excitotoxins and the development of motoneuronopathies in humans and primates.
2. Observations that elevated concentrations of glutamate are present in the plasma and/or cerebrospinal fluid (CSF) of patients who have died with ALS.
3. Observations of reduced contents of glutamate and aspartate in spinal cords and brains of patients who have died with ALS.
4. Evidence of decreased glutamate uptake by synaptosomes obtained from spinal cords and brains of ALS patients.
5. Autoradiographic evidence for altered numbers of EAA receptors in ALS patients and in a murine model of ALS.

In addition, a recent study has demonstrated that the gene encoding the receptor subunit GluR5 is found in the vicinity of the gene for familial ALS, raising the possibility that altered EAA receptors might be present in ALS patients (Eubanks *et al.*, 1993).

Intoxication with excitatory amino acids can lead to motoneurone dysfunction in humans

One of the best-described associations between intoxication with EAA and neurological sequelae has been the recent delineation of a clinicopathological syndrome arising after the consumption of mussels contaminated with domoic acid. Domoic acid is a neurotoxin with receptor binding properties similar to those of kainic acid. Ingestion of this EAA in humans produces headaches, confusion, seizures, myoclonus, loss of reflexes and coma, in addition to other systemic features (Teitlebaum *et al.*, 1990). Generalized weakness, hyperreflexia and extensor plantar responses can also be present, as can fasciculations and distal atrophy. Neuropathological examination of patients who died following intoxication have demonstrated neuronal loss, predominantly in the hippocampus and amygdala, in a distribution similar to that seen in experimental studies in animals involving kainate administration. Electromyographic evaluation of affected patients revealed spontaneous activity suggestive of acute denervation, which would sometimes last for at least 1 year after the intoxication. These findings were interpreted as being consistent with an acute, non-progressive neuronopathy involving anterior horn cells, or a motor neuropathy. No lesions were found within the motor nuclei of the brainstem or spinal cord in the one patient who was studied at autopsy (Teitlebaum *et al.*, 1990).

A relation between exposure to neurotoxins and motor system degeneration has also been suggested to occur in lathyrism (Spencer *et al.*, 1991). Lathyrism is a neurodegenerative disease affecting the motor system which has been associated with the ingestion of chickling peas (*Lathyrus sativus*) containing the neurotoxin β-*N*-oxalylamino-L-alanine (BOAA). A spastic paraplegia develops in humans after consumption of *Lathyrus* and feeding cynomolgus monkeys a diet of *Lathyrus sativus* produces a spastic paralysis similar to human lathyrism. The BOAA-activated conductance responsible for the toxicity is largely voltage-independent and resistant to *N*-methyl-D-aspartate (NMDA) antagonists (MacDonald & Morris, 1984). An interesting clinical observation is that after years of functional stability, some patients with lathyrism develop increasing deficit due to loss of lower motoneurone function. This clinical picture of progressive, lower motoneurone deficit resembles the progression of ALS.

Although a putative excitatory neurotoxin has been hypothesized to be involved in the Guamanian form of ALS (G-ALS), the neurotoxin has not been identified. The candidate neurotoxin, beta-*N*-methylamino-L-alanine (BMAA) (Spencer *et al.*, 1987), has not withstood critical scrutiny for involvement in G-ALS and it has not been detected in the plasma or CSF of North American ALS patients (Perry *et al.*, 1990).

These clinical observations of motor system dysfunction following intoxication with EAA or excitotoxins indicate that EAA are at least possible candidates in the pathophysiology of ALS. Studies in animals have demonstrated that EAA administration can induce degenerative neuronal changes in the CNS (Whetsell & Shapira, 1993). For example, intrathecal administration of EAA such as NMDA and kainate produce neuronal death in the spinal cord, where NMDA effects are found in both the dorsal and ventral grey matter whereas kainate effects predominate in the ventral horns of the spinal grey matter (Curtis & Malik, 1985; Hugon *et al.*, 1989b; Nag & Riopelle, 1990). In other studies, however, intrathecal administration of quisqualate (QUIS/AMPA) and kainate, but not NMDA, produced damage to spinal cord neurones (Urca & Urca, 1990).

Elevated concentrations of excitatory amino acids in the plasma and CSF of ALS patients

In ALS patients, abnormally high fasting plasma glutamate levels have been detected (Plaitakis & Caroscio, 1987). The source of the increased plasma glutamate is not clear, although an abnormality in the catabolism of glutamate has been postulated. Giving oral loads of glutamate to ALS patients produced slower declines in glutamate levels than in controls or patients with other neurological diseases (Plaitakis, 1990). These elevated glutamate levels do not correlate with the age of the ALS patients studied, or to disease severity (Plaitakis & Constantakakis, 1993). Other studies of plasma glutamate concentrations in ALS have not found abnormalities in the level of glutamate, aspartate or glycine (Cottell *et al.*, 1990; Perry *et al.*, 1990). These discrepancies have in some cases been attributed to patient selection (Plaitakis & Constantakakis, 1993). Defects in leucocyte glutamate dehydrogenase (GDH) activity have been observed by some investigators, possibly indicating a defect in the catabolism of glutamate (Hugon *et al.*, 1989a). However, these findings have not been detected by other workers (Malessa *et al.*, 1991).

In patients with sporadic ALS, concentrations of glutamate, aspartate, *N*-acetylaspartate and *N*-acetylaspartylglutamate have been reported to be increased in CSF (Rothstein *et al.*, 1990; Rothstein *et al.*, 1991). Other studies have been unable to replicate some of these findings (Perry *et al.*, 1990). Attention has been drawn to the difficulties inherent in the determination of CSF EAA levels, including the use of high-performance liquid chromatography and careful sample handling (Spink & Martin, 1991). Plasma cysteine concentrations have also been reported to be increased, with impaired sulphoxidation of toxic thiols (Heafield *et al.*, 1990); however, these observations can not be confirmed (Perry *et al.*, 1991a, b).

There are grounds, however, for believing that even if elevation of plasma or CSF glutamate occurs, it may not be directly linked to the pathogenesis of ALS. Firstly, glutamate is poorly transported across the blood–brain barrier, and high plasma glutamate concentrations need not be associated with high brain levels. Secondly, efficient glutamate re-uptake mechanisms are present in brain, especially in synaptic terminals and glia which maintain extracellular glutamate concentrations (but see below). Thirdly, in the event that glutamate levels were elevated in brain and spinal cord, it might be expected that glutamate effects would be the most pronounced in the hippocampal formation and orbitofrontal cortex where glutamate binding is considerably greater than in the ventral horn of spinal cord. These sites are rarely damaged in ALS. To explain the selectivity of motoneurone loss in ALS, Plaitakis has hypothesized that glycinergic input may potentiate NMDA-induced excitotoxic damage in the spinal cord (Plaitakis, 1990).

Reduced contents of amino acids in spinal cord and brain in ALS

Studies of autopsied brains and spinal cords from patients who have died with the sporadic form of ALS have demonstrated decreased glutamate contents in many brain areas, as well as in the cervical and lumbar regions of spinal cord (Perry *et al.*, 1987; Plaitakis *et al.*, 1988; Malessa *et al.*, 1991; Tsai *et al.*, 1991) In sporadic ALS, the glutamate content of cervical spinal cord is only about half the normal level and impressive reductions in glutamate contents are also detected in brain (Perry *et al.*, 1987). The biochemical changes observed in the sporadic form of ALS are dissimilar from those detected in the Guamanian form of ALS, suggesting that these two disorders may have different pathogeneses. Contents of aspartate are also reduced in the cervical and lumbar spinal cord in ALS (Plaitakis *et al.*, 1988) and taurine contents are increased in some brain and spinal cord regions (Perry *et al.*, 1987).

Studies of amino acid contents in the brains and spinal cords of *wobbler* mice, an animal model of ALS, have demonstrated slightly decreased contents of glutamate, aspartate and glycine in spinal cord, compared with normal control littermates (Krieger *et al.*, 1991). However, the abnormalities in the amino acid content of the brains of wobbler mice were dissimilar to those observed in the brains of ALS patients.

These remarkably consistent observations of changes in amino acid contents in ALS are compatible with the hypothesis that a reduction of these amino acids results from excesssive synaptic release of EAA, possibly leading to neuroexcitotoxic damage to motoneurones (Plaitakis *et al.*, 1988). However, the reductions in the contents of glutamate and aspartate might reflect only the loss of motoneurones and spinal cord neurones from any cause. In cat and rabbit spinal cord, the contents of glutamate, aspartate and glycine are high in the ventral horn (Graham *et al.*, 1967), and in experiments where the numbers of motoneurones and interneurones were diminished by aortic occlusion, decreased aspartate and glutamate contents were observed (Homma *et al.*, 1979). Alternatively, a loss of descending tracts might also account for some of these changes in amino acid contents. Spinal cord glutamate contents in rats are significantly reduced following intercollicular brainstem transection (D'Arcangelo & Brancati, 1990). Potentially, a loss of corticospinal tract axons from any primary cause could also produce similar effects.

Impaired glutamate uptake in CNS tissue from ALS patients

Defects in high-affinity sodium-dependent glutamate transport have been detected in synaptosomes from CNS tissue obtained from patients with ALS, compared with controls (Rothstein *et al.*, 1992). In ALS patients, the maximal velocity of transport for glutamate is reduced in synaptosomes from spinal cord, motor and somatosensory cortex, but not in other brain regions which are relatively uninvolved in the disorder. The affinity of the transporter for glutamate was unaffected. Potentially, defective transport of glutamate could lead to elevated extracellular concentrations of glutamate or other EAA and produce excitotoxicity (Rothstein *et al.*, 1992). However, this mechanism does not account for the selective loss of motoneurones found in ALS. It is also possible that the changes in transport velocity are secondary to cell loss rather than primary and may reflect the decreased transport abilities of the surviving cells.

Decreased numbers of NMDA receptors in ALS

The distribution of glutamatergic ligand binding sites in the human spinal cord has been examined using quantitative receptor autoradiography (Allaoua *et al.*, 1992; Jansen *et al.*, 1990; Shaw *et al.*, 1991). These studies have demonstrated that [^{3}H]glutamate, [^{3}H]AMPA and [^{3}H]kainate receptor binding are concentrated in lamina II, with lower levels in the other laminae of the spinal cord (Allaoua *et al.*, 1992; Jansen *et al.*, 1990). NMDA receptor binding sites are distributed widely in both the ventral and dorsal horns and have the highest binding densities in lamina II (Allaoua *et al.*, 1992; Jansen *et al.*, 1990; Shaw *et al.*, 1991). In the ventral horns multiple foci of high [^{3}H]MK-801/NMDA receptor binding sites have been observed in the lumbosacral segments of human spinal cord and these foci probably correspond to motoneurones (Shaw *et al.*, 1991). These receptor binding studies have been complemented by electrophysiological studies in rats which have demonstrated that mammalian motoneurones possess both NMDA and non-NMDA receptors; however, the precise distribution of the receptors on the motoneurone has not been determined (Konnerth *et al.*, 1990).

An autoradiographic study of EAA receptor binding in spinal cords from patients who died having ALS has demonstrated reductions in [^{3}H]TCP receptor binding in both the ventral and dorsal horns, with unchanged AMPA binding (Allaoua *et al.*, 1992). TCP is an NMDA receptor ion channel antagonist. These observations suggest that ALS might be associated with a selective loss of [^{3}H]TCP/NMDA receptors. Motoneurones possess both NMDA and non-NMDA receptors and activation of either receptor subtype might mediate excitotoxic injury. However, the selective reduction in NMDA receptors with sparing of AMPA receptors could occur as a consequence of increased amounts of a circulating agonist of NMDA receptors in the CNS of ALS patients. This agonist would be directed to NMDA-receptor-bearing motoneurones and other spinal cord cells and could produce receptor and motoneurone losses by excessive NMDA receptor stimulation. An alternative explanation is that there is a down regulation in NMDA receptor number in ALS, possibly due to the continued presence of increased amounts of an NMDA-like agonist in various spinal cord regions (Allaoua *et al.*, 1992). This possibility was supported by the observation that NMDA receptor losses were also observed in the dorsal horn. Since afferent pathways are usually spared in ALS, both clinically and pathologically, it appears likely that receptor regulation is abnormal in ALS (Allaoua *et al.*, 1992).

Autoradiographic studies of glutamatergic ligand binding sites in the spinal cord of wobbler mice have also demonstrated alterations in NMDA receptor binding (Krieger *et al.*, 1993a). In severely affected mice with clinical and morphological evidence of extensive motoneurone loss, significantly decreased binding of [^{3}H]MK-801/NMDA was found in both the dorsal and ventral horns. [^{3}H]Kainate receptor binding was significantly increased in both the dorsal and ventral horns with [^{3}H]CNQX, an AMPA receptor ligand, being unchanged. The finding of selective reductions of NMDA receptor binding is similar to the autoradiographic findings in ALS patients (Krieger *et al.*, 1993a).

An additional study evaluating NMDA binding sites in human spinal cord by quantitative autoradiography using [^{3}H]MK-801, an NMDA receptor ion channel antagonist, has confirmed some of these results and demonstrated that the binding of [^{3}H]MK-801 to NMDA receptors was reduced by 40–45% in the dorsal and ventral horns of spinal cords from patients who died with ALS, compared with controls (Krieger *et al.*, 1993c). No significant differences in affinity were observed between spinal cords from ALS patients or controls. Once again, reductions in [^{3}H]MK-801 binding were seen in both the dorsal and ventral horns of spinal cord, further supporting the possibility that NMDA receptors were abnormally regulated and were decreased independently of motoneurone loss. Further substantiation for abnormal regulation of NMDA receptors in ALS was reported recently (Krieger *et al.*, 1993b). Spinal cord sections from autopsy material were exposed to phorbol ester before incubation with [^{3}H]MK-801 to determine levels of NMDA binding. Phorbol ester treatment increased [^{3}H]MK-801 binding in both ALS and control tissue to almost identical levels of specific binding for both groups. The increased [^{3}H]MK-801 binding could be completely blocked by concurrent exposure of spinal cord sections to H-7, a general protein kinase inhibitor (Krieger *et al.*, 1993b). These results suggest that NMDA receptors in ALS spinal cord are decreased as a result of abnormal protein kinase/phosphatase enzyme activity independently of motoneurone degeneration.

Conclusions

It is not clear whether the observed abnormalities described above are a cause or a consequence of the neurone loss which occurs in ALS. For instance, the autoradiographic studies described above have suggested that changes in NMDA receptor number in ALS may reflect abnormal regulation of protein kinase activity. It is possible that abnormal protein kinase/phosphatase regulation may be an underlying process resulting in changes to receptor distributions in ALS and that changes in EAA are secondary to this dysregulatory process. Potentially, an abnormality in protein kinase/phosphatase regulation might be responsible for the impairments in motoneurone activities which are altered in ALS.

Acknowledgement

It is a pleasure to thank Professor Tom Sears for the opportunity to work in his laboratory during my graduate studies. I also wish to thank my colleagues at UBC: Drs Chris Shaw, Seung Kim, Shirley Hansen, Andy Eisen and the late T.L. Perry, Sr. This work was supported by grants from the British Columbia Health Research Foundation, the Canadian MRC and the University Hospital Foundation.

References

Allaoua, H., Chaudieu, I., Krieger, C., Boksa, P., Privat, A. & Quirion, R. (1992). Alterations in spinal cord excitatory amino acid receptors in amyotrophic lateral sclerosis patients. *Brain Research*, **579**, 169–172.

Choi, D.W. (1988). Glutamate neurotoxicity and diseases of the nervous system. *Neuron*, **1**, 623–634.

Cottell, E., Hutchinson, M., Simon, J. & Harrington, M.G. (1990). Plasma glutamate levels in normal subjects and in patients with amyotrophic lateral sclerosis. *Biochemical Society Transactions*, **18**, 283.

Curtis, D.R. & Malik, J. (1985). A neurophysiological analysis of the effect of kainic acid on nerves fibres and terminals in the cat spinal cord. *Journal of Physiology (London)*, **368**, 99–108.

D'Arcangelo, P. & Brancati, A. (1990). Distribution of *N*-acetylaspartate, *N*-acetylaspartylglutamate, free glutamate and aspartate following complete mesencephalic transection in rat neuraxis. *Neuroscience Letters*, **114**, 82–88.

Eubanks, J.H., Puranam, R.S., Kleckner, N.W., Bettler, B., Heinemann, S.F. & McNamara, J.O. (1993). The gene encoding the glutamate receptor subunit GluR5 is located on human chromosome 21q21.1–22 in the vicinity of the gene for familial amyotrophic lateral sclerosis. *Proceedings of the National Academy of Sciences, USA* , **90**, 178–182.

Graham, L.T. Jr, Shank, R.P., Werman, R. & Aprison, M.H. (1967). Distribution of some synaptic transmitter suspects in cat spinal cord. *Journal of Neurochemistry*, **14**, 465–472.

Heafield, M.T., Fearn, S., Steventon, R.H., Waring, R.M., Williams, A.C. & Sturman, S.G. (1990). Plasma cysteine and sulphate levels in patients with motor neurone, Parkinson's and Alzheimer's disease. *Neuroscience Letters*, **110**, 216–220.

Homma, S., Suzuki, T., Murayama, S. & Otsuka, M. (1979). Amino acid and substance P contents in spinal cord of cats with experimental hind-limb rigidity produced by occlusion of spinal cord blood supply. *Journal of Neurochemistry*, **32**, 691–698.

Hugon, J., Tabaraud, F., Rigaud M., Vallat, J.M. & Dumas M. (1989a). Glutamate dehydrogenase and aspartate aminotransferase in leukocytes of patients with motor neuron disease. *Neurology*, **39**, 956–958.

Hugon, J., Vallat, J.M., Spencer, P.S., LeBoutet, M.J. & Barthe, D. (1989b). Kainic acid induces early and delayed degenerative neuronal changes in rat spinal cord. *Neuroscience Letters*, **104**, 258–262.

Jansen, K.L.R., Faull, R.L.M., Dragunow, M. & Waldvogel, H. (1990). Autoradiographic localisation of NMDA, quisqualate and kainate receptors in human spinal cord. *Neuroscience Letters*, **108**, 53–57.

Konnerth, A., Keller, B.U. & Lev-Tov, A. (1990). Patch clamp analysis of excitatory synapses in mammalian spinal cord. *Pflügers Archiv*, **417**, 285–290.

Krieger, C., Lai, R., Mitsumoto, H. & Shaw, C. (1993a). The wobbler mouse: quantitative autoradiography of glutamatergic ligand binding sites in spinal cord. *Neurodegeneration*, **2**, 9–17.

Krieger, C., Perry, T.L., Hansen, S. & Mitsumoto, H. (1991). The *wobbler* mouse: amino acid contents in brain and spinal cord. *Brain Research*, **551**, 142–144.

Krieger, C., Wagey, R., Lanius, R.A. & Shaw, C.A. (1993b). Activation of PKC reverses apparent NMDA receptor reduction in ALS. *Neuroreport*, **4**, 931–934.

Krieger, C., Nagey, R. & Shaw, C. (1993c). Amyotrophic lateral sclerosis: quantitative autoradiography of [^{3}H]MK-801/NMDA binding sites in spinal cord. *Neuroscience Letters*, **159**, 191–194.

MacDonald, J.F. & Morris, M.E. (1984). Mechanism of neuronal excitation by L-2-oxalylamino-3-amino- and L-3-oxalylamino-2-amino-propionic acid. *Experimental Brain Research*, **57**, 158–166.

Malessa, S., Leigh, P.N., Bertel, O., Sluga, E. & Hornykiewicz, O. (1991). Amyotrophic lateral sclerosis: glutamate dehydrogenase and transmitter amino acids in the spinal cord. *Journal of Neurology, Neurosurgery and Psychiatry*, **54**, 984–988.

Nag, S. & Riopelle, R.J. (1990). Spinal neuronal pathology associated with continuous intrathecal infusion of *N*-methyl-D-aspartate in the rat. *Acta Neuropathologica*, **81**, 7–13.

Perry, T.L., Hansen, S. & Jones, K. (1987). Brain glutamate deficiency in amyotrophic lateral sclerosis. *Neurology*, **37**, 1845–1848.

Perry, T.L., Krieger, C., Hansen, S. & Eisen, A. (1990). Amyotrophic lateral sclerosis: amino acid levels in plasma and cerebrospinal fluid. *Annals of Neurology*, **28**, 12–17.

Perry, T.L., Krieger, C., Hansen, S. & Tabatabaei, A. (1991a). Amyotrophic lateral

sclerosis: fasting plasma levels of cysteine and inorganic sulfate are normal, as are brain contents of cysteine. *Neurology*, **41**, 487–490.
Perry, T.L., Krieger, C., Hansen, S. & Tabatabaei, A. (1991b). Cystine, sulfate and ALS. *Neurology*, **41**, 1851–1852.
Plaitakis, A. (1990). Glutamate dysfunction and selective motor neuron degeneration in amyotrophic lateral sclerosis: a hypothesis. *Annals of Neurology*, **28**, 3–8.
Plaitakis, A. & Caroscio, J.T. (1987). Abnormal glutamate metabolism in amyotrophic lateral sclerosis. *Annals of Neurology*, **22**, 575–579.
Plaitakis, A. & Constantakakis, E. (1993). Altered metabolism of excitatory amino acids, *N*-acetyl-aspartate and *N*-acetyl-aspartyl-glutamate in amyotrophic lateral sclerosis. *Brain Research Bulletin*, **30**, 381–386.
Plaitakis, A., Constantakakis, E. & Smith, J. (1988). The neuroexcitotoxic amino acids glutamate and aspartate are altered in the spinal cord and brain in amyotrophic lateral sclerosis. *Annals of Neurology*, **24**, 446–449.
Rothman, S.M. (1992). Excitotoxins: possible mechanisms of action. *Annals of the New York Academy of Sciences*, **648**, 132–138.
Rothstein, J.D., Kuncl, R., Chaudhry, V., Clawson, L., Cornblath, D.R., Coyle, J.T. & Drachman, D.B. (1991). Excitatory amino acids in amyotrophic lateral sclerosis: an update. *Annals of Neurology*, **30**, 224–225.
Rothstein, J.D., Martin, L.J. & Kuncl, R.W. (1992). Decreased glutamate transport by the brain and spinal cord in amyotrophic lateral sclerosis. *New England Journal of Medicine*, **326**, 1464–1468.
Rothstein, J.D., Tsai, G., Kuncl, R.W., Clawson, L., Cornblath, D.R., Drachman, D.B., Pestronk, A., Stauch, B.L. & Coyle, J.T. (1990). Abnormal excitatory amino acid metabolism in amyotrophic lateral sclerosis. *Annals of Neurology*, **28**, 18–25.
Shaw, P.J., Ince, P.G., Johnson, M., Perry, E.K. & Candy, J. (1991). The quantitative autoradiographic distribution of [^{3}H]MK-801 binding sites in the normal human spinal cord. *Brain Research*, **539**, 164–168.
Spencer, P.S., Allen, C.N., Kisby, G.E., Ludoph, A.C., Ross, S.M. & Roy, D.N. (1991). Lathyrism and western Pacific amyotrophic lateral sclerosis: etiology of short and long latency motor system disorders. *Advances in Neurology*, **56**, 287–299.
Spencer, P.S., Nunn, P.B., Hugon, J., Ludolph, A.C., Ross, S.M., Roy, D.N. & Robertson, R.C. (1987). Guam amyotrophic lateral sclerosis–Parkinsonism–dementia linked to a plant excitant neurotoxin. *Science*, **237**, 517–522.
Spink, D.C. & Martin, D.L. (1991). Excitatory amino acids in amyotrophic lateral sclerosis. *Annals of Neurology*, **29**, 110.
Teitlebaum, J.S., Zatorre, R., Carpenter, S., Gendron, D., Evans, A.C., Gjedde, A. & Cashman, N.R. (1990). Neurologic sequelae of domoic acid intoxication due to the ingestion of contaminated mussels. *New England Journal of Medicine*, **322**, 1781–1787.
Tsai, G.C., Stauch-Slusher, B., Sim, L., Hedreen, J.C., Rothstein, J.D., Kuncl, R. & Coyle, J.T. (1991). Reduction in acidic amino and *N*-acetylaspartylglutamate in amyotrophic lateral sclerosis CNS. *Brain Research*, **556**, 151–156.
Urca, G. & Urca, R. (1990). Neurotoxic effects of excitatory amino acids in the mouse spinal cord: quisqualate and kainate but not *N*-methyl-D-asparate induce permanent neural damage. *Brain Research*, **529**, 7–15.
Whetsell, W.O. Jr & Shapira, N.A. (1993). Neuroexcitation, excitotoxicity and human neurological disease. *Laboratory Investigation*, **68**, 372–387.
Young, A.B., Penney, J.B., Dauth, G.W., Bromberg, M.B. & Gilman, S. (1983). Glutamate or aspartate as a possible neurotransmitter of cerebral corticofugal fibres in the monkey. *Neurology*, **33**, 1513–1516.

Index

Page numbers in *italic* type refer to illustrations and tables.